Mathematics for business, finance and economics

An informed understanding of mathematical methods and the skills to make use of them are essential for the aspiring numerate manager of today. An overview of the methods and an awareness of their range and relevance are important parts of a modern education in business management, finance and economics. This comprehensive and user-friendly textbook provides a thorough introduction to mathematical concepts and methods used in the analysis of business management, finance and economics. Much of the coverage is also relevant for students of other social sciences at university level where a quantitative approach is employed.

The ten chapters of the book are each carefully designed with a graduated approach to lead students through from a basic level to more advanced concepts and applications, enabling both students and teachers to choose the level appropriate for their course. Topics covered include:

- **Linear equations and inequalities**: with applications in market demand and supply; linear programming; production planning
- **Functions, derivatives and optimization of functions**: with application to profit and sales revenue maximization; cost minimization; income from taxes; aggregate sales curves; finance
- **Optimization under constraint**: including the Lagrange multiplier and substitution methods; the Kuhn-Tucker conditions.

The text is 'software aware' and most chapters contain illustrative computer programs relevant to the material covered, without making prior knowledge or extensive use of computers a requirement. Student exercises and comprehensive worked solutions are provided throughout.

F. M. Wilkes is Professor of Business Investment and Management at the University of Birmingham. He is author/co-author of a number of other books including *Capital Budgeting Techniques* and *Management of Company Finance*.

Mathematics for business, finance and economics

F. M. Wilkes

London and New York

First published 1994
by Routledge
11 New Fetter Lane, London EC4P 4EE

Simultaneously published in the USA and Canada
by Routledge
29 West 35th Street, New York, NY 10001

© 1994 F. M. Wilkes

Printed and bound in Great Britain by
T.J. Press (Padstow) Ltd, Padstow, Cornwall

All rights reserved. No part of this book may be reprinted or
reproduced or utilized in any form or by any electronic,
mechanical, or other means, now known or hereafter
invented, including photocopying and recording, or in any
information storage or retrieval system, without permission in
writing from the publishers.

British Library Cataloguing in Publication Data
A catalogue record for this book is available from the British Library

Library of Congress Cataloging in Publication Data
A catalogue record for this book has been requested

ISBN 0-415-11488-8 (hbk)
ISBN 0-415-11489-6 (pbk)

Contents

Preface		ix
1 Introduction		1
1.1	The number system	2
1.2	Sets	8
1.3	The fundamental laws of arithmetic	16
1.4	Exponents	19
1.5	Number notation, precedence and scales of measurement	23
1.6	Summation and product notation	36
1.7	Sequences and series	41
1.8	BASIC programs	49
	Additional problems	53
	References and further reading	54
	Solutions to exercises	54
2 Straight Lines and Linear Equations		60
2.1	Introduction	61
2.2	Co-ordinates	61
2.3	Straight lines – preliminaries	68
2.4	Identifying straight lines and equations	73
2.5	Two applications	81
2.6	Point–slope and general form	84
2.7	Straight lines and simultaneous equations	87
2.8	Elementary row operations	99
2.9	Other linear systems in up to two variables	101
2.10	Linear systems of equations in more than two variables	104
2.11	Gaussian elimination	112
2.12	Identities	115
2.13	Further applications	117
2.14	BASIC programs	123
	Additional problems	126
	References and further reading	127
	Solutions to exercises	127

3 Linear Inequalities — 135
- 3.1 Introduction — 136
- 3.2 Linear inequalities in one variable — 138
- 3.3 Linear inequalities in two or more variables — 144
- 3.4 Convex solution sets — 149
- 3.5 Linear programming: introduction and graphical method — 151
- 3.6 Introduction to the simplex method — 163
- 3.7 The iterative procedure — 166
- 3.8 Sensitivity analysis — 175
- 3.9 Duality — 179
- Additional problems — 182
- References and further reading — 184
- Solutions to exercises — 184

4 Functions and Turning Points — 194
- 4.1 Functions — 195
- 4.2 Quadratic functions — 201
- 4.3 Cubics and quartics — 221
- 4.4 Polynomials — 229
- 4.5 Descartes's rule of signs — 230
- 4.6 Rational functions — 233
- 4.7 Functions: further considerations — 241
- 4.8 BASIC program — 256
- Additional problems — 259
- References and further reading — 260
- Solutions to exercises — 260

5 Slopes, Derivatives and Turning Points — 269
- 5.1 Introduction — 270
- 5.2 Slope and turning points — 270
- 5.3 An approach to the derivative — 272
- 5.4 The power function rule — 277
- 5.5 Differentiating polynomials — 283
- 5.6 The product and quotient rules — 290
- 5.7 The chain rule — 295
- 5.8 The inverse function rule — 302
- 5.9 Implicit differentiation — 304
- 5.10 Higher order derivatives — 307
- 5.11 Local maxima and minima — 310
- 5.12 Global maxima and minima — 316
- 5.13 Concavity, convexity and points of inflection — 325
- 5.14 BASIC program — 333
- Additional problems — 337
- References and further reading — 339
- Solutions to exercises — 340

6 Functions of More Than One Variable — 361
6.1 Introduction — 362
6.2 Linear functions of several variables — 364
6.3 Quadratic functions — 369
6.4 Slopes and first-order derivatives — 373
6.5 Higher order partial derivatives — 382
6.6 Local maxima and minima — 387
6.7 Saddle points — 392
6.8 Stationary values: résumé — 398
6.9 Introduction to constrained optimization — 402
6.10 The method of substitution — 408
6.11 Substitution: concluding remarks — 419
6.12 BASIC program — 420
Additional problems — 427
References and further reading — 428
Solutions to exercises — 428

7 Constrained Optimization with Lagrange Multipliers — 449
7.1 Constrained optimization – Lagrange multipliers (i) — 450
7.2 Interpretation of the Lagrange multiplier — 461
7.3 Sign restricted and bounded variables — 464
7.4 Lagrange multipliers (ii) Inequality constraints — 484
7.5 Lagrange multipliers (iii) Inequality constraints with sign requirements — 492
7.6 The Kuhn–Tucker conditions — 498
7.7 Economic application – multi-product monopoly — 510
7.8 Concluding remarks — 515
Additional problems — 515
References and further reading — 517
Solutions to exercises — 517

8 Integration — 543
8.1 Introduction — 544
8.2 Rules for integration — 547
8.3 Application to the marginal analysis of the firm — 552
8.4 Differential equations — 555
8.5 Integration by substitution — 567
8.6 The definite integral — 571
8.7 Numerical integration — 584
8.8 Concluding remarks — 591
Additional problems — 592
References and further reading — 593
Solutions to exercises — 594

viii Contents

9 Exponential and Logarithmic Functions — 613
- 9.1 Exponential functions: introduction — 614
- 9.2 The natural exponential function and its derivative — 622
- 9.3 Integration of natural exponential functions — 629
- 9.4 Natural logarithmic functions — 635
- 9.5 The derivative of natural logarithmic functions — 643
- 9.6 Elasticity — 648
- 9.7 Natural logarithmic functions and integration — 654
- 9.8 Integration by parts — 658
- 9.9 Logarithmic and exponential functions to bases other than e — 663
- 9.10 Aggregate sales curves — 671
- 9.11 BASIC programs — 675
- Additional problems — 678
- References and further reading — 680
- Solutions to exercises — 680

10 Matrices — 698
- 10.1 Introduction — 699
- 10.2 Some fundamentals — 700
- 10.3 Addition, subtraction and scalar multiplication — 709
- 10.4 Matrix multiplication — 713
- 10.5 Matrix inversion — 723
- 10.6 Simultaneous equations (i) — 731
- 10.7 Rank — 739
- 10.8 Higher order systems: determinants and the inverse matrix — 743
- 10.9 Simultaneous equations (ii) — 748
- 10.10 Cramer's rule — 752
- 10.11 Homogeneous systems — 756
- 10.12 Concluding remarks — 760
- Additional problems — 761
- References and further reading — 762
- Solutions to exercises — 763

Appendix: Optimization: Further Considerations — 786
- A1 Second-order conditions for unconstrained maxima and minima using determinants — 787
- A2 Quadratic functions: stationary values — 791
- A3 Constrained optimization: second-order conditions using determinants — 797
- A4 Optimization of quadratic functions of two variables with a linear equality constraint — 800

Index — 805

Preface

This book adopts a user friendly approach to the introduction of mathematical methods and concepts used in the quantitative analysis of business management, economics and finance. The claim of being on the side of the student reader is not a unique one, but this textbook has a number of distinctive features designed to maximize the accessibility of a fairly broad syllabus, and these are set out below.

While the book is designed to support a quantitative approach to the study of management methods, business finance and industrial economics, much of the text is also relevant for students of general business studies and other social sciences at university level where a quantitative approach is employed. Amongst the principal objectives of the book are the following:

1 to provide a thorough treatment of some of the most important and relevant mathematical methods in the fields of business management, finance and economics;
2 to provide a wide breadth of coverage in the area of constrained optimization;
3 to draw attention to and make use of computer programs and software packages that are now available;
4 to make the material accessible to the widest possible group of students and to develop individual confidence and a positive outlook in the use of the methods.

The principles and methods are described as far as possible in narrative terms, and intuitive as well as technical explanations are offered. Progression is graded throughout each chapter, each of which contains numerous examples and end of section exercises with comprehensively worked answers.

The approach has a number of other distinctive features. Differential calculus, including functions of several variables, and integration are first of all presented through polynomial and rational functions, with exponential functions introduced at a later stage. The purpose of this ordering is to allow the principles and essence of the methods to be conveyed with a minimum of distraction.

Space is taken throughout the book to point out historical or other intriguing aspects of topics. Wherever possible, formality is eschewed and intuitive understanding is emphasized. The level of difficulty is treated consistently, progressively increasing throughout each chapter, then easing down again for the start of the succeeding chapter.

The book places emphasis on constrained optimization with detailed discussion of both equality and inequality constraints and sign restricted variables. The opening chapter provides self-contained basic material and is designed to ensure that students have an essential and reasonably homogeneous background for the remainder of the text. A summary of objectives is given for each chapter.

Throughout, where applications assist in the development of the material, they are woven into the text, with further applications in separate sections. When a new concept is introduced, although one or two terms only will subsequently be used in the text, other terms that are used to describe the concept are given. In conjunction with this feature, the detailed index should enable reference usage.

The presentation of all topics in the book is computer and software aware. Attention is drawn to the capabilities, usefulness and relative ease of use of today's mathematical software packages – in particular *Derive*. Most chapters also offer one or more illustrative computer programs in BASIC, the majority of which will run under Microsoft's *QBasic* which is supplied with MS-DOS versions 5 or later. However, it is important to note that computing expertise is not a requirement for the effective use of this book.

The presumed minimal background for students using the book is UK GCSE or O Level mathematics. While those with additional or A Level mathematics will find familiar material, there will also be much that is new in either substance or approach, and the management and industrial economics orientation will give new significance and insights to the methods presented.

The overall structure of the book has been designed to allow for flexible usage depending upon the course requirements, the individual preferences of lecturers or the degree of modularity in a degree programme. Consistent with the progressive nature of the material, chapters are as self-contained as possible, having their own specific objectives, mini bibliography and worked answers to exercises.

There are several ways in which the book could be used within a business or social science degree programme. For example, an introductory half course or the mathematics component of a combined quantitative methods course might cover Chapters 1–5, 9 (part) and 10. Of course, not all parts of each chapter need to be covered or given equal emphasis. There are over 100 sections in the book as a whole and the emphasis given to each section or subsection – for example in respect of the use of computer programs – can be varied without loss of continuity. The book as a whole would constitute a relatively comprehensive syllabus for a full one-year course, or half courses spread over two years, in a quantitatively oriented programme.

Whether you are a student on a specialist degree programme or are taking quantitative options in this area I hope that you will find this textbook relevant, interesting and genuinely user friendly throughout.

I would like to express my thanks to Mr C. R. Barrett for his investment of time in commenting on the manuscript and to Mr R. W. Bailey for comment on particular sections. My thanks also to Rosemary Nixon of Routledge for helping to make the concept a practical reality. Remaining imperfections are, of course, solely my responsibility. I welcome feedback from readers, and if you have specific comments or suggestions for improvement, I can be reached on E-Mail as F.M.Wilkes@bham.ac.uk.

Chapter 1

Introduction

1.1	The number system	2
1.2	Sets	8
1.3	The fundamental laws of arithmetic	16
1.4	Exponents	19
1.5	Number notation, precedence and scales of measurement	23
1.6	Summation and product notation	36
1.7	Sequences and series	41
1.8	BASIC programs	49
	Additional problems	53
	References and further reading	54
	Solutions to exercises	54

This chapter presents some fundamental concepts and introduces notation, terminology and other preliminaries that will be drawn upon later in the book. The representation of numbers, their classification, scales of measurement and historical background are reviewed.

An introduction to sets in Section 1.2 is followed by a description of important laws in the arithmetic of real numbers. Laws observed by exponents are considered in Section 1.4. The widely useful summation or *sigma* notation is introduced in Section 1.6 and is employed in an introduction to sequences and series in the following section. The chapter concludes with some illustrative BASIC computer programs.

By the end of this chapter, you should be familiar with relevant rules for operations involving numbers and variables. You will also be able to use appropriate and efficient notation, understand relevant terminology and see some of the value of the use of computers.

1.1 THE NUMBER SYSTEM

The effective use of numbers in business and commercial transactions is an essential skill. The aspiring financial analyst or economist must begin with a grasp of the fundamental operations involving numbers, and a basic idea of the classification of numbers gives a useful starting point. Knowledge of the numbers necessary for everyday reckoning and enumeration arose in antiquity. The *counting numbers* are the positive whole numbers symbolized by the sequence of Arabic *numerals*:[1]

$$1, 2, 3, 4, 5, 6, 7, 8, 9, \ldots$$

amongst which it will be noted that zero is not included. The counting numbers are also called *natural numbers*, but it should be noted that in actual usage the natural numbers are sometimes taken to include zero. The counting numbers form part of the larger set of whole numbers, the *integers*:

$$\ldots -5, -4, -3, -2, -1, 0, 1, 2, 3, 4, 5, \ldots$$

The integers comprise all positive and negative whole numbers and zero. The negative integers can be defined as the result of the subtraction of the counting numbers from zero. The integers (think of *integral* meaning *whole*) can themselves be seen as part of a more general category – the *rational numbers*. A *rational* number is the *ratio* of two integers and may therefore be an integer or a fraction. When expressed in decimal form, a rational number will either *terminate* or *repeat*. For example:

$$17 \div 8 = 2.125$$

is a rational number which terminates, while

$$16 \div 7 = 2.285714\ldots$$

is a rational number which repeats *ad infinitum* the six digits after the decimal point. A rational number will terminate only if its denominator has no prime factors other than 2 or 5. *Irrational numbers* cannot be expressed as the ratio of two integers. The word 'irrational' is not used in its everyday sense of 'unreasonable' but rather as 'non *ratio*-nal'. It is believed that the first number to be identified as irrational (to the consternation of the Pythagoreans) was $\sqrt{2}$. Note that while $\sqrt{2}$ can be expressed through Pythagoras's theorem *in terms of* two integers (as shown in Figure 1.1) it cannot be expressed as their ratio.

The idea of rationality can be expressed in slightly different ways. For example, whenever the ratio of two numbers is a rational number, the numbers are said to be *commensurable*.[2] Thus the diagonal and side of the right triangle with sides 1, 1 and $\sqrt{2}$ are *incommensurable*. The commensurability of two quantities means that they are *in rational proportion* to one

Introduction 3

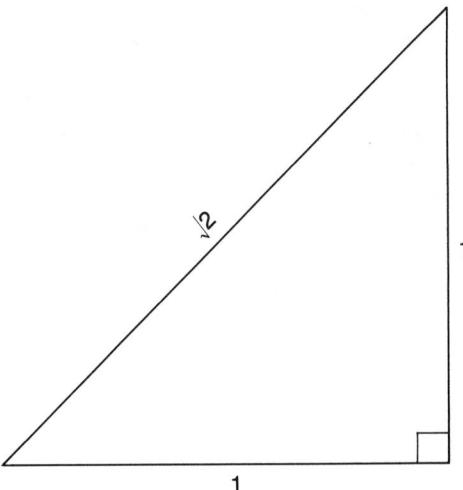

Figure 1.1

another – integral multiples of the same amount. While it may appear that irrational numbers are curiosities that are few and far between, this is not the case. In fact, in a sense that can be made precise, there are more irrational numbers than there are rational numbers. *Transcendental numbers* are irrational numbers which can never emerge as the solution to a polynomial equation (quadratics, cubics, quartics etc. – see Chapter 4) in which the coefficients are rational numbers. While this may seem rather obscure, numbers such as 'π' and 'e' are transcendental. Numbers which

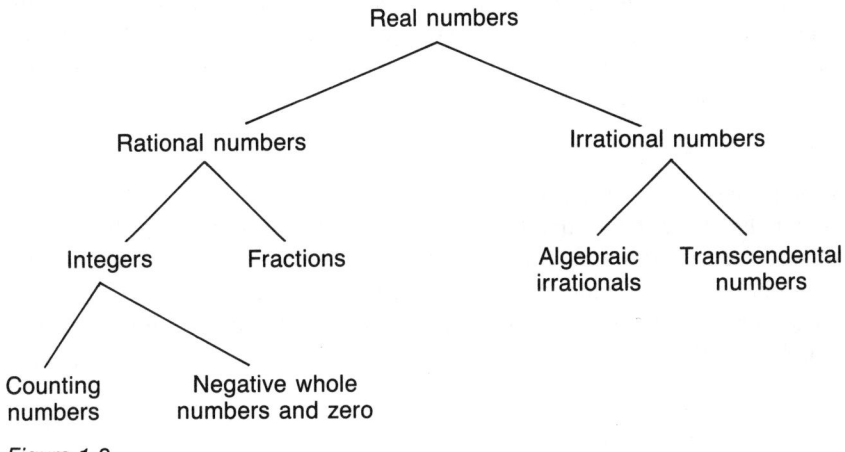

Figure 1.2

can appear as the roots[3] of polynomial equations in rational coefficients are called *algebraic* numbers. Algebraic numbers, which include all rationals, may be rational or irrational. The *real numbers* constitute the rational and irrational numbers taken together. The real numbers contain all possible decimal representations, and the system of real numbers can be set out in the form of a tree diagram as in Figure 1.2.

For centuries it was thought that the real numbers contained every meaningful possibility. However, the development of numbers beyond the counting numbers had arisen partly as a result of the need to solve certain types of equation – and in this respect a disconcerting gap remained. An equation such as

$$2x - 6 = 0$$

can be solved in terms of the counting numbers ($x = 3$), but an equation such as

$$2x + 6 = 0$$

finds no solution in the counting numbers. In order to obtain a solution ($x = -3$) the concept of number needs to be broadened to include the negative integers and, in general, zero. As late as the sixteenth century there were still mathematicians who were prepared to argue against the existence of negative numbers.[4] But further categories of number are still required. For example, an equation such as

$$2x - 5 = 0$$

cannot be solved in terms of the integers. The concept of number needs to be broadened still further to allow solution to an equation of this nature, one which can certainly arise out of everyday experience – as for example when dividing an inheritance of five hectares between two heirs. Clearly, there is a practical as well as a conceptual requirement for the rational numbers. But the inclusion of the rationals still did not go far enough, since there is no solution in rational numbers to the equation

$$x^2 - 2 = 0$$

which arises out of the geometry of right-angled triangles and might represent the straight line distance between one location and another spot a mile to the east and a mile to the north. It was in contexts such as this that irrational numbers came to be accepted. But still there remained equations for which there was no solution in either rational or irrational numbers. The simplest of these is

$$x^2 + 1 = 0$$

In the eighteenth century the great mathematician Euler (1707–83) introduced 'i' where

$$i = \sqrt{(-1)}$$

and suggested that i should be accepted as a new type of number.[5] The case in favour is a strong one, as i obeys all the normal rules of arithmetic and its use allows the solution of the remainder of the polynomial equations in rational coefficients. In this respect, i plays the role of an *ideal element* – something which is added to an existing system in order to complete the system and eliminate special cases.[6] Here, the special case would be those algebraic equations that cannot be solved in terms of the real numbers. Numbers of the form ki, in which k is a non-zero real number, are called *imaginary numbers*. The square root of *any* negative number can be written as a real multiple of i. The term 'imaginary' is unfortunate as it may give the impression that the numbers are somehow detached from reality. This is far from being the case and i is useful in a variety of practical contexts including economic growth models. *Complex numbers* have both a real and an imaginary component and are written in the form

$$a + bi$$

where a and b are real numbers. Examples of complex numbers are

$$5 + 4i, \quad 6 - 2i, \quad -3.5 + 7.7i, \quad \sqrt{2} + 0.4i$$

In the complex number $a + bi$, the product bi is said to be the imaginary part of the complex number, while a is the real part. Complex numbers usually arise in pairs which have the same real part and which differ in that the imaginary part is added in one member of the pair and subtracted in the other. These pairs are called *complex conjugates* and take the form

$$a + bi \text{ and } a - bi$$

For example the equation

$$x^2 - 6x + 13 = 0$$

has as its roots[7] the complex conjugates

$$x = 3 + 2i \text{ and } x = 3 - 2i$$

The term *conjugate* does not apply only to the case of complex roots. The roots of a quadratic equation in rational coefficients can also be described as 'conjugates' when the roots themselves are simply *irrational* rather than complex. Thus the roots of the equation

$$x^2 - 2 = 0$$

are the conjugate pair

$$+1.4142\ldots \text{ and } -1.4142\ldots$$

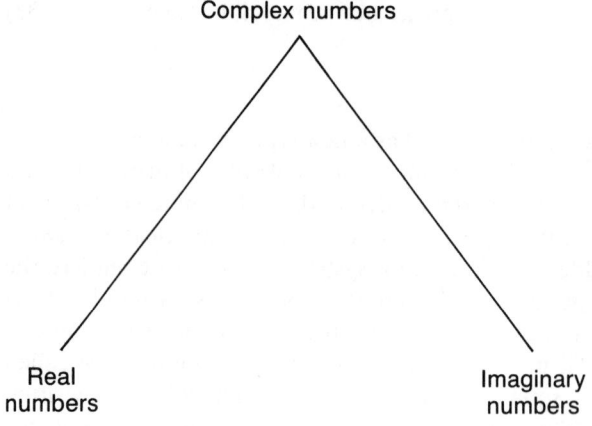

Figure 1.3

Two complex numbers $a + bi$ and $c + di$ are *equal* if and only if $a = c$ and $b = d$. Only when $a = 0$ and $b = 0$ is the complex number itself zero. Addition and subtraction of complex numbers is straightforward and proceeds as follows:

$$(a + bi) + (c + di) = (a + c) + (b + d)i$$
$$(a + bi) - (c + di) = (a - c) + (b - d)i$$

In other words, the addition or subtraction can be carried out on the real and imaginary parts separately. As examples consider

$$(7 + 5i) + (-3 + 8i) = 4 + 13i$$
$$(10 - 2i) - (8 - 5i) = 2 + 3i$$

Complex numbers include as special cases the real numbers (produced when $b = 0$) and the imaginary numbers (for which $a = 0$). Figure 1.2 can now be extended as shown in Figure 1.3.

The 'components' of the number system can be obtained by starting, where we began, with the counting numbers. The operation of subtraction amongst the counting numbers produces the integers, and division produces the rational numbers. The picture is completed by the extraction of roots, which operation produces the irrational and the complex numbers.

EXERCISES 1.1

1 For the following numbers:

$$7 \quad -4 \quad 3.2 \quad \pi \quad \sqrt{2} \quad \tfrac{1}{2} \quad 1/\sqrt{2} \quad e$$
$$0 \quad \sqrt{(-4)} \quad 4 + 2i \quad 10^{10}$$

Identify each number as being a

  (i) counting number
  (ii) integer
  (iii) rational number
  (iv) irrational number
  (v) transcendental number
  (vi) real number
  (vii) imaginary number
  (viii) complex number

In each case select the *narrowest* category to which the number belongs.

2 Give an example of each of the following:

  (i) a natural number
  (ii) a rational number
  (iii) an imaginary number
  (iv) a counting number
  (v) a complex number
  (vi) an integer
  (vii) a transcendental number
  (viii) an irrational number
  (ix) a real number

3 Give the results of the following operations of addition or subtraction involving complex numbers:

  (i) $8 + 4i + 6 + 2i$
  (ii) $7 - 3i + 2 + 6i$
  (iii) $-10 + 4i + 11 - 4i$
  (iv) $4i - (2 - 2i)$
  (v) $8 - 3i - (8 + 3i)$

4 True or false?

  (i) A quadratic equation cannot have a purely imaginary number as a solution.
  (ii) All transcendental numbers are irrational.
  (iii) Real numbers are a special case of complex numbers.
  (iv) Imaginary and complex numbers have no practical relevance.
  (v) The decimal representation of a rational number cannot continue indefinitely.

1.2 SETS

It is often convenient to refer to several items taken together. Such a collective entity is called a *set*, and sets have certain properties that it is useful to know. The items that make up the set are called the *elements* or *members* of the set. If an item x is a member of a set A we write

$$x \in A$$

This expression is read as

'x is a member of A'

or as

'x belongs to A'

Where it is necessary to make clear that an element does *not* belong to a set, the 'member of' symbol is struck out. Thus if y does not belong to A this is written as

$$y \notin A$$

The full membership of a set can be expressed in two ways. If the number of elements is not large, then all of the elements can be listed. Thus the set A may be

$$A = \{1, 2, 3, 4, 5, 6, 7\}$$

in which the convention of enclosing the list of elements within braces ('curly brackets') should be noted. The alternative way of identifying the set is to describe the properties that are required for membership of the set. Thus in the case of set A we could equally well have written

$$A = \{x: x \text{ is a counting number less than } 8\}$$

where the inside of the bracket reads:

'members of the set are x where x is a ...'

This description is sometimes called the *defining relation* for the set. The set A, which has a finite number of elements, is an example of a *finite set*. Sets which have the same number of elements are said to be *equivalent* even though their elements may differ. Thus the sets

$$\{1\ 2\ 3\} \quad \{5\ 6\ 7\} \quad \{X\ Y\ Z\}$$

are equivalent. Sets are said to be *equal* when the membership of the two sets is identical. Thus the sets

$$\{1\ 2\ 3\} \quad \{2\ 3\ 1\}$$

are equal; the order in which the elements are listed does not matter. A *subset* of a given set is an assemblage of some (or even all) of the elements of a set *and no other elements*. Thus for example

$B = \{1, 2, 3, 4\} \quad C = \{2, 4, 6\} \quad D = \{1, 3, 5, 7\}$
$E = \{5, 6, 7\} \quad F = \{3\}$

are all subsets of the set A. The subset relationship is represented by

$B \subseteq A$

which reads

'B is a subset of A'

If B is a *proper subset* of A, it will contain some, but not all, of the elements of A. This is indicated by the notation

$B \subset A$

But the set

$G = \{2, 4, 6, 8\}$

is *not* a subset of A since it contains an element which is not an element of A. The set G *not* being a subset of A is indicated by

$G \nsubseteq A$

A *superset* contains a given set as a subset. For example, the set A above is a superset of B, C, D, E and F but not of G. The set D is a superset of F. The set of counting numbers up to and including 10 is an example of a superset of the set A. The term 'superset' is quite often used in marketing when speaking of product improvements. If, for instance, version 4 of a piece of software is said to be a superset of version 3.5, it is meant that version 4 incorporates all of the features of version 3.5 and, hopefully, some useful additions! The concept of a subset is a useful way to link the different types of number. Counting numbers can be seen as a subset of the integers which in turn can be seen as embedded within the real numbers. The standard labelling of the types of number as sets is

N the positive integers (natural numbers)
I or Z the integers
Q the rational numbers
R the real numbers
C the complex numbers

'Being a subset of' is a relationship which can be applied successively. Thus in general, if

$F \subseteq E$ and $E \subseteq A$ then $F \subseteq A$

The subset relationship is a *transitive relation*. Transitivity, where it exists, is an important property of relations in economic theory. For example, in the theory of consumer behaviour, if among three possible choices of commodities option A is preferred to B and B is preferred to C, then if the preference relation is transitive, it can be inferred that A is preferred to C. There is a special set which is a subset of all other sets. This is the set containing no elements, the *null set*, and is represented by the symbol ∅. We shall see that the null or *empty* set is more than just a conceptual fine point. The *universal set*, U, consists of all the elements that are being studied in a particular context. It follows from this definition that, within that context, all sets are subsets of U. All of the subsets of a given set can now be classified. They fall into four categories as follows:

1 Each individual member of the set
2 All possible combinations of the members
3 The set itself
4 The null set

The number of subsets for a set with n elements is 2^n. The set of all subsets of a given set is called the *power set* and is sometimes written as 2^s. Thus the set A with seven elements has $2^7 = 128$ subsets, while a set with 30 elements will have a power set containing 1,073,741,824 members!

Operations on sets

We may wish to refer to the aggregate membership of two or more sets. This totality is called the *union* of the sets. The union of the sets E and F is written as

$E \cup F$

where the union symbol ∪ is read as *cup*. The operation of union can be illustrated using the sets E and F above:

$E \cup F = \{3, 5, 6, 7\}$

And as another example of union:

$B \cup C = \{1, 2, 3, 4, 6\}$

In this case note that where an element is included in both sets it appears once only in the union. The order of the elements is immaterial and it does not matter whether we write $B \cup C$ or $C \cup B$, the result is the same. Union is an operation that can include more than two sets. For example:

$C \cup E \cup F = \{2, 3, 4, 5, 6, 7\}$

The union of sets is the elements of the sets taken together; the *intersection* of sets is the elements that the sets have *in common*. The symbol for intersection is an inverted union symbol, ∩, which is read as *cap*. So:

$B \cap C$

refers to the elements that belong to *both* of the sets B and C. Thus:

$B \cap C = \{2, 4\}$

Note that where two sets have no elements in common, the intersection of the sets has no members, i.e. it is the null set. Thus for example

$F \cap C = \emptyset$

Sets such as F and C which have no elements in common are said to be *disjoint*. The intersection of more than two sets can be found. The result would be those elements to be found in all of the sets. For example, if the set

$H = \{2, 3, 4, 5\}$

then

$B \cap C \cap H = \{2, 4\}$

An expression involving operations on sets may well include both union and intersection. For example, to find

$E \cap \{B \cup C\}$

first work out $B \cup C$. We obtain

$B \cup C = \{1, 2, 3, 4, 6\}$

Now find the intersection of E with this union

$E \cap \{B \cup C\} = \{6\}$

Consider also

$\{B \cap C\} \cup \{D \cap E\}$

Finding the two bracketed terms first

$B \cap C = \{2, 4\}$

and

$D \cap E = \{5, 7\}$

with the result that

$\{B \cap C\} \cup \{D \cap E\} = \{2, 4, 5, 7\}$

At times we may wish to refer to all of the items under consideration that do *not* belong to a certain set. The all-inclusive nature of this concept of the *complement* of a set allows contradictions,[8] and it is preferable to consider complementarity as being *relative* to another set. This is not restrictive, since the other set could be the universe of discourse. Complementarity can be defined as a relation between any two sets, and any set and its complement are disjoint. We shall here be concerned only with the universal complement of a set – all those elements in the universal set that do not belong to A. The complement of the set A is written as

$$\bar{A} \text{ or } C\{A\} \text{ or } A'$$

For example if the universe of discourse was the counting numbers up to 10, then the complement of the set A would be the numbers 8, 9 and 10. This would be written as

$$\bar{A} = \{8, 9, 10\}$$

The *set difference* between two sets is the elements that are contained in one set but not in the other. The set difference between B and C (represented by $B - C$) is

$$B - C = \{1, 3\}$$

while

$$D - F = \{1, 5, 7\}$$

and

$$F - B = \emptyset$$

The *symmetric difference* is the set of elements that belong to *precisely one* of two given sets. The symmetric difference is therefore the union of the sets from which the intersection is excluded. The symmetric difference is denoted by the symbols $\ominus$ or ∇. Thus the symmetric difference of the sets B and C is

$$B \ominus C = \{1, 3, 6\}$$

A useful way to represent sets and the operations on them is the *Venn diagram* (after John Venn (1834–1923), the English logician and mathematician). In a Venn diagram, the universal set is represented by a square within which individual sets are shown as circles. Thus the Venn diagram representations of union, intersection and complementarity would appear as shown below. In Figure 1.4 (a) the union of the sets B and C is the shaded area covered by the two sets together. It is meaningful to consider the union of disjoint sets – which in Venn diagram representation would appear as two non-overlapping circles as shown in Figure 1.4(b). Figure 1.4(c) shows the union of a set with one of its proper subsets. This will be the set itself.

Introduction 13

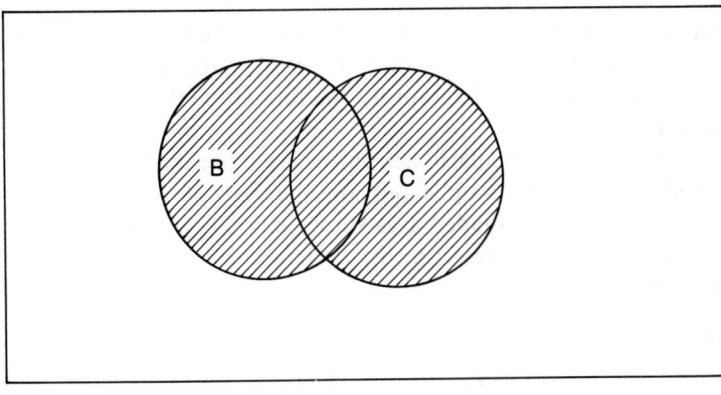
(a)

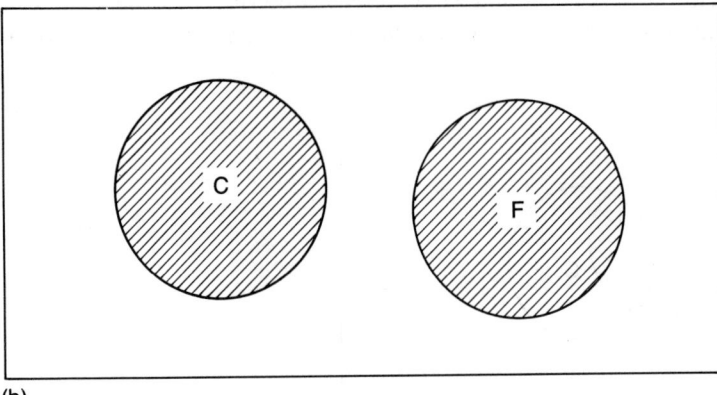
(b)

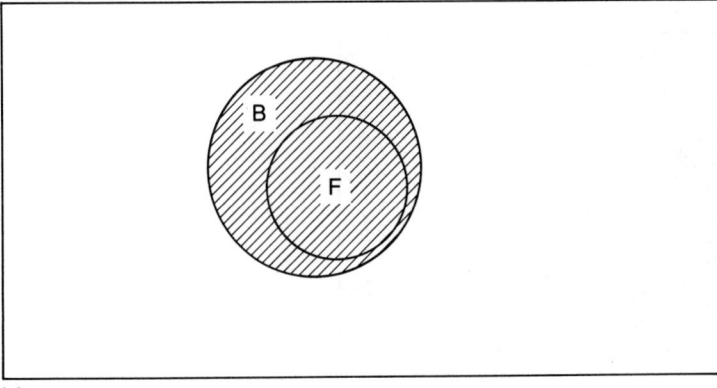
(c)

Figure 1.4

14 Mathematics for business, finance and economics

The intersection of two sets is shown as the overlap between the circles. It appears as the shaded lenticular region in Figure 1.5(a). The complement of a set D is shown in a Venn diagram by shading all of the universal set except D as in Figure 1.5(b).

Venn diagrams can be particularly helpful where more than two sets are involved and where there are complicated combinations of operations. For example, in Figure 1.6(a) the shaded area shows $B \cap C \cap H$ while in 1.6(b) the shaded area shows the intersection of the complement of C with the intersection of B and H, i.e.

$$\{B \cap H\} \cap C'$$

Venn diagrams are useful in *counting problems*, where it is required to establish the number of elements in a subset having certain properties in relation to the set or sets of which it is a part. For example, suppose that in Figure 1.6 the sets B, C and H represent subscribers to the *Financial Times*, *Time* magazine and *The Economist* respectively. The shaded area in

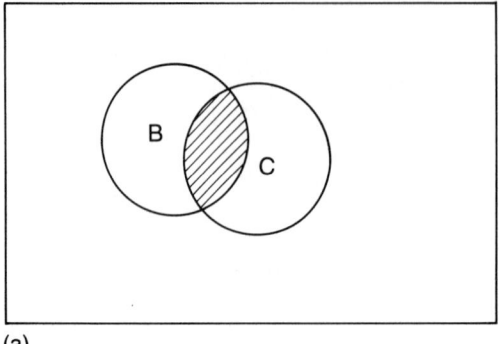

(a)

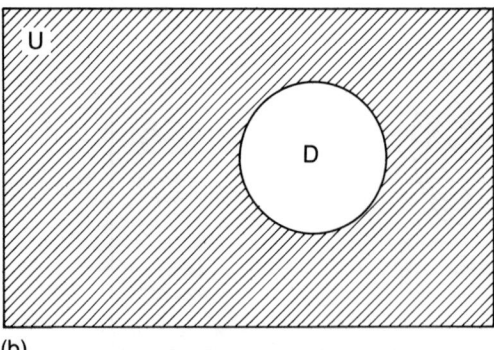

(b)

Figure 1.5

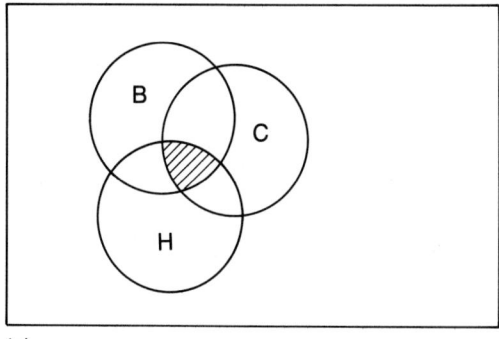
(a)

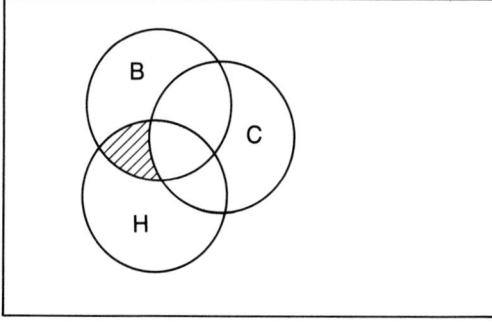
(b)

Figure 1.6

1.6(a) represents people subscribing to all three publications, while 1.6(b) shows those who subscribe to the *Financial Times* and *The Economist* but not to *Time* magazine.

EXERCISES 1.2

1 For the following sets:

$$A = \{1, 2, 3, 4, 6, 7, 8\} \quad B = \{2, 3, 4, 6\}$$
$$C = \{3, 4, 8\} \quad D = \{1, 4, 6, 8\} \quad E = \{6, 7, 8\}$$
$$F = \{2\} \quad G = \{1, 3, 5\} \quad H = \{1, 4, 6, 7\}$$
$$I = \emptyset \quad J = \{0\} \quad K = \{4, 6\}$$

(i) Which of the sets, if any, are *not* subsets of A?

(ii) Find:

(a) $D \cup E \cup F$

(b) $B \cap D$
   (c) $E \cap G$
   (d) $B \cap \{C \cup H\}$
   (e) $\{B \cap C\} \cup G$
   (f) $F \cup J$
   (g) $F \cup I$

   (iii) What is the total number of subsets of the set A?
   (iv) Which sets are supersets of K?

2 For the following sets:

   $A = \{2, 3, 4, 5, 6, 7, 8\} \quad B = \{2, 3, 4, 5\}$
   $C = \{3, 5, 7\} \quad D = \{2, 4, 6, 8\} \quad E = \{6, 7, 8\}$
   $F = \{4\} \quad G = \{1, 3, 5\} \quad H = \{3, 4, 5, 6\}$
   $I = \{3, 5\}$

   (i) Which sets are *not* proper subsets of A?
   (ii) Find:

      (a) $C \cup E \cup F$
      (b) $B \cap C$
      (c) $F \cap C$
      (d) $B \cap C \cap H$
      (e) $B \cap \{C \cup H\}$
      (f) $\{E \cap D\} \cup \{G \cap H\}$

   (iii) What are the following set differences:

      (a) $A - C$
      (b) $B - F$

   (iv) What is the symmetric difference, $G \ominus H$?
   (v) If the set A is the universe of discourse, what are the complements of:

      (a) B
      (b) D

   (vi) Which sets are *not* supersets of set I?

1.3 THE FUNDAMENTAL LAWS OF ARITHMETIC

The three fundamental laws in the arithmetic of real numbers are:

1 the commutative law

2 the associative law
3 the distributive law

These laws apply in a common-sense way to real numbers. The *commutative law* states that, for the operations of addition and multiplication, the order in which the numbers are written down is immaterial. For any two real numbers x and y the commutative law means that

(a) $x + y = y + x$
(b) $x.y = y.x$

Note that (a) does *not* apply to the operation of subtraction, which is not commutative (except for the trivial case of $x = y$) and that (b) does not apply to the operation of division – again with the trivial $x = y$ case excepted. The *associative law* states that where three or more numbers are written down the result of addition alone, or the result of multiplication alone, does not depend on the order in which the operations are carried out. Thus with three real numbers x, y and z, the associative law means that[9]

(a) $\quad x + y + z = (x + y) + z = x + (y + z) = (x + z) + y$
(b) $\quad x.y.z = (x.y)z = x(y.z) = y(x.z)$

Note that the associative law does not apply to the operations of subtraction and division. This point can be illustrated by means of a *counter-example*. Clearly:

$$(8 - 4) - 2 = 2 \neq 8 - (4 - 2) = 6$$

and

$$(8/4)/2 = 1 \neq 8/(4/2) = 4$$

The *distributive law* involves the operations of addition and multiplication together. This law shows how to expand a bracket which contains a sum and which is multiplied by another real number. For three real numbers x, y and z, the distributive law states that

$$x(y + z) = xy + xz \tag{1.1}$$

Thus

$$5(4 + 3) = 35 = 5(4) + 5(3)$$

Note that if, simultaneously, division and subtraction replace multiplication and addition, the distributive law does *not* apply. That is,

$$x/(y - z) \neq x/y - x/z$$

18 Mathematics for business, finance and economics

But if the only change is subtraction replacing addition in (1.1), the distributive law holds:

$$x(y - z) = xy - xz$$

The three laws apply to *sets*, in which context the operation of union replaces addition and, somewhat surprisingly, intersection replaces multiplication.[10] Thus for given sets B, C and D:

Commutative law: $\quad B \cup C = C \cup B$
$\qquad\qquad\qquad\quad B \cap C = C \cap B$

Associative law: $\quad B \cup C \cup D = \{B \cup C\} \cup D$
$\qquad\qquad\qquad\qquad\quad = B \cup \{C \cup D\}$
$\qquad\qquad\qquad\qquad\quad = \{B \cup D\} \cup C$

Distributive law: $\quad B \cap \{C \cup D\} = \{B \cap C\} \cup \{B \cap D\}$

When the laws are applied to the sets B, C and D of Section 1.3, the outcome is fairly obvious in the cases of the commutative and associative laws. To illustrate the distributive law result, begin with the left-hand side of the equation:

$$C \cup D = \{1, 2, 3, 4, 5, 6, 7\}$$

so that

$$B \cap \{C \cup D\} = \{1, 2, 3, 4\}$$

On the right-hand side:

$$B \cap C = \{2, 4\}$$

and

$$B \cap D = \{1, 3\}$$

so that

$$\{1, 3\} \cup \{2, 4\} = \{1, 2, 3, 4\}$$

EXERCISES 1.3

1 For real numbers a, b and c, which of the fundamental laws of arithmetic are involved in the following statements?

   (i) $(a + b)c = ac + bc$
   (ii) $ab = ba$
   (iii) $6 + (4 + 7) = (6 + 4) + 7$

2 Given the sets:

$$B = \{6, 7, 8, 9\} \quad C = \{4, 5, 6, 7\} \quad D = \{2, 3, 7, 8\}$$

find

(i) $B \cap \{C \cup D\}$
(ii) $\{B \cap C\} \cup \{B \cap D\}$

Which of the fundamental laws is illustrated by your results?

1.4 EXPONENTS

We have already made use of terms such as x^2, x^3 and x^n. We now take a closer look at variables raised to powers and the rules that apply. First of all, some definitions. In the term x^n, x is said to be the *base* while n is the *exponent*. It will be helpful to recall exactly what x^n is. It is x multiplied by itself n times:

$$x^n = x.x.x.x.x.x.x.x\ldots$$

There are a number of rules for handling exponents and we shall use the above simple definition to reinforce the rules in some simple examples. Here is the first of the rules.

> The *product rule* Where a term x^n is multiplied by a term x^m, the result can be written as
>
> $$x^n.x^m = x^{n+m}$$

The validity of the product rule can be confirmed by selecting particular values for n and m and setting out the multiple in full. Thus:

$$x^3.x^2 = (x.x.x)(x.x) = x^5$$

The product rule can be applied no matter how many individual terms of the form x^n are to be multiplied together. Thus, for example,

$$x^2.x^5.x^6 = x^{13}$$

Now consider the next of the rules.

> The *power rule* When x^n is itself raised to a power m, the result is x raised to the product of the two powers involved. Thus:
>
> $$(x^n)^m = x^{nm}$$

Again, this rule can be fixed in the mind by confirmation with a simple example. For instance:

$$(x^2)^3 = (x.x)(x.x)(x.x) = x^6$$

As with the product rule, somewhat more complicated cases can be dealt with under the power rule. We will consider three examples here. First of all consider the case of repeated raising to a power:

$$[(x^n)^m]^p = x^{nmp}$$

So that, for example,

$$[(x^2)^3]^4 = x^{24}$$

In the context of raising a bracketed term to a power, note also that given two variables x and y:

$$(xy)^n = x^n y^n$$

A special case here is where only one variable, x, is involved but x has a coefficient, a, which is other than unity. It is the case that

$$(ax)^n = a^n x^n$$

So, for example,

$$(2x)^3 = 2x.2x.2x$$
$$= 2^3 x^3 = 8x^3$$

Where there are two variables, each of which is already raised to some power

$$(x^m y^p)^n = x^{mn}.y^{pn}$$

It is again helpful in fixing this in one's mind to apply it to an example which is simple enough to work out at length. Thus:

$$(x^2.y)^3 = (x^2.y)(x^2.y)(x^2.y)$$
$$= x^6.y^3$$

Note that $(xy)^n$ is a special case in which $m = p = 1$.

> **The *quotient rule*** Where a term consists of the ratio of x raised to differing powers, the result is x raised to the difference of the powers. That is:
>
> $$\frac{x^n}{x^m} = x^{n-m}$$

For example,
$$\frac{x^5}{x^2} = \frac{x.x.x.x.x}{x.x} = x.x.x = x^3$$

The quotient rule is the product rule applied to the case where one of the powers is negative. There are several interesting special cases of the quotient rule:

(a) $m = n$ $\dfrac{x^n}{x^n} = x^{n-n} = x^0 = 1$

This instance of the rule provides a revealing demonstration of the fact that $x^0 = 1$.

(b) $m = n + 1$ $\dfrac{x^n}{x^{n+1}} = x^{-1} = \dfrac{1}{x}$

which shows up the fact that the reciprocal of x is x raised to the power of minus one.

(c) $n = 0$ $\dfrac{x^0}{x^m} = x^{0-m} = x^{-m} = \dfrac{1}{x^m}$

Since, as already seen, $x^0 = 1$ the quotient rule can be used to confirm that x raised to the power of $-m$ is the reciprocal of x^m.

Now consider *rational exponents*, firstly where the numerator is one and where the denominator is an integer. The simplest case is $x^{1/2}$. This is a convenient way to represent the square root of x, and includes both the positive and the negative value of the root. It should be noted that, by default, the representation of a square root by use of the *radical sign*, $\sqrt{}$, is taken to refer to the positive root only.[11] So, for example, $4^{1/2} = \pm 2$ while $\sqrt{4} = 2$.[12]

For an integer m, when x is raised to the power $1/m$ the result is the *m*th *root* of x. This is represented as $x^{1/m}$ while the equivalent representation using radical signs uses a superscript m thus: $\sqrt[m]{x}$. In algebraic manipulation it is always preferable to work with rational exponents rather than radical signs. Therefore, an early stage in rendering a term or expression more suitable for manipulation or simplification is the conversion of any radical signs to the rational exponent equivalent. The power rule for exponents can be used to confirm the fact that $x^{1/m}$ is the mth root of x. By definition, the *m*th root of a number is the value that when multiplied by itself m times produces the original number. Therefore, applying the rule,

$$(x^{1/m})^m = x^{m/m} = x^1 = x$$

As examples of the real roots corresponding to rational exponents consider the following:[13]

$$36^{1/2} = \pm 6$$
$$125^{1/3} = 5$$
$$256^{1/4} = \pm 4$$
$$243^{1/5} = 3$$
$$64^{1/6} = \pm 2$$

in which it will be noted that, whenever m is an even number, there will be two real roots of opposite sign. Now consider the case where the numerator of the rational exponent is a number other than one. If x is raised to the power n/m, the result is the mth root of x^n. That is:

$$x^{n/m} = \sqrt[m]{x^n}$$

For example, $x^{2/3}$ is the cube root of x squared:

$$x^{2/3} = \sqrt[3]{x^2}$$

Thus when $x = 8$, $x^{2/3} = 4$. As further examples involving rational exponents, x raised to the power $3/2$ is the square root of x cubed,

$$x^{3/2} = \sqrt{x^3}$$

so that if $x = 16$, then $x^{3/2} = \pm 64$. As a final example, $x^{3/4}$ is the fourth root of x cubed, so that if $x = 16$, $x^{3/4} = \pm 8$. The product, power and quotient rules also apply to the case of *irrational* exponents.[14] As an example involving the product rule consider

$$x^{\sqrt{2}} \cdot x^{\sqrt{2}} = x^{2\sqrt{2}}$$

Care may be required, as in this example, where the result, the sum of the exponents, is *not* x^2. As a further example of the product rule, note that

$$x^{\pi} \cdot x^{\pi} = x^{2\pi}$$

As an example applying the power rule, consider

$$(x^{\sqrt{2}})^{\sqrt{2}} = x^2$$

where in this case the result, requiring the product of the exponents, *is* x^2. Now consider an example with irrational exponents involving the quotient rule:

$$\frac{x}{x^{\sqrt{2}}} = x^{1-\sqrt{2}} = x^{-0.414\ldots}$$

In all the above cases, we have taken the base to be a variable and the exponent to be a constant. In other circumstances the reverse may be true, or indeed both base and exponent may be variables as in x^x. We shall not at this point overanticipate our discussion of exponential functions (which result when the exponent is variable) and will simply point to a couple of examples. Note that in

$$3^x = 81$$

the value of x emerges as 4, and in

$$-1024^x = -4$$

the value of x required to satisfy this equation is 0.2. Finally, note that exponents may themselves have exponents, allowing 'astronomical' numbers to be represented in a form involving just a few numerals. For example the number represented by 10 *raised to the power* 10^{10} would be one followed by one hundred zeros.

EXERCISES 1.4

1 Use the laws of exponents to simplify the following terms:

(i) $x^4 x^5$
(ii) $x^2 x^4 x^6$
(iii) $(x^3)^5$
(iv) $[(x^3)^5]^4$
(v) $(3x)^4$
(vi) $(x^3 y^4)^2$
(vii) x^7/x^4
(viii) x^7/x^9
(ix) $(x^{1/2})^6$
(x) $x^{\sqrt{3}} x^{\sqrt{3}}$
(xi) $(x^{\sqrt{3}})^{\sqrt{3}}$
(xii) $x^2/x^{\sqrt{2}}$

2 Evaluate:

(i) $25^{1/2}$
(ii) $216^{1/3}$
(iii) $8^{2/3}$
(iv) $16^{3/2}$
(v) $16^{0.25}$
(vi) $-243^{0.2}$

1.5 NUMBER NOTATION, PRECEDENCE AND SCALES OF MEASUREMENT

Notation and conventions regarding the representation of numbers are taken for granted in everyday arithmetic. But there is a wide variety of ways in which numbers can be represented and operated on. While numbers in themselves do not involve measurement, they can be, and of course frequently are, applied to measurement. In the context of measurement, comparisons made between numbers will depend upon the nature of the measurements that it is appropriate to employ. This section is subdivided under four heads:

1 Absolute values
2 Place value notation
3 Rules of precedence for arithmetic operations
4 Scale of measurement

1.5.1 Absolute values

The *sign* of a number can be very important – as when the number represents profit, an account balance, or the time remaining to complete a project. But there are other situations in which the sign of a number is of less significance than its *size*. One instance of this is when numerical methods are being used to approximate a value such as the yield on a project. What matters is the *absolute* accuracy of the approximation rather than whether it is an overestimate or an underestimate. In computing, the achievement of an agreeable visual format of output may require the separate representation of the magnitude of a number and its sign. The magnitude of a number regardless of sign is called the *absolute value* or *modulus* of the number. For some number x, the absolute value or modulus can be written as

$$\text{ABS}(x) \text{ or } |x|$$

the former notation is frequently used in computer programming, while the | | modulus symbol predominates elsewhere. When x is positive, the absolute value is the number itself. When x is negative, the modulus is minus x. Thus, for example,

$$|62| = 62 \text{ and } |-62| = 62$$
$$\text{ABS}(-100) = 100 \text{ and } \text{ABS}(-\pi) = \pi$$

Moduli have a number of properties. For two numbers a and b, which can be positive or negative, the following properties hold:[15]

(i) $|a+b| \leqslant |a| + |b|$

So, for example,

$$|6+4| = 10 = |6| + |4|$$

but

$$|6+(-4)| = 2 < |6| + |-4|$$

(ii) $|a-b| \geq ||a| - |b||$

For example:
$$|12-9| = 3 = ||12| - |9||$$
but
$$|-12 - 9| = 21 > ||-12| - |9|| = 3$$

(iii) $ABS(ab) = ABS(a)ABS(b)$

so that
$$ABS[(-6)5] = ABS(-6)ABS(5) = 30$$

In this section we have looked at the absolute values of individual numbers. The concept of absolute value also applies to functions, and *absolute value functions* are considered in Section 15 of Chapter 5.

1.5.2 Place value notation

The most common notation for numbers (here used by default) is *place value notation* using a *base* of ten. This is the *denary* system. But since no one format is the most suitable for all purposes, it is as well to be aware of the alternatives. The variations on the place value theme use the fact that any positive number can be written as

$$a_n b^n + a_{n-1} b^{n-1} + a_{n-2} b^{n-2} + \cdots + a_1 b^1 + a_0 b^0$$
$$+ a_{-1} b^{-1} + a_{-2} b^{-2} + \cdots$$

where b is the base or *radix* and the a_i are non-negative integers less than b. Therefore, in the denary system, in the number 8,459,067.123 each *digit* or *numeral* multiplies a power of 10 appropriate to its place in the number. The number is

$$8(10^6) + 4(10^5) + 5(10^4) + 9(10^3) + 0(10^2) + 6(10^1) + 7(10^0)$$
$$+ 1(10^{-1}) + 2(10^{-2}) + 3(10^{-3})$$

The numerals

0 1 2 3 4 5 6 7 8 9

are symbols, each representing a number, and are the Arabic numerals,[16] which came to Europe in the Middle Ages through translations of Arabic texts. Such was the superiority of this notation that the effect on European mathematics was dramatic. The base of the place value notation need not be 10. There are other useful possibilities. For example, *octal* numbers use a base of 8. Thus the number

   654.3

in octal represents

$$6(8^2) + 5(8^1) + 4(8^0) + 3(8^{-1})$$
$$= 428.375 \text{ (to base 10)}$$

Octal numbers are used to some extent in computing. Numbers to the base of 16 are widely used by computer programmers. These are *hexadecimal* numbers. Hexadecimal numbers use as numerals the digits 0 to 9 and the letters A B C D E and F. Thus in 'Hex' the number 3AE represents

$$3(16^2) + 10(16^1) + 14(16^0)$$
$$= 942 \text{ (to base 10)}$$

An important common property of both hexadecimal and octal numbers is the use of a base which is a power of 2. This makes for relatively easy conversion between hex and octal numbers and *binary* numbers which are of central importance in computing. Binary notation uses the base of 2 so that the only digits employed are 0 and 1 – the *bi*nary digi*ts* or *bits*. A sequence of bits (usually 8, sometimes 16) encodes a single piece of data – for example a number or an alphabetic or other character – and is called a *byte*. The binary number 111011.1 represents

$$1(2^5) + 1(2^4) + 1(2^3) + 0(2^2) + 1(2^1) + 1(2^0) + 1(2^{-1})$$
$$= 58.5 \text{ (to base 10)}$$

A distinction should be made between a binary number and a *binary coded decimal* in which the place values represent powers of 10 but binary digits replace the usual decimal numerals. Thus the binary coded form of the decimal number 579 would appear as

   0101   0111   1001

Binary coded decimals find their principal application in computing. There are other bases that find occasional use; examples are base 5 (quinary) and base 12 (duodecimal). In duodecimal notation the letters A and B or, alternatively, T and E are used to provide the additional numerals. It is interesting to note that the Babylonians used the base of 60. This is known as *sexagesimal* notation, and 60 is a convenient base for a number of arithmetic purposes as it has many factors. Fragments of a sexagesimal system can be seen today in the subdivision of hours into minutes and minutes into seconds.

   In some of the examples above, we have used a point '.' to indicate where the negative powers of a base begin. Within the decimal system, the use of this notation was pioneered by John Napier (1550–1617) the Scottish mathematician and the originator of logarithms. To be more precise, we have used *fixed point* notation, in which the number is written with the decimal point separating the fractional and integral parts of the number. In fixed point notation there are fixed numbers of digits before and after the decimal point. For example, if the numbers

0.0054321 and 126.72

were both expressed in fixed point notation with ten digits, five of which follow the decimal point, then the numbers would appear in the following way:

00000.00543 and 00126.72000

respectively. Fixed point notation (or an informal variant) is the most common, everyday usage of the decimal system. In *floating point* notation, a number is shown as a multiple of a base raised to a power. The ratio

$$\frac{250,000}{128}$$

with four places before and after the decimal point has as its quotient in fixed point notation the number

1953.1250

This number can be shown in floating point form in several ways. For example:

19.53125×10^2

or

0.1953125×10^4

or

195312.5×10^{-2}

or even

1953.125×10^0

with the 10^0 suppressed, the last instance being equivalent to fixed point form. The exponent can be positive or negative but cannot contain a decimal point. In floating point form the decimal point need not therefore appear in its 'proper', fixed form, place. The base, here 10, is as is appropriate to the context. Where the coefficient of the base-raised-to-a-power is as large as possible subject to still being less than the value of the base itself, the result is called *scientific* notation or *exponential notation*. In scientific notation the result of the division above would appear as

$$1.953125 \times 10^3$$

Note that in the standard output used by some computer software (for example QuickBasic), numbers with more than a certain number of significant figures would probably be presented in exponential notation but, instead of writing 0.05960464 as;

$$5.960464 \times 10^{-2}$$

the number may appear as

$$5.960464E^{-2}$$

where, quite simply, E^{-2} replaces 10^{-2}. So far, we have discussed notational schemes for representing individual numbers. There is also a variety of possible ways in which arithmetic operations on numbers can be represented. The everyday manner of writing the sum

$$2 + 2 = 4$$

is with the *operator*, here addition, the symbol for which is '+', placed between the *addends* or in general the *arguments* to which the operation applies. This arrangement of arguments and operator is called *infix* notation. By contrast *prefix* notation places the operator before the arguments. Thus the above sum would be written in prefix notation as

$$+2 \; 2 = 4$$

Prefix notation is also called *Polish* notation. In *postfix* notation the operator is positioned after the arguments and the sum would therefore appear as

$$2 \; 2 + \; = 4$$

Postfix notation is also called *reverse Polish* notation and is used internally by computers and in some scientific calculators.

1.5.3 Precedence of arithmetic operations

Rules of precedence govern the order in which operations are carried out. In the absence of such rules, in infix notation the expression

$$7 - 3 \times 4$$

could have two distinct values depending on which of the operations, subtraction or multiplication, was carried out first. Thus with subtraction performed first

$$7 - 3 \times 4$$
$$= 4 \times 4$$
$$= 16$$

whereas with multiplication performed first

$$7 - 3 \times 4$$
$$= 7 - 12$$
$$= -5$$

In fact multiplication takes priority over subtraction so that given this rule of precedence the unambiguous result of the calculation is -5. Arithmetic operations are carried out in the following order:

1 exponentiation
2 multiplication and division
3 addition and subtraction

Therefore the expression

$$7 - 3 \times 2^2$$
$$= 7 - 3 \times 4$$
$$= 7 - 12$$
$$= -5$$

Infix notation alone does not always give the most economical representation of a stated series of operations. For example consider the following set of instructions concerning four digits and three operators:

(a) subtract three from seven;
(b) add four to two;
(c) form the product of the numbers produced by (a) and (b).

The faithful execution of the instructions is secured most clearly by the use of *parentheses* (these are *round* brackets such as those enclosing this remark). With infix notation and parentheses, the expression is written as

$$(7 - 3) \times (4 + 2) = 24$$

which is more evident as an accurate statement of what is required than the bald sequence of symbols (or *string*):

$$7 \times 4 + 7 \times 2 - 3 \times 4 - 3 \times 2$$

Where parentheses are used, precedence is accorded to their contents. Parenthesized strings are evaluated first – they are treated as a whole.

30 Mathematics for business, finance and economics

Treating terms collectively in this manner is called *aggregation*. Normal rules of precedence apply within the parentheses. An advantage of reverse Polish notation (which is shared by prefix notation) is that there is no need for the use of parentheses. With reverse Polish notation, the instructions (a)–(c) can be written unambiguously as

$$7\ 3 - 4\ 2 + \times$$

From right to left the string reads: 'multiply the sum of 4 and 2 by the difference between 7 and 3'. Although unambiguous, ease of reading is not a notable feature of either prefix or postfix notation, and the legibility provided by parentheses is a worthwhile net gain. More complicated sets of instructions than (a)–(c) may require several levels of bracketing or *nested brackets*. Brackets have rules of precedence among themselves. The innermost brackets are evaluated first from which point working proceeds outwards to the highest level of bracketing. By convention, the innermost brackets are parentheses which are worked out before *square* brackets [ ] which are themselves evaluated before *braces* or 'curly' brackets { }. Occasional use is also made of a *vinculum*, a line ———— placed above or below the terms to be aggregated, as for example in

$$\overline{ax + by - c}$$

Where a vinculum is used, the terms so aggregated are evaluated first – even before parentheses. Where there is no other prioritization, the expression is evaluated by proceeding from left to right. As an example of an arithmetic calculation involving nested brackets consider the expression

$$4 + 2\{5[5 - 1] - [7(6 + 1) - 2(3 \times 2^3 - 8)]\} - 10$$

Taking the innermost (round) brackets from left to right

$$6 + 1 = 7 \text{ and since } 2^2 = 4 \text{ and } 3 \times 4 = 12 \text{ then } 3 \times 2^3 - 8 = 16$$

so the expression becomes

$$4 + 2\{5[5 - 1] - [7 \times 7 - 2 \times 16]\} - 10$$

The square brackets are

$$5 - 1 = 4 \text{ and } 49 - 32 = 17$$

so that the expression is now

$$4 + 2\{5 \times 4 - 17\} - 10$$

Within the braces

$$20 - 17 = 3$$

so that we are left with

$$4 + 2 \times 3 - 10$$

i.e.

$$4 + 6 - 10$$
$$= 0$$

We have been discussing the rules of precedence between the arithmetic operations of exponentiation, multiplication/division and addition/subtraction.[17] The word 'operator' (which, strictly speaking, means the *symbol* for the oper*ation*) can be used in other contexts where mathematical or logical operations are to be performed on variables, values, expressions or strings of letters or symbols. For example, the symbols:

$$= \quad < \quad \leqslant \quad > \quad \geqslant \quad \neq$$

are termed *relational operators*. There are also *logical operators* such as NOT, AND or OR. A *binary operation* applies to any *two* variables, values, strings, expressions, sets etc. For example the operation of taking the intersection of two sets is a binary operation. In its most general sense, the word 'operation' can apply to any *procedure* as applied, for example, to expressions or to strings of characters. Integration, for which the operator symbol is

$$\int$$

and differentiation are examples of operations in the wider sense.

1.5.4 Scales of measurement

We now consider the appropriate scale of measurement for numbers, given what it is that the numbers are supposed to represent. In economics, monetary variables such as prices, costs and revenues and physical variables such as output or consumption levels are measured on a *ratio* scale. A ratio scale has a fixed zero and both differences between, and ratios of, numbers are meaningful. We may therefore speak of profits increasing by £500,000, investment rising by DM2.4m, costs doubling or manufacturing output falling by 10 per cent. However, the units in which these variables can be measured are not unique. Prices could be measured in pence rather than pounds, profits in £k rather than pound units and investment in dollars rather than Deutschmarks. But all of these possible conversions have the common property of preserving ratios. For example, if profits double when measured in pounds, they also double in terms of £k and, given a steady exchange rate, they also double in Deutschmark terms. On a ratio scale the unit of measure is unique only up to an *identity transformation*. That is, if we are measuring a variable in units x, then so far as a model is concerned, units y would serve just as well, where

$$y = ax \text{ for any } a > 0$$

For example, if x is pounds and y is Deutschmarks then (at the time of writing) $a = 2.42$. If x is grammes and y is kilogrammes then $a = 0.001$. Any equilibrium conditions derived from the model (such as the equation of marginal revenue with marginal cost) are not affected, though of course the numerical values at which the equilibrium is reached will depend on the units chosen. Ease of use, convention or transactions being conducted in an international unit of account (e.g. the ECU) or a single national currency (usually the US dollar) are common grounds for selecting one unit over another.

It is not always appropriate or even meaningful to measure a variable against a ratio scale. An example from economics and finance is provided by the use of *utility functions* in consumer theory or portfolio theory where alternatives are to be ordered on the grounds of desirability and choices made on the basis of individual or collective preferences. For such purposes an *ordinal* scale is appropriate. When a variable is measured against an ordinal scale, not only can no significance be attached to ratios of numbers, but differences between numbers have no meaning apart from the *sign* of the difference. This is because, as its name suggests, an ordinal scale simply puts into rank order the sets of values of the arguments of the function. To illustrate, suppose that an investor prefers the combination (A) of 20 units of return and 100 units of risk[18] to combination B giving 30 units of return and 160 units of risk. This preference ordering would be reflected by a function f which attached a larger number to combination A than to combination B, i.e. a function f for which

$$f(20, 100) > f(30, 160)$$

If the values actually given by f are 55 and 54 respectively, that is fine. If the numbers attached to the combinations were 1,000,000 and 3 respectively or 1000.001 and 1000 that would serve just as well. It cannot be inferred that because there is 'only' a difference of 0.001 between the two values that there is not much to choose between the combinations. All that matters is that A gets a larger number than B and this may be all that can be said in this and a wide variety of other choice situations. In these circumstances, given any function f which accurately ranks sets of values of its arguments, then *any* transformation of the function f that preserved the ordering would do just as well. For a transformation to be *order preserving*, all that is required is that the transformation function has a positive slope[19] throughout the range of values to which it is to be applied. So for example

$$g = f^2 \text{ or } g = \sqrt{f}$$

would both meet this condition and represent the ordering just as well as *f*. For other purposes an ordinal scale may be too weak and a ratio scale too strong. An *interval scale* falls between these extremes. In an interval scale of measurement absolute differences between values are meaningful but ratios are not.[20] There is no unique zero for the scale. Temperature is measured on an interval scale in degrees Celsius or Fahrenheit. Dates are also measured on an interval scale. To speak of the delivery of stock being a week late makes sense, but no meaning is attached to *ratios* of dates. It is meaningful to speak of one temperature reading being 20 degrees higher than another, but it is nonsense to say that a day on which temperature reached 30 °C was twice as hot as one on which the maximum temperature was 15 °C. A variable measured on an interval scale is unique up to an *order preserving linear transformation*. This is defined as

$$y = ax + b \text{ where } a > 0$$

in which *x* represents the original units of measure and *y* is the transformed unit. The coefficient *a* must be positive for the transformation to be order preserving. The constant term *b* may be positive, negative or zero. Note that with $b = 0$ the identity transformation is produced. An example of such an order preserving linear transformation is provided by the well-known formula for converting degrees Celsius to degrees Fahrenheit:

$$F = 32 + 1.8C$$

In the *Theory of Games*, payoff functions are usually measured on an interval scale. In a two-player zero sum game, the gains of one participant are the losses of the other. In multi-player games, solution concepts often require payoffs to be jointly considered, transferred or combined. Money is often used to measure the payoffs, but an order preserving linear transformation would do just as well.

A *nominal* scale is even weaker than an ordinal scale. When a nominal scale is used, observations are simply put into categories without ordering or quantitative measurement. Official or commercial application forms often call for nominal scale information, for example gender, marital status, salutation, occupation etc. This information can be matched with other records such as purchasing decisions to form the basis of targeted mailings.

The *scales* of measurement discussed above do not stipulate particular *units* of measure. Financial variables may be measured in £, $, DM, yen etc. While the units of volume measure may be tonnes, barrels, litres etc. Within this diversity note that there are important business and economic performance indicators which take the form of ratios and so are independent of units of measure – percentages are *dimensionless* in this sense. For example, the fact that first quarter turnover of Acme PLC is already 35 per cent of last year's total does not depend on whether turnover is measured in £, £k or £m.[21] Rates of change such as percentage growth in gross domestic product or charges such as the interest rate on a loan provide further examples. An important concept in economics is *elasticity*. As a finite approximation, *elasticity of demand* is given by the ratio

$$\frac{\text{\% change in quantity demanded}}{\text{\% change in price}}$$

Demand elasticity is used as a measure of the responsiveness of quantity demanded to changes in price because, unlike the slope of the demand curve, elasticity is independent of the units in which price and quantity are measured.

It can be important to distinguish between *flow* variables and *stock* variables. A production function expresses output q in terms of the inputs employed in its manufacture. For example:

$$q = f(\text{labour, capital, materials})$$

All of these variables, labour, capital, materials and of course output itself, are measured as *rates of flow per unit time*. The time period (day, week, month or year) is selected as is convenient. Similarly, optimal production plans derived from linear programming models relate to resource availabilities and production levels that are expressed as rates of flow per unit time. In contrast, a stock variable, as its name suggests, relates to the total number or aggregate value of items in existence at some point in time. The number of cars on the road, a firm's finished goods inventory or the nation's currency reserves and available supply of capital goods provide examples of stock variables. Macroeconomic models in particular make use of the concept of stock variables. Where resources are limited it is important to distinguish between two types of resource – those resources for which unused supplies today can be carried over to tomorrow and those for which non-use of today's supply means that the productive potential is lost. The former category is called a *pool resource* and the latter a *non-pool resource*. Materials often represent a pool resource whereas labour is non-pool – unworked hours yesterday are not available today.

EXERCISES 1.5

1 Evaluate the following:

(i) $|3 + 7|$
(ii) $|-5|$
(iii) $|3 - 5|$

(iv) $|-5 \ -5|$
(v) $|-5|+|-5|$
(vi) $||-5|-|5||$
(vii) $|(-7)4|$

2 Express in decimal form:

(i) 345 (octal)
(ii) 234 (hexadecimal)
(iii) 1BE (hexadecimal)
(iv) 1011011 (binary)

3 Express the base 10 number 555 in the following forms:

(i) hexadecimal
(ii) octal
(iii) binary

4 (i) Convert the number 2AF from hexadecimal to binary.
(ii) Convert the number 1110110101 from binary to hexadecimal.

5 Express the number 965 as a binary coded decimal.

6 Express the ratio 31250/64 in

(i) fixed point form
(ii) scientific notation

7 Evaluate:

(i) $10 - 4 \times 23$
(ii) $(10 - 4)2^3$
(iii) $-6(9 - 5) - (1 - 5)[(12 - 9) - (3 - 8)]$
(iv) $50 - 0.5\{5[7 - 2] - [5(3 + 7) - 3(4 \times 2^3 - 7)]\} - 5$

8 Against which of the scales of measurement described are the following entities normally measured?

(i) Currency
(ii) Consumer satisfaction
(iii) Delivery dates
(iv) Volume of output
(v) Ethnic origin

9 Which of the following are independent of the units in which the underlying phenomena are measured?

(i) Inventory
(ii) Inflation
(iii) Interest rates
(iv) Capital investment

1.6 SUMMATION AND PRODUCT NOTATION

It is frequently necessary to refer to a list of numbers or expressions which are to be summed. The numbers might represent values of the same variable observed over time, such as the returns on an investment, or values of different variables measured in the same units at a given time – say the end of year balances of the budget centres in an organization. When considering an investment opportunity, the returns over the whole lifetime of the project, after appropriate weighting, will need to be aggregated. To illustrate, suppose that a project will generate returns in each of five successive periods. The returns could be represented as

$$x_1, x_2, x_3, x_4, x_5$$

The list itself could be referred to using a subscript notation as being a set comprising elements

$$x_i \text{ where } i = 1, 2, 3, 4, 5$$

and where x_i is the typical element. We may wish to know the sum of the returns, S. This could be set out in full as

$$S = x_1 + x_2 + x_3 + x_4 + x_5$$

Such an exhaustive listing would clearly not be usable if the number of elements is large. A more compact way to represent the sum S is as follows:

$$S = \sum_{i=1}^{i=5} x_i$$

This is *sigma notation* with Σ being the Greek capital letter sigma. Sigma can be read as an instruction to add all elements of the form x_i, the *terms* of the summation, starting from the subscript value indicated below the Σ, $i = 1$, the *lower limit of summation*, and concluding at the i value indicated above sigma – in this example $i = 5$, which is the *upper limit of summation* in this case. The subscript i is known as the *index of summation*. The use of the capital sigma in this way to indicate a summation is due to Euler. Sigma notation has the great merit of compactness and is also very flexible. For example, if we wanted to consider the returns after the project was up to speed and before it began to run down, we might wish to omit the first and last years from the sum. This is achieved simply by altering the limits of summation:

$$S = \sum_{i=2}^{i=4} x_i = x_1 + x_2 + x_3$$

If the upper and lower limits of summation are the same, just one element is involved. For instance

$$S = \sum_{i=2}^{i=2} x_i = x_2$$

A further special case is where all the terms in the summation are the same. The result will simply be the constant value multiplied by the number of terms that the summation contains, as, for example when each of the terms takes the constant value k:

$$\sum_{i=1}^{i=n} k = k + k + k + k + \cdots = nk$$

So, for example,

$$\sum_{i=1}^{i=6} 7 = 7 + 7 + 7 + 7 + 7 + 7 = 42$$

It is sometimes convenient to split a summation into two or more parts, e.g.

$$\sum_{i=1}^{3}(x_i + y_i) = \sum_{i=1}^{3} x_i + \sum_{i=1}^{3} y_i$$

in which x_i might represent the direct costs of a project in year i with y_i representing the indirect costs. This result is known as the *sum rule*. The result is valid if the addition sign in the above is replaced by the subtraction sign. Note that to achieve economy of notation, the '$i =$' is edited out of the upper limit of summation. A weighted sum may be required – as in investment appraisal where project returns are weighted by factors dependent upon the time of receipt of the return and the interest rate. For the moment simply note that a weighted sum such as

$$w_1 x_1 + w_2 x_2 + w_3 x_3$$

can be written as

$$\sum_{i=1}^{3} w_i x_i$$

A special case of a weighted sum arises when the weights are the same throughout – as for example in

$$\sum_{i=2}^{5} 4x_i^3 = 4x_2^3 + 4x_3^3 + 4x_4^3 + 4x_5^3$$

$$= 4 \sum_{i=2}^{5} x_i^3$$

Note that when, as here, each term in the summation is multiplied by a constant, it is possible to take that constant outside of the summation sign. The following are examples of the ways in which the notation is used:

$$\sum_{x=1}^{3} x = 1 + 2 + 3 = 6$$

which shows the sum of specific numbers rather than variables. In this case the terms are the index of summation itself. A convenient formula exists for finding the value of such a summation, given only the number of terms, the first term and the common difference between the terms.[22] More complicated expressions may require direct calculation. Consider

$$\sum_{x=1}^{20} (0.1x^3 - 2x^2 - 3x + 10)$$

While such a sum *could* be worked out manually, it would take a considerable time with significant probability of error. The sum results quickly from a BASIC[23] program as illustrated in Section 1.8. As a further example consider

$$\sum_{n=2}^{5} \frac{1}{n} = \frac{1}{2} + \frac{1}{3} + \frac{1}{4} + \frac{1}{5}$$

$$= \frac{30}{60} + \frac{20}{60} + \frac{15}{60} + \frac{12}{60}$$

$$= \frac{77}{60}$$

Summations may contain an *unlimited* number of terms. This arises when there is no upper limit and/or no lower limit to the index of summation. For instance the summation

$$\sum_{n=1}^{\infty} 2^{-n} = \frac{1}{2} + \frac{1}{4} + \frac{1}{8} + \frac{1}{16} + \frac{1}{32} + \cdots = 1$$

shows a *sum to infinity*. In this case there is a finite sum and the series is said to *converge*. A financial example is provided by securities such as Consols or War Loan, the returns on which are perpetual.[24] The sum of a finite number of terms of an infinite series (up to $n = 100$, say, in the case above) is known as a *partial sum*. A summation may be taken over more than one index. Two examples are provided by the problems of finding the variance of the yield on a portfolio of many securities, and of finding the variance of time taken to perform a sequence of tasks with random durations. Consider the double sum

$$\sum_{i=1}^{i=2} \sum_{j=1}^{j=2} (c_{ij} x_i x_j)$$

The most convenient way to work this sum is to take the first value of the index i and run j through all its values, then take the next value of i and again run the j index through all its values – and so on. Adopting this approach, the sum above, set out in full, is

$$c_{11} x_1 x_1 + c_{12} x_1 x_2 + c_{21} x_2 x_1 + c_{22} x_2 x_2$$
$$= c_{11}(x_1)^2 + (c_{12} + c_{21}) x_1 x_2 + c_{22}(x_2)^2$$

It is not necessary in general for the two indices to have the same upper and lower values, and there may be more than two indices over which to sum. The summation required may be the values of an expression which is a function of the values of the indices. For example, consider the double sum

$$\sum_{x=1}^{8} \sum_{y=1}^{7} (x^2 - xy - y^2 + 2x + 3y)^2$$

which is easily evaluated using a program such as that given in Section 1.8. Summation notation is widely useful in finance, economics, queuing theory, linear programming, discounted cash flow analysis and practically any application involving series.

Product notation

There are also circumstances when it is the *product* of the terms that is called for, as when finding the discount factor to apply to a cash flow element when year to year interest rates differ. In such cases *product notation* is required, with the symbol Π being used to indicate the product of terms in just the same way that Σ indicates the sum. Thus:

$$\prod_{k=1}^{k=n} x_k = x_1 . x_2 . x_3 \ldots x_{n-1} . x_n$$

represents the product of n terms x_k. The product may involve values of the index itself as in the *factorial* of a number. This is

$$\prod_{k=1}^{k=n} k = 1 \times 2 \times 3 \times 4 \times 5 \times \cdots \times n$$

$$= n!$$

where $n!$ is read as 'n factorial'. Thus when $n = 6$

$$\prod_{k=1}^{k=6} k = 720 = 6!$$

Section 1.8 includes a BASIC program for calculating the factorial of a number. Where there is no upper limit to the value of the index, as in

$$\prod_{k=1}^{\infty} x_k$$

we speak of an *infinite product*. It is important to note that the value of the product of an infinite number of terms is not necessarily itself infinite. For example:

$$\prod_{k=1}^{\infty} \frac{4k^2}{4k^2 - 1} = 1.570796\ldots = \frac{\pi}{2}$$

which is known as *Wallis's product*.[25]

EXERCISES 1.6

1 Evaluate:

(i) $\sum_{t=0}^{t=4} 2^t$

(ii) $\sum_{t=-2}^{t=2} (0.5)^t$

(iii) $\sum_{t=0}^{t=3} 3^t$

(iv) $\sum_{t=-2}^{t=1} 2^{-t}$

(v) $\sum_{t=-2}^{t=2} (0.2)^{-2t}$

2 Evaluate:

(i) $\prod_{k=1}^{k=7} k$

(ii) $\prod_{k=-4}^{k=4} k$

(iii) $\prod_{k=0}^{k=4} k!$

3 Write out in full:

(i) $\sum_{i=1}^{i=6} x_i$

(ii) $\sum_{j=0}^{j=3} kx_j$

(iii) $\sum_{i=1}^{i=2} [kx_i + (y_i)^2]$

(iv) $\sum_{i=1}^{i=4} w_i x_i$

(v) $\sum_{i=1}^{i=2} \sum_{j=1}^{j=2} v_i w_j x_i y_j$

4 Express concisely in sigma notation:

(i) $y_1 + y_2 + y_3 + y_4 + y_5$
(ii) $-x_0 - x_1 - x_2$
(iii) $2x_1 + 2x_2 + 2x_3 + 2x_4$
(iv) $w_0 x_0 + w_1 x_1 + w_2 x_2$
(v) $w_1 x_1 y_1 + w_1 x_2 y_1 + w_1 x_3 y_1 + w_2 x_1 y_2 + w_2 x_2 y_2 + w_2 x_3 y_2$

1.7 SEQUENCES AND SERIES

In finance, economics and business the decision-maker is frequently presented with sequences of numbers. The numbers may represent the quarterly interest payments on an investment, successive values of GDP or annual profit figures for the last decade. With their elements taken in succession, each of these collections of numbers represents a *sequence* — an ordered set of numbers that is either finite or *denumerable*.[26] The set of cash returns to a 20-year project,

$$R_1, R_2, R_3, \ldots, R_t, \ldots, R_{19}, R_{20}$$

is a sequence. The symbol R_t, indicating the return in year t, represents the typical member of the sequence and is called the *general term*. Where successive terms of a sequence are related in the same way, the sequence is called a *progression*. The way in which pairs of terms are related defines a particular kind of progression. Because of their importance in finance and economics, we shall be concerned with *arithmetic progressions* and *geometric progressions*. In an arithmetic progression, consecutive terms differ by a fixed amount. Thus, for example, the sequence

$$50, 60, 70, 80, 90, 100$$

is an arithmetic progression. So is the sequence

$$13, 10, 7, 4, 1, -2, -5, \ldots$$

The natural numbers also form a sequence as does the set of integers. In the first of the progressions above, the *common difference* between successive terms is 10 and the first term is 50. In the second example, the first term is 13 and the common difference is -3. In general, if the first term is represented by a and the common difference by d, any arithmetic progression containing n terms will have the form:

$$a, a+d, a+2d, a+3d, \ldots, a+(n-1)d \qquad (1.2)$$

The example progressions above have, respectively,

$a = 50, d = 10, n = 6$

and

$a = 13, d = -3, n = \infty$

Using the first example to check the value of the last term as given by (1.2), this should be

$50 + (6 - 1)10 = 100$

which is correct. A financial example of arithmetic progression is provided by *simple interest*. Under simple interest, the same fixed payment is made in each period regardless of the amount on deposit or the sum outstanding. Thus if £100 is banked on 1 January 1994 at 12.5 per cent simple interest per annum, the amount on deposit will increase by £12.50 each year, and a running record of these amounts makes up an arithmetic progression as follows:

1994	1995	1996	1997	1998	1999
100	112.5	125	137.5	150	162.5

The value of the investment after n years can be obtained by use of (1.2) but a little care is required. The value after one year is the *second* term in the progression and in general the value after n years is given by the $(n + 1)$th term in the progression: $a + nd$. So, since $a = 100$ and $d = 12.5$, the value after 20 years will be

$a + nd = 100 + 20(12.5)$
$ = 350$

A further illustration of the use of arithmetic progressions is provided by *straight line depreciation*. In this method, the book value of an asset in any year will be its book value of the preceding year less a constant amount. If an asset is initially valued at £50,000 and is written off over 10 years at £5,000 per year, its book value over time will form an arithmetic progression:

	Initial value	One year	Two years	Three years
Book value:	£50,000	£45,000	£40,000	£35,000

The nth term of this progression gives the book value after $(n - 1)$ years. Thus, since $a = 50,000$ and $d = -5000$, the book value after seven years (corresponding to term eight in the series) will be

$50,000 + (8 - 1)(-5000) = 15,000$

It is often useful to know the sum of a number of terms in a sequence. Such a sum is called a *series*. The sum of a sequence can be written concisely in sigma notation. So for example

$$\sum_{t=1}^{t=20} R_t$$

represents the sum of a sequence of 20 terms where the general term is R_t. Where the sequence has a finite number of terms the series is said to be a *finite series*; otherwise it is an infinite series. The number of terms is the *length* of the series. An important type of series is one which takes the following form:

$$a_0 + a_1 x + a_2 x^2 + a_3 x^3 + \cdots$$

This is a *power series*. Power series are sometimes used to give a more tractable representation of a function. A particularly important power series is the *Taylor series*. The sum S_n of an arithmetic progression containing n terms could be written out in full as

$$S_n = a + (a + d) + (a + 2d) + (a + 3d) + \cdots + [a + (n-1)d]$$

but it is possible to develop a formula for this sum by noting the result of adding the terms in a particular way.[27] If the first and last terms are added, the result will be

$$a + [a + (n-1)d] = 2a + (n-1)d$$

and if the second and second last terms are added

$$(a + d) + [a + (n-2)d] = 2a + (n-1)d$$

Similarly, if the third and third last term are added:

$$(a + 2d) + [a + (n-3)d] = 2a + (n-1)d$$

As may be surmised, the terms can be paired off so that the sum of all pairs is $2a + (n-1)d$. Since there are $n/2$ pairs all told, the total must be

$$S_n = \frac{n}{2} [2a + (n-1)d] \tag{1.3}$$

The formula (1.3) is a great timesaver! For example, to find the sum to 100 terms of the arithmetic progression

$$10, 15, 20, 25, 30, \ldots$$

note that $a = 10$, $d = 5$ and, of course, $n = 100$. Therefore, using the formula, the sum is

$$S_{100} = 50 [20 + (99)5]$$
$$= 27{,}750$$

For the progression

$$100, 95, 90, 85, \ldots$$

the value of a is 100 and $d = -5$. So, from (1.3), the sum to 31 terms will be

$$S_{31} = 15.5[200 + 30(-5)]$$
$$= 775$$

while the sum to 100 terms is

$$S_{100} = 50[200 + 99(-5)]$$
$$= -14,750$$

The formula can also be used in a different way to answer questions such as the following. What is the value of n that is required to make the sum to n terms reach zero for the progression

$$-200, -190, -180, -170, \ldots$$

In this case, $a = -200$ and $d = 10$. Therefore, the value of n is required for which

$$S_n = \frac{n}{2}[-400 + (n-1)10] = 0$$

i.e.

$$\frac{n}{2}[-410 + 10n] = 0$$

so, discounting the trivial case in which n itself is zero, the sum is zero when

$$-410 + 10n = 0$$

i.e. after 41 terms. A formula such as (1.3) which states the required sum in terms of known parameters is called a *closed form expression*. Where closed form expressions can be obtained they are extremely useful. Note that the formula (1.3) will work even if the number of terms is odd. In the case of odd n, the solitary middle term will be the $[(n+1)/2]$th term with a value of

$$a + \{[(n+1)/2] - 1\}d$$

which is one-half of $2a + (n-1)d$. Useful application of (1.3) is found in simple interest calculations. Suppose that someone is in the habit of borrowing £100 annually at 5 per cent simple interest. What would be the total interest that they would pay over a 20-year period? The interest payments in each year form an arithmetic progression as can be seen from the Σ row in Table 1.1.

Table 1.1

£100 invested at	Interest payable at				
	$t=1$	$t=2$	$t=3$	$t=4$	$t=5$
$t=0$	5	5	5	5	5
$t=1$		5	5	5	5
$t=2$			5	5	5
$t=4$				5	5
$t=5$					5
Σ	5	10	15	20	25

The total of interest payments over the 20-year period is found from (1.3) with the values $a=5$, $d=5$ and $n=20$. The result is

$$\frac{20}{2}[2(5) + 19(5)] = 1{,}050$$

A further example of the use of the formula is provided by a common system of royalty payments. Suppose that a firm of publishers have decided on a print run of 5000 copies of a particular book for which no reprint is envisaged. The book will sell at £10 per copy. Royalty payable to the author is expressed as percentages of the sales price and is calculated on a sliding scale as follows:

First 1000 copies	10%
Next 1000 copies	11.25%
Third 1000 copies	12.5%
Fourth 1000 copies	13.75%
Fifth 1000 copies	15%

If the book sells out, what will be the total royalty payment to the author? The royalty payments form an arithmetic progression in which the first term is 10 per cent of £10 (1000) = £1000. Each successive tranche is greater than its predecessor by 1.25 per cent of £10,000, so the common difference is £125. Application of formula (1.3) produces

$$S_n = \frac{5}{2}[2{,}000 + (5-1)125]$$

$$= 6{,}250$$

In a *geometric progression* it is the *ratio* of each term to the preceding term that is constant throughout. In other words, the terms differ by a constant multiplier. The following progressions are geometric:

5, 10, 20, 40, 80, 160, ...

128, 32, 8, 2, 0.5, 0.125, ...

Successive terms in each of these progressions take the following form:

$$a, ad, ad^2, ad^3, ad^4, ad^5, \ldots, ad^{n-1}$$

where the first term is a, the constant factor of difference between adjacent terms, the *common ratio*, is d and the nth term is ad^{n-1}. The number of terms in a geometric progression must be finite or else denumerable. In the first illustration, $a = 5$ and $d = 2$ while in the second case $a = 128$ and $d = 0.25$. In demography, if a population grows at a constant annual rate from a base level a at a given time, the annual population figures form a geometric progression. For example, if the annual rate of growth is 2 per cent, then $d = 1.02$, and if the initial population is 50,000,000, then the population figures year by year will form the progression

50,000,000   51,000,000   52,020,000   53,060,400   54,121,608 ...

The inexorable nature of proportionate growth is apparent from the progression, as is the fact that the year on year increase, although constant in percentage terms, gives rise to ever increasing *actual* additions to the population.[28] There are many examples of the use of geometric progressions in business and finance. In accounting, under the *declining balance* method of depreciation, the first term, a, represents the asset's original book value, the factor of difference, d (which must be less than one), gives the proportionate change in value from year to year, and the $(n + 1)$th term of the progression gives the value of the asset after n years. For example, if a car depreciates in value by 20 per cent each year, then the common ratio will be:

$$d = 1 - 0.2 = 0.8$$

and the value of the car after n years of depreciation is given by the $(n + 1)$th term of the geometric progression. Thus the value of a car bought initially for £9000 would be, after three years,

$$ad^n = £9000(0.8)^3 = £4608$$

In finance, geometric progressions are important in compound interest calculations and the sum to n terms is of particular importance. As was the case with arithmetic progressions, it is possible to obtain a closed form expression for the sum S_n:

$$S_n = a + ad + ad^2 + ad^3 + ad^4 + \cdots + ad^{n-1}$$

Note that

$$dS_n = ad + ad^2 + ad^3 + ad^4 + \cdots + ad^{n-1} + ad^n$$

from which it is clear that S_n and dS_n have all but their first and last terms in common. That is:

$$S_n - dS_n = a - ad^n$$

so

$$S_n(1-d) = a - ad^n = a(1-d^n)$$

so that

$$S_n = \frac{a(1-d^n)}{1-d} \tag{1.4}$$

Many instances of the use of (1.4) are found in business finance, where it is one of the most useful of all expressions. Here we give one example, and draw attention to a particular property of (1.4). An *annuity* consists of the receipt (or payment) of a fixed amount each year. The *present value* of an annuity is the lump sum to which the future receipts are at present equivalent.[29] If the *discount rate* that is appropriate is $100r$ per cent, then the present value of an annuity of £1 received at the end of each of n years will be

$$\frac{1}{1+r} + \frac{1}{(1+r)^2} + \frac{1}{(1+r)^3} + \frac{1}{(1+r)^4} + \cdots + \frac{1}{(1+r)^n}$$

$$= \sum_{t=1}^{t=n} (1+r)^{-t} \tag{1.5}$$

The constant difference in the annuity sum is $1/(1+r)$ and the first term, a, is also $1/(1+r)$. Use of this information in (1.4) produces

$$S_n = \frac{(1+r)^{-1}[1-(1+r)^{-n}]}{1-(1+r)^{-1}}$$

$$= \frac{1-(1+r)^{-n}}{r} \tag{1.6}$$

Expression (1.6) allows rapid calculation of the present value of any annuity. For example, with a discount rate of 10 per cent ($r = 0.1$) an annuity of £50 for six years would have a present worth of

$$£50 \frac{[1-(1+0.1)^{-6}]}{0.1} = £282.24$$

The important property of (1.4) is that if $d < 1$ the series has a bound on its value even if the number of terms is infinite – it is convergent. In a sum to infinity, the index of summation may be unlimited from above, from below, or in respect of either extreme, as with

$$\sum_{t=a}^{\infty} x_t \quad \text{or} \quad \sum_{-\infty}^{t=b} x_t \quad \text{or} \quad \sum_{-\infty}^{\infty} x_t$$

Where there is *no* finite limit to the sum to infinity, the series is said to be *divergent*. Divergence may mean that the sum $S_n \to \infty$ or that $S_n \to -\infty$ or that S_n oscillates. If in (1.4) the value of d is less than 1, it is clear that as n increases without limit, d^n approaches zero. This can be written as

$$\text{as } n \to \infty \quad d^n \to 0$$

So

$$\text{as } n \to \infty \quad S_n \to \frac{a}{1-d} \tag{1.7}$$

Relation (1.7) is important in financial calculations and can be used to give the present value of certain undated stocks (such as Consols). Progressions, sequences and series may be specified by some general rule or by the use of a *recursion formula*. For example, factorials lend themselves well to recursive definition in the following way:

$$n! = n(n-1)!$$

which, if n was 6, leads to the progression

$$6! = 6 \times 5! = 720$$
$$5! = 5 \times 4! = 120$$
$$4! = 4 \times 3! = 24$$
$$3! = 3 \times 2! = 6$$
$$2! = 2 \times 1! = 2$$
$$1! = 1 \times 0! = 1$$

in which it should be noted that the value of 0! is defined as 1. Note also that if recursion is used there must always be a *terminating condition*. In the case of factorials, termination occurs when 0! is encountered.

EXERCISES 1.7

1 Find the sum to n terms of the following progressions:

  (i) 50, 60, 70, 80, ... (where $n = 20$)
  (ii) 150, 135, 120, 105, ... (where $n = 15$)
  (iii) −1000, −950, −900, −850, ... (where $n = 25$)

2 For what value of n does the sum to n terms of the following progression reach zero?

$$-2{,}000{,}000, -1{,}920{,}000, -1{,}840{,}000, -1{,}760{,}000, \ldots$$

3 Find the eleventh term in the geometric progression:

   2.5, 5, 10, 20, ...

4 Find the sums of the following geometric series to the stated number of terms:

   (i) 1.5, 3, 6, 12, 24, ...
   (10 terms)
   (ii) 1024, 512, 256, 128, ...
   (12 terms)
   (iii) 1,000,000, 1,100,000, 1,210,000, 1,331,000, ...
   (7 terms)

5 Find the sum to infinity for:

   (i) 256, 64, 16, ...

   (ii) $\dfrac{100}{1.2}, \dfrac{100}{(1.2)^2}, \dfrac{100}{(1.2)^3}, \ldots$

1.8 BASIC PROGRAMS

For the first BASIC program consider a summation required in Section 1.6. This was

$$\sum_{x=1}^{20} 0.1x^3 - 2x^2 - 3x + 10$$

This sum is computed using a loop construction. One way of doing this is shown in Program 1.1.

Program 1.1

```
FOR x = 1 TO 20

    forthisx = 0.1*x^3 - 2*x^2 - 3*x + 10

    sumsofar = sumsofar + forthisx

NEXT x

PRINT sumsofar
```

Although efficiency could be improved, the program will deliver the goods in most dialects of BASIC.[30] Exponentiation is represented by a caret (^)

50 Mathematics for business, finance and economics

and multiplication by an asterisk (*). In this program, 'forthisx' and 'sumsofar' are labels for variables. Keystrokes could have been saved by using y and z as variable labels, but it is helpful when rereading an old program, or in trying to understand someone else's work, to have such descriptive variable labels. It is scarcely an issue here, but processing time could have been economized marginally by defining x as an integer variable. Program 1.1 uses a FOR–NEXT loop.[31] For each value of x, forthisx holds the corresponding value of the expression, and sumsofar then adds to its own previous value the current value of forthisx. When the calculations have been prepared for the last value of x, the loop is exited and the program prints the current value of sumsofar – which is the required total. This is −1760. Program 1.1 could be polished up with printed statements about what the program does, 'error trapping', and a better interface with the user. However, these matters are not our central concern. The second program relates to the double sum required in Section 1.6. This was

$$\sum_{x=1}^{8} \sum_{y=1}^{7} (x^2 - xy - y^2 + 2x + 3y)^2$$

which is computed by the following BASIC program.

Program 1.2

```
FOR x = 1 TO 8
      FOR y = 1 TO 7
            thiscomb = (x^2 - x*y - y^2 + 2*x + 3*y)^2
            sumsofar = sumsofar + thiscomb
      NEXT y
NEXT x
PRINT sumsofar
```

The program output shows that the value of the double sum is 41692. Note that the FOR–NEXT loop for the variable y lies within the loop for x. So for each value of x, y will run through its full range of values. Then the next value of x is taken, and so on. Loops can be nested within loops as required by the summation, up to the limits imposed by the software. The next program prints the factorial of a number (up to a limit determined by the size of number that the software can accommodate).[32] The program is as follows.

Program 1.3

```
INPUT 'Number for which the factorial is required'; m
   IF m = 0 THEN
         z = 1
   ELSEIF m < 0 THEN
         z = 0
   ELSE
         z = 1
         FOR i = 1 TO m
         z = z*i
         NEXT i
   END IF
   PRINT z
```

This program will run in QBasic (which comes with MS-DOS 5) or Microsoft QuickBasic. The program first asks (through the INPUT statement) for the number for which the factorial is required; this number is labelled m. Provision is made for the fact that 0! = 1. If the user inputs a negative number, the program will print zero as the result.[33] An 'IF...THEN...ELSE' block construction is used. In the output, factorials up to 10 appear in conventional place value notation without separating commas. The factorial of 10 is shown as

  3628800

For numbers greater than 10, the factorial of the number is printed (with seven figure accuracy) in exponential notation. For 34! the result appears as

  2.958328E + 38

which is of course

  2.958328×10^{38}

In Section 1.6 we saw that *Wallis's product* approached the value $\pi/2$ as the number of terms increased indefinitely. The following program uses this fact to obtain an estimate of the value of π.

Program 1.4

```
INPUT 'Number of terms'; n
prodsofar# = 1
FOR k = 1 TO n
    numerator# = 4*k^2
    denominator# = numerator# - 1
    ratio# = numerator#/denominator#
    prodsofar# = prodsofar# *ratio#
NEXT k
pi# = 2*prodsofar#
PRINT 'Estimate of π by Wallis's product for';
PRINT n; 'terms is'; pi#
```

This program will run in Microsoft QuickBasic or MS-DOS QBasic. It will also run under GW-BASIC if line numbers are added, which can be done by typing in the first line as

    10 INPUT 'Number of terms'; n%

and the second line as

    20 prodsofar# = 1

and so on. The program again uses a FOR–NEXT loop, but this time with a running *product* being calculated. The user must first input the number of terms for which the product is to be evaluated. The '#' symbol specifies *double precision*; the variable 'ratio#' is calculated to more decimal places than would be a variable labelled 'ratio'. Double precision also allows the storage of larger numbers, an important consideration in this program if the number of terms in the series is to be large. The program output states the number of terms and the estimate that emerges. As is evident from the results shown in Table 1.2, Wallis's product makes a slow approach to the value $\pi/2$, over 100,000 terms being required to achieve a value of π accurate to the fifth decimal place!

In this initial review chapter, we have considered aspects of notation, the measurement and representation of numbers and sets and sequences of numbers have been introduced. It is now time to build on these foundations. In Chapter 2 this process begins with a study of linear equations and linear systems and their relevance to business and economic decision making.

Table 1.2

Number of terms	Estimate of π
10	3.06770
50	3.12608
100	3.13379
500	3.14002
1000	3.14081
5000	3.14144
10000	3.14151
25000	3.14156
50000	3.14158
100000	3.14158
200000	3.14159

ADDITIONAL PROBLEMS

1 What is the present value of:

  (i) A sum of £100 to be received eight years hence, under a discount rate of 8 per cent?
  (ii) A debt of £250 to be paid in five years' time, with a discount rate of 12 per cent?

2 What is the present value of the cash flow:

  $t = 1$ $t = 2$ $t = 3$
  100   130   170

  at

  (i) 10 per cent discount?
  (ii) 20 per cent discount?

3 Find the present value of an annuity of £100 for:

  (i) 5 years at 20 per cent interest;
  (ii) 10 years at 5 per cent interest.

4 The population of a country is growing at the rate of 3 per cent per annum. The population at present stands at 100,000,000. What will the population be in:

  (i) Four years' time?
  (ii) Ten years' time?

5 Draw up a table showing the population in 10 years' time and 20 years' time given an initial population of 70,000,000 and annual growth rates of:

 (a) $2\frac{1}{2}\%$ (b) 5%

6 Find the book value of a car after five years, bought initially for £10,000, under:

 (i) Straight line depreciation at £1,000 per year;
 (ii) Declining balance depreciation at 10 per cent per year..

REFERENCES AND FURTHER READING

1 Borowski, E. J. and Borwein, J. M. (1989) *Dictionary of Mathematics*, Collins.
2 Boyer, C. B. (1968) *A History of Mathematics*, Wiley.
3 Chiang, A. C. (1984) *Fundamental Methods of Mathematical Economics* (Third Edition), McGraw-Hill.
4 Clapham, C. (1989) *The Concise Oxford Dictionary of Mathematics*, Oxford University Press.
5 Gottfried, B. S. (1986) *Programming with Basic* (Third Edition), McGraw-Hill.
6 Hart, W. L. (1966) *College Algebra* (Fifth Edition), Heath.
7 Hollingdale, S. (1989) *Makers of Mathematics*, Penguin.
8 Samuels, J. M., Wilkes, F. M. and Brayshaw, R. B. (1990) *Management of Company Finance* (Fifth Edition), Chapman and Hall.

SOLUTIONS TO EXERCISES

Exercises 1.1

1 The number 7 is a *counting number*; -4 is an *integer*; 3.2 is *rational*; π is *transcendental*; $\sqrt{2}$ is *irrational*; $\frac{1}{2}$ is a *rational* number; $1/\sqrt{2}$ is an *irrational* number; e is *transcendental*; 0 is an *integer*; $\sqrt{(-4)}$ is an *imaginary number*; $4 + 2i$ is a *complex number*; 10^{10} is an *integer*. Note that 10^{10} could have been classified as a counting number, although some patience would be required to reach it! *Natural number* is a narrower category than the integers, but was not included in the list.

2 (i) Any whole number, positive, negative or zero; e.g. 20.
 (ii) Any number that can be expressed as the ratio of two integers, e.g.

 $3.142857142857142857\ldots = 22/7$

 (iii) Any real number that *cannot* be expressed as the ratio of two integers, e.g. $\sqrt{2}$.

(iv) An irrational number that cannot be the root of any polynomial in rational coefficients, e.g. π.
(v) Any real multiple of i (where $i = \sqrt{(-1)}$), e.g. $-15i$.
(vi) A complex number is a number that may have real and imaginary components, e.g. $17 - 24i$. The complex numbers include the real numbers and the imaginary numbers as special cases.

3 (i) $14 + 6i$
  (ii) $9 + 3i$
  (iii) 1
  (iv) $-2 + 6i$
  (v) $6i$

Note from (v) and (iii) that the difference between or the sum of two complex numbers could be either (a) imaginary, when the real parts cancel, or (b) real, when the imaginary parts cancel.

4 (i) False. For example, the roots of the equation
$$x^2 + 1 = 0$$
are $\pm i$.
  (ii) True.
  (iii) True.
  (iv) False.
  (v) False. For example, $10/3 = 3.333333\ldots$.

Note, with regard to case (iv), that many models involve parabolas which do not cross or touch the x axis, and which therefore have complex roots.

Exercises 1.2

1 (i) G and J
  (ii) (a) $\{1, 2, 4, 6, 7, 8\}$
      (b) $\{4, 6\}$
      (c) $\emptyset$
      (d) $\{3, 4, 6\}$
      (e) $\{1, 3, 4, 5\}$
      (f) $\{0, 2\}$
      (g) $\{2\}$
  (iii) $2^7 = 128$
  (iv) A, B, D and H

Note re (i) and (iii) that the null set is a subset of all sets.

2 (i) G is not a subset of A at all (since it contains the element 1 which is not a member of A). A is not a *proper* subset of itself.

(ii) (a) {3, 4, 5, 6, 7, 8}
   (b) {3, 5}
   (c) ∅
   (d) {3, 5}
   (e) {3, 4, 5}
   (f) {3, 5, 6, 8}
(iii) (a) {2, 4, 6, 8}
   (b) {2, 3, 5}
(iv) {1, 4, 6}
(v) (a) {6, 7, 8}
   (b) {3, 5, 7}
(vi) D, E and F

Exercises 1.3

1 (i) The *distributive* law.
   (ii) The *commutative* law.
   (iii) The *associative* law.

2 (i) {6, 7, 8}
   (ii) {6, 7, 8}

Note that the equality of (i) and (ii) illustrates the validity of the distributive law in terms of sets.

Exercises 1.4

1 (i) x^9 (ii) x^{12} (iii) x^{15} (iv) x^{60} (v) $81x^4$
  (vi) $x^6 y^8$ (vii) x^3 (viii) x^{-2} (ix) x^3 (x) $x^{2\sqrt{3}}$
  (xi) x^3 (xii) $x^{2-\sqrt{2}}$

2 (i) ±5 (ii) 6 (iii) 4 (iv) 64 (v) ±2
  (vi) −3

Exercises 1.5

1 (i) 10 (ii) 5 (iii) 2 (iv) 10 (v) 10
  (vi) 0 (vii) 28

2 (i) 229 (ii) 564 (iii) 446 (iv) 91

3 (i) 22B (ii) 1053 (iii) 1000101011

4 (i) 1010101111 (ii) 3B5

5 1001 0110 0101

6 (i) 488.28125 (ii) 4.8828125×10^2

7 (i) −22 (ii) 48 (iii) 8 (iv) 20

8 (i) Ratio scale (ii) Ordinal scale
  (iii) Interval scale (iv) Ratio scale (v) Nominal scale

9 Inflation and interest, expressed as percentage rates.

Exercises 1.6

1 (i) 31 (ii) 7.75 (iii) 40 (iv) 7.5
  (v) 651.0416

2 (i) 5040 (ii) 0 (iii) 288

Note that in the case of (iii), $0! = 1$.

3 (i) $x_1 + x_2 + x_3 + x_4 + x_5 + x_6$
  (ii) $kx_0 + kx_1 + kx_2 + kx_3$
       $= k(x_0 + x_1 + x_2 + x_3)$
  (iii) $kx_1 + y_1^2 + kx_2 + y_2^2$
        $= k(x_1 + x_2) + y_1^2 + y_2^2$
  (iv) $w_1 x_1 + w_2 x_2 + w_3 x_3 + w_4 x_4$
  (v) $v_1 w_1 x_1 y_1 + v_1 w_2 x_1 y_2 + v_2 w_1 x_2 y_1 + v_2 w_2 x_2 y_2$

4 (i) $\sum_{i=1}^{i=4} y_i$ (ii) $-\sum_{i=0}^{i=2} x_i$ (iii) $2 \sum_{i=1}^{i=4} x_i$

  (iv) $\sum_{i=0}^{i=2} w_i x_i$ (v) $\sum_{i=1}^{i=2} \sum_{j=1}^{j=3} w_i x_j y_i$

Exercises 1.7

1 (i) $a = 50$, $d = 10$, $n = 20$, so $S_n = 2900$
  (ii) $a = 150$, $d = -15$, $n = 15$, so $S_n = 675$
  (iii) $a = -1000$, $d = 50$, $n = 25$, so $S_n = -10,000$

2 $n = 51$

Note: The first 25 terms are negative, the twenty-sixth is zero. The next 25 terms are positive and cancel out their negative counterparts.

3 Since $a = 2.5$, $d = 2$ and $n = 11$, the nth term is

$$ad^{n-1} = 5120$$

4 (i) 1534.5 (ii) 2047.5 (iii) 9,487,172

5 (i) $341.\overline{33}$ (ii) 500

Additional problems

1. (i) $\dfrac{100}{(1.08)^8} = 54.03$

   (ii) $\dfrac{250}{(1.12)^5} = 141.86$

2. (i) $\dfrac{100}{1.1} + \dfrac{130}{(1.1)^2} + \dfrac{170}{(1.1)^3}$

   $= 90.9091 + 107.4380 + 127.7235$
   $= 326.0706$

   (ii) $\dfrac{100}{1.2} + \dfrac{130}{(1.2)^2} + \dfrac{170}{(1.2)^3}$

   $= 83.3333 + 90.2778 + 98.3796$
   $= 271.9907$

3. (i) 299.06 (ii) 772.17

4. (i) $100,000,000(1.03)^4 = 112,550,881$
   (ii) $100,000,000(1.03)^{10} = 134,391,638$

5.
	10 years	20 years
$2\tfrac{1}{2}\%$	89,605,918	114,703,151
5%	114,022,624	185,730,839

Note that, with 5 per cent growth, the population size shows a two-and-a-half-fold increase over 20 years.

6. (i) £5000 (ii) £5905

NOTES

1. Further discussed in Section 1.6 below.
2. The wider aspect of commensurability is that the quantities concerned can be expressed in common units.
3. A *root* is a value satisfying a given equation.
4. For example the French geometer and algebraist Viete (1540–1603).
5. Euler also introduced summation notation (see Section 1.6) and the symbols π and e. The equation $e^{i\pi} + 1 = 0$, which links five fundamental constants, is called *Euler's formula* (see Chapter 9).
6. See Borowski and Borwein (1989).
7. Try confirming the fact that $3 + 2i$ *is* a solution by working out the equation for this value of x. Whenever i^2 is encountered, substitute -1.
8. For example Russell's paradox.
9. With bracketed operations being carried out first. See Section 1.5.3 below.
10. We shall also refer to the laws in respect of operations on matrices in Chapter 10.

11 This said, in applied work the positive root is usually taken for granted. The radical sign is inconvenient for algebra and its use is best confined to arithmetic operations. We shall, however, use radicals occasionally for completeness.
12 Using radicals, the negative root can be indicated as $-\sqrt{4} = -2$.
13 Complex roots are not stated here.
14 For a technical discussion of irrational exponents see Hart (1966).
15 Inequalities such as $\leq$ 'less than or equal to' are discussed in detail in Chapter 3.
16 Which may, in fact, have arisen in India.
17 Also included amongst arithmetic operations are negation, modulo arithmetic and integer division.
18 The unit of return may be the yield on a portfolio and the unit of risk the variance of yield. Note that, other things being equal, less risk would be preferred to more.
19 The function should be *isotonic*.
20 However, *ratios of differences* are meaningful on an interval scale.
21 When a change of units also changes the information base, then a different statistic may result. This may occur if sales were measured in terms of *volume* rather than *value*.
22 The formula sums an arithmetic series to n terms.
23 Or another language such as Pascal, Fortran or C.
24 The 'war' in War Loan was the Napoleonic War.
25 A program using Wallis's product to estimate the value of π is given in Section 1.8.
26 A set may have an unlimited number of members but if the members of the set can be matched, one to one, with the natural numbers the set is *denumerable*. In contrast, the set of all points between (say) zero and one cannot be put into correspondence with the natural numbers and is not denumerable.
27 Assume for what follows that the value of n is even. The result holds good if n is odd.
28 Proportionate growth is further explored in Chapter 9.
29 For further discussion of annuities and present values see Samuels, Wilkes and Brayshaw (1990).
30 Line numbers used in older versions of BASIC are omitted here. Where necessary, as with GWBasic, begin each line with a number greater than for the previous line, followed by a space.
31 Other loop constructions are available in some versions of BASIC and other languages. Examples are: DO WHILE and DO UNTIL.
32 This is 34! in Microsoft QuickBasic v 4.5.
33 A more sophisticated response – such as printing a message and asking the user to try again – could be included.

Chapter 2

Straight Lines and Linear Equations

2.1	Introduction	61
2.2	Co-ordinates	61
2.3	Straight lines – preliminaries	68
2.4	Identifying straight lines and equations	73
2.5	Two applications	81
2.6	Point–slope and general form	84
2.7	Straight lines and simultaneous equations	87
2.8	Elementary row operations	99
2.9	Other linear systems in up to two variables	101
2.10	Linear systems of equations in more than two variables	104
2.11	Gaussian elimination	112
2.12	Identities	115
2.13	Further applications	117
2.14	BASIC programs	123
	Additional problems	126
	References and further reading	127
	Solutions to exercises	127

In this chapter you will learn important concepts concerning straight lines and linear equations. Sections 2.1–2.7 introduce fundamental ideas and Sections 2.8–2.12 build on this material to introduce more advanced concepts and methods. Sections 2.5 and 2.13 consider specific applications and Section 2.14 provides BASIC programs related to this chapter.

By the end of the chapter, you should be able to plot lines with given equations, express the lines in the most suitable form of equation, and find the equation of any line with two known points or other known properties. You will have covered material useful in linear programming, the calculus and other contexts and will have laid foundations which can be built upon in later chapters.

You will also be able to manipulate and solve systems of linear simultaneous equations in two, three or more variables. You will have examined a number of linear models, learnt of their use in business and economics, and seen how the models themselves can be adapted to accommodate changing circumstances.

2.1 INTRODUCTION

The simplest ideas are usually the most valuable, and turn out to be the ones most widely used. This is only partly due to the fact that simpler concepts are likely to be more widely known. It also results from the fact that fundamental techniques are likely to be of practical value more often than their relatively sophisticated cousins. This principle is nowhere more evident than in the case of straight lines and linear equations which are widely used throughout business studies and economics. Amongst the variety of applications that we shall examine here and in Chapter 3 are

Breakeven analysis
Depreciation
Stock control
Forecasting
Investment income generation
Market demand and supply
Linear programming and machine utilization

Linear expressions recur frequently in these and other contexts throughout later chapters. But we begin with the representation of algebraic expressions in diagrammatic form. It is of enormous value to be able to show equations in a readily assimilable way and also to have the option of using algebraic techniques on a problem originally conceived in geometrical terms.

2.2 CO-ORDINATES

In Chapter 1 we considered the system of real numbers, all of which can be matched with points on the *real line*, a straight line of unlimited extent. This is done by first establishing an arbitrary point on the line to represent zero, and marking off a distance, also discretionary, on one side of the zero point to represent unity. With the real line laid out horizontally, points to the right of zero are designated as corresponding to positive numbers – this again is a conventional choice. Having established the unit interval, all of the positive integers follow as the unit distance is repeatedly added on the right. By the same process the negative integers are established to the left of the zero point. Rational numbers are located by subdivision of the unit intervals while the irrational numbers are located by geometric construction. The real line with the construction for $\sqrt{2}$ is shown in Figure 2.1.

The real line contains all of the real numbers with no gaps remaining unfilled. There is an exact one-to-one correspondence between the points on the real line and the real numbers which are the *co-ordinates* of the points. The real line is a single *co-ordinate axis* which can be used to make comparisons between numbers in terms of algebraic size, the larger numbers being to the right. For example, 3 is to the right of 2, and -2 is to the right

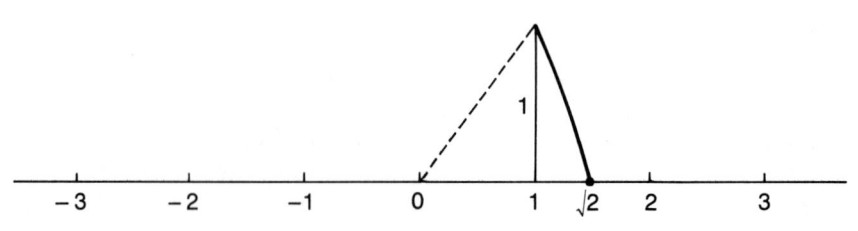

62 Mathematics for business, finance and economics

Figure 2.1

of -3. The line can also be used to show intervals (the numbers lying between specified limits) and other subsets of the real numbers. All of this work relates to a single dimension.

The value of using co-ordinates is greatly expanded when a second real line is introduced at right angles to the original real line and passing through the zero point. The two lines are now termed *co-ordinate axes* and the resulting system of *rectangular* or *Cartesian* coordinates allows all points in two dimensions (the plane) to be uniquely labelled and located by the values of the two co-ordinates. It should be noted that the points on the plane could be located by reference to *any* intersecting pair of straight lines as axes. When the lines intersect at right angles, we speak of a *Cartesian* co-ordinate system. The x axis is sometimes referred to as the *abscissa* and the y axis as the *ordinate*. The co-ordinates represent an *ordered pair* of numbers in which the horizontal or x co-ordinate value is always stated first and the vertical or y co-ordinate value is stated second. These arrangements are a matter of convention rather than logical necessity. It is important to note that the two axes do not have to be measured to the same scale or in the same physical or monetary units. In Figure 2.2 the rectangular co-ordinate system is shown and specific points P_1, P_2, P_3 and P_4 are indicated. The co-ordinates of the points are

$$P_1(2,1) \quad P_2(4,3) \quad P_3(-2,1) \quad P_4(2,-3)$$

The first co-ordinate value represents the distance to be moved from zero in the x direction (taking account of sign) and the second co-ordinate represents the vertical displacement from zero (i.e. against the y co-ordinate axis). Thus the co-ordinates will be written as (x,y) with subscripts as needed to identify particular points. In what follows it should be noted that no results would be invalidated by switching the labelling of the axes or by reversing the positive and negative sides of either axis.

Having identified some particular points in the plane, three questions naturally arise:

1 What is the straight line distance between any two points?
2 What is the slope of the line segment connecting the points?

Straight lines and linear equations 63

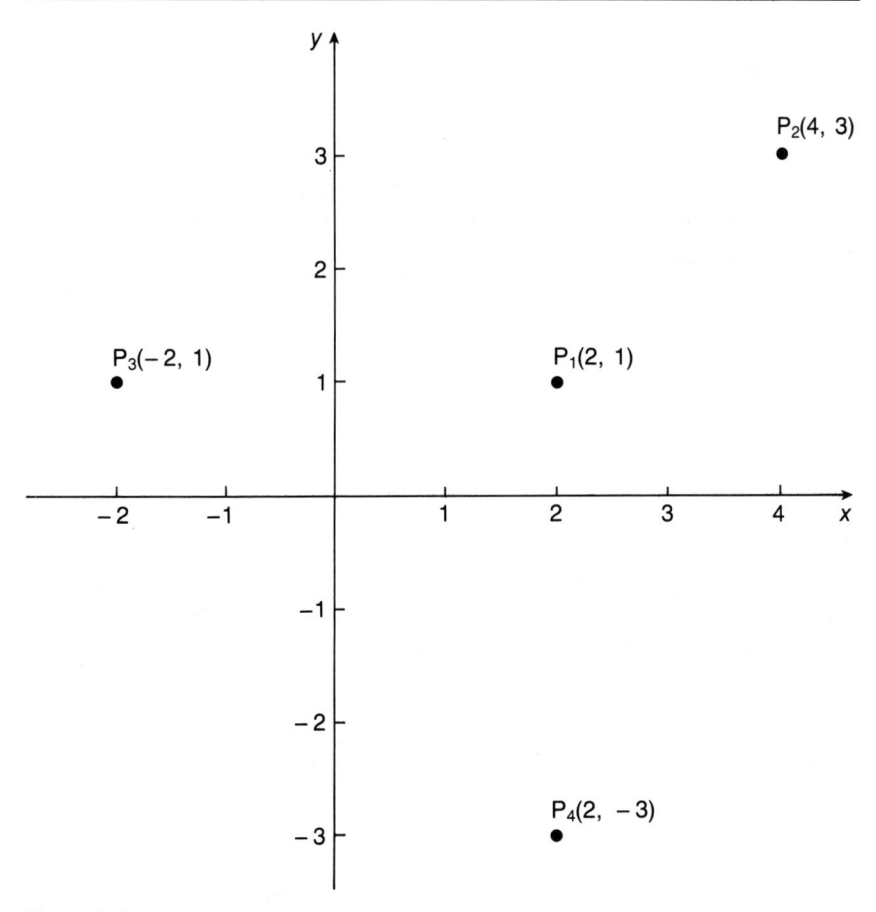

Figure 2.2

3 What is the algebraic expression for the line passing through two particular points?

In this section we shall address questions 1 and 2. Question 3 is considered in the following section. The straight line distance between any two points with given co-ordinates can be found by the use of Pythagoras's theorem. For example the distance between P_1 and P_2 is

$$d = \sqrt{(4-2)^2 + (3-1)^2}$$
$$= \sqrt{8}$$

while the straight line distance separating the points P_4 and P_3 is

$$d = \sqrt{[2-(-2)]^2 + (-3-1)^2}$$
$$= \sqrt{(32)}$$

In general, the *distance formula* for the straight line distance between any two points in the plane with co-ordinates (x_2, y_2) and (x_1, y_1) is

$$d = \sqrt{(x_2 - x_1)^2 + (y_2 - y_1)^2} \qquad (2.1)$$

Two special cases are included within equation (2.1). In the case where the two points are on the same vertical line, the formula simplifies to give the difference between the y co-ordinates, and where the two points are on the same horizontal line the distance is simply the difference in the x co-ordinate values. The full results for distances between the points are

Points	Distance
P_1P_2	2.828427
P_1P_3	4
P_1P_4	4
P_2P_3	6.324555
P_2P_4	2
P_3P_4	5.656854

All lines other than vertical lines have a defined *slope*. The slope of a straight line connecting any two points is the ratio of the change in y values to the change in x values. The slope of the line joining P_2 and P_1 is

$$\text{slope} = \frac{3-1}{4-2} = +1$$

while the slope of the line connecting P_3 and P_4 will be

$$\text{slope} = \frac{-3-1}{2-(-2)} = -1$$

In general, if the co-ordinates of the two points are (x_1, y_1) and (x_2, y_2), then the slope will be

$$\text{slope} = \frac{y_2 - y_1}{x_2 - x_1}$$

It does not matter which of the points is taken to be the 'first' point. The result is the same for distance and slope. This is because in the case of distance the differences are squared, while in the case of slope both the numerator and the denominator of the slope ratio are reversed in sign if the order of the two points is changed. In any slope calculation consistency must be observed as to which point is used as the 'first' point in taking the values for x and y. Note that the slope of the P_1P_3 line is zero as there is no change in the y value, while the slope of the P_1P_4 line is *undefined* since division by zero would be involved (the line is vertical). The full results for the slopes are

Points	Slope
P_1P_2	1
P_1P_3	0
P_1P_4	undefined
P_2P_3	0.333333
P_2P_4	3
P_3P_4	−1

The slope of a line depends on the units of measurement along the vertical and horizontal axes.[1] For example, if the units of measurement along the y axis were originally pounds sterling and if these units are now changed to pence, with no change in the units of measure along the x axis, the slope of all lines is increased a hundredfold. Slope is not a dimensionless concept, which is why in certain applications (such as demand curves) it can be better to work in terms of *proportionate* changes in the variables rather than the absolute changes involved in measures of slope. The two co-ordinate axes divide the plane into four sections called *quadrants*, labelled as shown in Figure 2.3.

The first quadrant, where neither variable is negative, is often called the *positive quadrant*. In the first and third quadrants the two variables have the same sign while in the second and fourth quadrants the signs are

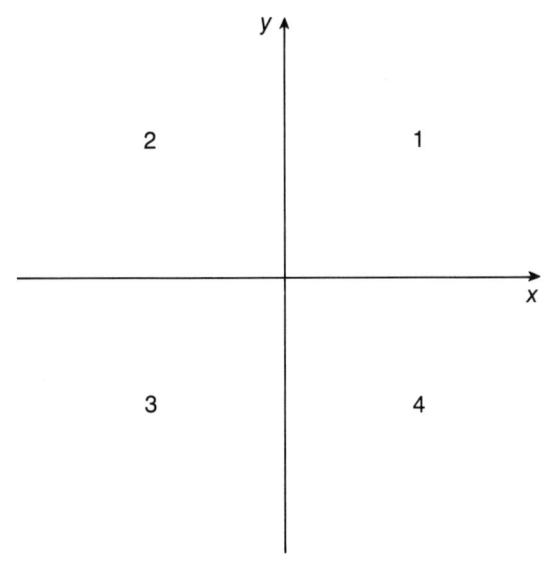

Figure 2.3

66 Mathematics for business, finance and economics

opposite. Points in three-dimensional space can be identified uniquely if a third, z, axis is introduced at right angles to the x, y plane. This is usually shown as in Figure 2.4.

Note that in Figure 2.4 there is a right angle between each pair of axes – as at the corner of a cube. The co-ordinate system divides three-dimensional space into eight *orthants*. Practical work is usually confined to the positive orthant. Although manual sketch diagrams in three dimensions can sometimes be worthwhile, precise geometrical work is not normally attempted by hand. There are several computer software packages that will produce excellent 'wire-frame' diagrams of functions in three dimensions. It is worth investigating such packages, the use of which can enhance understanding of both general concepts and particular functions.[2]

Points in three space have *co-ordinate triples* (x, y, z). The Pythagorean distance formula extends to this case so that the distance d between two points P_2 and P_1 in three-dimensional space can be thought of as the corner to corner diagonal through a cube and where d is given by

$$d = \sqrt{(x_2 - x_1)^2 + (y_2 - y_1)^2 + (z_2 - z_1)^2}$$

For example, the straight line distance between the points

$P_1(8, 9, 10)$ and $P_2(6, 5, 4)$

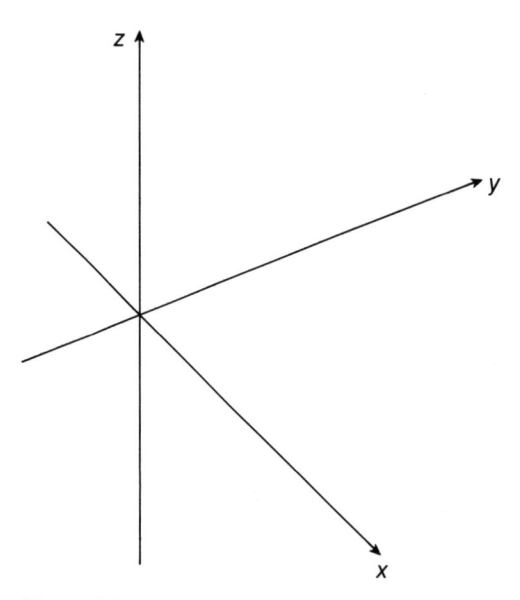

Figure 2.4

is given by
$$d = \sqrt{(8-6)^2 + (9-5)^2 + (10-4)^2}$$
$$= \sqrt{4 + 16 + 36}$$
$$= \sqrt{56}$$
$$\approx 7.48$$

The concept of slope in three-dimensional space is more complicated in that it is necessary to specify the direction of movement. We shall take up this question in Chapter 6. In the next section, we consider linear equations and the straight lines that are the geometrical counterparts of the equations.

EXERCISES 2.2

1 Given the following points:

$P_1(2, 5)$ $P_2(5, 2)$ $P_3(3, -2)$
$P_4(-2, -2)$ $P_5(-1, 3)$

find the distance between the following pairs of points:

(i) P_1 and P_2 (ii) P_2 and P_3 (iii) P_1 and P_5
(iv) P_1 and P_4 (v) P_3 and P_5

2 For the points P_1 to P_5 in exercise 1, find the slopes of the straight lines connecting

(i) P_1 and P_5 (ii) P_1 and P_2 (iii) P_1 and P_3
(iv) P_5 and P_2 (v) P_4 and P_1

3 Suppose that the point $P_6(2, 2)$ is added to the five points above. What can be said about the slopes of the lines connecting the points

(i) P_6 and P_2 (ii) P_6 and P_1

(iii) P_6 and the origin

4 Find the straight line distance between the following pairs of points in three-dimensional space:

(i) $P_1(9, 5, 8)$ and $P_2(5, 3, 4)$
(ii) $P_1(3, 6, 12)$ and $P_2(8, 6, 6)$
(iii) $P_1(7, -2, -8)$ and $P_2(8, -1, 1)$.

2.3 STRAIGHT LINES – PRELIMINARIES

The equation of a straight line can be written in several ways. The most convenient of these is often the *slope–intercept form*:

$$y = mx + c \qquad (2.2)$$

In (2.2) the value of m is the slope and the value of c is the intercept – the point where the line cuts the vertical (y) axis. In Figure 2.5, the line L has *negative slope* ($m < 0$) and *positive intercept* ($c > 0$) while line M has *positive slope* ($m > 0$) and *negative intercept* ($c < 0$).

Line L might represent the equation

$$y = -0.5x + 10$$

while line M might represent the equation

$$y = 1.2x - 5$$

Horizontal lines (parallel to the x axis) have zero slope ($m = 0$) and therefore have the form

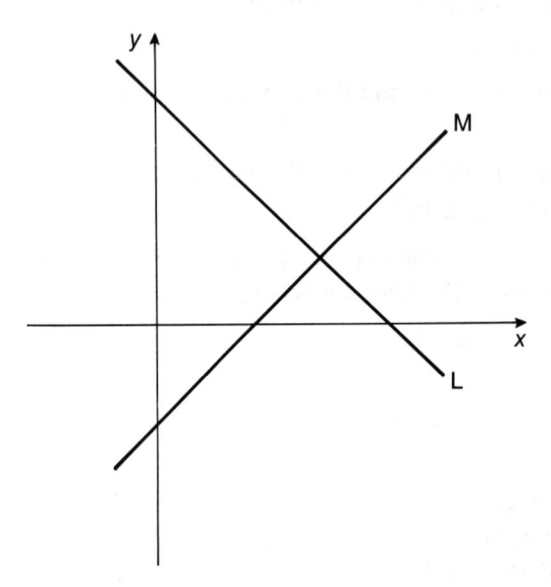

$$y = \text{constant} \qquad (2.3)$$

Figure 2.5

Straight lines and linear equations 69

So the straight line defined by the equation

$$y = 10$$

is a horizontal line 10 units above the x axis while

$$y = -15$$

is a horizontal line 15 units below the x axis. As we shall see, even this most simple form of equation has its uses. *Vertical lines* have an *undefined slope*.[3] For such lines, it is the x value that does not change, and in the ratio that would otherwise define the slope value, division by zero would be involved. Since this is not permitted, the slope is undefined. A vertical line can be written as

$$x = \text{constant} \qquad (2.4)$$

Thus $x = -10$ defines a vertical line that cuts the x axis 10 units to the left of the origin. Equations (2.3) and (2.4) themselves have special cases of interest. The lines defined by

$$y = 0 \text{ and } x = 0$$

are the equations of the x axis and the y axis respectively. Equation (2.2) has a further important special case – where the value of c and the intercept is zero, the straight line going through the origin. An equivalent way of expressing this is to observe that, when the value of c is zero, x and y *always remain in the same proportion*, the factor of proportionality being m, as evidenced by the fact that (2.2) can now be written as

$$\frac{y}{x} = m$$

Returning to the general case of the straight line in slope–intercept form

$$y = mx + c$$

there are two *parameters*, m and c, each of which can be independently selected and which together uniquely identify a particular straight line. If the value of c is varied while m remains constant a *family of lines* is generated, with the common value of slope as the familial characteristic. Some members of the family of lines given by $m = +0.5$ are shown in Figure 2.6.

If y represents costs and x represents output level, then a family of lines defined by slope is created if fixed costs c, are changed while unit variable cost, given by m, remains the same. Note that lines belonging to the same family defined by slope are parallel. Thus the straight lines defined by

$$y = 4x + 10$$
$$y = -7 + 4x$$

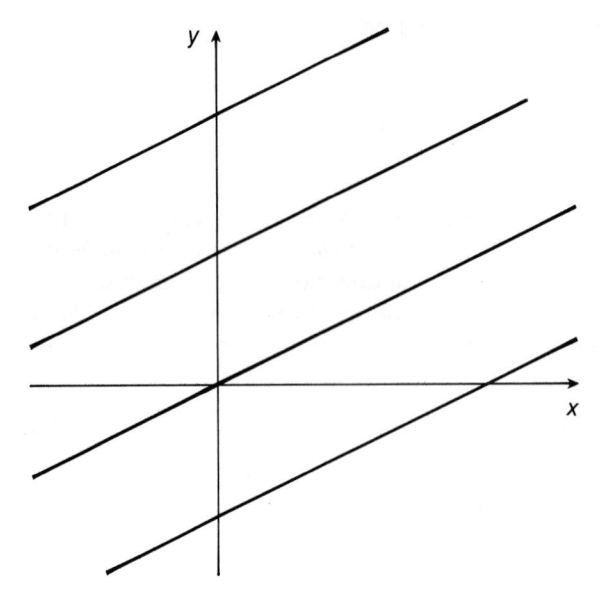

Figure 2.6

are parallel. A family of lines can also be defined by the common characteristic of passing through a given point. Such lines are also said to be *concurrent*. Figure 2.7 shows members of the family of lines passing through the point $(0, c)$.

Note that in Figure 2.7 one of the lines passing through the point $(0, c)$ is the y axis itself (having the equation $x = 0$). In the context of a linear cost function, concurrent lines would be generated if fixed costs c remained constant while unit variable cost m changed. Finally, note that it is not necessary that the common point is on an axis for the lines to be concurrent. The point could be anywhere in the x, y plane. As we have seen, lines which belong to the same family *defined by slope* are parallel. In contrast, *perpendicular* lines intersect at right angles and, except for the case of horizontal and vertical lines, have slopes of m and $-1/m$ respectively (i.e. where $m \neq 0$). Perpendicular lines are also described as *normal* or *orthogonal*. So the equations of two perpendicular straight lines can be written as

$$y = mx + c$$

and

$$y = \frac{-1}{m} x + k$$

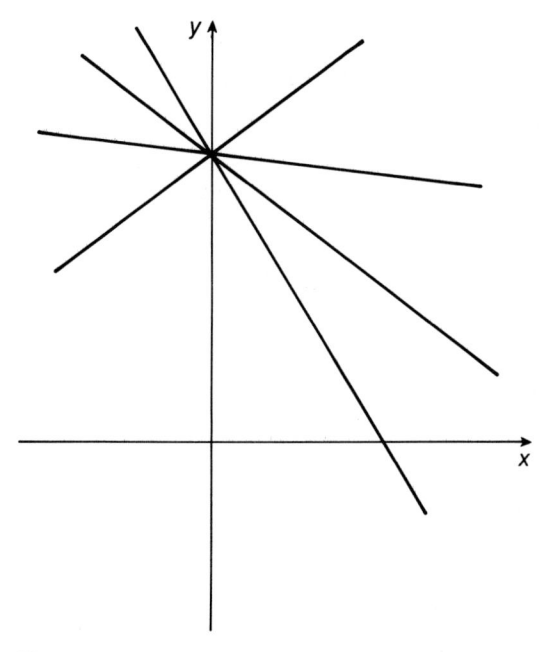

Figure 2.7

where c and k are the y intercepts of the lines. Thus, for example, the lines

$$y = 2x - 4$$

and

$$y = 3 - 0.5x$$

are perpendicular. These lines are shown in Figure 2.8.

Note that $y = 3 - 0.5x$ is not the *only* straight line that is perpendicular to the line $y = 2x - 4$. Any straight line with equation in the form

$$y = -0.5x + k$$

is perpendicular to the line $y = 2x - 4$. Thus the lines

$$y = 100 - 0.5x$$

and

$$y = -0.5x$$

are also orthogonal to $y = 2x - 4$ (and to *any* line having a slope of 2).

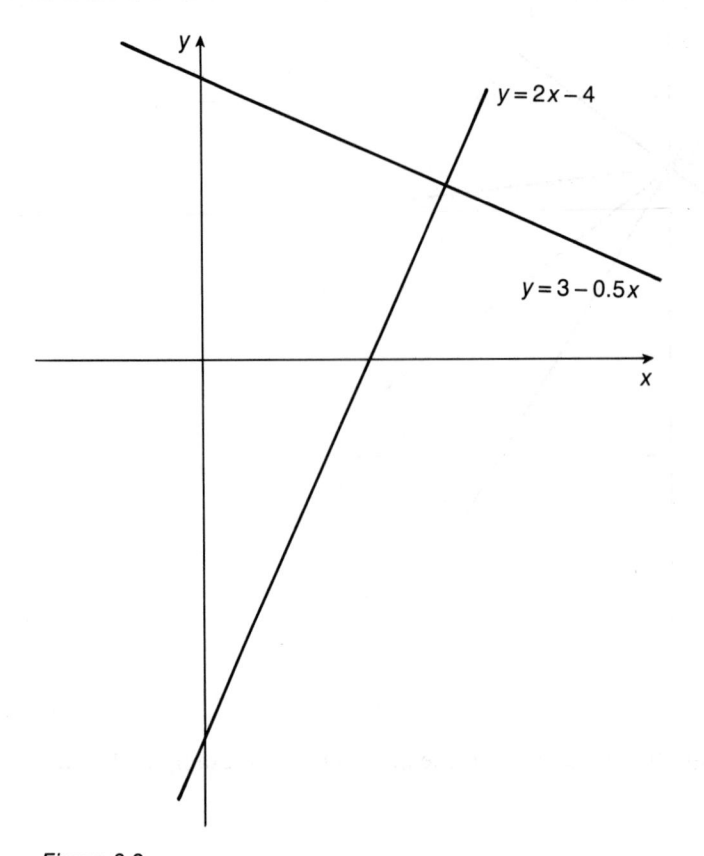

Figure 2.8

Similarly, any line the equation for which satisfies

$$y = 2x + c$$

is orthogonal to the line $y = 3 - 0.5x$. Another way the relationship of orthogonality can be expressed is that *the product of the slopes of perpendicular lines is minus one.*[4] One important use of orthogonal lines is in mathematical programming, where movement from a current position in a direction perpendicular to the contours of the function being optimized represents movement in the direction of *steepest ascent* and therefore the fastest rate of increase of the objective.

EXERCISES 2.3

1 What can be said about the slope of the straight lines:

   (i) $y = 3x + 5$
   (ii) $y = 10 - 0.5x$
   (iii) $y = 50$
   (iv) $y = \sqrt{2}x$
   (v) $x = 25$

2 For the following straight lines:

   (i) $y = 5x + 4$
   (ii) $y = 4x + 5$
   (iii) $y = 10x + 4$
   (iv) $y = 5x - 17$
   (v) $y = -x + 4$
   (vi) $y = 5x$
   (vii) $y = 8x$

   Identify those lines which are concurrent, and those which are parallel.

3 Which of the following lines are orthogonal:

   (i) $y = 5x + 10$
   (ii) $y = 0.2x - 0.1$
   (iii) $y = 20 - 0.2x$
   (iv) $y = -5x + 6$
   (v) $y = -5x$

2.4 IDENTIFYING STRAIGHT LINES AND EQUATIONS

To use straight lines effectively in business and economic applications, some basic skills are necessary. In relation to an individual straight line these include:

1 obtaining the equation of a straight line from a literal description of the relationship between the variables;
2 plotting a straight line with a given equation;
3 finding the equation of a line with known slope going through a given point;
4 finding the equation of a line going through two known points;
5 establishing on which side of a line a given point lies.

Consider these skills in turn.

Obtaining a linear equation from a literal description

The starting point for building many models in business applications is frequently a literal description of some part of the overall problem, and it is important to be able to make use of information in this form. For example, suppose that the following statements are made:

(a) 'Total costs are the sum of fixed and variable costs. Variable costs accrue at the constant rate of £6 per unit produced, while inescapable costs are £10,000.'
(b) 'The maximum price that can be charged for the product is £500 less 10p for each unit sold.'
(c) 'Sales revenue is price times quantity. The market currently determines price at £10 regardless of the number of units that we sell.'

The above statements correspond to linear relationships described by the following equations:

(a) $y = 10,000 + 6x$

   (in which y is total costs and x is production volume);

(b) $y = 500 - 0.1x$

   (in which y is price and x is quantity sold);

(c) $y = 10x$

   (where y is sales revenue and x is volume of sales).

Plotting a straight line with a given equation

One reason why this is a useful skill is that a sketch diagram of relationships is a great aid to understanding and a useful way to prompt interesting questions. A unique straight line is identified if any two distinct points on the line are known. The two given points are then connected and the line is extended beyond each of them. Given the equation of the line, the two points that are usually the most convenient to find are the intercepts with the co-ordinate axes. There is one exception which we shall encounter shortly, but consider for the moment the line with the equation

$$y = -4x + 20$$

The point where the line crosses the y axis is easiest to establish. This is obtained by setting $x = 0$ in the equation and solving for y. The result is $y = 20$ and the line is known to pass through the point $(0, 20)$. Next, the x intercept is identified by setting $y = 0$ and solving for x. The result is

$$0 = -4x + 20$$

so that

$$4x = 20$$

i.e.

$$x = 5$$

So the point $(5, 0)$ is on the line. Note that the intercept with the x axis represents a solution of the equation

$$mx + c = 0$$

and the x intercept will therefore occur at the point $x = -c/m$. Having identified two points, the straight line that passes through them can be drawn in. For the present example this is shown in Figure 2.9.

The exceptional case mentioned earlier is where the straight line passes through the origin so that the x and y intercepts are the same. In this case a further point is generated by inserting a convenient value of x into the

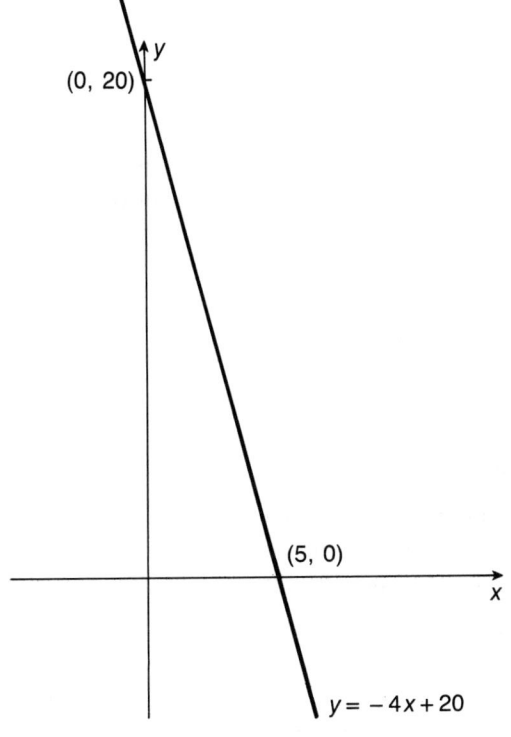

Figure 2.9

equation and the point with the resulting co-ordinates is then joined to (0, 0). For example if the line has the equation

$$y = -4x$$

a convenient value of x is $x = 1$ for which the corresponding value of y is -4. To identify the line the origin should be connected with the point $(1, -4)$. Note that lines which pass through the origin will have points in only two of the four quadrants.

Equation of a line with known slope and given point

The method for finding the equation of a straight line with known slope passing through a given point is a useful skill in cases where the rate of change is known and where a single observation of corresponding x and y values is available. The method will be explained through an example. Suppose that it is known that a line has slope $+2$ and passes through the point $(3, 5)$. The coefficients in the straight line equation must fit the data given. Thus in the slope–intercept 'template'

$$y = mx + c$$

the fact that the slope is 2 produces

$$y = 2x + c$$

so the one remaining unknown is the value of c. The co-ordinates of the point on the line ($x = 3, y = 5$) inserted into the template mean that

$$5 = 2(3) + c$$
$$5 = 6 + c$$
$$c = -1$$

so that the equation of the line is

$$y = 2x - 1$$

For further practice, we shall find the equations of the following:

(a) the line with slope -3, passing through the point $(3, 8)$;
(b) the line with zero slope, passing through the point $(3, 8)$;
(c) the line having undefined slope, and passing through the point $(3, 8)$.

In example (a), the equation must take the form

$$y = -3x + c$$

and, given that the point $(3, 8)$ must satisfy the equation, c must be such that

$$8 = -3(3) + c$$

so that

$$c = 8 + 9$$
$$= 17$$

and the full equation is therefore

$$y = -3x + 17$$

In (b) the slope is zero, so the equation of the line must read

$$y = 0x + c$$

in which the value of x is immaterial (as the line is horizontal). With the given point $(4, 5)$, the value of c must be such that

$$5 = 0(4) + c$$

i.e.

$$c = 5$$

so that the equation of the line is simply

$$y = 5$$

In example (c) the line is vertical. We can get at the equation of the line by realizing that for vertical lines it is the value of y that is immaterial (x will always be the same). In other words,

$$x = 0y + c$$

which, with the given point $(5, -6)$, means that

$$5 = 0(-6) + c$$

i.e.

$$c = 5$$

and so

$$x = 5$$

Equation of a line through two known points

Finding the equation from two known points is a useful skill when a relationship is known to be linear and where two observations of pairs of x and y values have been made. If two points on the straight line are specified, the co-ordinates of these points inserted into the straight line equation produce two simultaneous equations in m and c, one equation corresponding to each point. The equations are then solved for m and c. Methods for solving simultaneous linear equations will be discussed in more

78 Mathematics for business, finance and economics

detail later in the chapter; for now we shall consider two examples – solved by the *elimination procedure*. The problem is as follows:

What line is determined by the two points (2, 13) and (−1, 4)?

The two parameters of the line, m and c, must fit the data given. Thus with

$$y = mx + c$$

the two points produce, respectively,

$$13 = m(2) + c$$

i.e.

$$13 = 2m + c$$

and, for the second point,

$$4 = m(-1) + c$$

i.e.

$$4 = -m + c$$

If, term by term, the second equation is subtracted from the first, c disappears leaving:

$$13 - 4 = [2m - (-m)] + (c - c)$$

which simplifies to

$$9 = 3m$$
$$m = 3$$

The value $m = 3$ can now be inserted into either of the equations with the result that $c = 7$. The equation of the line is therefore

$$y = 3x + 7$$

Now consider a further example. Find the equation of the straight line passing through the points

(3, 9) and (−1, −15)

The equations produced are

$$9 = 3m + c$$

and

$$-15 = -m + c$$

Subtraction now produces:

24 = 4m

m = 6

and substitution of this value into either of the original equations yields the result that c = −9, so that the equation of the line is

y = 6x − 9

Note that this solution procedure is particularly convenient in these circumstances, as the coefficient of c will always be unity in both equations.

Establishing on which side of a line a point lies

The ability to determine on which side of a given line a point lies is important when the equation of the line represents a limit − for example

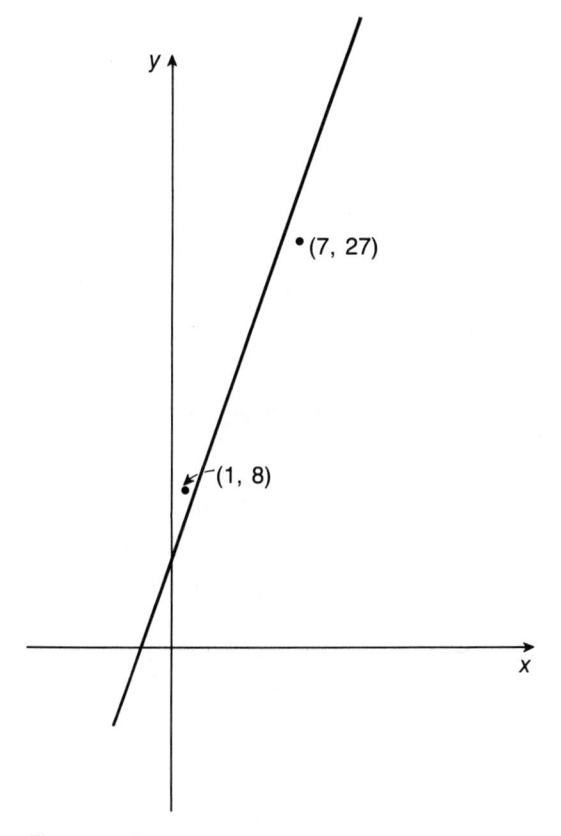

Figure 2.10

80 Mathematics for business, finance and economics

the maximum amount of resource available – or a measure of performance, for example return for given risk in portfolio theory. It will most often be necessary to know whether the y co-ordinate of the point is greater or less than that given by the equation. Thus given the point (1, 8) and the line

$$y = 4x + 2$$

the insertion of the x co-ordinate value of 1 into the equation produces $y = 6$ so that the point is *above* the line. In contrast the point (7, 27) is below the line since the y value given by the equation when $x = 7$ is 30. The two points and their relation to the line are illustrated in Figure 2.10. To set this process in an economic context, suppose that x represents output and that y represents total costs of production (when resources are used efficiently) within the current technology. Thus an output of 7 units produced at a cost of £27 will be super-efficient and thus must correspond to either an improvement in production technology or a reduction in resource prices.

The slope–intercept form of equation will most often be the preferred algebraic expression of a straight line in applications where there is a clear *subject of the equation* or *dependent variable*, i.e. when the value of y clearly follows from the value of x. Some important applications are considered in the following section.

EXERCISES 2.4

1 Find the equations of the following lines:

(i) Slope = 4, passing through the point (5, 25)
(ii) Slope = −5, passing through the point (18, 10)
(iii) Horizontal, passing through the point (7, 10)
(iv) Vertical, passing through the point (8, 4).

2 (i) For a particular straight line passing through the origin, if x changes, then y changes by half the amount but in the opposite direction. What is the equation of the line?
(ii) What is the equation of the line that passes through the points (2, 10) and (−1, 1)?

3 (i) What is the equation of the straight line passing through the points (1, 7) and (2, 11)?
(ii) For a given line, if x changes in value, then y changes by the same amount in the opposite direction. The line passes through the point (4, 6). What is the equation of the line?

4 (i) What is the equation of the straight line passing through the points (2, −1) and (4, 9)?
(ii) What is the equation of the straight line passing through the point (1, 3) and which is orthogonal to the line $y = 4 − 0.5x$?

5 Given the straight line

$$y = 5x - 2$$

establish whether the following points are on, above or below the line:

(i) (2, 9)
(ii) (3, 13)
(iii) (−2, −11)
(iv) (1, 2)

2.5 TWO APPLICATIONS

In this section we consider the use of straight lines in the prediction of company income and in monitoring the level of stock. In *forecasting* with simple *linear regression*, a straight line of best fit is calculated using observations of corresponding values of two variables. Conventionally the equation is written in the form

$$Y = a + bX$$

The intercept term a and the slope coefficient b are set at values which together match the data as closely as possible. With X as the independent variable, there should be grounds for supposing that changes in X actually bring about changes in the dependent variable Y; there should be a causative relation between the variables. For example, X may represent income and Y expenditure, or X may be the outlay on sales promotion while Y represents total sales revenue. Quite often the independent variable represents time, either abstract time ($t = 0, 1, 2, \ldots$) or calendar date. To illustrate, suppose that observations of the level of a company's turnover (sales revenue) have been related to time in the following manner:

$$Y = 45 + 3.8X$$

where Y is the turnover (in £m) and $X = 0$ represents the year 1982, $X = 1$ represents 1983 and so on. The regression equation can be used to give a single figure forecast of turnover in future years *on the assumption that the underlying relationship is unchanged*. For example, to generate a forecast value for 1996, a value of $X = 14$ should be inserted into the equation, with the result that the value of Y predicted is £98.2m. For 1997, the forecast value of Y passes the hundred million mark for the first time − at £102m. A value of $X = -3$ inserted into the equation would indicate, assuming that the relationship was valid at that time, that the company's turnover in 1979 was £33.8m. The regression line is shown in Figure 2.11.

If the regression equation was estimated on the basis of observations from 1982 to 1995 inclusive, then the predictions for 1996 and 1997

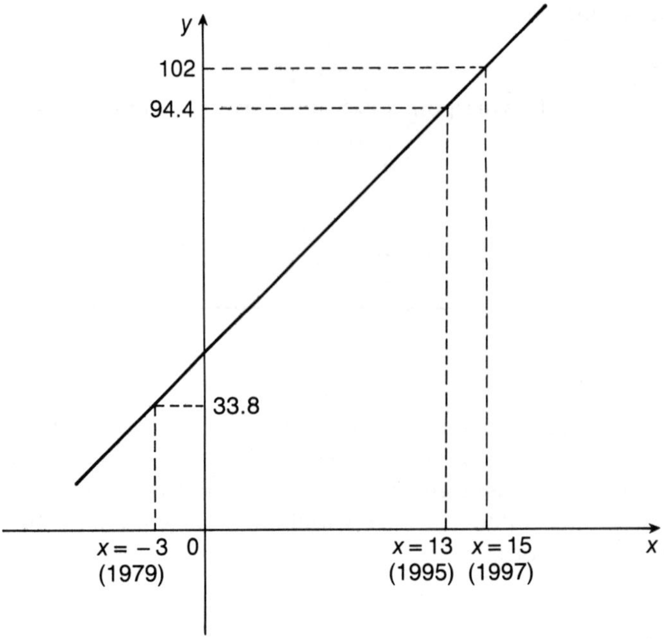
y
102
94.4
33.8
x = −3
(1979)
0
x = 13
(1995)
x = 15
(1997)
x

Figure 2.11

represent *extrapolation* (as indeed does the estimation of the value for 1979). Use of the equation to generate values within the period 1982–1995 represents *interpolation*. These cases are distinguished as there is generally more confidence in the model structure being valid if the X value stays within the range of sample values. The further that X goes outside of this range, the less is the likelihood that the model structure still applies.

Inventory control

Consider the classical model of *inventory* or *stock control*. In this model, a single item of stock is withdrawn from inventory at a known and constant rate. Suppose that a department has an opening inventory of 525 reams of paper which are used at the rate of nine reams per day. Because of the delay between ordering and delivery (the *lead time*) orders for replenishment of stock are placed when stock on hand falls to 150 units. The relationship between inventory level I and the number of working days, t, is linear with the intercept term being the opening inventory and where the slope is given by the rate of withdrawal from stock. Thus

$$I = 525 - 9t$$

is the equation of stock against time. The equation can be used to find the

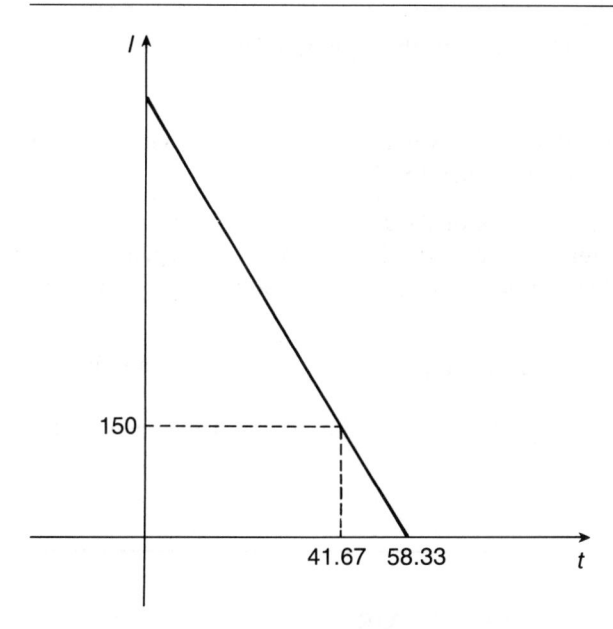

Figure 2.12

number of days before an order for replenishment is placed. This will be the value of t such that

$$I = 525 - 9t = 150$$

Therefore

$$9t = 375$$

so that

$$t = 41.67$$

and so the order for replenishment should be placed during the forty-second working day. The graph of inventory against time is shown in Figure 2.12.

Note that (ignoring any practical problems involving fractions of days) a *stockout* would occur if lead time exceeded $150/9 = 16.67$ days. The level of stock which triggers a re-order is known as the *re-order level* (150 here).

EXERCISES 2.5

1 A company's profits in year t are forecast by linear regression to be $\pi(t)$ where

$$\pi(t) = 80 + 5t$$

where $t = 0$ corresponds to 1992.

(i) What value of profit is given by the equation for

   (a) 1994   (b) 1997

(ii) Assuming that the relationship was valid at the time, what would be the estimate of the profit figure for 1989?

2 A company has an opening stock of 1000 units of an item, the stock level being depleted by 40 units per day's trading. An order for replenishment of stock is placed when the level of inventory is sufficient to cover six days' demand.

   (i) Write down an equation showing stock level against the number of days trading.
   (ii) Given the current policy, on what day should a replenishment order be placed?

2.6 POINT–SLOPE AND GENERAL FORM

The *point–slope* form of equation can be written as

$$y - y_1 = m(x - x_1) \tag{2.5}$$

where in (2.5) (x_1, y_1) are the co-ordinates of a specific point which it is known lies on a line of slope m. For example, suppose that we require the equation of the line of slope 2 passing through the point $(1, 5)$. Putting these co-ordinates and an m value of 2 into equation (2.5) produces the result

$$y - 5 = 2(x - 1) \tag{2.6}$$

which could then be rearranged into slope–intercept form as

$$y = 2x + 3 \tag{2.7}$$

Note that while the point–slope form is as in (2.5) the 'details' as in (2.6) will depend on the point selected, but these should rearrange into the same slope–intercept expression. For example, suppose that instead of the point $(1, 5)$ another point on the same line – say $(-5, -7)$ – had been given. Insertion of the data into equation (2.5) produces

$$y - (-7) = 2[x - (-5)]$$

i.e.

$$y + 7 = 2(x + 5) \tag{2.8}$$

which rearranges into

$$y = 2x + 3$$

as before. The point–slope form arises from the fact that, given one point (x_1, y_1) on a particular line, any other unspecified point (x, y) on the same line must be such as to result in the same slope given by

$$\frac{y - y_1}{x - x_1} = m \tag{2.9}$$

Indeed, equation (2.9) is itself known as the *gradient form* of the equation of the straight line through the point x_1, y_1 with slope m. The situation is illustrated in Figure 2.13, where the slope m is (as always for straight lines) the ratio of the change in y and the change in x between any two points on the line.

The point–slope form (2.5) represents cross multiplication from (2.9). Straight line equations can also be written in *general form*. This is

$$Ax + By + C = 0 \tag{2.10}$$

so that the line $y = 2x + 3$ would appear in general form as

$$2x - y + 3 = 0 \tag{2.11a}$$

or equally well as

$$-2x + y - 3 = 0 \tag{2.11b}$$

In (2.10) the equation is not asymmetric – no one variable is singled out for special treatment as the dependent variable (the *subject of the equation*).

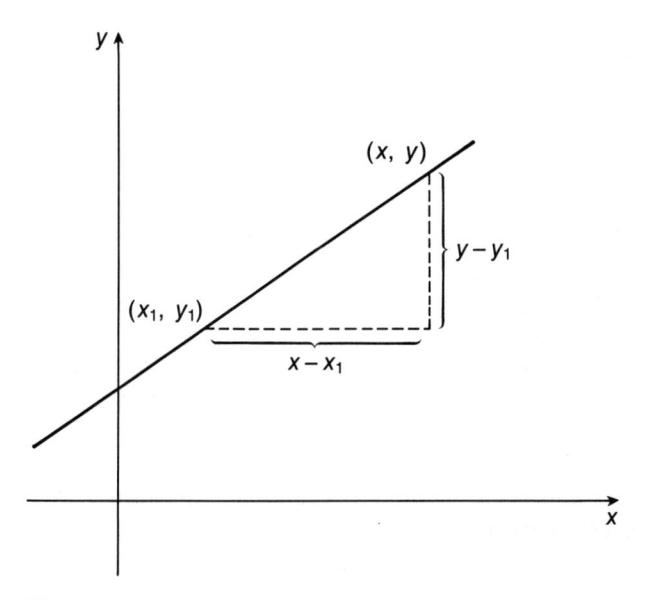

Figure 2.13

The general form (2.10) is *the general equation of the first degree in two variables*. One advantage of general form is that it extends readily to linear expressions involving more than two variables. For example, a linear relationship between three variables x, y and z can be written as

$$Ax + By + Cz + D = 0 \tag{2.12}$$

This produces a plane surface in three dimensions. A *straight line* in three spaces can be seen as the intersection of two planes. A variant of the general form is that in which all of the variable terms are on the left-hand side of the equation (the *LHS*) and the constant term is on the right-hand side (*RHS*) as in

$$ax + by = c$$

This is frequently the manner in which linear equations arise (e.g. as constraints) either individually or simultaneously. The equation written in slope intercept form as

$$y = 2x + 3$$

would appear as

$$2x - y = -3$$

or as

$$-2x + y = 3$$

If the equation is presented in this form it is less obvious on which side of the line a particular point lies. For example, with the line given by

$$-2x + y = 3$$

does the point with co-ordinates $(-4, -6)$ lie above or below the line? In fact the point is below the line since when $x = -4$ is substituted into the equation the resulting value of y is -5. Since this value is greater than the y co-ordinate of -6, the point lies below the line. Equations in which the constant term and no variable terms are on the right-hand side frequently occur in problems involving resource limitations. For example, suppose that x and y represent the number of units purchased of each of two types of security, the prices of which are £5 and £4 per unit respectively. If a budget of £2000 is to be invested, the linear equation that results is

$$5x + 4y = 2000$$

In this context, a point above the line would represent expenditure over budget, while a point below the line would correspond to an underspend.

EXERCISES 2.6

1 Use the point–slope form to express the equations of the lines:

   (i) with slope 3 passing through the point $(5, 8)$;
   (ii) with slope -6 passing through the point $(8, 12)$;
   (iii) with slope -0.5 and y intercept at -20;
   (iv) with a $45°$ slope passing through the origin.

2 Express in general form, the equation of the straight line that passes through the points $(5, 5)$ and $(-5, 35)$.

3 A straight line having slope 4 passes through the point $(5, 25)$. Express the equation of the line in

   (i) point–slope form;
   (ii) slope–intercept form;
   (iii) general form.

4 For the line given by

$$-5x + 2y = 20$$

is the point P $(8, 32)$ located on, above or below the line?

2.7 STRAIGHT LINES AND SIMULTANEOUS EQUATIONS

Given two straight lines L and M exactly *one* of the following relationships must hold:

1 L and M have *all* of their points in common;
2 L and M have *none* of their points in common;
3 L and M have exactly *one* point in common.

These mutually exclusive alternatives are graphed in Figure 2.14. In Figure 2.14(a) L (broken line) and M (dotted line) are superimposed. For practical purposes the lines are the same although their equations may arise in such a way that this is not obvious at the outset. For example, the equation of the line L might be

   L: $y = 7.5 - 0.25x$

while the line M might arise in the following form:

   M: $2x + 8y - 60 = 0$

Of course, it must be possible to convert, by permissible operations, the expression for L into the expression for M and vice versa.

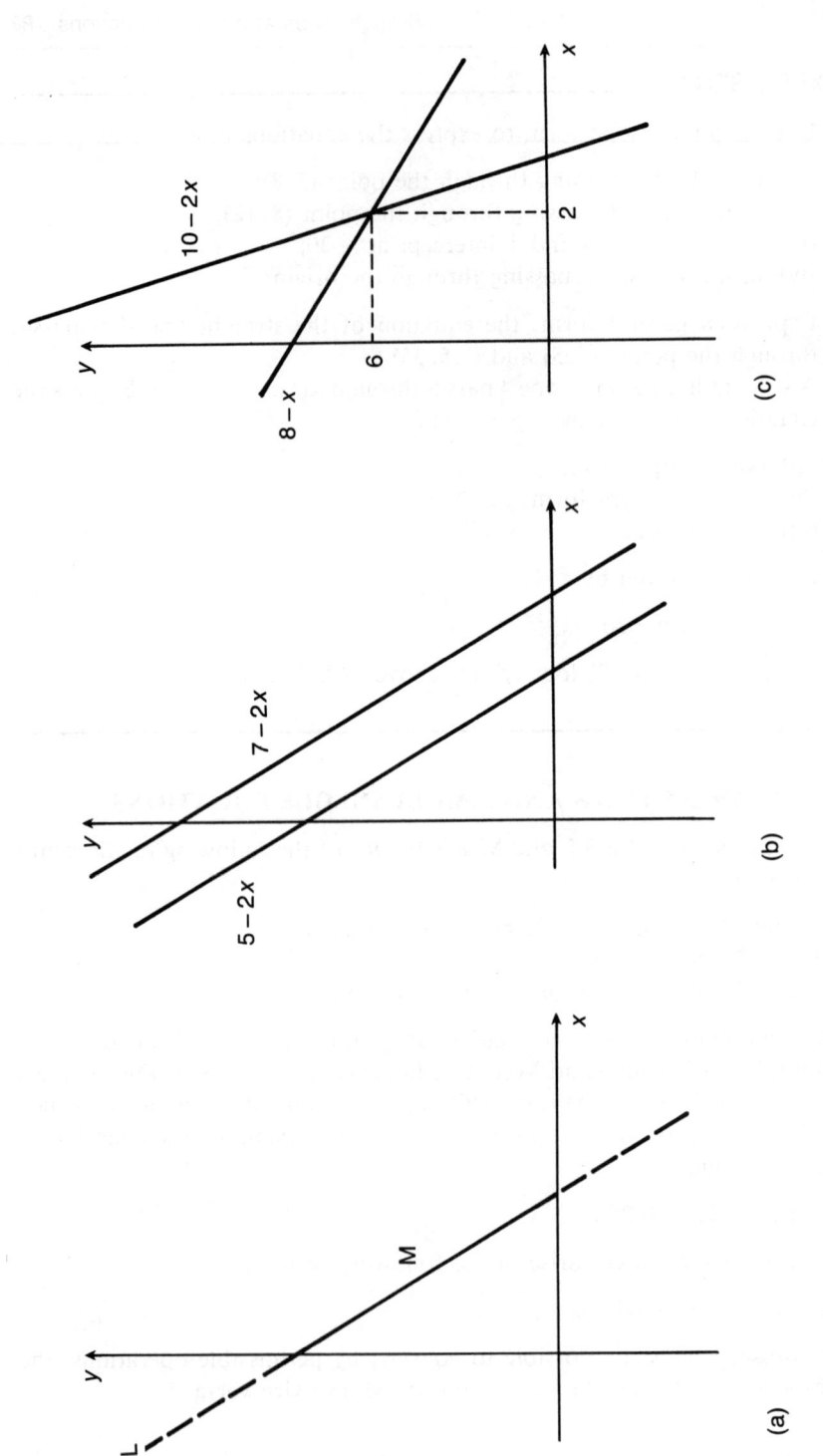

Figure 2.14

In Figure 2.14(b) the lines are parallel. In slope–intercept form the m value is the same but the c values are different. The lines might be

L: $y = -2x + 5$
M: $y = -2x + 7$

In Figure 2.14(c) the lines have different slopes. This is the important distinction between this case and cases (a) and (b). Straight lines of different slope must cross once and once only. For example, the equations of the lines in Figure 2.14(c) might be

L: $y = -2x + 10$
M: $y = -x + 8$

The one point that the lines have in common – when both equations are satisfied – is the point (2, 6). This point of intersection of the lines corresponds to the simultaneous solution of the equations that define the lines. For a unique solution of linear equations to exist, the equations have to be both *independent* and *consistent*. Two equations are independent if it is not possible to obtain one equation from the other by multiplication through by a non-zero constant.[5] Thus cases (b) and (c) represent lines with independent equations. Consistent equations are those which can be simultaneously satisfied. Thus cases (a) and (c) represent lines with consistent equations. Clearly then, only case (c) corresponds to equations which are both independent *and* consistent. Table 2.1 summarizes the properties of each of the three cases above.

The three cases can also be distinguished in terms of set theory if the lines are thought of as sets of points. The point(s) in common, if any, can be seen as the intersection of the two sets. Table 2.2 shows how the three possibilities can be presented.

Solving simultaneous equations

There are several ways in which systems of linear equations can be solved. In Chapter 10, methods using matrices and determinants are covered. Later

Table 2.1

Case	Independent	Consistent	No. of solutions
(a)		*	Infinite
(b)	*		0
(c)	*	*	1

Table 2.2

Case	L ∩ M	Remarks
(a)	L or M	Same m, same c
(b)	∅	Same m, different c
(c)	(x, y)	Different m

in this chapter we consider Gaussian elimination, but at this point we shall introduce two simple methods:

1 *substitution*
2 *elimination*

Substitution

The method will be introduced in the context of finding a solution to the following two simultaneous equations:

$$3x + y = 11 \qquad (2.13)$$
$$x + 2y = 12 \qquad (2.14)$$

Note the form in which the equations are expressed. This is often the manner in which the equations arise and it is convenient for solution. First express (2.13) with y as the subject of the equation:

$$y = 11 - 3x \qquad (2.15)$$

Now *substitute* the right-hand side of (2.15) into (2.14) in place of y. The result is

$$x + 2(11 - 3x) = 12$$
$$x + 22 - 6x = 12$$
$$-5x = -10$$
$$x = 2$$

Now substitute 2 for x in equation (2.15). The result is

$$y = 11 - 3(2)$$
$$y = 5$$

The solution of the simultaneous equations (2.13) and (2.14) is the unique pair of values $x = 2$, $y = 5$. This outcome could equally well have been obtained by the initial use of (2.14) to express x in terms of y as follows: from (2.14)

$$x = 12 - 2y$$

which when substituted into (2.13) produces

$$3(12 - 2y) + y = 11$$

so that

$$36 - 6y + y = 11$$
$$36 - 5y = 11$$
$$-5y = -25$$

and therefore

$$y = 5$$

and from either (2.13) or (2.14), $x = 2$. Consider another example in the use of the substitution approach. Suppose that the equations are expressed in general form as follows:

$$7x + 4y - 51 = 0$$
$$-3x + 7y + 48 = 0$$

From the first of the equations

$$4y = 51 - 7x$$

i.e.

$$y = 12.75 - 1.75x$$

which, when substituted into the second equation, gives

$$-3x + 89.25 - 12.25x + 48 = 0$$

so that

$$15.25x = 137.25$$

Therefore

$$x = 9$$

and y then emerges as

$$y = 12.75 - 1.75(9)$$

so that

$$y = -3$$

Elimination

In the method of elimination, one of the equations is first multiplied through by a constant chosen to be such that the coefficients of x and y become the same in each equation. Thus if (2.13) is multiplied throughout by 2 the result is

$$6x + 2y = 22$$

If (2.14) is subtracted from the modified (2.13), the terms in y will cancel, eliminating y, and one equation in x alone results. In detail:

$$6x + 2y = 22$$
$$x + 2y = 12$$
$$\overline{}$$
$$5x = 10$$
$$x = 2$$

Substitution of $x = 2$ into either equation will produce $y = 5$. The result could equally well have been obtained by multiplication of (2.14) by 3 to give equal coefficients of x in each equation. In this way, x would have been eliminated as follows:

$$3x + y = 11$$
$$3x + 6y = 36$$
$$\overline{}$$
$$-5y = -25$$
$$y = 5$$

In this case the minor inconvenience of the negative values following subtraction could have been avoided by subtraction of (2.13) from the modified (2.14). The two equations are graphed in Figure 2.15 and the solution values of x and y are indicated.

Once $x = 2$ has been obtained substitution of this value into either equation is equivalent in geometric terms to moving up the line $x = 2$ until it hits either equation and then moving parallel to the x axis to establish the value of y as shown by the arrows. The diagram confirms that it does not matter into which equation $x = 2$ is substituted; nor does it matter whether y is obtained first rather than x. We now consider four exercises in the use of the methods of substitution and elimination. These exercises will be followed by illustrations from finance and management.

Exercise 1

$$5x - 2y = 26$$
$$-x - 4y = 8$$

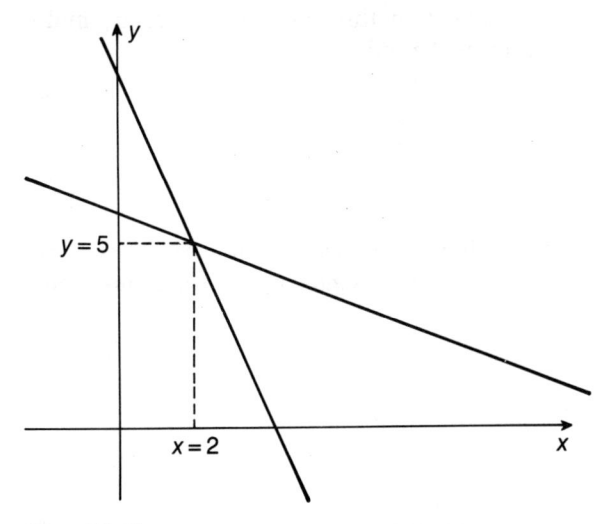

Figure 2.15

Here the use of substitution is suggested by the second equation in which x appears with unit (negative) coefficient. Rearrangement produces

$$x = -4y - 8$$

which when substituted into the first equation results in

$$5(-4y - 8) - 2y = 26$$

so that

$$-20y - 40 - 2y = 26$$
$$-22y = 66$$

Therefore $y = -3$ and from the rearranged second equation

$$x = -4(-3) - 8$$
$$= 4$$

In the light of the value of y here, recall that there is nothing to prevent negative values from arising as the solution to systems of simultaneous linear equations.

Exercise 2

$$3x + 8y = 64$$
$$-2x + 2y = 16$$

The use of elimination is convenient in this case. In particular, multiplication of the second equation by 4 produces the result

$$3x + 8y = 64$$
$$-8x + 8y = 64$$
$$\overline{}$$
$$11x = 0$$

so that $x = 0$. Insertion of this value into either equation produces $y = 8$. In this case the intersection of the lines represented by the equations occurs on the y axis.

Exercise 3

$$3x + y = 11$$
$$6x + 2y = 24$$

Again using elimination here, multiplying the first equation by 2 and subtracting the second results in

$$6x + 2y = 22$$
$$6x + 2y = 24$$
$$\overline{}$$
$$0 = -2$$

This impossible requirement to equate two different constants means that the equations are *inconsistent* – they have no simultaneous solution. The lines represented by the equations are parallel.

Exercise 4

$$3x + y = 11$$
$$-12x - 4y = -44$$

Using substitution in this case, from the first equation

$$y = 11 - 3x$$

and insertion into the second results in

$$-12x - 4(11 - 3x) = -44$$

so that

$$-12x - 44 + 12x = -44$$

Therefore

$$0 = 0$$

which means that the two equations are not independent (the second is -4 times the first). So there is an infinite number of solutions in this case, but recall that not *anything* will do in terms of x and y values – 'most' points are not solutions. Only those points having co-ordinates satisfying the relationship $y = 11 - 3x$ will fit. All other points are off the line. Now consider two practical illustrations of the use of simultaneous equations, firstly in the field of personal financial management and then in an industrial context.

Financial management example

On retirement, a company manager expects to receive a pension plus a lump-sum payment of £60,000. The manager seeks to obtain an annual income of £7000 by investing the lump sum in loan stock yielding 13 per cent and a deposit account yielding 9 per cent. What amounts should be invested in each case?[6] The problem produces a pair of simultaneous equations which can be solved by elimination or substitution to give the required amounts. If x represents the sum invested in the loan stock and y represents the amount placed on deposit, then assuming nothing is held as cash we have

$$x + y = 60,000$$

The requirement of an annual income of 7000 produces the equation

$$0.13x + 0.09y = 7000$$

In this example the method of substitution is more convenient. From the equation governing the total sum to be invested we obtain

$$y = 60,000 - x$$

which, upon substitution into the annual income equation, produces the result

$$0.13x + 0.09(60,000 - x) = 7000$$
$$0.04x + 5400 = 7000$$
$$0.04x = 1600$$

and so

$$x = 40,000$$

So the value of y is

$$y = 20,000$$

Two-thirds of the sum available should be placed in the loan stock and one-third put on deposit to generate the required return.[7] Practical considerations such as brokerage or management charges not represented in this model could be included in an expanded model. But even the simple form has potential. For example, nothing stands still in the real world, circumstances change – interest rates being a case in point. Suppose in the context of this example that interest rates are reduced by one percentage point. What should be the response of our investor, given his or her stated objectives? First note that, if no changes are made, the investor's income will suffer since

$$0.12(40,000) + 0.08(20,000)$$
$$= 4800 + 1600$$
$$= 6400$$

To retain the original level of income, the equations now become

$$x + y = 60,000$$
$$0.12x + 0.08y = 7000$$

and, again using substitution, from the first equation

$$y = 60,000 - x$$

Insertion of this relationship into the second equation produces

$$0.12x + 0.08(60,000 - x) = 7000$$

and so

$$0.12x + 4800 - 0.08x = 7000$$

from which

$$0.04x = 2200$$
$$4x = 220,000$$

so that

$$x = 55,000$$

and the investor would be faced with a fairly radical adjustment of the portfolio with £55,000 to be invested in loan stock and just £5000 remaining on deposit. Now suppose that interest rates are reduced by another percentage point. The equations are now

$$x + y = 60,000$$
$$0.11x + 0.07y = 7000$$

Using elimination for a change, multiplying the first equation by 0.07 results in the system:

$$0.07x + 0.07y = 4200$$
$$0.11x + 0.07y = 7000$$
$$-0.04x = -2800$$

so that

$$4x = 280{,}000$$

i.e.

$$x = 70{,}000$$

which would require that

$$y = -10{,}000$$

Now while the algebra does not rule out this outcome, an implicit assumption in the use of this model would be that neither x nor y is negative. So unless the investor has a somewhat unlikely facility to *borrow* at 7 per cent, the income target can no longer be achieved even if the entire lump sum was invested in the potentially more lucrative (and higher risk) security.

Multi–process production

Simultaneous linear equations often arise in *blending problems*. While these problems usually call to mind physical processes such as oil refining, in fact the financial example just considered can be seen as a problem of 'blending' investments to produce the requisite financial outcome. In an altogether different context suppose that a firm can make a product in either or both of two processes. The first process (in which the amount produced will be represented by x) is the more economical but it is a heavy user of machine time which is severely constrained. Each unit produced in the first process requires 10 minutes of machining time, of which a total of 8200 minutes is available. Each unit produced in the second process (y) requires only 4 minutes of machining time. Total production in the two processes combined must be 1000 units. Clearly with the financial pressure to make as much as possible in the first process, no machining time should be left unused. Thus the two requirements are

$$10x + 4y = 8200$$
$$x + y = 1000$$

Using elimination and multiplying the machine time equation by 4 results in

$$10x + 4y = 8200$$
$$4x + 4y = 4000$$
$$\overline{}$$
$$6x = 4200$$

Therefore

$$x = 700 \text{ and } y = 300$$

So 300 units must be produced in the second process. As more machine time becomes available, production in the second process will fall. For example, if 9400 minutes of machine time were available, first process production would rise to 900 units with just 100 units remaining to be produced in the second process.

EXERCISES 2.7

1 Solve the following systems of simultaneous equations using elimination:

(i) $4x + 3y = 110$
$2x + y = 50$

(ii) $5x - y = 20$
$3x + 2y = 12$

(iii) $6x - 5y = 16$
$4x + 2y = 0$

(iv) $0.4x + 5y = 7$
$2x - 0.5y = 4.4$

2 Solve the following simultaneous equations using the method of substitution:

(i) $2x + y = 20$
$7x + 3y = 66$

(ii) $x - 11y = 40$
$2x - 7y = 35$

3 Solve the following simultaneous equations using the method of your choice:

(i) $y = 4x - 35$
$2x + y = 25$

(ii) $x + 25 = 3.5y$
$x = 4.5y - 33$

4 Where possible, find solutions to the following systems of simultaneous equations. Where there is not a unique solution, state whether there are *any* solutions and the reason why the situation occurs:

(i) $0.4x + 0.2y = 3.6$
$0.1x + 0.3y = 2.9$

(ii) $-0.5x + y = 9.5$
$x - 2y = -19$

(iii) $2x - y = -3$
$-2x + 0.5y = -0.5$

(iv) $x - 2y = -9$
$-2x + 4y = 19$

2.8 ELEMENTARY ROW OPERATIONS

In solving a linear equation system, the need will often arise to change an equation into a more convenient form without altering the information it contains, or to combine or interchange equations in such a way that the solution to the system as a whole is unaffected. These permissible changes are called *elementary row operations*[8] and can take the following forms:

1. the multiplication, throughout, of any equation by any non-zero constant (either positive or negative);
2. the addition, to any equation in the system, of a constant multiple of any other equation in the system;
3. the interchange of the positions of two equations in the system.

The repeated application of 2 means that any linear combination (a weighted average) of equations in the system can be added to any equation in the system. In fact, 1 can be seen as a special case of 2 where multiples of an equation are added to itself. To see that the elementary row operations are legitimate, consider some examples. As we have already seen, the system

$$3x + y = 11$$
$$x + 2y = 12$$

has the solution $x = 2$, $y = 5$. So also does the system

$$3x + y = 11$$
$$7x + 4y = 34$$

In this system, the second equation of the original system has been replaced by the original second equation plus twice the first equation. The following system also has the solution $x = 2$, $y = 5$:

$$x - 3y = -13$$
$$x + 2y = 12$$

In this case the first equation of the original system has been replaced with the first less twice the second equation. As a final example, the system

$$x - 3y = -13$$
$$7x + 4y = 34$$

also has $x = 2$, $y = 5$ as its solution. In this case both of the substitutions used earlier have been made. The elementary row operations generate, as it were, a family of systems of two linear equations having the same point as the solution. Any two distinct equations passing through the point (2, 5) can be operated on by elementary row operations to produce *all* linear equations passing through (2, 5). The situation is graphed in Figure 2.16 which shows the four equations used.

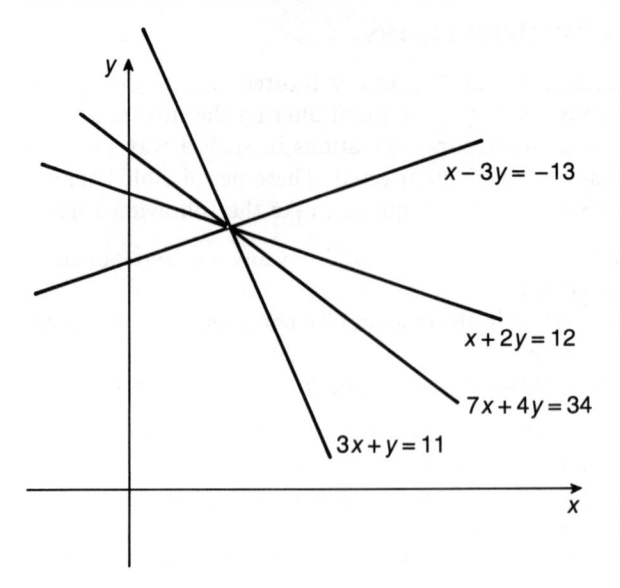

Figure 2.16

Note that any *three* or more of these equations in two variables have the solution (2, 5). In general, for a system of three equations in two variables to have a solution, any one of the equations must be obtainable from the other two by elementary row operations. These points are further developed in the following section. An understanding of elementary row operations is essential for the method of Gaussian elimination (discussed in Section 2.10). Elementary row operations are also implicit in the Simplex method of linear programming (considered in the following chapter) and which is closely related to Gaussian elimination. However, before concluding this section, it should be noted that there are further operations on equations that are possible. For example, both sides of an equation can be raised to the same power. Consider the equation

$$2x - 1 = x + 2$$

The same value of x (3) will satisfy

$$(2x - 1)^2 = (x + 2)^2$$

but note that the operation of squaring has introduced another 'solution': $x = -1/3$, which is *not* valid for the original linear equation. Normally, the problem solver would be moving from the more complicated quadratic to the linear form. But in such a step, a possible solution will be *lost*. However, the 'lost' solution can be retrieved once the other solution has been obtained.

2.9 OTHER LINEAR SYSTEMS IN UP TO TWO VARIABLES

While linear systems of two equations in two unknowns arise commonly, so also do other arrangements. Cases where the number of equations and unknowns is greater than two are introduced in the following section. Here we consider possibilities involving no more than two variables, specifically the cases of:

1 one equation in one unknown
2 one equation in two unknowns
3 whole number requirements

One equation in one unknown

The simplest case of all is that of a single linear equation in one unknown. But even this least complicated framework will lead to an important point. Suppose that in a particular case

$$2x + 7(x - 4) + 3(6 - 2x) - 20 = 20$$

This simplifies to

$$3x - 30 = 0$$

so that

$$x = 10$$

In this case, the single equation has but a single solution. Now suppose that

$$(5x - 103) + 9(2x + 5) + 7(10 - 3x) = 2(6 + x)$$

Elimination of the brackets and simplification of the left-hand side produces

$$2x + 12 = 12 + 2x$$

with the result that

$$0 = 0$$

which, although accurate, is hardly enlightening. Whenever the result of simplification is $0 = 0$, the original expression represents an *identity* – a relationship which *all* values of the variables satisfy – rather than an equation which has to be solved for the specific values of the variables for which it is true. Identities are discussed in more detail in Section 2.11. Now suppose that the equation is

$$6(2x + 10) + x = 5(20 + 3x) - 2(x - 10)$$

which reduces to

$$13x + 60 = 120 + 13x$$

i.e.

$$60 = 120$$

The requirement to equate two unequal constants means that the original expression cannot be satisfied for *any* values of x. The two parts of the expression are inconsistent (they plot as parallel lines against x) and produce a contradiction. If this occurs in practice (assuming error-free calculation) it may mean that the model has been misspecified or that the desired solution, e.g. an equilibrium point, does not exist. But this outcome can itself be useful information. The problem of finding the x intercept of a straight line produces a single equation in one variable:

$$mx + c = 0$$

which solves for $x = -c/m$, a unique value except for the cases where $m = 0$, $c = 0$ (where the line is the x axis) and $m = 0$, $c \neq 0$ (where no solution exists). The value of $x = -c/m$ represents a *zero* of the function

$$y = mx + c$$

and is said to be the *root* of the equation (any x value for which the corresponding y value is zero).

One equation in two unknowns

The case of two or more variables in one linear equation also brings out some interesting points. Take as an example the equation

$$5x + 2y = 24 \qquad (2.16)$$

This equation has an infinite number of solutions, but this does not mean that *any* pair of values will satisfy the equation. Unlike an identity, *some* x and y combinations are excluded. The equation imposes a restriction that not all pairs of values will satisfy. Only those pairs corresponding to points on the line defined by the equation will be solutions. Some of the non-solutions can be usefully categorized, for example in this case there are no solutions in quadrant three (where x and y are both negative). In the case of lines which go through the origin, then apart from the origin itself, x and y must always be the same sign for a line with positive slope or always of opposite sign where the slope is negative. In finding solutions to (2.16) either x or y may be freely chosen and a specific value of the other variable results. There is one *degree of freedom* in the setting of variables. In one linear equation in n variables there are $n - 1$ degrees of freedom.

Table 2.3

x	y
0	12
2	7
4	2

Whole number requirements

At times only solutions satisfying a number of practical conditions may be acceptable – e.g. those in non-negative integers. In the case of equation (2.16) there are three such solutions as shown in Table 2.3.

An equation with integral coefficients such as (2.16) in which only whole number solutions are meaningful is *Diophantine*.[9] Models expressed in rational coefficients in which only integral, sign restricted, solutions are acceptable are quite common in practice, and more difficult to solve. Such problems usually arise in conjunction with an inequality relationship – such as a budget constraint – and their solution may require the use of *integer programming* methods.[10] Problems involving sign restricted variables in linear systems are considered in Chapters 3 and 7.

EXERCISES 2.9

1 Find the value of x which satisfies the following equations:

(i) $5x + 3(x - 8) + 4(12 - 0.5) + 16 = 100$
(ii) $20 - 4x + 3(2x - 5) + 5(4x + 68) = 2(4x + 5) + 110$

2 Comment on the following equations:

(i) $3(5x + 10) - (8x + 50) = -5(8 - 2x) - 3x$
(ii) $6(4x + 2) - 7(3x + 1) \stackrel{.}{=} 8(5 + 3x) - 3(7x + 11) - 2$

3 Given the equation

$$8x + 5y = 40$$

which of the following are solutions, or are consistent with solutions to the equation:

(i) the point $x = 2.5$, $y = 4$
(ii) the point $x = 0$, $y = 8$
(iii) the point $x = -2$, $y = 11.2$
(iv) the point $x = 5$, $y = 8$
(v) points on the line

$$0.5x + 0.3125y - 2.5 = 0$$

(vi) points on the line

$$y = 8 - 1.6x$$

(vii) points on the line

$$20x + 12.5y = 50$$

2.10 LINEAR SYSTEMS OF EQUATIONS IN MORE THAN TWO VARIABLES

When a decision-maker has more than two controls or courses of action, models will involve more than two variables. Our purpose here is not to detail methods of solution, but to describe conditions in which solutions exist, and the number of solutions to be expected. There will be a *unique* solution if there are the same number of consistent equations as there are unknowns and if these equations are independent (no equation being a weighted average of the other equations). Such a system with one solution is said to be *determinate*. An example of a three-equation, three-unknown system (of the kind that we shall need to solve in Chapter 7) is the following:

$$3x + 2y - z = 0$$
$$z + x + y = 34$$
$$z - x + y = 26$$

The solution to such systems can be obtained in several ways, e.g. by use of Gaussian elimination (described in the following section), matrix inversion or Cramer's rule (discussed in Chapter 10). Here we shall use substitution or a combination of substitution and elimination. From the first equation state z in terms of x and y:

$$z = 3x + 2y$$

Substitution for z in the second and third equations produces

$$3x + 2y + x + y = 34$$
$$3x + 2y - x + y = 26$$

which simplifies to

$$4x + 3y = 34$$
$$2x + 3y = 26$$

which is a system of two equations in two variables, to which elimination or substitution can be applied. Elimination rapidly produces the result:

$$x = 4 \text{ and } y = 6$$

and insertion of these values into the expression for z gives the result $z = 24$.
In summary, the approach used above is as follows:

1 make use of one of the equations to state one of the variables explicitly in terms of the other two;
2 substitute for the chosen variable in the remaining two equations;
3 use either substitution or elimination to solve the 2×2 system that is left.

As a further example of this approach, consider the system

$$3x - 2y + 4z = 110$$
$$2x + 2y - 3z = -30$$
$$-2x + 6y + 4z = 220$$

The most convenient substitution is to use the third equation to state x in terms of y and z:

$$x = 3y + 2z - 110$$

which, when substituted into the first equation, gives

$$7y + 10z = 440$$

and in the second equation the result is

$$8y + z = 190$$

So we now need to solve the 2×2 system

$$7y + 10z = 440$$
$$8y + z = 190$$

in which substitution is an efficient option. From the second of the remaining equations

$$z = 190 - 8y$$

which when substituted into the first equation gives

$$-73y = -1460$$

so that

$$y = 20$$

and, by back substitution, first into the relationship between y and z

$$z = 190 - 8(20)$$
$$= 30$$

and then into the relationship stating x in terms of y and z

$$x = 3(20) + 2(30) - 110$$
$$= 10$$

The complete solution to the system is therefore

$$x = 10 \quad y = 20 \quad z = 30$$

By a similar process, it can be confirmed that the solution to the 3×3 system

$$5x + 4y - z = 0$$
$$z + 2x + 2y = 20$$
$$z + x - y = 15$$

is

$$x = 2 \quad y = 1 \quad z = 14$$

Of course, not all 3×3 systems will have unique solutions. For example, in

$$2x + 3y + 4z = 90$$
$$4x + 6y + 8z = 150$$
$$5x + 2y + 3z = 100$$

the left-hand side of the second equation is twice that of the first, but the right-hand side is 150 rather than 180. These equations are therefore inconsistent and the system has no solution.[11] However, if the right-hand side of the second equation had been 180, the equations would have been consistent but not independent (the second equation now being twice the first) and the system would have a continuum of solutions.[12] Similarly, a system that begins as one of two equations in three variables will not have a unique solution. For example, if the equation

$$z - x + y = 26$$

had been omitted from the system solved earlier on, we should have been left with

$$4x + 3y = 34$$

and, of course,

$$z = 3x + 2y$$

for which there is a continuum of solutions. However, as previously noted, not just *any* values of x, y and z will do. If, as here, the two equations are independent and consistent, the solution points will lie along a straight line. In general, a consistent system in which there are more variables than there are independent equations is said to be *indeterminate*. An example of an indeterminate system is illustrated in Figure 2.17.

The case illustrated in Figure 2.17 is one of three variables and two independent and consistent equations, and shows the two planes corresponding to the equations intersecting in the positive orthant. The planes

Straight lines and linear equations 107

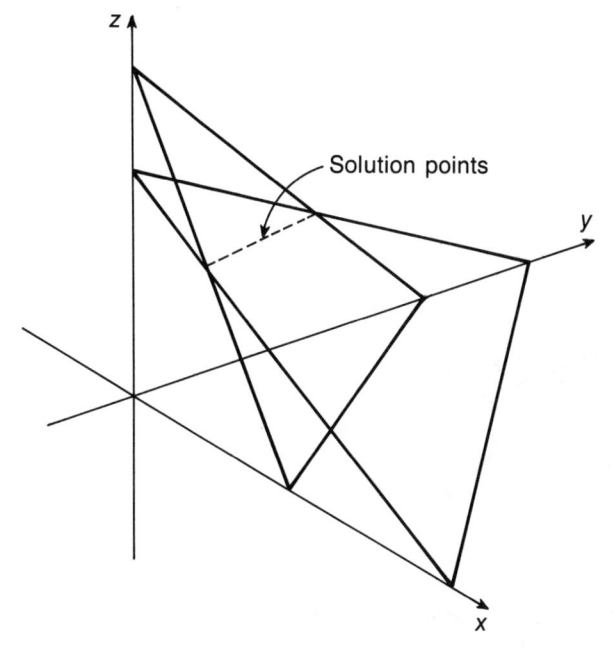

Figure 2.17

are not limited to the positive orthant but only those parts of the planes are illustrated. The points common to both planes, lying on a straight line, are shown. If a third equation is added, and if this is independent of the original two equations and consistent with them, the third plane that this equation produces will intersect the line common to the first two planes at a single point. The co-ordinates of this point will be the values of the variables which uniquely solve the three-equation system.

A system of three *independent* equations in two variables cannot be consistent and therefore cannot have a solution. In general a system in which there are more independent equations than there are variables is said to be *overdeterminate*. In the three-equation, two-variable case, the values of x and y which satisfy any two of the equations will contradict the third, as for example in

$$2x + y = 13$$
$$x + y = 9$$
$$x + 2y = 11$$

This system is graphed in Figure 2.18 where it is evident that whichever line is last to be drawn will not pass through the intersection of the other two lines. The system can be made determinate by the deletion of one of the

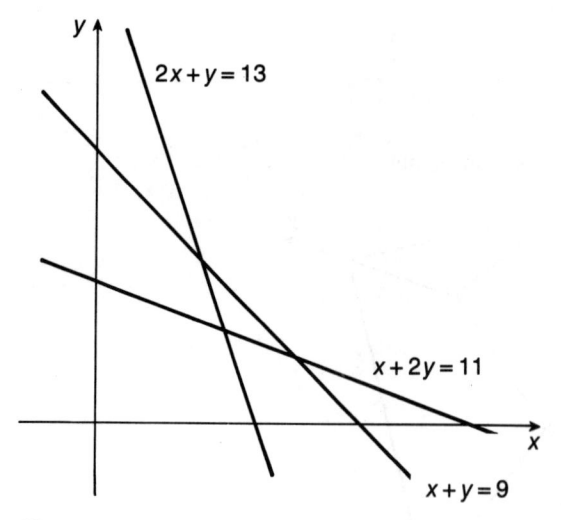

108 Mathematics for business, finance and economics

[Figure showing three lines: $2x+y=13$, $x+2y=11$, and $x+y=9$ on x-y axes]

Figure 2.18

equations or by modification of at least one of the equations to include a term in an additional variable.

Note that there are circumstances when a system of three equations in two variables *can* have a unique solution. A minor amendment to the inconsistent case just considered produces such a system:

(i) $2x + y = 13$

(ii) $x + y = 8$

(iii) $x + 2y = 11$

for which the solution is

$x = 5$ and $y = 3$

This solution is produced by any *two* of the three equations, and although in any such system that has a unique solution the equations must be *consistent*, they cannot be *independent*. This means that any one of the equations can be obtained by combining the other two. In the system above, for example, the second equation can be formed by adding (i) and (iii) and dividing the result by 3.[13] The interaction between the equations is shown in Figure 2.19.

A system of two equations in three variables will not produce a unique solution – it must be indeterminate. But provided that the equations are

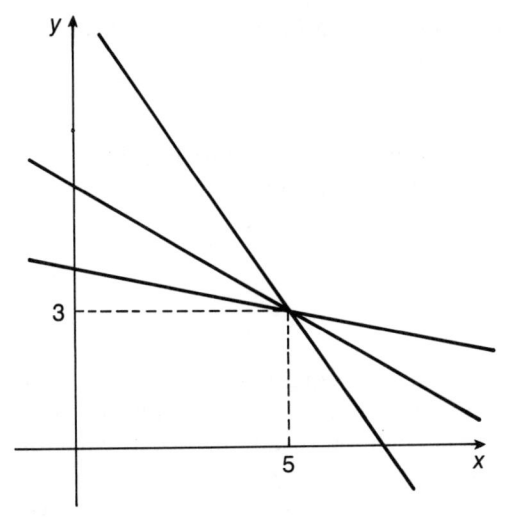

Figure 2.19

independent and consistent, a 'solution' for two of the variables can be given in terms of the third.[14] For example consider

(i) $\quad 5x + y + 4z = 100$

(ii) $\quad 2x + 2y - 8z = 64$

From the first equation, y can be stated in terms of x and z as

$$y = 100 - 5x - 4z$$

which on substitution into (ii) produces

$$2x + 200 - 10x - 8z - 8z = 64$$

so that

$$-8x - 16z = -136$$

or

(iii) $\quad x = 17 - 2z$

and substitution for x into (ii) produces

(iv) $\quad y = 15 + 6z$

Acceptable values for x and y do not correspond to a unique point but rather form a set of points – the *solution set*.[15] Definite values for x and y can only be obtained by specifying a value for z, or by the addition of a third equation consistent with (i) and (ii) and independent of them. For example, if we specify that $z = 4$ then

$x = 9$ and $y = 39$

emerge as the solution. The same unique solution results if either of the following equations is added to the original system:

$x + 2y + 5z = 11$

or

$3x - 4y + z = 67$

Note that both of the above additional equations include the point $(9, -9, 4)$ and any one of the family of planes associated with this point would have done equally well.[16] The addition of any particular equation says more about the necessary relationships between x, y, and z than does the selection of an individual value for z.[17] The usefulness of this arrangement in which x and y are stated in terms of z is that when one of the variables, say z, is determined outside of the system (perhaps by the government) the corresponding values of the in-system variables can be easily worked out. Variables which are determined outside a system are said to be *exogenous*, while variables determined within the system are *endogenous*. Treating some variables as exogenous is one way in which an otherwise indeterminate system can be made to work out. The equations in which each of the endogenous variables is stated in terms of the exogenous variables are said to be *reduced form equations*. Thus (iii) and (iv) above are reduced form equations.

Economics provides a number of examples of systems with fewer equations than unknowns. For example, in barter economies a system involving n goods can be solved for the $n - 1$ exchange ratios between $n - 1$ of the goods and an arbitrarily selected nth good or *numeraire*. If 'accounting money' is introduced, and a specific price is set for the numeraire, definite equilibrium prices result for the remaining $n - 1$ goods. In management science, an example of an indeterminate system occurs in the *transportation simplex* or *modified distribution method* for solving transportation problems. The method requires the establishment of values for $m + n$ unknowns from $m + n - 1$ equations as part of the process of the evaluation of unused routes. But since it is the *relative* cost of solutions that matters, one of the unknowns is set at a convenient arbitrary value (which then cancels out in the relative evaluations). Finally, as we have seen, a linear equation in three variables such as

$z = ax + by + c$

produces a *plane* in three-dimensional space. But how is a straight *line* in three dimensions to be represented in algebraic rather than geometric terms? One way is to note that a straight line in three space satisfies *two* linear equations

$$z = a_1x + b_1y + c_1$$

and

$$z = a_2x + b_2y + c_2$$

which do not represent parallel or identical planes. Straight lines in three space may also be specified by *parametric equations*[18] or by *vector equations*.

EXERCISES 2.10

1 Use substitution to solve the following systems of three linear equations in three unknowns:

(i) $\quad 4x + 3y - z = 23$
$\quad\quad z + 2x + 5y = 23$
$\quad\quad z + x - y = 6$

(ii) $\quad 5x - 2y - z = 0$
$\quad\quad 7z - 10x - 4y = 10$
$\quad\quad z + 5x + 8y = 70$

(iii) $\quad 8x + 5y - z = 0$
$\quad\quad z - 3x - 2y = 74$
$\quad\quad z + x - 5y = 90$

(iv) $\quad 4x + 5y - z = 0$
$\quad\quad z - x + y = 30$
$\quad\quad z + 2x + 2y = 40$

2 What can be said regarding solutions to the following systems and the relationships between the equations?

(i) $\quad 3x + 2y + 4z = 45$
$\quad\quad 6x + 4y + 8z = 95$
$\quad\quad 2x + 5y + 3z = 50$

(ii) $\quad 2x + y + z = 35$
$\quad\quad 3x + y + z = 40$
$\quad\quad 4x + 2y + 2z = 70$

(iii) $2x + 3y + 5z = 115$
$4x + 2y - 2z = 10$
$4x + 4y + 3z = 105$

3 State x and y in terms of z for the following systems:

(i) $x + 4y - 5z = 40$
$8x + 2y + 5z = 20$

(ii) $2x + 3y + 2z = 54$
$4x + 7y + 2z = 116$

2.11 GAUSSIAN ELIMINATION

One of the most efficient ways to solve simultaneous linear equations is the method of *Gaussian elimination*. The method uses the fact that the solution of the system, if there is one, is unaffected by the use of elementary row operations. The method will be explained through numerical examples starting with the three-variable, three-equation system

$$2x + 5y + z = 46$$
$$x + 3y + 2.5z = 45$$
$$2x + 4.5y + 4z = 74$$

First divide the first row by 2 in order to produce a unit coefficient for x,[19] giving

$$x + 2.5y + 0.5z = 23$$

For each subsequent row subtract a multiple of this row that eliminates x. For the second row subtract the modified first row as it stands, giving

$$0.5y + 2z = 22$$

and subtracting twice the new first row from the third gives

$$-0.5y + 3z = 28$$

So the system is now

$$x + 2.5y + 0.5z = 23$$
$$0.5y + 2z = 22$$
$$-0.5y + 3z = 28$$

The first row will remain as it now is, and the process is repeated for the second and subsequent rows. Multiplying the new second row by 2 gives

$$y + 4z = 44$$

and adding one-half of this to the third row gives

$$5z = 50$$

Now complete the process by dividing the third row by 5 (there are no subsequent rows to deal with) to give a unit coefficient for z:

$$z = 10$$

So the system is now

$$x + 2.5y + 0.5z = 23$$
$$y + 4z = 44$$
$$z = 10$$

For a system with a unique solution, the value of the last variable will be stated in the last row,[20] with the values of the remaining variables being obtained by back substitution. Putting $z = 10$ into the second equation results in

$$y + 40 = 44$$

so $y = 4$. Now substituting $y = 4$ and $z = 10$ into the first equation gives

$$x + 10 + 5 = 23$$

so that $x = 8$ and the complete solution to the system is

$$x = 8 \quad y = 4 \quad z = 10$$

This is the method of Gaussian elimination with back substitution, a very efficient method for solving simultaneous linear equations. *Gauss–Jordan elimination* or *Jordan elimination* is an elaboration of Gaussian elimination where, when *any* row has been divided through to give a 1 as its first non-zero coefficient, a multiple of this row is subtracted from preceding as well as subsequent rows to eliminate the corresponding variable from all equations. So, in our case, having got the second row to

$$y + 4z = 44$$

and as before adding one-half of it to the third row, in Gauss–Jordan elimination we now also subtract 2.5 times this row from the first to eliminate y in all equations but the second, giving

$$x - 9.5z = -87$$
$$y + 4z = 44$$
$$5z = 50$$

The third equation is now divided by 5, and 9.5 times the result is added to the first equation and 4 times the result is subtracted from the second equation to give the neat result

$$x = 8$$
$$y = 4$$
$$z = 10$$

In comparative terms, Gaussian elimination with back substitution calls for less work than Gauss–Jordan elimination and is usually preferred. Now consider another example. We have already encountered the system

$$x - y + z = 15$$
$$2x + 2y + z = 20$$
$$5x + 4y - z = 0$$

Using Gaussian elimination, the first equation needs no division, and x is eliminated from the second and third equations by subtraction of twice the first equation and 5 times the first equation respectively with the result

$$x - y + z = 15$$
$$4y - z = -10$$
$$9y - 6z = -75$$

Dividing the second equation by 4 and subtracting 9 times the result from the third equation gives the last two equations as

$$y - 0.25z = -2.5$$
$$-3.75z = -52.5$$

Division of the third row by -3.75 now gives

$$z = 14$$

and the full system is

$$x - y + z = 15$$
$$y - 0.25z = -2.5$$
$$z = 14$$

Back substitution gives

$$y - 0.25(14) = -2.5$$

Therefore

$$y = 1$$

Substituting the obtained values of z and y into the first equation gives

$x - 1 + 14 = 15$

$\therefore x = 2$

and the solution to the system is therefore:

$x = 2 \quad y = 1 \quad z = 14$

Gaussian elimination is an efficient means of solving simultaneous linear equations and can also be used for inverting matrices. It also underlies the simplex method of linear programming.

EXERCISES 2.11

1 Use Gaussian elimination to solve the following system:

$x + 3y + 2z = 29$
$2x + 8y + 6z = 78$
$3x + 7y + 7z = 85$

2 Solve the following system using Gaussian elimination:

$2x + y + 2z = 19$
$4x + 3y + 6z = 37$
$2x + 2y + 3z = 26$

2.12 IDENTITIES

An *identity* is an equation which is true for *all* values of the variables involved. For this reason it is sometimes called an *identical equation* as distinct from a *conditional equation* – the form we have been dealing with up to now – which is true for only certain values of its variables.[21] For an equation to be an identity, it must be possible to cancel out all terms involving variables leaving only the equation of two equal constants. Where it is necessary to distinguish identities from equations, the symbol $\equiv$ is used in place of the equation symbol $=$. Identities may involve expressions in several variables, and they arise in economics, finance and management science. Any manipulation permitted for an equation can also be applied to an identity. Thus elementary row operations could be carried out on identities. In economics the best known examples of identities are those involved in national income accounting. Let Y represent the value of gross domestic product. One way of measuring Y is to take the total value of household incomes. Let C represent the total expenditure by households on consumption, while S represents the level of savings – defined as that part of income which is not spent on consumption. With these definitions, the identity that results is

$$Y \equiv C + S$$

Now GDP can also be defined and measured from the expenditure side as

$$Y \equiv C + I$$

where C is consumption expenditure as before and I represents the total expenditure on investment goods. In theory the two measures of income should be the same (in practice an adjustment is necessary to make them equal as a result of measurement imperfections). As a result of the accounting definitions, it follows that

$$Y \equiv C + S$$
$$\equiv C + I$$

Thus

$$S \equiv I$$

This identity states that *actual* savings always equal *actual* investment. Note that this does not mean that the levels of *intended* savings and investment are always equal – the equality is a property of equilibria only. In working with models of national income accounting of this kind, the identity sign is not always used. This will not lead to any errors, but care is necessary in drawing conclusions.

In production planning, x and y might represent daily production levels of two goods manufactured by a firm. A feasible production plan (represented by a pair of values of x and y) will not call for the use of more resources than are currently available to the firm each day. If each unit of good x requires 20 minutes of machine time and each unit of good y requires 13 minutes of machine time, then the total machine time requirement for any production plan is given by

$$20x + 13y$$

Suppose that on any day a maximum of 480 minutes of machine time is available. The amount of unused or *slack* machine time will be s where s is defined by

$$s = 480 - 20x - 13y$$

The overall allocation of machine time manufacturing usage and idle time will then be stated by the identity

$$20x + 13y + s \equiv 480 \tag{2.17}$$

In practice, the identity sign is usually replaced by the equality sign in expressions like (2.17), but it is important not to overlook altogether the fact that an identity is involved. Relationships such as (2.17) which link resource requirements and availability are important in solving production planning problems by *linear programming*.[22]

2.13 FURTHER APPLICATIONS

Here we consider two further illustrations of the use of straight lines in business and economics:

1 breakeven analysis
2 market equilibrium

First consider the business application.

Breakeven analysis

The widely used *breakeven analysis* uses simultaneous equations in a simple way to determine the level of output at which revenue (income) for a period covers costs (outgoings). Cost is given by

$$C = F + V$$

where C represents total cost, F is fixed costs (overheads such as rent, lease payments, interest charges) and V represents variable costs (wages, materials, fuel etc.). Breakeven analysis takes variable costs to be proportional to output level, q. Thus

$$V = bq$$

in which the constant, b, represents the unit variable cost of production. It then follows that

$$C = F + bq$$

The firm's total revenue, R, will also be proportional to output level:

$$R = pq$$

in which p represents product price (i.e. *unit revenue*, here taken to be constant). Breakeven output level is achieved at the point where total revenue covers total costs, i.e. where

$$R = C$$
$$pq = F + bq$$
$$(p - b)q = F$$

118 Mathematics for business, finance and economics

so that

$$q^* = \frac{F}{p-b} \qquad (2.18)$$

where q^* is the breakeven output level. The model is graphed in Figure 2.20, in which q_c represents plant capacity.

In Figure 2.20, it is clear that for the revenue and cost lines to cross in the positive quadrant product price must be strictly greater than unit variable cost. If this is not the case, no breakeven output level exists. The relationship of q^* to q_c is also important. If the breakeven output level is a very high proportion of plant capacity, the financial viability of the enterprise will be very much at the mercy of minor changes in parameter values such as unit variable cost. For a numerical example let $P = £10$, $b = £6$, $F = £10,000$ and $q_c = 4000$ units of output. Substitution in formula (2.18) yields

$$q^* = \frac{10,000}{10-6} = 2500$$

so that the breakeven output level represents usage of the plant at 62.5 per cent capacity. Note that if price had been originally set at $p = 8$, while this value of unit revenue covers unit variable cost, the theoretical breakeven level $q^* = 5000$ is not reached by the time that the firm runs into the capacity limit. The breakeven formula can also be used to deduce the

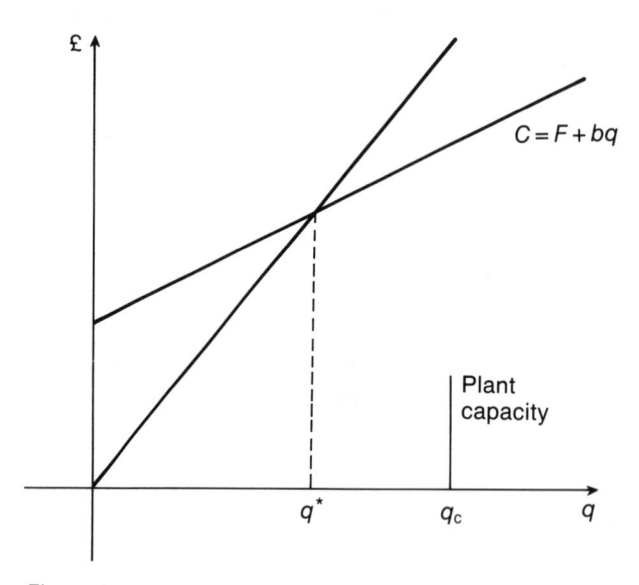

Figure 2.20

consequences of changes in the parameter values p, b and F. For example suppose that interest rate changes, a wage agreement and inflation of raw material costs result in new values of $b = £6.8$ and $F = £10,400$. Two questions of interest would be the following

(a) What would be the new breakeven output level?
(b) What new level of price would leave the original breakeven output level unchanged?

To address (a) substitute the changed values of the parameters into (2.18), producing

$$q^* = \frac{10,000}{3.2} = 3250$$

so that at the original price the output level required to break even increases by 30 per cent. As regards (b), the new level of price required to maintain the original breakeven output level will be p where

$$2500 = \frac{10,400}{p - 6.8}$$

so that

$$2500p - 17,000 = 10,400$$

$$p = \frac{27,400}{2500}$$

$$= £10.96$$

Note from the formula that equal proportionate changes in F, b and p leave the breakeven output level unchanged.

Market equilibrium

Analysis of the market for a good often makes use of linear supply and demand curves relating price to quantity supplied or demanded. A market is in equilibrium at a price where the quantity of the good that consumers wish to purchase is the same as the quantity that the price stimulates producers to offer. In other words, the market reaches equilibrium where the supply and demand equations are solved simultaneously. Suppose that, in the market for a particular product, supply and demand are given by the linear equations

Demand: $p = 1000 - 5q$
Supply: $p = 200 + 3q$

in which p is price and q is quantity. The equations define the *demand* curve[23] and the *supply curve* respectively. The market is in equilibrium for values of p and q satisfying both equations. Thus

$$1000 - 5q = 200 + 3q$$

so that

$$800 = 8q$$

Therefore

$$q = 100$$

and from either the supply or demand equation

$$p = 500$$

which is the *equilibrium price* or *market clearing price*. At a price above equilibrium, supply exceeds demand, a condition known as *excess supply*. For example, at a price p of 650, suppliers wish to offer 150 units but consumers would only be prepared to take 70 units so excess supply is 80 units. In contrast, at a price below the equilibrium level, demand exceeds supply, a situation known as *excess demand*. For example, at a price p of 350, consumers wish to purchase $q = 130$ units while producers would only be willing to supply $q = 50$ units. Figure 2.21 graphs the situation.

In Figure 2.21, the equilibrium position is E, and the excess supply and excess demand disequilibria are illustrated. This model can be expanded in a variety of ways. For example, suppose that a tax of £t per unit is imposed

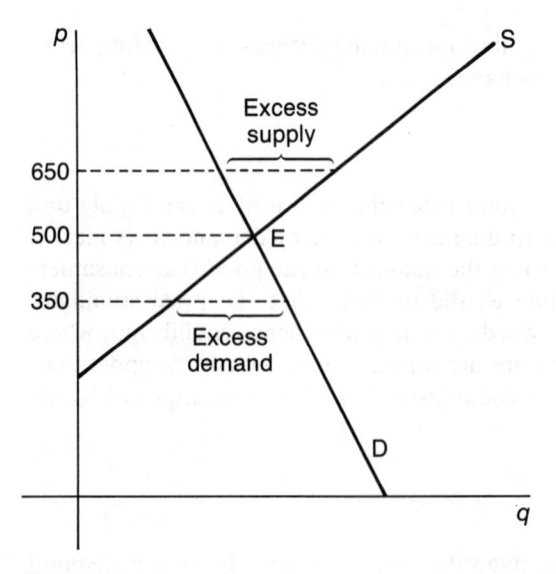

Figure 2.21

on this product by the government.[24] This will have the effect of shifting up the supply curve by the amount t at all points. This is because, for any value of q, the consumer must now for each unit pay the price given by the supply curve plus the government tax of £t. Various questions arise. For instance, does the presence of the tax mean that the equilibrium price will rise by the amount of the tax? Analysis shows that this will *not* be the case. For example, suppose that the tax levied was $t = 40$. Market equilibrium is now given by

$$1000 - 5q = 200 + 3q + 40$$

so that

$$760 = 8q$$

and therefore

$$q = 95 \quad \text{and} \quad p = 525$$

so the market clearing price has gone up by £25 rather than the full £40 of the tax. Note that while consumers are paying an extra £25 per unit for the product, the firm is receiving £15 less. These relative magnitudes, 25 and 15, depend on the slopes of the supply and demand curves, and give an indication of the comparative burden of the tax borne by consumers and producers. A further question is the level of tax that would maximize the revenue raised – a problem best addressed by calculus methods in a later chapter.

EXERCISES 2.13

1 A firm has fixed costs of £4800 per week. It makes a single product, for which the unit variable cost of production is £3. The product sells at a price of £5. Plant capacity is 3600 units per week.

   (i) Find the breakeven output level.
   (ii) What percentage of plant capacity is utilized at the breakeven output level?
   (iii) If fixed cost changes to £5400 and variable cost becomes £3.25, what will be the new breakeven output level?
   (iv) With the new cost data of part (iii), what selling price would leave the original breakeven output level unchanged?

2 A firm's monthly fixed costs are £13,500. The firm makes a single product, for which the unit variable cost is £8. The product sells at a price of £12. Plant capacity is 6750 units per month.

   (i) Find the breakeven output level.

(ii) What percentage of plant capacity is utilized at the breakeven output level?

(iii) Find the effect on breakeven output level of the following changes:

(a) Cost and selling price both increase by £1.
(b) Cost and selling price both increase by $12\frac{1}{2}$ per cent.

(iv) With other data at their original values, what percentage reduction in unit variable costs would be needed in order to reduce the breakeven output level to 2250 units?

3 The market demand curve for a product is given by

D: $p = 200 - q$

and the supply curve is given by

S: $p = 50 + 0.5q$

(i) Find the excess demand at a price p of 80.
(ii) Find the excess supply at a price p of 130.
(iii) Find the equilibrium (market clearing) price and quantity.
(iv) Now suppose that the government imposes a tax of t per unit supplied, so that the new, tax-inclusive, supply curve is

S: $p = 50 + 0.5q + t$

Assuming market clearing, find:

(a) The government's receipts from the tax (T) for $t = 15$ and $t = 30$.
(b) What would be the costs to the exchequer of a *subsidy* of 7.5 per unit supplied?

4 The market demand curve for a product is given by

D: $p = 120 - 0.4q$

and the supply curve with existing plant is given by

S_1: $p = 0.6q + 20$

(i) Find the market clearing price and quantity.

Suppose that new manufacturing plant is installed that results in the supply curve

$$S_2: \quad p = 0.1q + 40$$

(ii) From what level of output does the new plant result in a lower supply price than that resulting from the old plant?
(iii) With the new plant, what is the excess demand or supply at the old equilibrium price?
(iv) Find the equilibrium price and quantity with the new supply curve.
(v) What would be the excess demand or supply at the new equilibrium price using the old plant?

2.14 BASIC PROGRAMS

The first program finds the straight line distance between two points in two dimensions and gives the slope of the line segment connecting the points.

Program 2.1

```
CLS

PRINT 'Program finds the distance between two points'
PRINT 'and the slope of the line connecting them.'
PRINT

PRINT 'Press any key to continue with the program...'

DO

LOOP WHILE INKEY$ = ''

DO

    CLS

    INPUT 'X-coordinate of first point '; x1
    INPUT 'Y-coordinate of first point '; y1

    PRINT

    INPUT 'X-coordinate of second point '; x2
    INPUT 'Y-coordinate of second point '; y2

    PRINT

    distance = SQR((x1 - x2)^2 + (y1 - y2)^2)

    PRINT 'The distance between the two points is '; distance
    PRINT

    IF x2 = x1 THEN

            PRINT 'Slope undefined.'
            PRINT

    ELSE

            slope = (y2 - y1)/(x2 - x1)

            PRINT 'Slope of line connecting the points is '; slope
            PRINT

    END IF

    PRINT 'Do another? Type y and press enter to repeat...'

    PRINT

    INPUT answer$

LOOP WHILE answer$ = 'y' OR answer$ = 'Y'

END
```

Straight lines and linear equations 125

The program uses two DO...LOOP constructions. The first of these holds the initial message on the screen until the user presses a key. The second loop is the main part of the program and is repeated if the user presses the 'y' key. The program runs in QuickBasic or QBasic.

The second program solves a system of two linear equations in two unknowns.

Program 2.2

```
CLS

PRINT 'The format of the equations is:'
PRINT 'ax + by = c'
PRINT 'dx + ey = f'
PRINT

INPUT 'Coefficient of x in first equation'; a
INPUT 'Coefficient of y in first equation'; b
INPUT 'RHS of first equation'; c

INPUT 'Coefficient of x in second equation'; d
INPUT 'Coefficient of y in second equation'; e
INPUT 'RHS of second equation'; f

PRINT

det = a*e – b*d

IF det = 0 THEN

    IF a*f = c*d THEN

        PRINT 'Equations are dependent – many solutions'
        PRINT

    ELSE

        PRINT 'Equations are inconsistent – no solutions'

    END IF

ELSE

    x = (c*e – b*f)/det

    y = (a*f – c*d)/det

    PRINT 'Solution is...'
    PRINT

    PRINT 'x = '; x; TAB(20); 'y = '; y

END IF

END
```

The program allows for the possibility of dependence or inconsistency between the equations. The variable labelled 'det' is the determinant of the coefficient matrix.[25] The program runs in QuickBasic and QBasic and, with the insertion of line numbers, will run in GW-Basic.

ADDITIONAL PROBLEMS

1. The relationship between a company's annual revenue from sales (Y) and time (X) is estimated as

    $Y = 90 + 7.6X$

    in which Y is expressed in units of £m and $X = 0$ represents the year 1982.

    (i) Use the equation to forecast sales revenue for:

        (a) 1992   (b) 1994

    (ii) If the relationship remains valid, in what year should sales revenue first exceed £200m?

    (iii) If the same relationship obtained prior to 1982, what is the retrospective estimate from the equation for sales revenue in 1979?

2. A retailer's stock of a certain commodity at the start of the first trading day of the year is 720 units. Demand for the commodity is constant at the rate of 12 units per day. Demand is met by withdrawal from stock.

    (i) State the relationship of inventory level (I) to time (t) in the form of a linear equation.

    (ii) Find the time at which a stockout would occur in the absence of replenishment.

    (iii) A re-order is triggered when the stock level falls to 175 units. On which trading day will this occur?

3. A retired executive has a lump sum of £70,000 which he wishes to invest so as to produce an annual income of £8000. Options have been narrowed down to two possible investments: a share portfolio yielding 15 per cent per annum and a building society deposit account yielding 10 per cent per annum.

    (i) What sum should be invested in each option?

    (ii) Now suppose that there is a general fall in interest rates and the return on the alternatives became $12\frac{1}{2}$ per cent for the share portfolio and $7\frac{1}{2}$ per cent for the building society deposit account. How much should be switched from the building society account to the share portfolio to maintain the annual income at £8000?

REFERENCES AND FURTHER READING

1. Anderson, D. R., Sweeney, D. J. and Williams, T. A. (1991) *An Introduction to Management Science* (Sixth Edition), West Publishing.
2. Borowski, E. J. and Borwein J. M. (1989) *Dictionary of Mathematics*, Collins.
3. Chiang, A. C. (1984) *Fundamental Methods of Mathematical Economics* (Third Edition), McGraw-Hill.
4. Hart, W. L. (1966) *College Algebra* (Fifth Edition), Heath.
5. Mizrahi, A. and Sullivan, M. (1988) *Mathematics for Business and Social Sciences* (Fourth Edition), Wiley.
6. Rich, A., Rich, J. and Stoutemyer, D. (1989) *Derive: User Manual* (Third Edition), Soft Warehouse.
7. Weber, J. E. (1982) *Mathematical Analysis: Business and Economic Applications* (Fourth Edition), Harper and Row.
8. Wilkes, F. M. and Brayshaw, R. E. (1986) *Company Finance and its Management*, Van Nostrand Reinhold.

SOLUTIONS TO EXERCISES

Exercises 2.2

1. (i) $\sqrt{(5-2)^2 + (2-5)^2} = \sqrt{(18)}$
   (ii) $\sqrt{(20)}$
   (iii) $\sqrt{(13)}$
   (iv) $\sqrt{(65)}$
   (v) $\sqrt{(41)}$

2. (i) Slope $= \dfrac{5-3}{2-(-1)} = \dfrac{2}{3}$
   (ii) -1
   (iii) -1
   (iv) $-1/6$
   (v) $7/4$

3. (i) Zero (horizontal).
   (ii) Undefined (vertical).
   (iii) Slope is $+1$.

   Note: In the case of (i) the slope formula would involve division by zero.

4. (i) $d = [(9-5)^2 + (5-3)^2 + (8-4)^2]^{1/2}$
   $= [16 + 4 + 16]^{1/2}$
   $= [36]^{1/2}$
   $= 6$

128 Mathematics for business, finance and economics

(ii) $d = [(3-8)^2 + (6-6)^2 + (12-6)^2]^{1/2}$
$= [25 + 0 + 36]^{1/2}$
$= [61]^{1/2}$
≈ 7.81

(iii) $d = [(7-8)^2 + (-2-1)^2 + (-8-1)^2]^{1/2}$
$= [1 + 9 + 81]^{1/2}$
$= [91]^{1/2}$
≈ 9.54

Exercises 2.3

1 (i) The slope is 3.
  (ii) The slope is -0.5.
  (iii) Zero slope – the line is horizontal.
  (iv) Slope = $\sqrt{2}$.
  (v) Undefined slope – the line is vertical.

2 The lines (i), (iv) and (vi) are parallel.
  The lines (i), (iii) and (v) are concurrent.
  The lines (vi) and (vii) are concurrent.

3 The lines (i) and (iii) are orthogonal.
  The lines (ii), (iv) and (v) are orthogonal.

Exercises 2.4

1 (i) $y = 4x + 5$
  (ii) $y = -5x + 100$
  (iii) $y = 10$
  (iv) $x = 8$

2 (i) $y = -0.5x$
  (ii) $y = 3x + 4$

3 (i) $y = 4x + 3$
  (ii) $y = 10 - x$

4 (i) $y = 5x - 11$
  (ii) $y = 2x + 1$

5 (i) Above the line.
  (ii) On the line.
  (iii) Above the line.
  (iv) Below the line.

Straight lines and linear equations 129

Exercises 2.5

1. (i) (a) Here, $t = 2$ and so $\pi(t) = 90$; (b) here, $t = 5$ and so $\pi(t) = 105$.
   (ii) In this case $t = -3$, so that from the equation $\pi(t) = 65$.

2. (i) $I = 1000 - 40t$

   where I is the level of inventory and t is the number of days trading.

   (ii) Stock falls to the predetermined re-order level on completion of t days trading such that

   $$I = 1000 - 40t = 240$$

   so $t = 19$.

Exercises 2.6

1. (i) $y = 3x - 7$
   (ii) $y = -6x + 60$
   (iii) $y = -0.5x - 20$
   (iv) $y = x$

2. $y + 3x - 20 = 0$ or $-y - 3x + 20 = 0$

3. (i) $y - 25 = 4(x - 5)$
   (ii) $y = 4x + 5$
   (iii) $4x - y + 5 = 0$ or $-4x + y - 5 = 0$

4. The point is above the line.

Exercises 2.7

1. (i) $x = 20$, $y = 10$
   (ii) $x = 4$, $y = 0$
   (iii) $x = 1$, $y = -2$
   (iv) $x = 2.5$, $y = 1.2$

2. (i) $x = 6$, $y = 8$
   (ii) $x = 7$, $y = -3$

3. (i) $x = 10$, $y = 5$
   (ii) $x = 3$, $y = 8$

   Note: Substitution is the more convenient method in these cases, since rearrangement is not required.

4. (i) Unique solution: $x = 5$, $y = 8$.
   (ii) Infinite number of solutions (which must, however, satisfy $y = 9.5 + 0.5x$). The 'two' equations describe the same line.

(iii) Unique solution: $x = 2$, $y = 7$.
(iv) No solutions. The equations are inconsistent; they generate parallel lines.

Exercises 2.9

1 (i) $x = 10$
   (ii) $x = -12.5$

2 (i) The equation simplifies to the following statement:
$$-20 = -40$$
Therefore, the equation cannot be satisfied for *any* values of x.

   (ii) The equation simplifies to the statement
$$0 = 0$$
Therefore, the equation represents an identity and will be satisfied for any value of x.

3 Cases (i), (ii) and (iii) all satisfy the equation, but case (iv) does not. Cases (v) and (vi) specify precisely the same relationship between x and y as does the original equation, but case (vii) corresponds to a line that is parallel to the original, and it is therefore inconsistent with solutions to the given equation.

Exercises 2.10

1 (i) $x = 5$, $y = 2$, $z = 3$
   (ii) $x = 4$, $y = 5$, $z = 10$
   (iii) $x = 10$, $y = 8$, $z = 120$
   (iv) $x = 2$, $y = 4$, $z = 28$

2 (i) The system is inconsistent (see first and second equations) and there are no solutions.
   (ii) The first and third equations of this system are not independent and there is therefore not a *unique* solution to the system. Either the first and second equations or the second and third equations can be used to express any two of the variables in terms of the third variable.
   (iii) Here there is a unique solution:
$$x = 5 \qquad y = 10 \qquad z = 15$$
So in this case the three equations are both independent and consistent.

3 (i) $x = -z$, $y = 10 + 1.5z$
   (ii) $x = 15 - 4z$, $y = 8 + 2z$

Exercises 2.11

1 Using the first equation to eliminate x from the second and third equations produces

$$x + 3y + 2z = 29$$
$$2y + 2z = 20$$
$$-2y + z = -2$$

from which the second equation is used to eliminate y from the third equation, which is then divided through by 3. The result of these operations is

$$x + 3y + 2z = 29$$
$$y + z = 10$$
$$z = 6$$

from which back substitution produces the solution of the system as

$$x = 5 \qquad y = 4 \qquad z = 6$$

2 Using the first equation to eliminate x from the second and third equations produces

$$x + 0.5y + z = 9.5$$
$$y + 2z = -1$$
$$y + z = 7$$

from which the second equation is used to eliminate y from the third equation, which is then divided through by -1 with the resulting equations

$$x + 0.5y + z = 9.5$$
$$y + 2z = -1$$
$$z = -8$$

Using back substitution the solution of the system is

$$x = 10 \qquad y = 15 \qquad z = -8$$

Exercises 2.13

1 (i) Breakeven output level is at

$$q = \frac{4800}{5 - 3} = 2400 \text{ units}$$

(ii) Two-thirds of plant capacity.

132   Mathematics for business, finance and economics

(iii) New breakeven output level is at

$$q = \frac{5400}{5 - 3.25} = 3086 \text{ units}$$

(iv) New value of price p is such that

$$\frac{5400}{p - 3.25} = 2400$$

so that $p = 5.5$.

2   (i) Breakeven output level is at

$$q = \frac{13{,}500}{12 - 8} = 3375 \text{ units}$$

(ii) 50 per cent of plant capacity.
(iii) (a) Breakeven output level is unchanged.
(b) New breakeven output level is at

$$q = \frac{13{,}500}{13.5 - 9}$$
$$= 3000$$

(iv) 25 per cent reduction required.

3   (i) Excess demand of 30 units.
(ii) Excess supply of 45 units.
(iii) $q = 100$, $p = 100$
(iv) (a) $T = 1350$ when $t = 15$
$T = 2400$ when $t = 30$
(b) Cost of subsidy (at $t = -7.5$) is 787.5.

4   (i) $p = 80$, $q = 100$
(ii) $q \geqslant 44$
(iii) 300 units of excess supply.
(iv) $q = 160$, $p = 56$
(v) 100 units of excess demand.

Additional problems

1   (i) (a) £166m
        (b) £181.2m
    (ii) 1997
    (iii) £67.2m

2   (i) $I = 720 - 12t$
    (ii) 60 days.
    (iii) On the forty-sixth trading day.

Straight lines and linear equations 133

3 (i) £20,000 in the share portfolio and £50,000 in the building society.
   (ii) £35,000 should be moved from deposit to the securities.

NOTES

1 Where there *are* units of measurement. $y = f(x)$ does not necessarily involve units.
2 One such package, *Derive*, has been regularly used in the preparation of material for this book.
3 The slope in this case is informally referred to as *infinite*.
4 For the purpose of this definition, neither of the lines should be vertical.
5 Plus, of course, any necessary rearrangement of the terms.
6 The reason that not all the lump sum is placed in the higher yielding security is *risk*. Where an appropriate measure of risk is available, this can be included in a more sophisticated model.
7 Further examples of the use of linear equations in an investment context are given in Wilkes and Brayshaw (1986).
8 The rows correspond to the rows of matrices. See *elementary matrix operations* in Chapter 10.
9 After Diophantus of Alexandria, a third-century mathematician.
10 For an introduction to integer programming, see Anderson, Sweeney and Williams (1991).
11 The inconsistency would be revealed during a solution attempt by a requirement to equate two different constants.
12 Had the dependence not been detected at the outset, at some point during attempted solution $0 = 0$ would have emerged.
13 This is not the only way of looking at the mutual dependence of these equations. For example (i) is obtainable by multiplying equation (ii) by 3 and subtracting (iii). Dependence between equations in a linear system is considered in Chapter 10.
14 But not necessarily *any* two of the variables. For example the system $5x + y + 4z = 100$; $2x + 2y + 8z = 64$ can be solved for x and y in terms of z, but not y and z in terms of x. In fact x must be 17 for the system to be consistent.
15 Solution sets are discussed in relation to inequalities in Chapter 3.
16 Except, of course, one of the two planes already utilized.
17 Which, pedantically, could be said to be equivalent to the addition of the equation $z = 4$.
18 See Hart (1966).
19 If the coefficient of x is zero in the first equation, exchange the position of this equation with one in which the coefficient of x is non-zero. This switch is an elementary row operation which leaves the solution of the system unchanged.
20 If in eliminating y from the last equation z also disappears, then if the right-hand side is not also zero the equations are *inconsistent*. If the right-hand side value *is* zero, the equations are not independent and there are infinitely many solutions.
21 These values are the roots of the equation. For example, the conditional equation $3x^2 - 72 = 6x$ is true only for $x = 4$ and $x = -6$.
22 Introduced in Chapter 3.

23 Many factors influence the demand for a product. The *Marshallian demand curve* is the best known example where 'own price' is expressed as a function of quantity.
24 A tax of this nature is known as a *specific sales tax*. In the UK, *excise duties* take this form, on petrol for example.
25 Determinants are introduced in Chapter 10.

Chapter 3

Linear Inequalities

3.1	Introduction	136
3.2	Linear inequalities in one variable	138
3.3	Linear inequalities in two or more variables	144
3.4	Convex solution sets	149
3.5	Linear programming: introduction and graphical method	151
3.6	Introduction to the simplex method	163
3.7	The iterative procedure	166
3.8	Sensitivity analysis	175
3.9	Duality	179
	Additional problems	182
	References and further reading	184
	Solutions to exercises	184

This chapter explains the nature of linear inequalities, the operations that may be carried out on them and some of their many areas of application. Sign requirements on decision variables are introduced in the context of inequality relationships involving several variables.

The second part of the chapter builds on the basic knowledge of linear inequalities acquired in earlier sections to introduce linear programming problems. Both graphical and simplex solution procedures are described.

By the end of the chapter, you will have learnt to simplify individual and simultaneous linear inequalities to obtain the set of solutions. You will also be able to solve linear programming problems involving several variables, understand the scope of sensitivity analysis and be able to interpret the solution obtained.

3.1 INTRODUCTION

In linear models it is highly desirable that relationships between variables are, if possible, expressed as equations. This is because equations are highly convenient for purposes of manipulation and solution. However, in business, economic and financial problems first descriptions of the relationships in the model frequently involve statements that certain quantities can be equal only in the extreme – for example expenditure and budget. Indeed, the most fundamental of all economic laws, the *law of scarcity*, is an inequality and states that the material wants of a society, W, are greater than the total volume of goods and services, Q, that the economy can produce in order to satisfy these wants. This elemental truth is stated as a *strict inequality* thus:

$$W > Q \tag{3.1}$$

In (3.1) the symbol $>$ means '*is strictly greater than*'. In any hypothetical society in which the law of scarcity did not apply, the study of economics would be reduced to an abstract exercise. In management science, a frequently encountered inequality relationship is that between the quantity of a resource needed by a production plan and the amount of that resource available. In Section 2.10, we considered an example in which x and y represented the daily output levels of two products made by a firm. Each unit of the first product requires 20 minutes of machine time, while each unit of the second product requires 13 minutes of machine time. There are 8 hours of machine time available each day, and the amount of machine time called for in any production plan should not exceed the available total. This can be stated as an inequality:

$$20x + 13y \leqslant 480 \tag{3.2}$$

where in (3.2) the symbol $\leqslant$ means '*less than or equal to*' which has exactly the same meaning as '*should not exceed*' or '*not greater than*'. 'Less than or equal to' is an example of a *weak inequality* where the equation of the two sides is permissible as well as the allowable inequality. Weak inequalities arise much more commonly than strict inequalities and they are also much more convenient. Whenever there are budgetary constraints, limited physical resources, minimum targets or variables which cannot be negative, a weak inequality is defined. Other inequality symbols in common use are

$<$ '*strictly less than*'

and

$\geqslant$ '*greater than or equal to*'

The symbol $<$ could be read as '*neither greater than nor equal to*', while $\geqslant$ could equally well be read as '*not less than*'. If any of the symbols are struck out, the opposite meaning to that of the original symbol is taken. For example, the symbol $\not>$ represents '*not greater than*' (and is equivalent to $\leqslant$), while $\not\leqslant$ means '*not less than or equal to*' (i.e. greater than). Most importantly the struck-out equality symbol $\neq$ has the meaning '*not equal to*'. Incidentally, in BASIC the relation 'not equal to' is represented by a combination of the 'greater than' and 'less than' strict inequality symbols, < >, and may be thought of as meaning necessarily greater than or less than. Taken together, the equality and inequality relationships are called *relational operators*, and in computer programming the relational operators are used to compare collections of symbols ('*strings*') as well as numbers. In total there are six possible relationships:

Relational operator	Meaning
=	Equal to
$\neq$ or < >	Not equal to
<	Strictly less than
>	Strictly greater than
$\leqslant$ or <=	Less than or equal to
$\geqslant$ or >=	Greater than or equal to

The notion of '*approximately equal to*' is also a useful concept, although what is meant by 'approximately' has to be interpreted in context. The symbol is $\approx$ and a common usage is in rounding numbers. Thus we may write

$$3.1389 \approx 3.14$$

In the next section, we consider the process of rearrangement of inequalities into more convenient forms, and the operations that can be performed upon them.

EXERCISES 3.1

1 What is the appropriate inequality symbol for the following literal descriptions of a relationship:

   (i) not greater than;
   (ii) strictly greater than;
   (iii) neither less than nor equal to;
   (iv) not equal to.

2 Replace the following symbols with simpler alternatives:

   (i) $\not<$ (ii) $\not>$ (iii) $\not\leqslant$ (iv) $\not\geqslant$

3.2 LINEAR INEQUALITIES IN ONE VARIABLE

Model building and analytical work involving inequality relationships frequently require the rearrangement, simplification or manipulation of inequalities. Any such re-expressions must preserve the relationships involved. The following operations are permissible and may be carried out on an inequality:

1 Addition or subtraction of *the same* constant term to or from each side of an inequality.
2 Multiplication or division of both sides of the inequality by a *positive* constant.
3 Multiplication of both sides of an inequality by a *negative* number and *reversal of the direction of the inequality*.

Consider these operations in turn.

Addition of a constant

Thus if it is required that

$$x < 10$$

the ability to add a constant to both sides means that it is valid to write

$$x + 4 < 14$$

or, since the constant may be any real number, it will also be true that

$$x - \pi < 10 - \pi$$

The constant may in fact be an unknown, as in

$$x + b < 10 + b$$

All that is required is that both sides of the inequality are varied by the same amount.

Multiplication by a positive constant

Given the inequality

$$x \leqslant 10$$

after multiplication of both sides by (say) 20, it will be true that

$$20x \leqslant 200$$

or, after division by 20, it will be true that

$$0.05x \leqslant 0.5$$

and it would be equally valid to write

$$\pi x \leqslant 10\pi$$

Or, if the original relationship had been

$$-0.5y \geqslant 100$$

then after multiplication by 4 it will be true that

$$-2y \geqslant 400$$

and for any value x known to be positive

$$-0.5xy \geqslant 100x$$

Multiplication by a negative constant/reversal of direction

The validity of this rule is most clearly illustrated by the comparison of two real numbers. For example:

$$6 < 7$$

but when this inequality relationship is multiplied through by -1 the valid statement is

$$-6 > -7$$

since, between negative numbers, the value further to the right on the real line[1] is the greater of the two. Thus with an initial statement

$$2x - 4 \geqslant x - 3$$

multiplication by -1 throughout the inequality produces the equivalent statement that

$$-2x + 4 \leqslant -x + 3$$

while

$$2 - 2x \leqslant 4x - 10$$

when divided by -2 becomes

$$x - 1 \geqslant 5 - 2x$$

or, for any value b known to be negative,

$$b(2 - 2x) \geqslant b(4x - 10)$$

A linear inequality involving one unknown can often be rearranged so that the range of values of the unknown which satisfy the relationship are more clearly seen. For example, given that

$$2x - 3 \leqslant x + 5$$

the *solution set* can be found — the set of values of x for which the relationship is true. The solution set if not empty will be infinite. This is shown in the relationship above, where the addition of 3 to both sides produces

$$2x \leqslant x + 8$$

and subtraction of x from both sides produces

$$x \leqslant 8$$

which is the solution set in this case. Consider another example. Suppose that the initial inequality is

$$10 - 2x \leqslant 22 + x$$

Subtraction of 10 from both sides results in

$$-2x \leqslant 12 + x$$

Subtraction of x from both sides gives

$$-3x \leqslant 12$$

and division by -3 produces the solution set as

$$x \geqslant -4$$

However, suppose that the original relationship had been

$$6 - 4x + 5(x - 2) \geqslant 3(x - 2) - (2x - 3)$$

This expression simplifies to

$$x - 4 \geqslant x - 3$$

and to be satisfied would therefore require that

$$0 \geqslant 1$$

The solution set in this case is the empty set ∅. This inequality relationship therefore cannot be satisfied by any value of x. In contrast, if the original inequality relationship had been

$$10(x + 5) - 3(2x + 5) \geqslant -2x + 40 - 2(5 - 3x)$$

this simplifies to

$$4x + 35 \geqslant 4x + 30$$

in which all of the terms in x are seen to cancel, leaving the 'requirement' that

$$5 \geqslant 0$$

which means that the original relationship is true *for all* values of x. This is so, since it is a truthful relationship that did not really involve x at all, and therefore cannot *not* be true for any x. So far, we have considered cases involving weak inequalities. Similar principles apply to strict inequalities (the reader may rework the above examples using $<$ and $>$ in place of $\leq$ and $\geq$ respectively). In either case, when just one inequality is involved the resulting solution set, if not empty, is defined by an upper or lower bound on x. When more than one relationship must be satisfied, there may be both upper and lower bounds on x. Consider the *simultaneous inequalities*

(a) $\quad 2x - 3 \leq x + 5$
(b) $\quad x + 1 \leq 2x - 3$

We have already seen that (a) can be simplified to $x \leq 8$. Requirement (b) can be rearranged as $x \geq 4$, so that taken together the solution set is

$$4 \leq x \leq 8$$

which is a solution set that is bounded from both above and below. Before proceeding further, an important distinction between linear equations and linear inequalities should be drawn out. As inequalities, (a) and (b) can be satisfied simultaneously and produce a non-empty solution set. But if (a) and (b) were *equations* they would be inconsistent – as must always be the case for independent equations in a single variable. But inconsistency *can* occur with inequalities too. For example the system:

(c) $\quad 20 - 3x \leq 10 - x$
(d) $\quad 7x - 12 \leq 3 + 2x$

requires that $x \geq 5$ (from (c)) and simultaneously that $x \leq 3$ (from (d)). Thus the solution set is null. To further point up the distinction between simultaneous equations and inequalities, there is no limit to the number of distinct linear inequalities that can simultaneously hold. For example suppose that

(e) $\quad 2x - 3 \leq x + 5$
(f) $\quad 3x - 4 \leq 2x + 2$
(g) $\quad x + 1 \leq 2x - 3$
(h) $\quad 1 - 2x \leq -x - 4$

The *individual* solution sets of the inequalities are

(e) $\quad x \leq 8$
(f) $\quad x \leq 6$
(g) $\quad x \geq 4$
(h) $\quad x \geq 5$

Note that (e) and (f) represent *upper bounds* on x. The more restrictive of the two is (f) which is termed the *least upper bound* or *supremum*. If the least upper bound, (f), is satisfied, then (e) must also hold. In general, if the least upper bound (LUB) is fulfilled, *all* the other upper bounds will be satisfied of necessity and do not therefore need to be referred to. We can then concentrate on a single numerical value – the LUB. Lower bounds are given by (g) and (h), and of these, (h) is the *greatest lower bound* or *infimum*. If (h) is satisfied, (g) must also be fulfilled. Of all the lower bounds, only the greatest lower bound need be explicitly considered. This is because if the greatest lower bound (GLB) is satisfied, all of the other lower bounds will be automatically fulfilled. Therefore, for a system of simultaneous inequalities, the solution set will be the range between these tightest limits. That is,

$$GLB \leqslant x \leqslant LUB$$

which in this case means that

$$5 \leqslant x \leqslant 6 \tag{3.3}$$

Solution sets of this kind, defined between the infimum and the supremum, often result when the stability of the solution to a model is being investigated in relation to the values taken by certain key parameters. Work of this nature is called *sensitivity analysis*, and the range of values of a variable for which a solution holds good (expressed as in (3.3)) is sometimes called a *tolerance interval* or *range of feasibility*. A solution set for one variable can be identified as a set of points on the real line – a section of it – which may or may not include the end points. A range such as (3.3) which involves weak inequalities at both ends of the interval is said to be a *closed interval* on the line as the end points, 5 and 6, are included in the set. This is shown as in Figure 3.1(a), the solid circles at the limits of the interval indicating

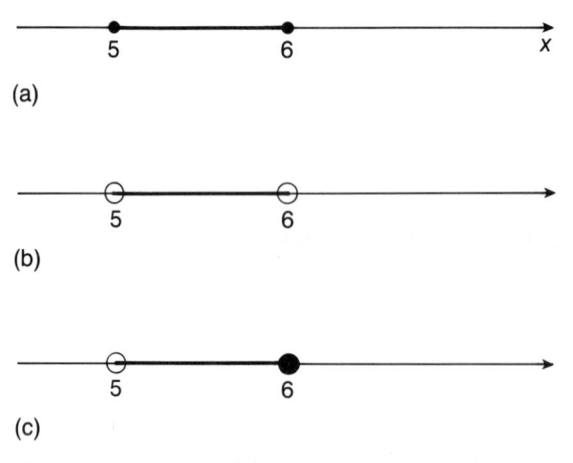

Figure 3.1

that the end points are included. If the solution set (3.3) had involved strict inequalities at both ends, i.e.

$$5 < x < 6$$

then an *open interval* on the real line would have been defined. This would appear as shown in Figure 3.1(b) with the end points as 'hollow' circles – the convention to signify that they are not included in the interval.

An *open–closed* interval is defined if the lower bound is *not* included, but the upper bound *is* within the range. This would be the case if the solution set had been

$$5 < x \leqslant 6$$

which would have appeared as shown in Figure 3.1(c). Obviously, when this situation is reversed as for

$$5 \leqslant x < 6$$

then the solution set defines a *closed–open* interval on the real line. Following the exercises we can now extend consideration of inequality relationships to those involving several variables.

EXERCISES 3.2

1 Find the solution set for each of the following linear inequalities:

(i) $4x - 8 \leqslant 2x + 12$
(ii) $10x + 8 \geqslant 12x - 9$
(iii) $6x + 25 \geqslant 3x + 10$
(iv) $4(5 + 2x) \leqslant 20 + 6x$

2 Comment on the solution sets in the following cases:

(i) $6(x - 10) + 3(25 - x) \leqslant 7(4 + 2x) - (11x + 14)$
(ii) $7(x - 10) + 5(16 - x) \leqslant 6(x - 8) - 4(x - 15)$

3 Find the solution set for each of the following systems of inequalities:

(i) $3x + 5 \leqslant 6x - 10$
    $x - 15 \leqslant 33 - 3x$
(ii) $4x - 20 \leqslant 3x + 10$
    $3x + 15 \leqslant 8x - 30$
(iii) $6x - 5 \geqslant 5x + 5$
    $7x - 5 \leqslant 2x + 20$

4 Find the solution set for each of the following systems of linear inequalities:

(i) $3x + 2 \leqslant 2x + 10$
$3x - 4 \leqslant 4x - 8$
$5x + 6 \leqslant 4x + 12$
$6 - x \leqslant 2x - 3$

(ii) $x + 20 \geqslant 3x + 6$
$4x + 12 \leqslant 8x + 4$
$x + 2 \geqslant 11 - 2x$
$5x - 5 \leqslant 4x + 4$

(iii) $2x + 6 \leqslant 4x + 2$
$4 - x \leqslant x - 2$
$5x - 1 \leqslant 4x + 8$
$3x + 2 \leqslant x + 16$

(iv) $5x + 5 \leqslant 4x + 14$
$3x - 2 \leqslant x + 12$
$x + 3 \leqslant 2x + 1$
$2 - x \leqslant x - 4$.

3.3 LINEAR INEQUALITIES IN TWO OR MORE VARIABLES

Linear inequalities usually involve more than a single variable as, for example, when the condition relates to an expenditure limit on resources. Consider the inequality

$$2x + y \leqslant 13$$

The solution set is all points on or below the line defined by the equality part of the relation and is shown in Figure 3.2.

In Figure 3.2, the inequality part of the relationship defines the permitted (hatched) side of the line. The solution set is bounded only by the line and extends without limit below the line. The solution set in this case is called a *half space* or *half plane*, and the set could be literally described as 'the half space on or below the line $y = 13 - 2x$'. Now consider the weak inequality

$$x + 3y \leqslant 19$$

The solution set is as shown in Figure 3.3.

In Figures 3.2 and 3.3, there is no exclusion of negative values for x and y. In practice, however, it is often the case that only non-negative values of the variables are meaningful – as for example with production levels.

Linear inequalities

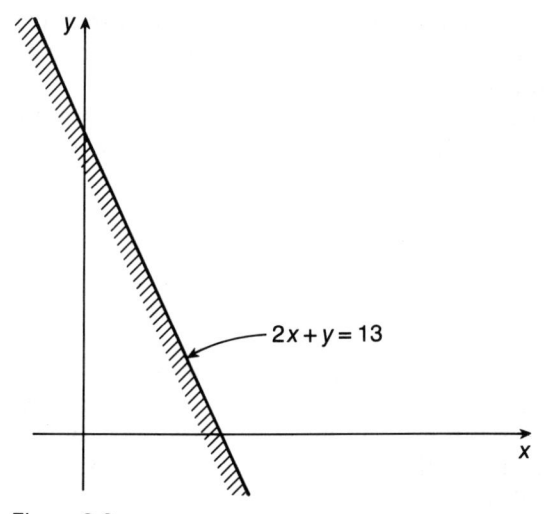

Figure 3.2

$2x + y = 13$

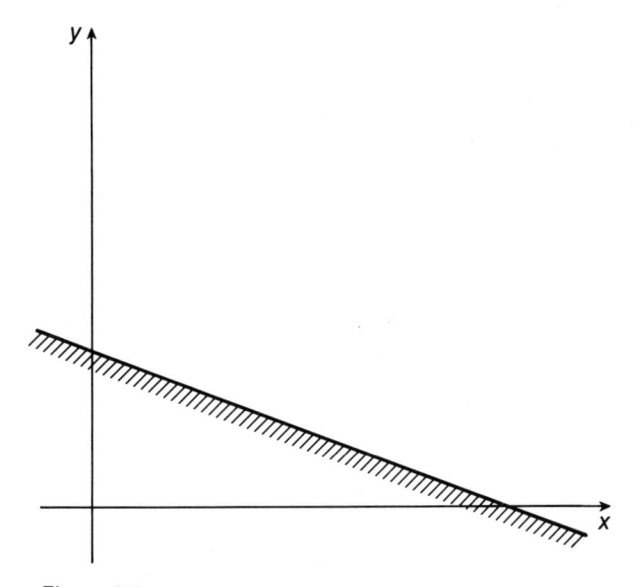

Figure 3.3

Negative values of the variables are excluded by the introduction of the *sign requirements*

$$x \geq 0 \quad \text{and} \quad y \geq 0$$

in addition to the existing inequalities. In practice, it is almost always the case that sign requirements will take the form of *non-negativity* requirements (rather than *non-positivity* requirements). These conditions restrict the solution set to that part of the original set lying in the positive quadrant as shown in Figures 3.4(a) and 3.4(b).

Sign requirements are no different in character from any other weak inequality. The non-negativity requirements on x and y could equally well have been written in the following manner:

$$1x + 0y \geqslant 0$$

and

$$0x + 1y \geqslant 0$$

respectively. The equality part of the requirements – as in any other equation – will produce a straight line in the x, y plane. In this case, the lines correspond to the y and x axes respectively. Now consider the system produced by the two linear inequalities and the sign requirements taken together. The full system is then

$$2x + y \leqslant 13$$
$$x + 3y \leqslant 19$$
$$x \geqslant 0, \ y \geqslant 0$$

The solution set for this system is illustrated in Figure 3.5.

In Figure 3.5, the solution set is the perimeter and interior of the polygon OCEB. Solution sets of this form will be considered in Section 3.5. For the moment, note that the shape of the solution set for simultaneous linear

(a)   (b)

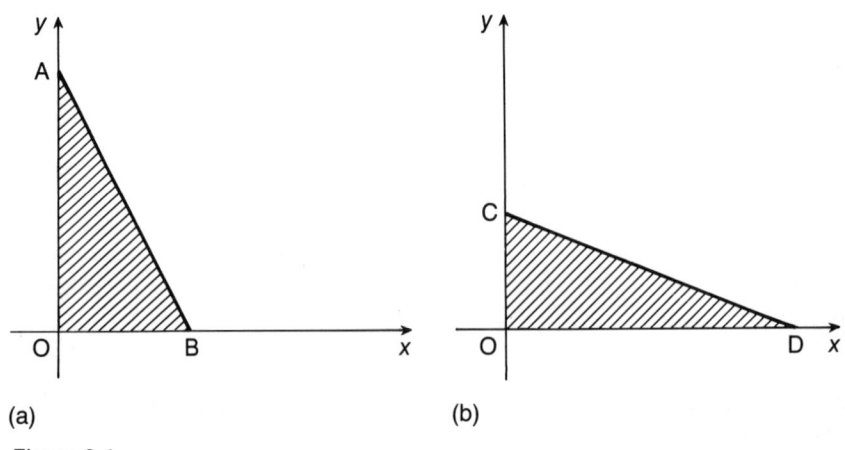

Figure 3.4

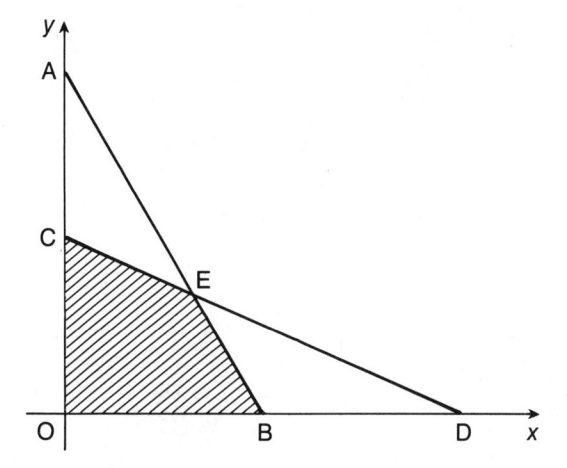

Figure 3.5

inequalities can vary considerably. Figures 3.6(a)–(e) illustrate the solution sets to the following systems:

(a) $2x + y \leq 13$ (b) $2x + y \leq 13$
       $x + 3y \geq 19$ $4x + 2y \geq 20$
                               $x \geq 0$

(c) $2x + y \geq 13$ (d) $x + 3y \leq 19$
       $x + 3y \leq 19$ $3x + 9y \geq 57$
       $y \geq 2$

(e) $2x + y \geq 180$
       $3x + 6y \leq 540$
       $x + y = 100$
       $x \geq 0$ $y \geq 0$

Note that in Figure 3.6(a) no sign restriction on x has been stipulated. In Figure 3.6(b) the solution set is that part of the *strip* between the parallel lines that does not involve negative values of x. In Figure 3.6(c) the solution set is the hatched triangle which, as it happens, does not include any negative values of x even though these have not been expressly ruled out. In Figure 3.6(d), the solution set is the straight line, representing the only way that the requirements can be simultaneously fulfilled – as strict equalities. In Figure 3.6(e), where the requirements include an equality relationship, the solution set is that part of the line given by the equation that is included in the region defined by the inequalities. This is the line segment AB as shown. In practice, a constraint set of this nature is not

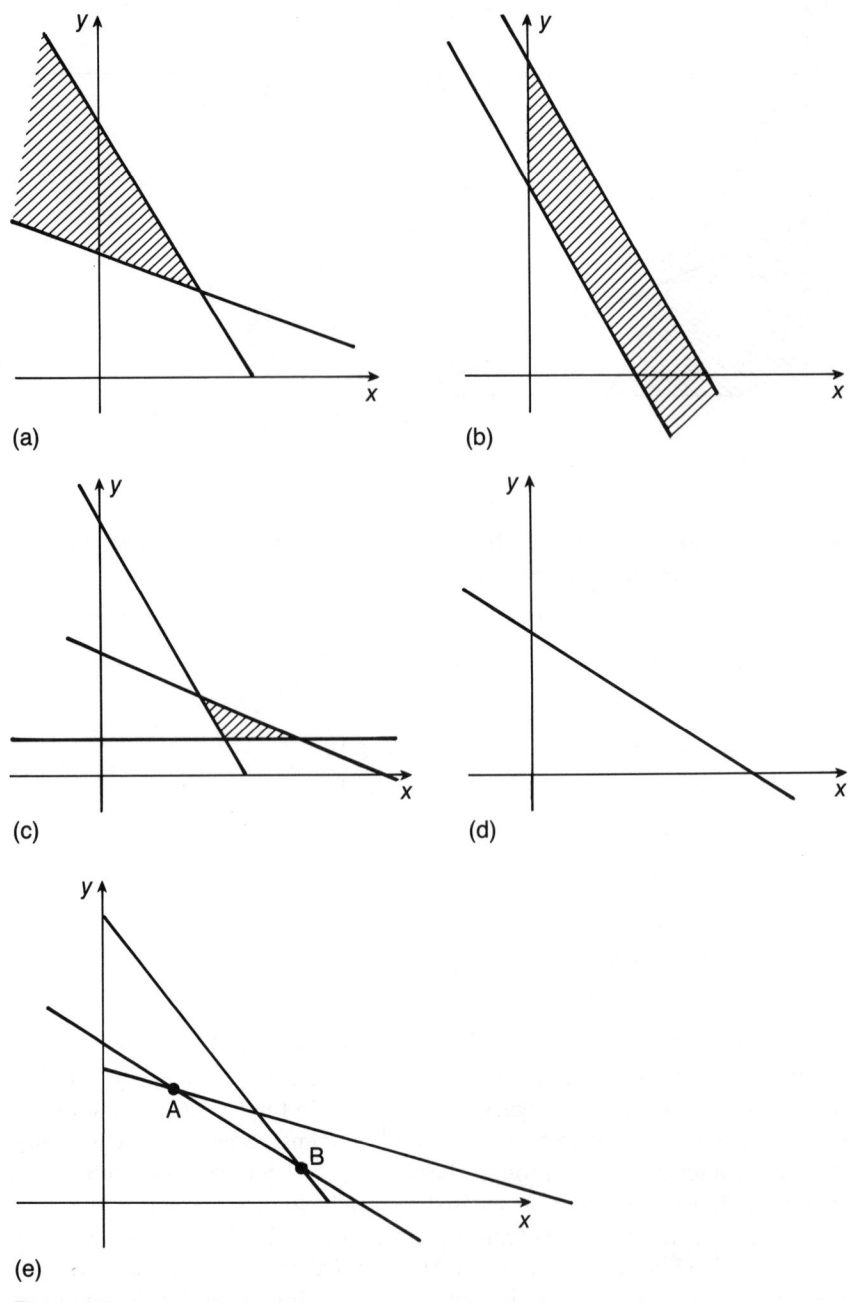

Figure 3.6

uncommon. For example, in batch production problems, x and y might represent the amounts produced of a uniform product by different processes or in different plants. The inequalities would represent resource availability constraints while the strict equality $x + y = 100$ would represent a customer order for 100 units. Simultaneous inequalities in several variables do not necessarily have solutions. For example, if in case (d) above the inequalities involved had been

$$x + 3y \leqslant 18$$
$$3x + 9y \geqslant 57$$

then the intersection of the individual solution sets is empty – being on conflicting sides of the parallel lines. Where the equality part of the relationships produces non-parallel lines, the solution set may still be empty if sign requirements are present. For example, the system

$$2x + y \leqslant 18$$
$$x + 2y \geqslant 40$$
$$x \geqslant 0, \quad y \geqslant 0$$

restricted as it is to the positive quadrant by the sign requirements, has no solution.[2]

EXERCISES 3.3

1 Describe the nature of the solution set in the following cases:

(i) $2x + y \leqslant 25$

(ii) $x + y \leqslant 10$
$x \geqslant 0 \quad y \geqslant 0$

(iii) $5x - 2y \geqslant 20$
$y \geqslant 2.5x - 8$

(iv) $2x + 4y \geqslant 40$
$5x + 9y \leqslant 80$
$x \geqslant 0 \quad y \geqslant 0$

(v) $x + 2y \leqslant 120$
$x + y = 80$
$x \geqslant 0 \quad y \geqslant 0$.

3.4 CONVEX SOLUTION SETS

Whatever the shape of the solution set generated by the system of simultaneous linear inequalities, it will always have one very important property – that of *convexity*. In two dimensions, none of the interior angles of the solution set polygon should be *reflex* (over 180°) and hence the boundary of the feasible region should not be a reflex polygon. That is, the feasible region will not be *re-entrant* − a convex set is one without 'indentations'. Clearly, in Figure 3.7 since the angle at point A is reflex, this could not result from a system of simultaneous linear inequalities. Convexity means that for *any* two points P and Q in the set, all points on a straight line connecting P and Q must also be in the set. Thus all of the solution sets of Figure 3.6 are convex. However, the set illustrated in Figure 3.7 is *not* convex since, because of the inward pointing vertex, it is possible to select points within the set (e.g. P and Q as shown) such that some of the points on the line segment connecting them are outside the set.

Convexity of the feasible region is an important property in problems where the optimal value of a function is sought for points lying within the feasible region. If the function to be optimized satisfies certain conditions, then if the feasible region is convex we can be sure that a point that is better than any other point in its locality will be the best overall.[3] An important concept in relation to convex sets is that of the *extreme point*. An extreme point in a convex set does not lie on a line segment joining any two other points in the set. Extreme points are important in the simplex method of linear programming. This method is called an *adjacent extreme point solution procedure*, since it works its way to the best position by moving between successively better adjacent extreme points. A non-convex set cannot be produced by linear inequalities which must all be *simultaneously* satisfied. Each linear inequality produces a solution set which is convex

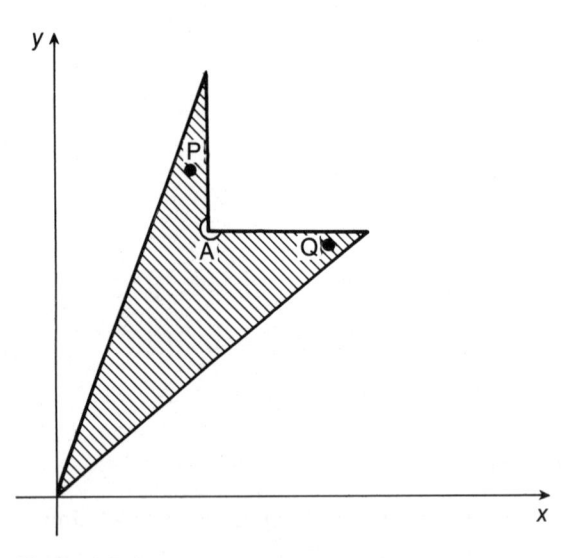

Figure 3.7

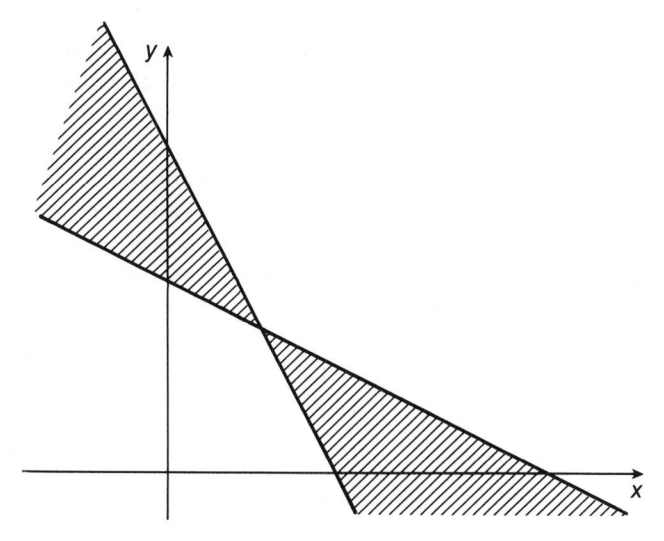

Figure 3.8

(a half plane in the two-dimensional case as we have seen) and the intersection of convex sets must itself be a convex set. But a non-convex feasible region involving only linear inequalities *could* be produced if there are *alternative* constraint systems. For example suppose that the permitted values of x and y must satisfy either

(i) $\quad 2x + y \leqslant 3$
   $\quad\quad x + y \geqslant 2$

or

(ii) $\quad 2x + y \geqslant 3$
   $\quad\quad x + y \leqslant 2$

The feasible region produced by this system is shown in Figure 3.8.

While Figure 3.8 illustrates a possible situation in practice, if such a problem *did* arise it could be addressed as two *sub-problems* corresponding to the alternative – and convex – regions defined by (i) and (ii) separately. The means employed would almost certainly involve linear programming to which we now turn.

3.5 LINEAR PROGRAMMING: INTRODUCTION AND GRAPHICAL METHOD

Convexity is important in problems that involve finding the maximum or minimum of a linear function subject to the condition that the values of its independent variables lie in the solution set of a system of linear inequalities. This is the situation in *linear programming* problems. If the solution set is convex, any point which is the best point in its own locality is the best point overall: *a local optimum must be a global optimum*. This property means that highly efficient solution procedures can be employed. Consider an example. A firm can produce two products X and Y in amounts x and y. These output levels are the firm's *decision variables*. Typically, management will wish to set production levels (choose values of x and y) so as to maximize some objective; for example, make as much profit as possible. Suppose that the most convenient time frame for the problem is one day. The accounts department estimate that each unit of X produced makes £2 profit while each unit of Y makes £5. Thus the firm's objective is to choose values of x and y so as to maximize profit z, where

$$z = 2x + 5y \tag{3.4}$$

Equation (3.4) is the firm's *objective function*. Since (3.4) is a linear function, unless there are some restrictions on the values taken by the decision variables, the objective function will have no finite maximum. Restrictions are provided in practice by the fact that resources (both physical and financial) that are needed to produce X and Y are not available in limitless amounts. If each unit of X produced requires 4 labour hours and each unit of Y requires 3 hours, a total daily availability of 48 labour hours defines the following inequality:

$$4x + 3y \leqslant 48 \tag{3.5}$$

where (3.5) is known as a *constraint*. The left-hand side of (3.5) gives the total labour time requirement of any *production plan* (a pair of values of x and y). For feasibility, the production plan must not call for more labour time than is currently available. Suppose also that each unit of X requires 1 kilogramme of a certain material and each unit of Y requires 2 kilogrammes. An overall daily availability of 22 kilogrammes of material defines a second weak inequality as a constraint:

$$x + 2y \leqslant 22 \tag{3.6}$$

In practice there may be a variety of constraints relating to other *scarce* resources such as floorspace, machine time or finance at various times; but we shall limit ourselves to two constraints here. A crucial word in this example is *each*. It is this word that tells us that the problem is linear. If the resources required to produce a further unit of a product depended on the amount already produced, a much less tractable problem would be defined. In addition to the resource constraints (3.5) and (3.6) there will normally be a prohibition on negative values of the decision variables, the sign requirements of earlier acquaintance. The complete problem in algebraic form is then to

maximize $\quad z = 2x + 5y$
subject to $\quad 4x + 3y \leqslant 48$
$\quad\quad\quad\quad\quad x + 2y \leqslant 22$
$\quad\quad\quad\quad\quad x \geqslant 0 \quad\quad y \geqslant 0$

A problem of this size can be solved either by use of an *algorithm* or by graphical means. An *algorithm* is a step by step procedure whereby either the solution is reached in a finite number of steps or it is shown that no solution exists. The principal algorithm in use is the *simplex method*, which uses the fact that the solution set to the constraints is convex. The simplex method involves the repetition of a cycle of calculations (*iteration*) in tabular form, and will be explained in later sections. Microcomputer software using the method is available and can be used to solve problems with many variables and constraints.[4] We begin by outlining a graphical approach which will illustrate the main features of the problems. The first step is to draw a graph of the constraints and the solution set they produce. This is shown in Figure 3.9.

The solution set is the interior and perimeter of the polygon OABC. The solution set is known as the *feasible set*, *feasible region* or *opportunity set*. In a production context it is also referred to as the *feasible production set*, and the outer edge of the set, the line segments connecting A, B and C, is the *production possibility frontier*, representing the feasible production

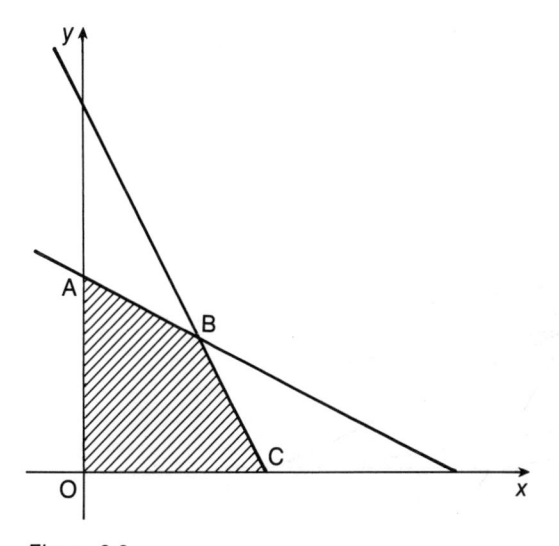

Figure 3.9

combinations which, with the given resource levels, are *Pareto optimal*. Pareto optimality, an important concept in economics, means that the output of one of the products cannot be increased without that of the other being reduced. Here we shall refer to Figure 3.9 as the *production possibility diagram*. Which of the points in OABC maximizes the value of the objective function? Prior to linear programming, a variety of rules of thumb and conventional nostrums were employed to give answers (usually imperfect) to problems of this kind. One of these prescriptions was

1 Produce as much as possible of the most profitable product(s).

but the chief yardstick was often *equipment utilization* as reflected in this 'rule':

2 Make sure that the production plan adopted makes full use of the scarce resources.

Rule 1 requires more detailed spelling out to be made operational and 2 will not always prove to be feasible. More than this, neither rule, despite their plausibility, can be relied on to produce the best position (although either may, coincidentally, be satisfied at the optimum). The chief shortcoming of rule 1 is that it does not take sufficient account of the constraints and relative resource levels. Rule 2 for its part fails to take account of the objective function! A correct solution procedure must take due cognizance of *both* aspects of the problem. The provision of the technical means to do

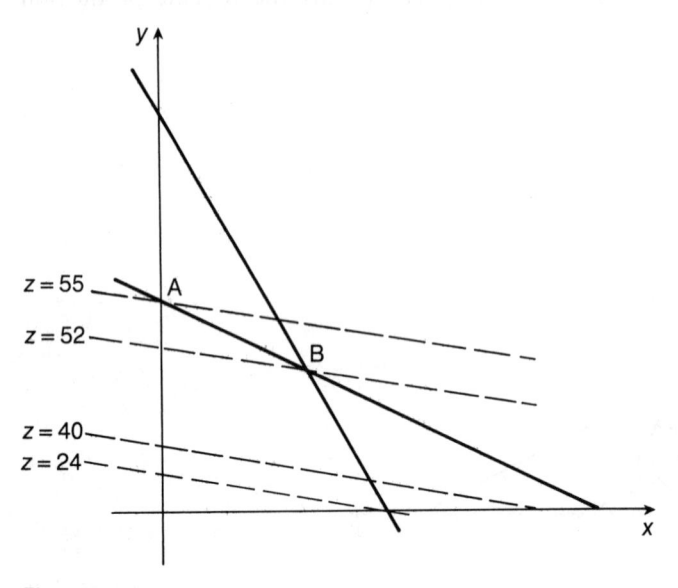

Figure 3.10

this was the great advance made by linear programming. In the graphical method, simultaneous consideration is achieved by introducing the objective function into the production possibility diagram. A *contour* of the objective function is a line such that all x, y combinations on the line correspond to the same value of z. The sense of 'contour' is the same as in an ordnance survey map with z representing height and x and y corresponding to the easterly and northerly directions respectively. In a profits maximizing context, the contours are also known as *iso-profit* lines. In Figure 3.10, the contours for $z = 24$, $z = 40$, $z = 52$ and $z = 55$ are shown as broken lines.

The further a contour is from the origin, the greater is the corresponding value of z. The point within OABC lying on the highest contour is the optimal solution to the linear programming problem. The graphical procedure is implemented by first drawing in a specimen contour through a convenient point – usually one of the intercepts. This could be C, for example, where the labour constraint line cuts the x axis. At this point, $y = 0$, so that from the constraint

$$4x = 48$$

and therefore $x = 12$. Insertion of $x = 12$ and $y = 0$ into the objective function gives $z = 24$. The y intercept of the contour is then found by solving for the value of y which gives $z = 24$ when x is zero. Thus:

$$24 = z = 2(0) + 5y$$

from which it follows that

$$y = 4.8$$

With the intercepts obtained, the sample contour can be drawn in the diagram and used as a guide for other contours. A value of z greater than 24 produces a contour parallel to the original, but further out from the origin. For example the $z = 40$ iso-profit line has its intercepts at $(20, 0)$ and $(0, 8)$. The specimen contour can now be moved upwards and to the right until there is just one point in common with the feasible region – in this case point A. This will be the profit maximizing position. In linear programming problems, the optimum point will always be situated at one of the corners of the feasible region, i.e. at an extreme point.[5] In other words, in two-variable problems, it will be located at a place where two of the constraint lines intersect. Note that one of the intersecting lines may be the x or y axis, representing the sign requirements satisfied as equalities. Once an optimum has been identified on the diagram, the approximate values of x and y could be read from the axes. But accurate values are preferable, and these can be obtained by solving the *intersecting* constraints as equations (these will be the *only* requirements that hold in equality form at this point). Thus at A, where the materials and x sign requirements both hold as equalities,

$$x + 2y = 22$$
$$x = 0$$

from which $y = 11$ which in turn produces $z = 55$, the highest possible level of profit. There is another value at the maximum position which is of interest. With $x = 0$ and $y = 11$ only 33 labour hours are required. There are therefore 15 hours of labour time unused or *slack*. As the optimum slack in the labour constraint amounts to over 30 per cent of the total availability, rule 2 is clearly seen not to apply. The concept of slack can be taken further. It is possible to re-express the original weak inequality constraints as equations by the introduction of *slack variables*, one to each constraint. For the first constraint we can define a slack variable s_1 as the amount by which the labour requirement falls short of the maximum availability. That means that s_1 is *defined as* the difference between the right-hand and left-hand sides of the constraint. That is:

$$s_1 = 48 - (4x + 3y)$$

Thus with the slack variable s_1 included specifically in the labour constraint this becomes[6]

$$4x + 3y + s_1 = 48$$

Similarly, with s_2 representing any slack, the materials constraint becomes

$$x + 2y + s_2 = 22$$

For the original constraints to be satisfied, the slack variables *cannot be negative*. Thus $s_1 \geqslant 0$ and $s_2 \geqslant 0$ need to be added to the sign requirements of the problem. The advantage gained by the introduction of slack variables is that equations have more desirable properties than inequalities. However, there is a price to be paid too! Every inequality transformed into an equation by the use of a slack variable means that the problem size is increased to the extent of

(a) one more variable;
(b) an additional sign requirement.

This increase in size is a price well worth paying in problems where the graphical approach cannot be used. We shall see later on, and again in Chapter 7, that an important relationship exists at the optimum between a slack variable and the value at the margin of the resource to which it corresponds. Returning to the solution of the current problem by graphical means (for which slack variables are not required) it is worth examining what would have happened had the production plan been set so as to make full use of both resources. The only point in the feasible region which achieves simultaneous full resource utilization is point B. The profit made here is found by first solving for the x and y values and then substituting into the objective function. The requirements fulfilled as equalities at B are

$$4x + 3y = 48$$
$$x + 2y = 22$$

Using the elimination method to solve for x and y gives

$$4x + 3y = 48$$
$$4x + 8y = 88$$
$$\overline{}$$
$$-5y = -40$$

so that

$$y = 8 \quad \text{and} \quad x = 6$$

which values produce $z = 52$. Thus £3 less profit is made at B than at the optimum. In other words there is an *opportunity loss* of £3, the difference between the profit levels at A and B. This is the price paid for the condition that all scarce resources be fully used. Although rule 1 is fulfilled at the optimum in the present example, this is not always the case. For example, suppose that the unit profit on x had been £4 rather than £2. The objective function would then have been

$$z = 4x + 5y$$

Contours through C, A and B are shown in Figure 3.11. The highest contour is now through B with $z = 64$. In comparison with A, 3 units less of y are produced at B, but the crucial economic point is this: the relative reduction of 3 units in y production allows a 6 unit rise in x output. This two to one *trade-off* or *rate of substitution* between x and y which was not worthwhile when the unit profit of x was £2 *is* worthwhile when x's unit profit is £5.

The slope of the relevant section of the production possibility frontier (given by the materials constraint between A and B) shows the rate of substitution that is possible between the products so that available resource levels are not exceeded. The slope of the iso-profit contours shows the rate of substitution between the products that will maintain a given level of profit. If at any point a more favourable rate of substitution exists (for movements in either direction) in terms of resources than for the maintenance of profit, the point in question cannot be optimal. Note that at each of the corner points A, B and C there are *two* such rates of substitution on the resource side, dependent on the direction of movement. At A and C the rates of substitution of y for x are ∞ and 0 respectively. The optimal

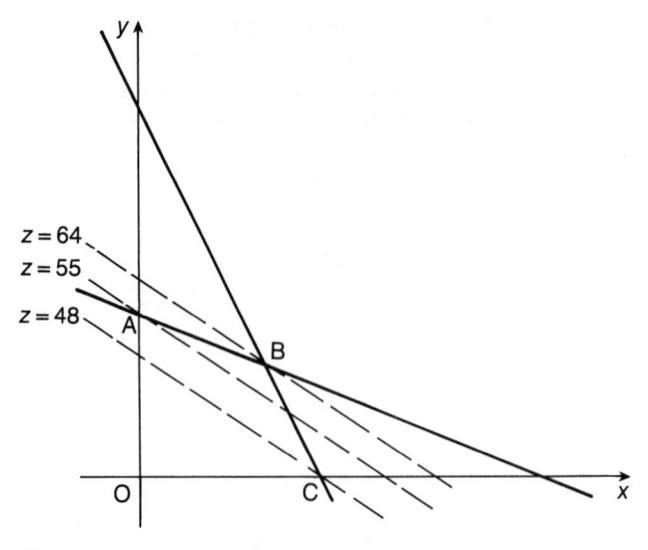

158 Mathematics for business, finance and economics

Figure 3.11

position depends on the relative slopes of the iso-profit lines and the constraints. If the objective function contours in absolute terms, have lesser slope than either constraint, then point A will be optimal; if they are steeper than both constraints C is best, and if intermediate in slope, B is the superior position. Consider the values of the slopes in the present example. For the labour constraint given

$$4x + 3y = 48$$

y can be made the subject of the equation by writing

$$3y = 48 - 4x$$

so that

$$y = 16 - 1.\overline{33}x$$

from which it is seen that the slope in absolute terms is $1.\overline{33}$. For the materials constraint, we have

$$x + 2y = 22$$

and to show up the slope we express y in terms of x as

$$2y = 22 - x$$

so that

$$y = 11 - 0.5x$$

Thus the modulus of the slope is 0.5. With the original unit profit figures, the objective function was

$$z = 2x + 5y$$

which can be written as

$$5y = z - 2x$$

i.e.

$$y = 0.2z - 0.4x$$

where z takes a constant value (e.g. 24, 40) for any particular contour. The absolute value of the slope is 0.4 which is less than either constraint – hence A is optimal. With the constraints in explicit form, note that the production possibility frontier (PPF) is an example of a piece-wise defined function:[7]

$$\text{PPF}(x) = \begin{cases} 11 - 0.5x & 0 \leqslant x \leqslant 6 \\ 16 - 1.\overline{33}x & 6 \leqslant x \leqslant 12 \end{cases}$$

So long as the iso-profit contours have a slope different from that of any constraint, one of the corner points of the feasible region will be *uniquely* optimal (if both products showed a loss then the optimum would be at zero). This is an important result, and it means that, of all the points in any feasible region, *only* the corner points need to be considered. If the slope of the objective function contours is the same as that of one of the constraints, then the optimum will be *non-unique*; the highest contour will be superimposed on the constraint edge with the same slope and two corner points will be equally good, as will all points in-between them. For example, suppose that

$$z = 8x + 6y$$

Then the contours would be as shown in Figure 3.12 and B, C and intermediate points produce $z = 96$.

In practice, non-uniqueness at the optimum is a *desirable* property, since the decision-maker is offered a costless choice between the equally profitable joint optima. Now consider another example:

maximize $z = 12x + 2y$

subject to $\quad 4x + y \leqslant 800$
$\quad\quad\quad\quad\quad 2x + 3y \leqslant 900$
$\quad\quad\quad\quad\quad x \leqslant 180$
$\quad\quad\quad\quad\quad x \geqslant 0 \quad y \geqslant 0$

In this problem, the third constraint represents an upper bound on the variable x (it may correspond to some practical sales limit on the product). The problem is graphed in Figure 3.13 and the feasible region is OABCD.

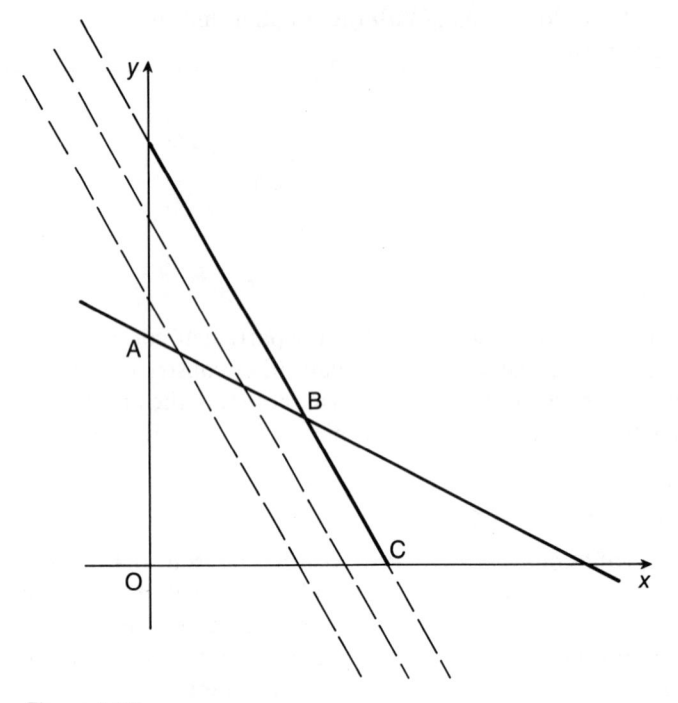

Figure 3.12

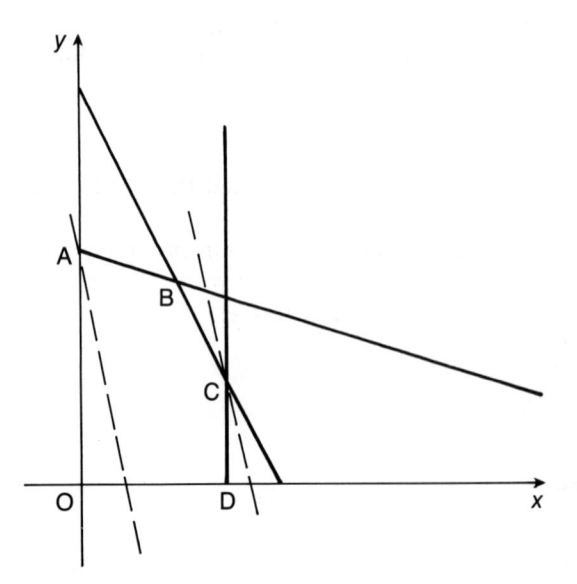

Figure 3.13

In Figure 3.13, a specimen contour is shown through point A. The optimum is at point C where

$$4x + y = 800$$
$$x = 180$$

from which it rapidly follows that $y = 80$ and that the value of the objective function is maximized at $z = 2320$. At point C, there are 300 units of slack in the second constraint. At point B, which corresponds to full resource utilization, the level of profit is $z = 2200$, revealing an opportunity loss of £120 in comparison with the optimum. Now consider a problem involving *minimization* of the objective function. The problem is

minimize $z = 3x + 5y$

subject to $\quad x + 4y \geq 60$
$\qquad\qquad\;\; 2x + y \geq 50$
$\qquad\qquad\;\; x \geq 0 \quad y \geq 0$

Notice that in this problem the constraints take the form of $\geq$ weak inequalities. In a problem in which the objective is to minimize (such as with a function expressing costs) and where, as in the present case, the coefficients of the decision variables are positive, the inequalities will be of this form. The problem is graphed in Figure 3.14.

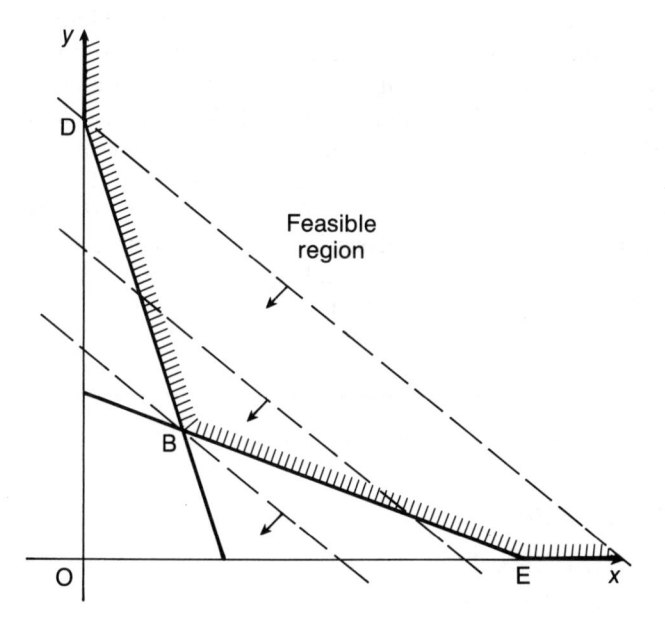

Figure 3.14

The feasible region here is the area bounded only from below by DBE and the axes. A specimen contour is shown through point D, with the arrow indicating the direction of improvement. The optimum occurs at B where

$$x + 4y = 60$$
$$2x + y = 50$$

which solve for

$$x = 20, \quad y = 10$$

and the minimum value of z is 110. The advantage of using linear programming – and in particular the simplex method – is that the program output (or the result of hand calculations in smaller problems) will provide more information than simply the optimal values of the decision and slack variables. If a resource is in short supply at the optimum (i.e. its constraint holds as an equality) management will certainly seek to secure more of the resource and achieve a relaxation in the constraint. In this way more profit would be obtained. But the extra resources will only be secured at a price. The optimum position can be investigated to reveal[8] whether or not this price is worth paying. The most that is worth paying is the increase in the objective function if a further unit of resource was available or if an upper bound was relaxed. These *marginal values*, *dual values* or *shadow prices*, as they are known, are crucial information for management in deciding whether a possible increase in resource is worthwhile. If a competitive company finds that its achievement of objective is restricted by a binding constraint, it will be someone's job to see that such constraints are made non-binding! More detailed discussion of dual values can be found in Section 3.7 following coverage of the simplex method. We can also find how the optimal production plan would respond to changes in the profitability of the products as well as variations in the resource levels. This is management information of great utility and is found through *sensitivity analysis* of the optimum. Sensitivity analysis in small problems can be carried out within the context of the graphical approach, but is more efficiently conducted using the simplex method.

EXERCISES 3.5

1 Solve the following linear programming problem using the graphical method:

maximize $F = 8x_1 + 10x_2$

subject to $\quad 3x_1 + 2x_2 \leqslant 54$
$\qquad\qquad\quad 2x_1 + 5x_2 \leqslant 69$
$\qquad\qquad\quad x_1 \geqslant 0 \quad x_2 \geqslant 0$

Indicate the location of the optimum and state the optimal values of the decision variables and the objective function.

2 With the constraints given in exercise 1, indicate the optimum position(s) and find the optimal values of x_1, x_2 and F for the following objective functions:

(i) $F = 8x_1 + 5x_2$
(ii) $F = 3x_1 + 10x_2$
(iii) $F = 8x_1 + 20x_2$

3 Use the graphical method to solve the following linear programming problem:

maximize $F = 35x_1 + 25x_2$

subject to $\quad 4x_1 + 3x_2 \leqslant 92$

$\quad\quad\quad\quad\quad x_1 + x_2 \leqslant 38$

$\quad\quad\quad\quad\quad x_1 \leqslant 20$

$\quad\quad\quad\quad\quad x_2 \leqslant 20$

$\quad\quad\quad\quad\quad x_1 \geqslant 0 \quad x_2 \geqslant 0$

4 Solve the following minimization linear programming problem using the graphical method:

minimize $F = 7x_1 + 3x_2$

subject to $\quad x_1 + 2x_2 \geqslant 55$

$\quad\quad\quad\quad\quad 4x_1 + x_2 \geqslant 80$

$\quad\quad\quad\quad\quad x_1 \geqslant 0 \quad x_2 \geqslant 0$

3.6 INTRODUCTION TO THE SIMPLEX METHOD

The simplex method will be introduced in the context of the following problem:

maximize $z = 6x_1 + 5x_2$

subject to $\quad 4x_1 + x_2 \leqslant 800$

$\quad\quad\quad\quad\quad 2x_1 + 3x_2 \leqslant 900$

$\quad\quad\quad\quad\quad x_1 \leqslant 180$

$\quad\quad\quad\quad\quad x_1 \geqslant 0 \quad x_2 \geqslant 0$

Suppose that x_1 and x_2 are the levels of two goods that can be produced by a firm; £6 and £5 are the respective unit profits of the products; 800 and 900 represent the amounts of material and labour time available per day;

4 and 2 are the requirements of these resources per unit of x_1 produced; 1 and 3 are the per unit resource requirements for x_2; and 180 is a limit imposed on the production of x_1 for other reasons. The problem is to find the daily production levels for the two goods that maximize daily profit. Begin by introducing a slack variable into each constraint to express the problem in terms of equations rather than inequalities:

maximize $z = 6x_1 + 5x_2$

subject to $\quad 4x_1 + x_2 + s_1 = 800$

$\qquad\qquad 2x_1 + 3x_2 + s_2 = 900$ (3.7)

$\qquad\qquad x_1 + s_3 = 180$

$\qquad\qquad x_1 \geqslant 0 \quad x_2 \geqslant 0 \quad s_1 \geqslant 0 \quad s_2 \geqslant 0$

Apart from the sign requirements, (3.7) is a system of three linear equations in five unknowns – the two original *structural variables*, x_j, augmented by the three slack variables, s_i. In general, therefore, a unique solution is not to be expected. But it turns out that things can be narrowed down a great deal. The simplex method rests on the fact that

No feasible point can be better than all of the corner points.

Only the extreme points can be unique optima, and even in the case of non-uniqueness there are always at least two optimal extreme points. So nothing is lost by restricting consideration to corner points. This decided, we now seek common properties of these points that can be used by a solution procedure, and address the question of how an optimum can be sought from among the extrema. First take a tour of the corners in the current problem to see what characteristics stand out. In graphing the feasible region, Figure 3.15 identifies at each corner those variables which are positive; variables not included in the list at each corner are zero.

The values of the variables, including the slack variables, at each feasible corner point constitute a *basic feasible solution* to the problem. A distinguishing feature of basic feasible solutions in linear programming problems is that *there are no more positive variables than there are constraints*. In fact there will usually be *the same* number of positive variables as constraints.[9] So in the current problem, the basic feasible solution corresponding to each corner point will have *three* positive variables. There is a further important point about the basic feasible solutions in Figure 3.15. Each basic feasible solution has (two) *adjacent extreme points*. In moving from any basic feasible solution to an adjacent solution, the set of positive variables changes by *one* element. For example, in moving from B to C, s_3 drops out and is replaced by s_2. In moving from O to D, s_3 drops out and is replaced by x_1. Thus provided that the number of positive variables remains at three and the constraints (3.7) remain satisfied, dropping one

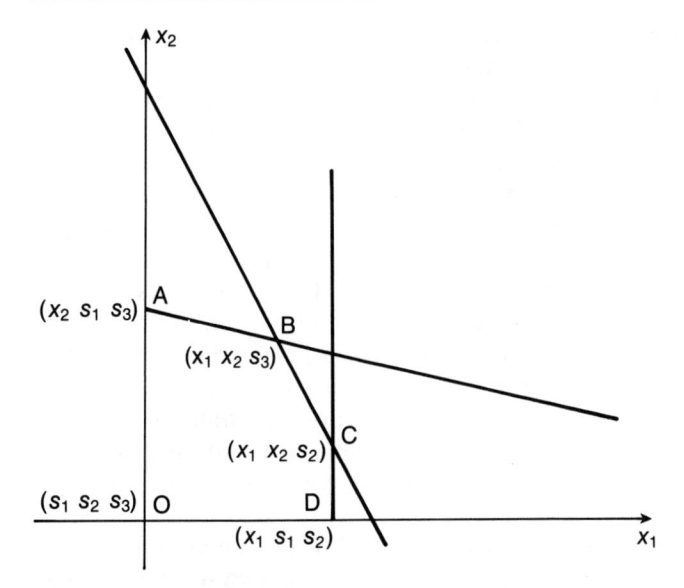

Figure 3.15

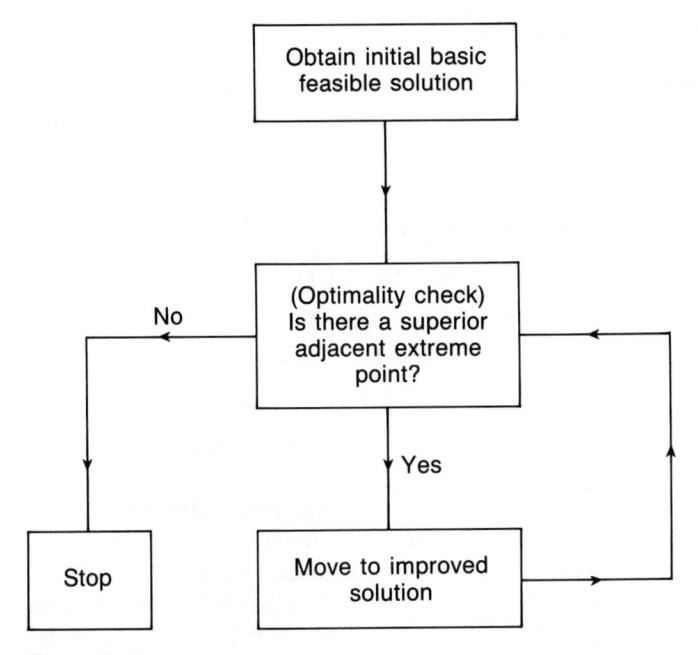

Figure 3.16

variable *to* zero and raising one variable *from* zero is equivalent to a move to an adjacent corner in the feasible set.

The simplex method uses the convexity property of the feasible region and the fact that the objective function is linear. In linear programming, if a particular basic feasible solution is not optimal, then one of its neighbours *must* represent an improvement. Also, when a linear function is being maximized over a convex region with a finite number of extreme points, to reach the optimum it is only necessary that each step from some starting point increases the value of the objective.[10] These properties make it possible to move from one basic feasible solution to an improved adjacent extreme point and to continue in this myopic fashion and be confident of reaching the optimum in a finite number of steps. We must still, of course, find a way of obtaining some basic feasible solution with which to start, and a means of assessing the improvement potential of neighbouring solutions. But the broad logic is now established and Figure 3.16 sets this out in the form of a flow diagram.

In Figure 3.16 the first box is sometimes called the *initialization phase*, while the loop represents the *iterative phase* of the simplex method. We can now begin to set out the detailed workings in tableau form.

3.7 THE ITERATIVE PROCEDURE

Consider the constraints as augmented by the slack variables. This is set out again below, but in this case each variable, including the slack variables, is shown in the objective function and in each constraint even if its coefficient is zero:

maximize $z = 6x_1 + 5x_2 + 0s_1 + 0s_2 + 0s_3$

subject to
$4x_1 + x_2 + 1s_1 + 0s_2 + 0s_3 = 800$
$2x_1 + 3x_2 + 0s_1 + 1s_2 + 0s_3 = 900$ (3.8)
$1x_1 + 0x_2 + 0s_1 + 0s_2 + 1s_3 = 180$
$x_1 \geq 0 \quad x_2 \geq 0 \quad s_1 \geq 0 \quad s_2 \geq 0 \quad s_3 \geq 0$

The next step is to arrange the information of (3.8) in a tabular format. This process is begun in Table 3.1.

In Table 3.1 there is one row identifying the variables, above which are the coefficients of the variables in the objective function and below which are the constraint coefficients. So all of the coefficients of each variable are in the same column. The next task is to identify a basic feasible solution with which to start and to add the description of this solution to the table. Where, as here, the origin is a feasible solution, this is taken as the initial basic feasible solution even though it may well be the worst solution! This likelihood is outweighed by the convenience of an easily obtained initial

Table 3.1

6	5	0	0	0
x_1	x_2	s_1	s_2	s_3
4	1	1	0	0
2	3	0	1	0
1	0	0	0	1

basic feasible solution, and the iterative phase of the simplex method brings about rapid improvement.[11] Table 3.2 adds the description of the origin.

In Table 3.2, the central column on the left lists the variables that are positive in the current basis. All other problem variables not shown here (the x_j in this case) are at zero level in this basic feasible solution. The origin consists of the slack variables set at values equal to the right-hand sides of their respective constraints. These values are shown to the right of the solution variable labels in the *solution column*. The column at the left shows the objective function coefficients of variables in the current solution.[12] The next task is to determine the value of the objective function at the current basic feasible solution, and to evaluate improvement potential – if any. The value of the objective function, z, is the sum of the products of the objective function coefficient of each variable and its corresponding level in the solution. This is a trivial exercise at the origin:

$$z = (0)800 + (0)900 + (0)180 = 0$$

The current value of z is inserted in the table directly below the solution column. We must now ascertain if improvement is possible. Recall that improvement can be sought via movement between adjacent basic feasible solutions differing by just one variable in the positive group. In the simplex tableau, this translates into evaluation of the consequences for the objective function if individual variables, currently zero, were introduced into the solution. We can conduct this investigation in terms of *unit level* changes in the values of possible incoming variables. While the introduction of a new variable to unit level will not usually represent a move all the way to a neighbouring basic feasible solution, it *will* show the constant *per unit* effect on z. The idea is to seek the variable giving the greatest per unit

Table 3.2

			6	5	0	0	0
			x_1	x_2	s_1	s_2	s_3
0	s_1	800	4	1	1	0	0
0	s_2	900	2	3	0	1	0
0	s_3	180	1	0	0	0	1

improvement, and then move to the adjacent basic feasible solution containing that variable.

Suppose that we set x_1 at unit level.[13] If $x_1 = 1$ this affects the variables in the original solution. In fact, the column of coefficients under x_1 in Table 3.2 (4, 2 and 1) shows by how much the respective variables in the current solution must be *reduced* in order to make way for one unit of x_1.[14] Thus s_1 must go down by 4, s_2 must decrease by 2 and s_3 must go down by 1.[15] These necessary reductions would affect the value of the objective function were it not for the zero coefficients of the slack variables. In later solutions not all variables will have zero objective function coefficients. The effect on z of the accommodating adjustments of variables in the current solution will be the sum of the products of entries in the x_1 column and the objective function coefficients of the solution variables:

$$4(0) + 2(0) + 1(0) = 0 \tag{3.9}$$

In return for this we get 1 unit of x_1 worth, as the head of the x_1 column reminds us, £6. So the *net decrease* in the objective function as a result of the introduction of 1 unit of x_1 is

$$4(0) + 2(0) + 1(0) - 6 = -6$$

i.e. there is a net *improvement* of 6 per unit of x_1 introduced. The number -6 is entered in a new row called the *net evaluation row*, the $z_j - c_j$ row or a variety of other names.[16] A negative number in the net evaluation row represents improvement potential and shows that the current solution is not optimal.[17] A net evaluation row number is found for each possible incoming variable. For example, the net evaluation or net evaluation row number for x_2 is

$$1(0) + 3(0) + 0(0) - 5 = -5$$

We shall refer to variables in the current solution as *basic variables* and those which are currently zero as *non-basic variables*. The basic variables in any solution will have $z_j - c_j$ values[18] of zero. The complete first simplex tableau is shown in Table 3.3.

Table 3.3

			6	5	0	0	0
			x_1	x_2	s_1	s_2	s_3
0	s_1	800	4	1	1	0	0
0	s_2	900	2	3	0	1	0
0	s_3	180	1	0	0	0	1
value of z →		0	-6	-5	0	0	0 ←

net evaluation row

The variable with the 'most negative' net evaluation number is now selected. This produces the greatest per unit improvement in the objective function and is the *entering variable*. The task now is to introduce the entering variable, x_1 here, to the greatest possible extent. This results in a move to an adjacent extreme point – an improved basic feasible solution, in this case point D. Notice from Figure 3.15 that, in moving from O to D, x_1 enters the group of positive variables and s_3 drops out. This *leaving variable* is found from the simplex tableau in the following way. Form the ratios of current solution values to the coefficients (rates of substitution) in the entering variable (x_1) column. These ratios show the maximum extent to which x_1 can be introduced without making the corresponding basic variable negative. Results are as follows.

s_1 row: ratio = 800/4 = 200
s_2 row: ratio = 900/2 = 450
s_3 row: ratio = 180/1 = 180

The smallest ratio identifies the leaving variable, s_3. The selection of s_3 to leave the solution means that the sign requirements are satisfied. The other basic variables will still be positive and the other non-basic variables will remain at zero level in the new solution with x_1. The tableau numbers in the row of the leaving variable constitute the *pivotal row*. The numbers in the entering variable column make up the *pivotal column*. The number at the junction of the pivotal row and column is the *pivotal number*. It is handy to identify the pivotal row and column as shown in Table 3.4.

The new solution corresponding to the introduction of x_1 and the removal of s_3 can now be built up. This will supersede the 'old' solution of Table 3.4. A new tableau will be formed directly below the description of the old solution. The first elements put in place are the labels of the variables in the new solution, and their corresponding objective function coefficients. The variables s_1 and s_2, which remain at a positive level in the new solution, retain their positions of first and second rows in the new tableau. The entering variable x_1 takes the position occupied by the leaving

Table 3.4

			6	5	0	0	0	
Pivotal column			x_1	x_2	s_1	s_2	s_3	
0	s_1	800	4	1	1	0	0	
0	s_2	900	2	3	0	1	0	
0	s_3	180	1	0	0	0	1	← Pivotal row
		0	−6	−5	0	0	0	

variable s_3, and it is in this row that the first new values are entered. The *new pivotal row* is the row in the same position in the new tableau as the pivotal row was in the old one. The elements in the new pivotal row are the old pivotal row values divided by the pivotal number. Since in this case the pivotal number is one, the new pivotal row values are the same as the old pivotal row values. With these numbers entered, the tableau is as shown in Table 3.5.

All the remaining numbers are found in the following manner. Start with the new level of s_2. Call this value the *new number*, and the number in the same position in the old solution (the original level of s_2) the *old number*. Now use the formula

new number = old number − *corresponding pivotal column number* × *corresponding new pivotal row number*

The corresponding new pivotal row number is the number in the new pivotal row in the same column as the new number. The corresponding pivotal column number is the number in the pivotal column in the same row as the old number. Thus, using the formula:

new value of $s_2 = 900 - 180 \times 2 = 540$

The formula means that from each row of the tableau is subtracted a constant multiple of another row (the new pivotal row). Since this represents an *elementary row operation*, the solution set of the system of linear equations[19] is unchanged and the problem has not been distorted in any way. The creation of the new pivotal row is itself an elementary row operation. The formula is now used to find the rate of substitution between s_2 and x_2 in the new solution. This value was 3 in the old solution, so therefore

new number = $3 - 0 \times 2 = 3$

Table 3.5

			6	5	0	0	0
			x_1	x_2	s_1	s_2	s_3
0	s_1	800	4	1	1	0	0
0	s_2	900	2	3	0	1	0
0	s_3	180	1	0	0	0	1
		0	−6	−5	0	0	0
0	s_1						
0	s_2						
6	x_1	180	1	0	0	0	1

The new rate of substitution between s_2 and x_1 is

new number = $2 - 1 \times 2 = 0$

In this case the corresponding pivotal column number is the old number itself. We can now fill in the entire s_2 row. It is

0 s_2 540 0 3 0 1 −2

Note the negative rate of substitution between s_2 and s_3. This means that s_2 is an *increasing* function of s_3; as s_3 is introduced s_2 will go up.[20] For all numbers in the s_1 row, the corresponding pivotal column number is 4. For example, the new level of s_1 itself is

new value of $s_1 = 800 - 180 \times 4 = 80$

The rate of substitution between s_1 and x_2 is

new number = $1 - 0 \times 4 = 1$

The remaining numbers in the new s_1 row are found in a similar fashion, and the complete new s_1 row is

0 s_1 80 0 1 1 0 −4

To complete the first iteration the value of the solution and the new net evaluation row are calculated. The value of the objective function is

$z = 0(80) + 0(540) + 6(180) = 1080$

The net evaluation row number under x_2 is the sum of the products of rates of substitution in the x_2 column and the corresponding (same row) objective function coefficients of variables in the solution, less the objective function coefficient of x_2, i.e.

$1(0) + 3(0) + 0(6) - 5 = -5$

while the net evaluation row number under s_3 is

$-4(0) \quad -2(0) + 1(6) - 0 = 6$

The tableau in full is shown in Table 3.6.

The new tableau describes point D, and a complete iteration of the simplex method has been carried out. A profit of 1080 is made, but a negative net evaluation number under x_2 means that the solution cannot be optimal; the variable x_2 should be introduced. In the diagram this corresponds to a move from point D to point C. The new pivotal column (set by the −5 net evaluation row number) and the new pivotal row (given by the smallest of the ratios 80/1, 540/3 and 180/0) are shown in Table 3.6.

This procedure is repeated until a solution is found with no negative numbers in the net evaluation row. Before carrying out the next iteration it is worth pointing out a number of short cuts that can help speed up the

172 Mathematics for business, finance and economics

Table 3.6

			6	5	0	0	0
			x_1	x_2	s_1	s_2	s_3
0	s_1	800	4	1	1	0	0
0	s_2	900	2	3	0	1	0
0	s_3	180	1	0	0	0	1
		0	−6	−5	0	0	0
0	s_1	80	0	1	1	0	−4
0	s_2	540	0	3	0	1	−2
6	x_1	180	1	0	0	0	1
		1080	0	−5	0	0	6

calculations. We saw that an 'old number' for which either the corresponding pivotal column number or the corresponding new pivotal row number is zero will be unchanged in the new tableau. Next, note that for basic variables − those already in the solution − the rates of substitution take the form of a one at the junction of the variable's 'own' row and column and zeros elsewhere in the column. A consequence of this is that basic variables always have zero net evaluation row numbers under their own column heading. Now draw up the solution at point C. This is shown in Table 3.7.

The value −14 in the net evaluation row of Table 3.7 shows that this solution is not optimal. This time s_3 must be introduced,[21] corresponding to a move to point B in Figure 3.15. For once, the pivotal element is not unity, so that the new pivotal row will differ from the pivotal row through division by the pivotal number, 10. The new pivotal row is

s_3 30   0   0   −0.3   0.1   1

The full solution corresponding to point B is shown in Table 3.8.

Since there are no negative numbers in the net evaluation row, the solution is optimal. This causes the iterative loop to be exited and is the

Table 3.7

			6	5	0	0	0
			x_1	x_2	s_1	s_2	s_3
5	x_2	80	0	1	1	0	−4
0	s_2	300	0	0	−3	1	10
6	x_1	180	1	0	0	0	1
		1480	0	0	5	0	−14

Table 3.8

			6	5	0	0	0
			x_1	x_2	s_1	s_2	s_3
5	x_2	200	0	1	-0.2	0.4	0
0	s_3	30	0	0	-0.3	0.1	1
6	x_1	150	1	0	0.3	-0.1	0
		1900	0	0	0.8	1.4	0

stopping rule. The optimal solution sets x_1 at 150, x_2 at 200, uses the two resources to the full (s_1 and s_2 are zero) and has x_1 at 30 units below its upper bound. The construction of a simplex tableau is equivalent to finding a solution to a set of simultaneous equations. In each tableau, a number of variables[22] (two here) is set to zero, and the constraints as equations are solved for the values of the remaining variables (three here). The process of moving from one tableau to the next is an efficient way of moving from one solution to another.

A useful feature of the simplex method is that it provides valuable information on the marginal valuation of resources. For example, the number 0.8 in the final net evaluation row under slack variable s_1 shows by how much the objective function z would increase if a further unit of resource 1 was obtained on the existing terms. For example, if the existing price of materials[23] is £3 per unit, then the *dual value* of £0.8 shows the maximum addition to the current price that it would be worth paying in order to obtain an additional unit of the resource. So if further units of material can be obtained for less than £3.8, profit will increase. Similarly, the dual value (or *shadow price* or *marginal value*) of £1.4 for resource 2 (found under s_2 in the net evaluation row) shows the maximum premium over the existing unit cost of labour that would be worth paying for extra units at the margin. With (say) a wage rate of £5 per hour for the existing 900 hours, the maximum overtime wage rate would be

£5 + £1.4 = £6.4 per hour

The dual values at the optimum for all resources can be found in the final net evaluation row under the slack variables. They work in the way that they do because having one extra unit of resource is equivalent, in objective function terms, to having the original level of resource but being able to set the corresponding slack variable to −1. Because the problem is linear, net evaluation row values apply to increases or decreases in the non-basic variables. So 0.8 also tells us that if s_1 could be *reduced* by 1 unit (i.e. made −1) then the objective function would *increase* by 0.8. Hence the interpretation of this figure as a dual value.

Dual values have a valuable use in assessing the desirability of producing new products. Firms with an existing product range are usually able to produce other products using similar technology. Dual values can be used to give an initial yes/no answer about the relative profitability of a new line. For example, suppose that with the data of Table 3.8 the firm *could* produce a third product requiring 2 units of material (the first resource) and 4 units of labour per unit produced. The unit profit on the third product is £7. Given the objective of profits maximization, should the product be introduced? One way to answer this question is to rework the problem again from scratch, with a new x_3 column in the simplex tableau with the data

$$\begin{matrix} 7 \\ x_3 \\ 2 \\ 4 \end{matrix}$$

But this is not necessary. If 1 unit of the third product is to be produced, there will be 2 units less of material available for current uses. Each unit withdrawn from use in the manufacture of x_1 and x_2 causes profit to go down by the dual value £0.8. Similarly, each unit of labour withdrawn from the production of x_1 and x_2 to make space for x_3 causes profit to go down by £1.4. In compensation for these losses, the third product would return a unit profit of £7. A simple reckoning tells us whether this is enough. The sum of products of dual values and the per unit resource requirements of the third product is the *profit producing potential* (in current uses) of the resources that the third product requires. This is

$$2(0.8) + 4(1.4) = 7.2$$

So the net effect on profit of producing 1 unit of the third product, after allowing for consequential reductions in x_1 and x_2, is a reduction in profit of £0.2. This figure

$$2(0.8) + 4(1.4) - 7 = 0.2$$

is the *opportunity loss* on 1 unit of the third product.[24] If a product shows any opportunity loss, it should not be produced. If the product had shown an opportunity *profit*, then it should be brought into the solution.[25] Note that although in accounting terms the third product shows the highest profit per unit (7 versus 6 and 5), given its consumption of scarce resources, it is *not profitable enough*. Note the opportunity loss of precisely zero on products that *are* included in the optimal plan:[26]

$$4(0.8) + 2(1.4) - 6 = 0$$
$$1(0.8) + 3(1.4) - 5 = 0$$

Furthermore, if each dual value is multiplied by the total amount of its corresponding resource or upper bound constraint, the number achieved is equal to the value of the objective function. That is:

$$800(0.8) + 900(1.4) + 180(0) = 1900 = z$$

This is always the case in linear programming. Dual values in a sense give an *average* as well as a marginal valuation of resources. But care must be exercised in this interpretation. A dual value of zero for a resource does *not* mean that it is worthless throughout as well as at the margin. If a resource necessary for production is reduced to zero, profit will also go to zero. The dual values can be seen as *apportioning* profit among the constraints binding at the optimum. This property can also be used as a partial check on workings: multiplication of the dual values in *any* solution by the respective right-hand side levels should give the current value of z. There are many technical refinements in the simplex method, but our considerations will be limited to the above discussion of maximization problems.[27]

EXERCISES 3.7

1 Use the simplex method to solve the following problem:

maximize $F = 2x_1 + 5x_2$

subject to $\quad 5x_1 + 3x_2 \leq 100$

$\qquad\qquad 4x_1 + 7x_2 \leq 140$

$\qquad\qquad x_1 \geq 0, \quad x_2 \geq 0$

2 Solve the following linear programming problem using the simplex method:

maximize $F = 8x_1 + 10x_2$

subject to $\quad 3x_1 + 2x_2 \leq 54$

$\qquad\qquad 2x_1 + 5x_2 \leq 69$

$\qquad\qquad x_1 \geq 0 \quad x_2 \geq 0$

3.8 SENSITIVITY ANALYSIS

The optimal solution in any linear programming problem depends on the values of the objective function coefficients, the right-hand side levels and the coefficients of the xs in the constraints. *Sensitivity analysis* or *post-optimality analysis*[28] examines the effect on the optimum of changes in parameters such as objective function coefficients. Sensitivity analysis can determine the range of variation in a parameter for which the existing

176 Mathematics for business, finance and economics

solution remains optimal and feasible. Here, the focus will be on analysis of the objective function coefficients, where optimality is the issue. An equivalent analysis can be conducted in terms of the right-hand side of the constraints, when feasibility of the current solution is the concern.

Consider the objective function coefficient of x_1. This was 6 originally, but now refer to this figure as π_1 and find the range of values of π_1 for which the current solution is optimal. The key to the procedure is the fact that the net evaluation row numbers must remain non-negative if the solution is to stay optimal.[29] The net evaluation row numbers will be expressed in terms of π_1, and each expression, when required to be non-negative, will give a bound on π_1. The narrowest interval produced by these bounds is the *tolerance interval, range of optimality* or *allowable range* for the parameter. To carry out the sensitivity analysis on π_1, first replace the specific value 6 by the unknown π_1 in the final tableau. With the two non-zero net evaluation row numbers (those under the slack variables s_1 and s_2) indicated by * and +, the tableau is then as shown in Table 3.9.

For this solution to remain optimal, the value of π_1 must be such that both * and + are non-negative numbers. Calculation of these net evaluation row numbers in the usual way gives the conditions

$$* = -0.2(5) - 0.3(0) + 0.3(\pi_1) - 0 \geqslant 0$$
$$+ = 0.4(5) + 0.1(0) - 0.1(\pi_1) - 0 \geqslant 0$$

From the * condition, it emerges that

$$-1 + 0.3\pi_1 \geqslant 0$$

so that the requirement on π_1 is

$$\pi_1 \geqslant 10/3$$

From the requirement on +, we have

$$2 - 0.1\pi_1 \geqslant 0$$

and therefore it is necessary that

$$\pi_1 \leqslant 20$$

Table 3.9

			π_1	5	0	0	0
			x_1	x_2	s_1	s_2	s_3
5	x_2	200	0	1	-0.2	0.4	0
0	s_3	30	0	0	-0.3	0.1	1
π_1	x_1	150	1	0	0.3	-0.1	0
		1900	0	0	*	+	0

So the *tolerance interval* for the objective function coefficient of x_1 is

$$10/3 \leq \pi_1 \leq 20 \tag{3.10}$$

The tolerance interval allows management to be advised that, not only is it the best plan to produce 150 units of x_1 and 200 units of x_2 if the unit profit on x_1 is precisely 6, but the plan *remains* optimal so long as π_1 is in the range given by (3.10). Information in this form is particularly valuable to management when the true value of a parameter is uncertain, or where the value is volatile. A similar analysis can be carried out on the objective function coefficient of x_2. In this case, the x_1 objective function coefficient returns to the value 6, and the conditions on the net evaluation row numbers are

$$* = -0.2\pi_2 - 0.3(0) + 0.3(6) - 0 \geq 0$$
$$+ = 0.4\pi_2 + 0.1(0) - 0.1(6) - 0 \geq 0$$

These conditions yield the tolerance interval

$$1.5 \leq \pi_2 \leq 9$$

So long as π_2 lies within this range, and provided other parameters are unchanged, the current solution stays optimal. Uncertainty may surround the values of *both* π_1 and π_2. In this case a *joint* sensitivity analysis can be conducted. Proceeding along the same lines as in the *single parameter analysis* above, the conditions on * and + that result are

$$* = -0.2\pi_2 - 0.3(0) + 0.3\pi_1 - 0 \geq 0$$
$$+ = 0.4\pi_2 + 0.1(0) - 0.1\pi_1 - 0 \geq 0$$

from which limits on the *relative* sizes of π_1 and π_2 emerge. These are, from *,

$$\pi_2 \leq 1.5\pi_1$$

and from +

$$\pi_2 \geq 0.25\pi_1$$

which can be combined in the form

$$0.25\pi_1 \leq \pi_2 \leq 1.5\pi_1 \tag{3.11}$$

and (3.11) is the tolerance interval for joint variation of π_1 and π_2. The results of this analysis are shown in Figure 3.17.

As long as the values of π_1 and π_2 fall within the shaded region, the original solution remains optimal. Note that when π_1 takes the value 6, the original tolerance interval for π_2 emerges as a line segment in the diagram. The results of the single-parameter analyses are special cases of the two-parameter results.

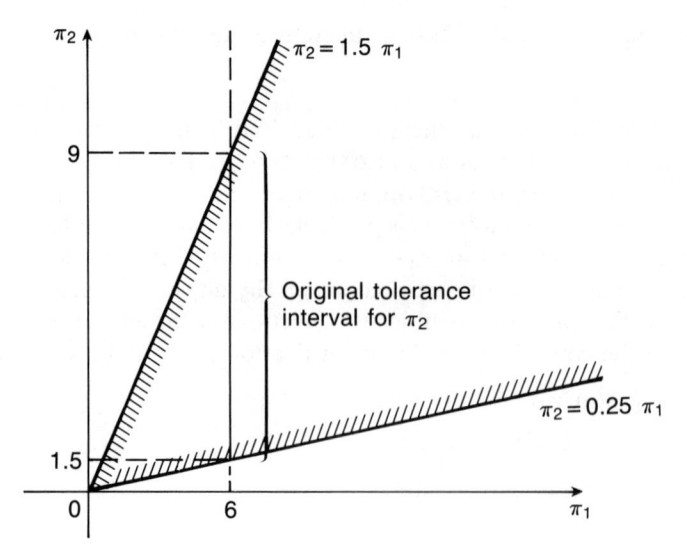

Figure 3.17

EXERCISES 3.8

1 Use the simplex method to solve the following problem:

maximize $F = x_1 + x_2$

subject to $\quad x_1 + 2x_2 \leqslant 100$

$\qquad\qquad 2x_1 + x_2 \leqslant 110$

$\qquad\qquad x_1 \geqslant 0 \qquad x_2 \geqslant 0$

State the optimal values of the dual variables and carry out individual and joint sensitivity analyses on the objective function coefficients.

2 Solve the following problem using the simplex method and conduct a joint sensitivity analysis on the objective function coefficients.

maximize $F = 4x_1 + 6x_2$

subject to $\quad 2x_1 + 5x_2 \leqslant 1700$

$\qquad\qquad 4x_1 + 3x_2 \leqslant 2000$

$\qquad\qquad x_1 \geqslant 0 \qquad x_2 \geqslant 0$

3.9 DUALITY

Every linear programming problem is associated in a symmetrical relationship with a *dual problem*. For example, for the problem

maximize $F = 10x_1 + 12x_2$

subject to $\quad 4x_1 + 2x_2 \leqslant 76$
$\quad\quad\quad\quad\;\; 3x_1 + 5x_2 \leqslant 85$
$\quad\quad\quad\quad\;\; x_1 \geqslant 0 \quad x_2 \geqslant 0$

the dual is

minimize $G = 76y_1 + 85y_2$

subject to $\quad 4y_1 + 3y_2 \geqslant 10$
$\quad\quad\quad\quad\;\; 2y_1 + 5y_2 \geqslant 12$
$\quad\quad\quad\quad\;\; y_1 \geqslant 0 \quad y_2 \geqslant 0$

The maximization problem is usually called the *primal problem* and the minimization problem is the *dual problem*, although the problems are duals of each other.[30] The dual problem uses all the information of the primal problem but in a different arrangement. Note the following.

1. The coefficients of the objective function of the primal problem become the constraint levels of the dual problem.
2. The right-hand side values of the primal problem become the objective function coefficients in the dual problem.
3. Columns of constraint coefficients in the primal problem become rows of constraint coefficients in the dual.
4. The direction of the constraint inequalities in the dual problem is the reverse of that in the primal problem.
5. The type of extreme value sought for the dual problem is opposite to that of the primal.

There is no asymmetry between the problems. For example, if the words 'primal' and 'dual' are interchanged throughout 1–5, the statements remain valid. The primal and dual problems are graphed in Figures 3.18(a) and 3.18(b).

In Figure 3.18(a), the feasible region for the primal problem is OABC. The optimum is at point B, with the broken line indicating the optimal contour of F. In Figure 3.18(b), the feasible region for the dual problem is the open area bounded from below by E'B'D'. The optimal solution is at point B', which produces a minimum of the dual objective function G. Each basic feasible solution in the primal problem has a corresponding basic, but not necessarily feasible, solution in the dual. Indeed, it turns out that in all linear programming problems there is only *one* basic feasible solution to the primal problem that corresponds to a basic feasible solution

180 Mathematics for business, finance and economics

(a)

(b)

Figure 3.18

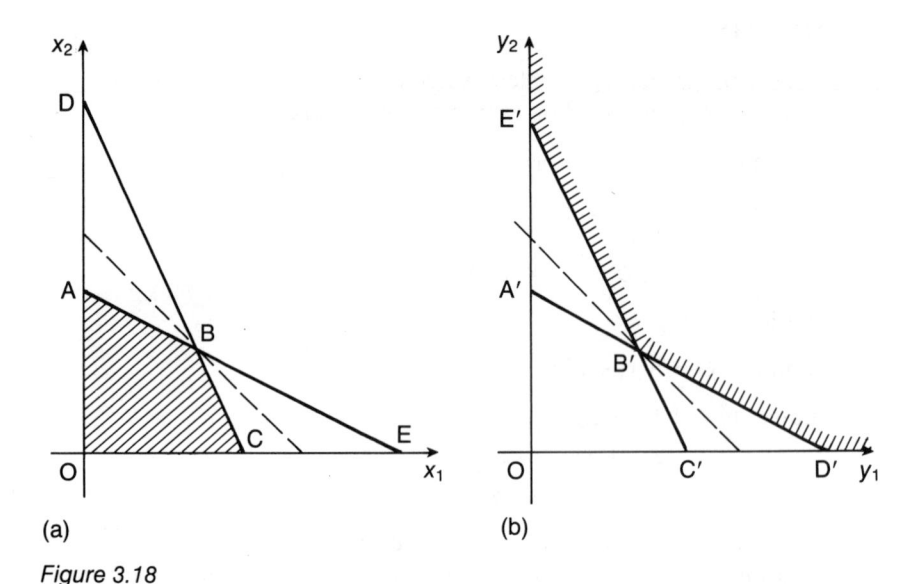

in the dual. What is more, these corresponding basic feasible solutions are the respective optima!

Apart from the intriguing symmetrical relationship between the primal and the dual, why is the dual problem worth studying? First of all, the dual structural variables y_i turn out to be the *dual values* of the primal problem. So the y_i are the net evaluation row numbers under the slack variables in the primal problem. Because of this, the dual is significant for economic interpretation and in sensitivity analysis. Second, just as the optimal values of the dual variables can be found in the solution to the primal problem, so the optimal values of primal variables are found in the net evaluation row of the optimal solution to the dual. This may be important for hand computation or in some extremely large linear programming problems, since the problem with fewer constraints is generally easier to solve. Table 3.10 shows the optimal tableau for the primal problem. Starting from the origin, this is obtained in the usual manner after two iterations.

Table 3.10

			10	12	0	0
			x_1	x_2	s_1	s_2
10	x_1	15	1	0	5/14	−2/14
12	x_2	8	0	1	−3/14	4/14
		246	0	0	1	2

The maximum value of the primal objective function F is achieved by setting x_1 to 15 and x_2 to 8. The optimal values of y_1 and y_2 in the dual are read off from the final net evaluation row of the primal problem in Table 3.10. As may be confirmed by graphical means, the dual objective function is minimized for $y_1 = 1$ and $y_2 = 2$, at which point the value of the dual objective function $G = 246$ is precisely the same as the primal objective function F. Equality of optimal values of the primal and dual objective functions is true for all linear programming problems. Feasible solutions of the dual problem have values of G greater than or equal to all values of F for feasible solutions to the primal problem. The equality holds only at the optimum.

Linear programming belongs to a wider class of constrained optimization problems but is the most important example of such problems, proving to be of enormous value in practice. There are many linear programming packages suitable for, or adaptable to, a wide range of problems from small-scale manufacturing, through finance and investment to overall corporate modelling. The size of problems that can now be solved is impressive. Modern mainframe computers can solve linear programming problems with *thousands* of variables and constraints. There is also a wide choice of user friendly software handling up to a hundred variables and constraints for use on personal computers. In principle, *all* economic and business problems ultimately involve constraints, and in most cases best, rather than adequate, solutions are preferred. So it is not surprising that linear programming methods have found widespread use in practice.

EXERCISES 3.9

1 Set out the duals to the following problems:

(i) maximize $F = 5x_1 + 4x_2$

subject to $\quad 2x_1 + 3x_2 \leqslant 20$
$\quad\quad\quad\quad\quad 4x_1 + x_2 \leqslant 15$
$\quad\quad\quad\quad\quad x_1 \geqslant 0 \quad x_2 \geqslant 0$

(ii) maximize $F = 13x_1 + 7x_2$

subject to: $\quad 3x_1 - 2x_2 \leqslant 10$
$\quad\quad\quad\quad\quad\; 2x_1 + 4x_2 \leqslant 12$
$\quad\quad\quad\quad\quad\; x_1 \geqslant 0 \quad x_2 \geqslant 0$

2 Express the duals to the following problems:

(i) minimize $F = 18x_1 + 20x_2$

subject to $\quad 2x_1 + x_2 \geqslant 6$

$\qquad\qquad\quad x_1 + 3x_2 \geqslant 5$

$\qquad\qquad\quad x_1 \geqslant 0 \qquad x_2 \geqslant 0$

(ii) minimize $F = 50x_1 + 40x_2$

subject to $\quad x_1 + 5x_2 \geqslant 30$

$\qquad\qquad\quad 3x_1 + x_2 \geqslant 20$

$\qquad\qquad\quad x_1 \geqslant 0 \qquad x_2 \geqslant 0$

ADDITIONAL PROBLEMS

1 A manufacturing company can make two products, each of which requires time on a cutting machine and a finishing machine. Relevant data are as follows.

	Product 1	Product 2
Cutting hours (per unit)	2	1
Finishing hours (per unit)	3	3
Unit profit (£)	6	4
Maximum sales (units per week)	200	200

The number of cutting and finishing hours available per week are 390 and 810 respectively. No other resources are in limited supply, but their costs have been allowed for in the unit profit figures.

Making use of the simplex method:

(i) With the objective of profits maximization, how much of each product should be produced?

(ii) What are the dual values of the scarce resources at the optimum?

(iii) A possible third product has a unit profit of £5 and would require 2 hours of cutting time and 2 hours of finishing time. Given the profits maximization objective, should this product be made?

(iv) Determine the maximum range of variation in each of the unit profit figures for the original solution to remain optimal.

2 Excel Limited can manufacture four products. Only three resources are in limited supply. Each product does not require every type of scarce

resource. Resource consumption (per unit) and financial details for the products are as follows.

	Product			
	A	B	C	D
Packaging labour	2	5	0	0
Machine time	4	1	0	0
Skilled labour	0	0	3	4
Unit profit	3	9	5	8

The monthly resource availabilities are as follows.

Resource	Hours available
Packaging labour	8000
Machine time	4000
Skilled labour	6000

(i) Determine the monthly production levels that would maximize profits.
(ii) What is the minimum unit profit at which the sale of product B would be worthwhile?
(iii) What is the dual value of skilled labour?

Note that the use of the simplex method is not essential in this problem.

3 Jewel Enterprises at present manufacture a range of four products. The resource consumption, financial and current production data are as follows.

	Product			
	1	2	3	4
Current production	1000	900	750	250
Unit profit	14	24	15	20
Machine hours (per unit)	1	1	1.5	0.5
Materials (per unit)	1.5	2.5	1	2

Both machine hours and materials are in limited supply and are fully used in the present production plan. A linear programming study has revealed that the dual values of the two resources are £4 for machine hours and £9 for materials.

(i) What is the opportunity loss on the existing production plan as a whole?
(ii) What are the profits maximizing production levels?

REFERENCES AND FURTHER READING

1 Anderson, D. R., Sweeney, D. J. and Williams, T. A. (1991) *An Introduction to Management Science* (Sixth Edition), West Publishing.
2 Dennis, T. L. and Dennis, L. B. (1991) *Management Science*, West Publishing.
3 Hillier, F. and Lieberman, G. J. (1990) *Introduction to Operations Research* (Fifth Edition), McGraw-Hill.
4 Mizrahi, A. and Sullivan, M. (1988) *Mathematics for Business and the Social Sciences* (Fourth Edition), Wiley.
5 Taha, H. A. (1992) *Operations Research* (Fifth Edition), Macmillan.

SOLUTIONS TO EXERCISES

Exercises 3.1

1 (a) $\leqslant$ (b) $>$ (c) $>$ (d) $\neq$ or $<>$
2 (a) $\geqslant$ (b) $\leqslant$ (c) $>$ (d) $<$

Exercises 3.2

1 (i) $x \leqslant 10$
  (ii) $x \leqslant 8.5$
  (iii) $x \geqslant -5$
  (iv) $x \leqslant 0$

2 (i) The solution set is empty. The identity simplifies to the statement $1 \leqslant 0$.
  (ii) All values of x satisfy the inequality. The terms in x cancel, leaving the statement $10 \leqslant 12$.

3 (i) $5 \leqslant x \leqslant 12$
  (ii) $9 \leqslant x \leqslant 30$
  (iii) The solution set is empty. The inequalities require that, simultaneously, $x \geqslant 10$ and $x \leqslant 5$.

4 (i) $4 \leqslant x \leqslant 6$
  (ii) $3 \leqslant x \leqslant 7$
  (iii) $3 \leqslant x \leqslant 7$
  (iv) $3 \leqslant x \leqslant 7$

Exercises 3.3

1 (i) The half space on or below the line $y = 25 - 2x$.
  (ii) The interior and boundary of the triangular region defined by the line segment $y = 10 - x$ between and including the points $(0, 10)$ and $(10, 0)$ and the x and y axes.
  (iii) The *strip* bounded from above by the line $y = 2.5x - 8$ and from below by $y = 2.5x - 10$.

Linear inequalities 185

(iv) The solution set is empty. Note that if the sign requirement on x was removed, it would then be possible to satisfy the two constraints.
(v) The line segment bounded by the points (80, 0) and (40, 40).

Exercises 3.5

1 For this problem, the optimal solution is located at the intersection of the two constraints. This is shown as point B in Figure A1. Optimal values of the decision variables and the objective function are

$x_1 = 12$, $x_2 = 9$ and $F = 186$

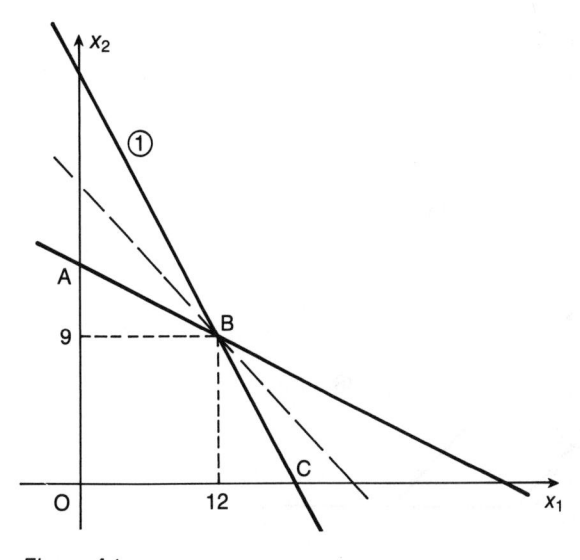

Figure A1

2 (i) The optimal solution is at point C in Figure A1, where the values of the decision variables are

$x_1 = 18$ $\quad$ $x_2 = 0$

and $F = 144$.

(ii) Point A is the optimum position here. At this point, the decision variable values are

$x_1 = 0$ $\quad$ $x_2 = 13.8$

The objective function stands at $F = 138$.

186 Mathematics for business, finance and economics

(iii) In this case, the contours of the objective function are parallel to the second constraint. As a result, there is a non-unique optimum. The corner points A or B, or any point in-between, will produce the highest value of $F = 276$.

3 In this problem the second of the 'resource' constraints is redundant — as can be seen from Figure A2. The first constraint always sets tighter limits for both x_1 and x_2, and the feasible region OABCD is determined by the first constraint and the upper bounds. The optimum position is at point C, where

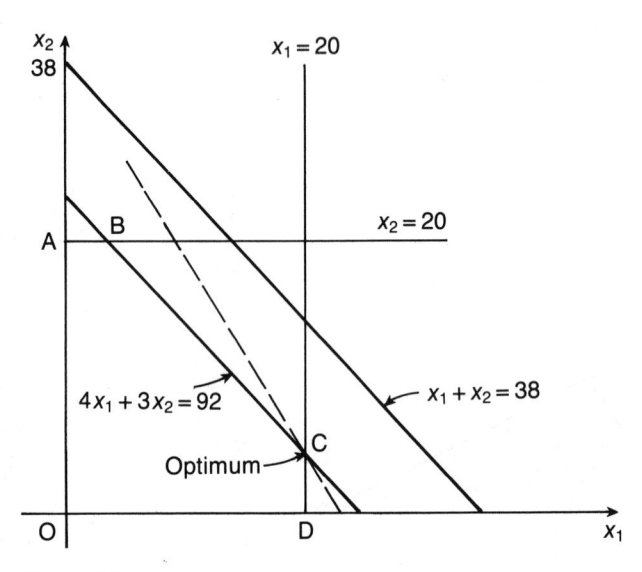

$$x_1 = 20, \qquad x_2 = 4 \quad \text{and} \quad F = 800$$

Figure A2

4 The problem is graphed in Figure A3. The feasible region is the open area bounded from below by the axes and ABC. Point B is optimal, being in contact with the lowest objective function contour. At point B the values of the decision variables and the objective function are

$$x_1 = 15, \qquad x_2 = 20 \quad \text{and} \quad F = 165$$

Linear inequalities 187

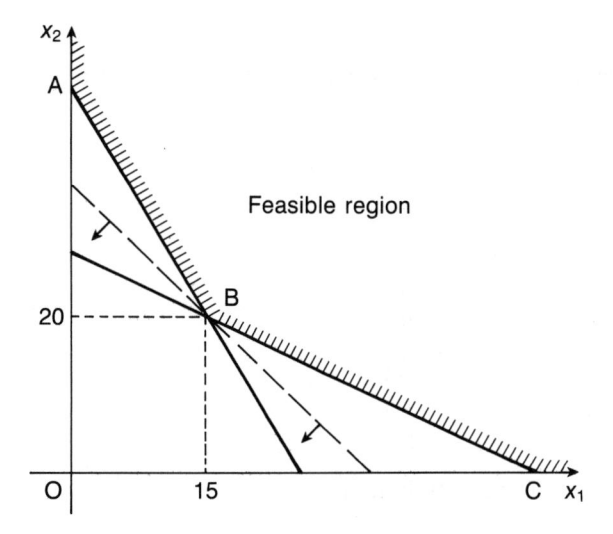

Figure A3

Exercises 3.7

1 The workings are as follows:

			5	3	0	0
			x_1	x_2	s_1	s_2
0	s_1	100	5	3	1	0
0	s_2	140	4	7	0	1
		0	-2	-5	0	0
0	s_1	40	23/7	0	1	$-3/7$
5	x_2	20	4/7	1	0	1/7
		100	6/7	0	0	5/7

2 The workings are as follows:

			8	10	0	0
			x_1	x_2	s_1	s_2
0	s_1	54	3	2	1	0
0	s_2	69	2	5	0	1
		0	-8	-10	0	0
0	s_1	26.4	2.2	0	1	-0.4
10	x_2	13.8	0.4	1	0	0.2
		138	-4	0	0	2
8	x_1	12	1	0	$0.\overline{45}$	$-0.\overline{18}$
10	x_2	9	0	1	$-0.\overline{18}$	$0.\overline{27}$
		186	0	0	$1.\overline{81}$	$1.\overline{27}$

which confirms the optimal solution obtained by the graphical method in Exercises 3.5, number 1.

Exercises 3.8

1 Two iterations are required. The simplex workings are as follows:

			1	1	0	0
			x_1	x_2	s_1	s_2
0	s_1	100	1	2	1	0
0	s_2	110	2	1	0	1
		0	-1	-1	0	0
0	s_1	45	0	3/2	1	$-1/2$
1	x_1	55	1	1/2	0	1/2
		55	0	$-1/2$	0	1/2
1	x_2	30	0	1	2/3	$-1/3$
1	x_1	40	1	0	$-1/3$	2/3
		70	0	0	1/3	1/3

Thus the solution values at the optimum are

$$x_1 = 40, \quad x_2 = 30, \quad s_1 = 0, \quad s_2 = 0, \quad F = 70$$

The dual values are 1/3 for each constraint. Using π_1 and π_2 to represent the objective function coefficients of x_1 and x_2 respectively, if π_1 alone varies, then the two non-zero net evaluation row numbers, under s_1 and s_2, should both remain non-negative and produce the following requirements on π_1:

$$2/3 - \pi_1/3 \geqslant 0 \text{ so that } \pi_1 \leqslant 2$$
$$-1/3 + 2\pi_1/3 \geqslant 0 \text{ so that } \pi_1 \geqslant 0.5$$

Thus the tolerance interval for π_1 is

$$0.5 \leqslant \pi_1 \leqslant 2$$

For the individual analysis on π_2, the net evaluation row numbers produce the following requirements:

$$2\pi_2/3 - 1/3 \geqslant 0 \text{ so that } \pi_2 \geqslant 0.5$$
$$-\pi_2/3 + 2/3 \geqslant 0 \text{ so that } \pi_2 \leqslant 2$$

Thus the tolerance interval for π_2 is

$$0.5 \leqslant \pi_2 \leqslant 2$$

With π_1 and π_2 both variable, the non-negativity of the net evaluation row numbers requires that

$$2\pi_2/3 - \pi_1/3 \geqslant 0$$

and

$$-\pi_2/3 + 2\pi_1/3 \geqslant 0$$

which together imply that

$$0.5\pi_1 \leqslant \pi_2 \leqslant 2\pi_1$$

2 The simplex workings are as follows.

			4	6	0	0
			x_1	x_2	s_1	s_2
0	s_1	1700	2	5	1	0
0	s_2	2000	4	3	0	1
		0	−4	−3	0	0
6	x_2	340	0.4	1	0.2	0
0	s_2	980	2.8	0	−0.6	1
		2040	−1.6	0	1.2	0
6	x_2	200	1	0	4/14	−2/14
4	x_1	350	0	1	−3/14	5/14
		2600	0	0	12/14	8/14

Using π_1 and π_2 to represent the objective function coefficients of x_1 and x_2 respectively, the requirement that the net evaluation row numbers be non-negative means that (multiplying by 14 throughout)

$$4\pi_2 - 3\pi_1 \geqslant 0 \quad \text{so that} \quad \pi_2 \geqslant 0.75\pi_1$$
$$-2\pi_2 + 5\pi_1 \geqslant 0 \quad \text{so that} \quad \pi_2 \geqslant 2.5\pi_1$$

which, taken together, give the joint restriction

$$0.75\pi_1 \leqslant \pi_2 \leqslant 2.5\pi_1$$

Exercises 3.9

1 (i) minimize $G = 20y_1 + 15y_2$
subject to $2y_1 + 4y_2 \geqslant 5$
$3y_1 + y_2 \geqslant 4$
$y_1 \geqslant 0 \quad y_2 \geqslant 0$

(ii) minimize $G = 10y_1 + 12y_2$
subject to $3y_1 + 2y_2 \geqslant 13$
$-2y_1 + 4y_2 \geqslant 7$
$y_1 \geqslant 0 \quad y_2 \geqslant 0$

Note: The labelling of the objective function and the dual variables is, of course, arbitrary. The sign requirements are an essential part of the problem formulation.

190　Mathematics for business, finance and economics

2　(i) maximize $G = 6y_1 + 5y_2$
　　 subject to　　$2y_1 + y_2 \leqslant 18$
　　　　　　　　　 $y_1 + 3y_2 \geqslant 4$
　　　　　　　　　 $y_1 \geqslant 0$　　$y_2 \geqslant 0$

　　(ii) maximize $G = 30y_1 + 20y_2$
　　　 subject to　　$y_1 + 3y_2 \leqslant 50$
　　　　　　　　　　 $5y_1 + y_2 \leqslant 40$
　　　　　　　　　　 $y_1 \geqslant 0$　　$y_2 \geqslant 0$

Note: Recall that the duals to minimizing linear programming problems are maximization problems. The problems are duals of each other.

Additional problems

1　Noting that the upper bound constraint on variable x_1 is redundant, the simplex workings are as follows:

			6	4	0	0	0
			x_1	x_2	s_1	s_2	s_3
0	s_1	390	2	1	1	0	0
0	s_2	810	3	3	0	1	0
0	s_3	200	0	1	0	0	1
		0	−6	−4	0	0	0
6	x_1	195	1	1/2	1/2	0	0
0	s_2	225	0	3/2	−3/2	1	0
0	s_3	200	0	1	0	0	1
		1170	0	−1	3	0	0
6	x_1	120	1	0	1	−1/3	0
4	x_2	150	0	1	−1	2/3	0
0	s_3	50	0	0	1	−2/3	1
		1320	0	0	2	2/3	0

　(i) The optimal production levels are 120 for product 1 and 150 for product 2, giving total profits of £1320.

　(ii) Dual values are: cutting hours, £2; finishing hours, £2/3.

　(iii) Carrying out the opportunity loss calculation on the third product,

$$2(2) + 2(2/3) - 5 = 1/3$$

　　so the third product *should* be brought into the production plan.

Linear inequalities 191

(iv) Letting the unit profits on the two products be π_1 and π_2 respectively, the tolerance intervals are

$$4 \leq \pi_1 \leq 8$$

and

$$3 \leq \pi_2 \leq 6$$

2 While the simplex method *could* be used to solve this problem, it is also possible to make use of the graphical method despite the fact that there are four decision variables. This is because the problem divides into two sub-problems. Letting the amounts produced be x_1, x_2, x_3 and x_4, the two sub-problems involved are

maximize $3x_1 + 9x_2$

subject to $\quad 2x_1 + 5x_2 \leq 8000$

$\quad\quad\quad\quad\quad 4x_1 + x_2 \leq 4000$

$\quad\quad\quad\quad\quad x_1 \geq 0 \quad\quad x_2 \geq 0$

and

maximize $5x_3 + 8x_4$

subject to $\quad 3x_3 + 4x_4 \leq 6000$

$\quad\quad\quad\quad\quad x_3 \geq 0 \quad\quad x_4 \geq 0$

(i) From the separate graphical solution of the two problems:

$x_1 = 0, \quad x_2 = 1600, \quad x_3 = 0, \quad x_4 = 1500.$

The profit made is £26,400.

(ii) This unit profit is that which would give the objective function contours the same slope as the second constraint. This is $\pi_2 = 3/4$.

(iii) The dual value of skilled labour can be found by re-solving the second sub-problem with a right-hand side value of 6001. This would allow a 1/4 increase in x_4 which translates into an increase of £2 in the objective function. This is the requisite dual value.

3 (i) The opportunity loss on the existing production plan is obtained by first finding the opportunity loss per unit of each product. These values are then multiplied by the existing production levels and summed. Given the dual values, the opportunity losses are as follows:

product 1 $1(4) + 1.5(9) - 14 = 3.5$
product 2 $1(4) + 2.5(9) - 24 = 2.5$
product 3 $1.5(4) + 1(9) - 15 = 0$
product 4 $0.5(4) + 2(9) - 20 = 0$

Therefore, the opportunity loss on the production plan as a whole is

$$1000(3.5) + 900(2.5) = 5750$$

(ii) Clearly, given the opportunity losses, production of products 1 and 2 should be discontinued and the released resources applied to products 3 and 4. But in order to find the new production levels of products 3 and 4, we need to know the total amounts of machine hours and materials available. Since the existing plan exhausts the supply of these resources, the total amounts available must be the sum of the existing production levels multiplied by the per unit resource requirements. Thus the machine hours available are

$$1000(1) + 900(1) + 750(1.5) + 250(0.5) = 3150$$

and material availability must be

$$1000(1.5) + 900(2.5) + 750(1) + 250(2) = 5000$$

Given the cessation of production of products 1 and 2, the problem therefore becomes

$$\text{maximize } F = 15x_3 + 20x_4$$

$$\text{subject to} \quad 1.5x_3 + 0.5x_4 \leqslant 3150$$

$$x_3 + 2x_4 \leqslant 5000$$

$$x_3 \geqslant 0 \quad x_4 \geqslant 0$$

which can be solved by the graphical method to give

$$x_3 = 1520, \quad x_4 = 1740, \quad F = 57{,}600$$

Note: In part (ii), given that we already know that both dual values are positive at the optimum, the resources must both be fully used, so that the constraints can be solved as strict equalities.

NOTES

1 And therefore, in the case of negative numbers, having the *lesser* absolute value.
2 A sketch of the system without the sign requirements shows that the solution set lies entirely within the second quadrant.
3 This point is taken up again in Chapter 7.
4 For example, *The Management Scientist* package accompanying Anderson, Sweeney and Williams can handle problems with up to 75 variables and up to 25 constraints.
5 The optimal point may not be unique, but all optima will give the same value of the objective function. In linear programming, the optimal solutions must always include corner points.

6 Note that given the definition of s_1 this equation should strictly be an *identity* since by definition s_1 always takes the value necessary to make both sides the same. However, in practice the identity sign is not used in this context.
7 Further description of piece-wise defined functions can be found in the next chapter.
8 Without any additional work being needed if the simplex method is employed.
9 In *degenerate* solutions there are fewer positive variables, owing to coincidental limitations imposed by constraints.
10 In linear programming we are not confronted with the possibility of an interminable sequence of ever-decreasing improvements.
11 If the origin is not feasible, the problem must be adapted in order to generate an initial basic feasible solution.
12 There are several minor variations of this tabular layout and in the terminology used to describe its components.
13 In geometric terms this corresponds to a move along the x_1 axis away from the origin and towards D.
14 This is fairly obvious here given that 4, 2 and 1 are constraint coefficients and that the s_i represent unused resources. The property will also obtain in each subsequent table produced by the simplex method.
15 These values (4, 2 and 1) are *rates of substitution* between the potential incoming variable and variables in the current solution.
16 z_j is the value obtained in (3.9) for incoming variable x_j, while c_j is the objective function coefficient of x_j.
17 Aside from the case when the optimal solution is *degenerate*, a technicality that will not concern us here.
18 As may be verified by carrying out the workings.
19 The equations are the constraints expressed in equality form with the addition of slack variables.
20 This effect is a result of x_1 being reduced by 1 unit for each unit increase in s_3.
21 Not because it is of any value in its own right, but because it will allow a more beneficial balance to be struck between the profit productive x_j.
22 In a problem with m constraints (and slack variables) and n structural (x) variables, n of the structural and slack variables are set to zero in any basic solution.
23 A figure already taken into account and netted out in obtaining the objective function coefficients. The unit profit figures of £6 and £5 assume that all units of material used are bought at a specified price (here we have taken this to be £3).
24 Had x_3 been included in the simplex tableau, this (0.2) would have been its net evaluation row number.
25 The current solution can be used as a starting point, but we shall not pursue the technical details.
26 If there was either an opportunity loss or a profit, the present production plan could not be optimal.
27 For more detailed technical material see Hillier and Lieberman (1990).
28 Some authors use the term *post-optimality analysis* in a wider sense. See for example Hillier and Lieberman (1990).
29 Only optimality is in question here. The feasible region is unaffected by changes in the objective function coefficients.
30 The primal and dual formulations can be seen as viewing different aspects of the same problem.

Chapter 4

Functions and Turning Points

4.1	Functions	195
4.2	Quadratic functions	201
4.3	Cubics and quartics	221
4.4	Polynomials	229
4.5	Descartes's rule of signs	230
4.6	Rational functions	233
4.7	Functions: further considerations	241
4.8	BASIC program	256
	Additional problems	259
	References and further reading	260
	Solutions to exercises	260

In this chapter, you will study the properties of functions widely used in economics, business studies and management science. The principal focus is on polynomials, with particular attention being given to the simpler forms of polynomial – quadratics and cubics. Section 4.6 deals with functions which are the ratio of polynomials and the chapter concludes with a review of relevant special properties possessed by some functions.

By the chapter's end, you will be able to solve quadratic equations and mixed systems of one linear and one quadratic equation. You will have learnt the principles of curve sketching and be able to use Descartes's rule of signs to establish the maximum number of positive roots of an equation. You will have considered business and economic applications of quadratic, cubic and rational functions.

4.1 FUNCTIONS

Economics, finance and management science deal principally with phenomena which are *measurable*. For example, prices, quantities, interest rates, profits and rates of taxation all fall into the category of quantifiable variables, the relationships between which the economist seeks to understand. The object of the understanding may be to predict outcomes or to influence, or ideally to optimize, the values taken by the variables over which control can be exercised. Where an optimization model is used, the objective will be to achieve the best value of some measure of performance or effectiveness. For example, a firm may wish to maximize its profits, or a contractor may wish to minimize the completion time of a project. The concept of a *function* arises in most theoretical or empirical studies of the relationships between variables whether or not the relationships are initially expressed in symbolic terms. Formally, a *function* is a relation between two sets that associates a unique element of the second set with each element (or n-tuple of elements [1]) of the first set. In the present context we shall consider a function to be a rule which links together a given number x and another *single* number: the *function* of x. 'Function of x' is written as

$f(x)$

and is usually read as 'f of x'. If the rule represented by $f(x)$ produces more than one value for a given x, as for example when a square root is taken, the rule in strict terms does not satisfy the definition of a function – which requires a unique outcome. But the formal requirement can often be met by focusing on a particular part or branch of the relation.[2] The rule or rules for the function could in principle be set out in literal terms as a series of specified operations to be carried out on x to obtain the number $f(x)$. For example, in a particular case the instructions could be to 'double x and add 5' which would be written in function notation as

$f(x) = 2x + 5$

or to 'square x and subtract 10' which would appear as

$f(x) = x^2 - 10$

Clearly, however, a function such as

$f(x) = x^5 - 8x^4 + 7x^3 - 2.5x^2 + 2x - 4$

would take some spelling out verbally. The $f(x)$ notation for a function was introduced by *Leonhard Euler* (1707–83) and is very convenient in much of the calculus. Alternative notation uses another symbol in place of $f(x)$, the default character being y. In the function

$y = x^2 - 3x - 10$ (4.1)

x is the *independent variable* or *argument* and y is the *dependent variable* or the *subject* of the equation. In graphing a function, the convention is to measure the dependent variable against the vertical axis. The classification of variables as independent or dependent fits naturally when there is a cause and effect relationship between x and y. Where no such relationship exists, the *explicit* form (4.1) may still be used for convenience, although the *implicit* expression of the relationship between the variables

$$x^2 - 3x - 10 - y = 0$$

makes no distinction as to dependence, and removes the asymmetric treatment of x and y. Where a causal relationship exists, x might be referred to as the *decision variable* or *control variable* in a management science context or, principally in macroeconomics, as an *instrument* if its value can be varied, if only between limits, by the decision-maker. The subject of the equation, y, would be referred to as the *objective* while the function itself (which it may be desired to optimize) is called the *objective function*.

There may be limits on the values which the dependent variable can take. These limits may be defined by the context of a particular application. For example, output may be limited from above by plant capacity and from below by the need to at least satisfy the most important customers. In other cases, the variable x may have no meaning if negative, or it may be impossible to carry out the operations required by the function for certain values of x. The set of permissible values of x is called the *domain* of the function. The set of values of y which result when the function is applied to values of x in the domain is called the *range* of the function. For the relationship between x and y to satisfy the definition of a function, each x value in the domain must produce a *unique* value of y within the range. But it is allowable for different values of x in the domain to result in the same y value in the range. The functional relationship can be represented as an input–output process which can be illustrated in the following way:

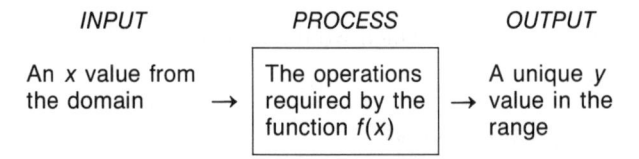

INPUT	PROCESS	OUTPUT
An x value from the domain $\rightarrow$	The operations required by the function $f(x)$	$\rightarrow$ A unique y value in the range

Now consider some examples illustrating different possibilities for the domain and range of functions. Three cases are shown in Figure 4.1. In the first of the cases, the function is

$$y = x^2$$

for which the domain is unlimited (minus infinity, shown as $-\infty$, to plus infinity). Real numbers must produce a non-negative square so that the

Functions and turning points 197

(a) Domain: $-\infty < x < +\infty$, $y = x^2$, Range $y \geq 0$

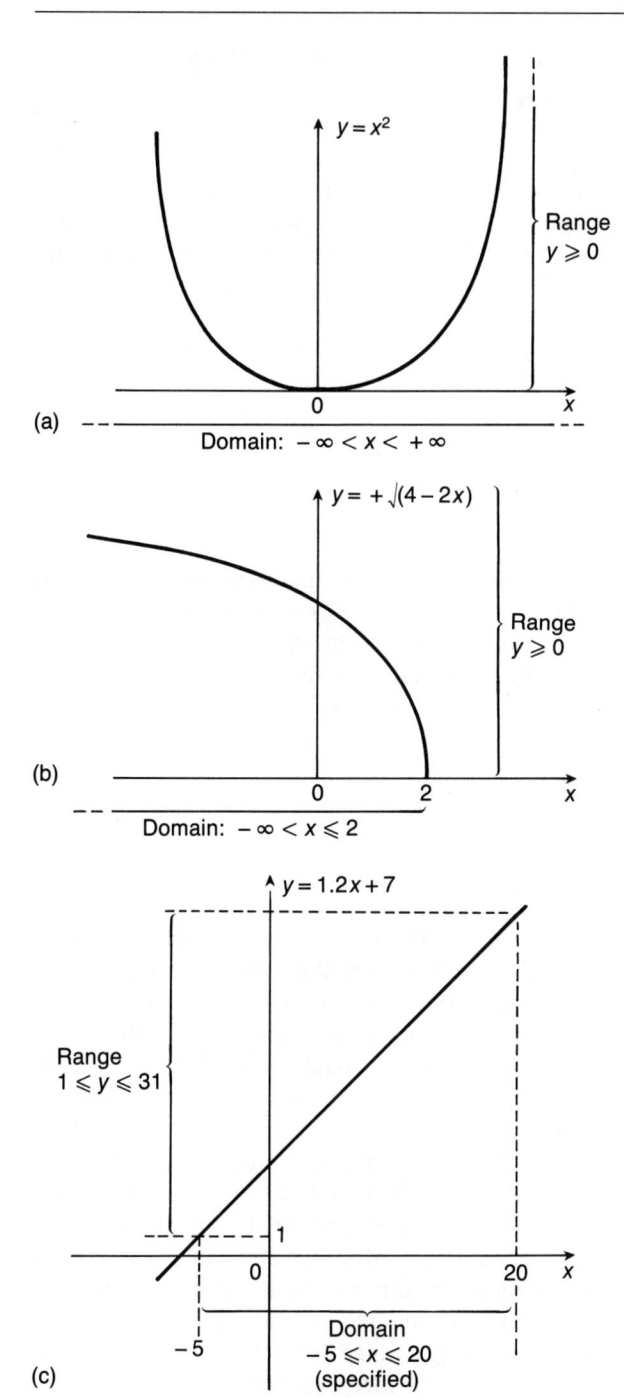

(b) Domain: $-\infty < x \leq 2$, $y = +\sqrt{(4-2x)}$, Range $y \geq 0$

(c) $y = 1.2x + 7$, Range $1 \leq y \leq 31$, Domain $-5 \leq x \leq 20$ (specified)

Figure 4.1

range is restricted to all positive numbers and zero. This can be shown as

Domain: $-\infty < x < \infty$
Range: $y \geqslant 0$

This case is illustrated in Figure 4.1(a). The domain is the real numbers for which the operation of squaring can be performed – i.e. all of the real numbers. Since the square of a real number cannot be negative, the range of the function is restricted, by its own nature, to the positive real numbers and zero. In the second of the cases, the function is

$$y = +\sqrt{(4-2x)}$$

(meaning that only the positive square root is involved in the function). The domain and range are

Domain: $-\infty < x \leqslant 2$
Range: $y \geqslant 0$

This case is illustrated in Figure 4.1(b). The operation required by the function cannot be performed on numbers in excess of $+2$. Consequently, in this case it is the domain which is restricted by the form of the function. The range will also be limited by the function, being the set of non-negative real numbers. In the third case, the function is

$$y = 1.2x + 7$$

and the domain and range are

Domain (specified): $-5 \leqslant x \leqslant 20$
Range: $1 \leqslant y \leqslant 31$

This case is illustrated in Figure 4.1(c). In this case, a restricted domain has been 'artificially' imposed. This may reflect the realistic values in the context of a particular application. For example, x might be a change in stock level. If the current inventory is 5 units, then the fall in stock cannot be greater than 5. If the maximum storage capacity was 25 units, then the rise in stock cannot exceed 20 units. When the domain of a function is restricted in this way, if the function is always increasing or always decreasing,[3] the endpoints of the *range* that result can be obtained by substituting the endpoints of the domain into the function. In practice there is often a restricted or target range for the output of the function – as for example when a y value must be achieved within specific limits of accuracy. For example, a non-profit organization aims to break even on average but in any year may wish its excess of income over outgoings to be within the range $-£10,000$ to $+£10,000$. A target range for a function will in turn imply a restricted domain for the independent variable.

Functions and turning points 199

We have considered linear functions in Chapter 2 and in later sections quadratics and higher degree functions will be considered. But before turning to quadratics, there is an interesting type of function that can be used to model realistic situations and can be well illustrated with the use of linear expressions. In a *piecewise-defined* function, the nature of the relationship between the variables changes at certain points within the domain. One example of a piecewise-defined function is the following:

$$f(x) = \begin{cases} 0 & \text{if } x = 0 \\ 2x + 10 & \text{for } 0 < x \leqslant 15 \\ x + 30 & \text{for } x > 15 \end{cases}$$

This function is graphed in Figure 4.2.

In Figure 4.2, disks indicate points where the form of the function changes. An unfilled disk indicates a point where the value of $f(x)$ is not included in the range. Thus with the uppermost line segment, the endpoint at $x = 15$ is excluded. Although the piecewise-defined function looks strange, it nevertheless fulfils the requirements for being a function. Indeed, far from being contrived, it is a form which corresponds to many practical situations in business and industry. For example, the function may represent a manufacturer's long-term costs of production for various output levels. The firm may have a choice between two plant sizes or may choose not to operate at all. If the decision is made not to produce the product, then long-term costs will be zero. Where production does occur,

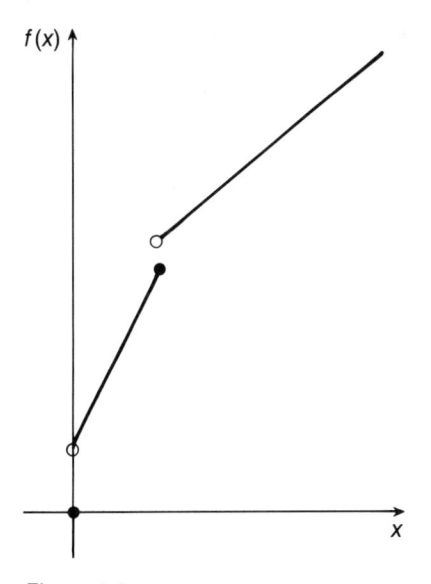

Figure 4.2

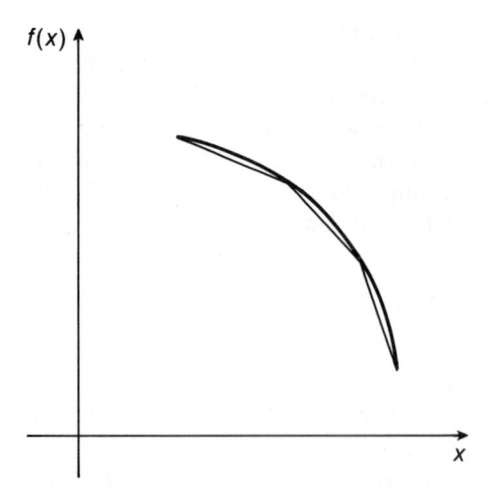

200 Mathematics for business, finance and economics

Figure 4.3

the $2x + 10$ component of the function shows the costs arising from the use of the smaller of two possible plant sizes that the manufacturer may employ. This plant size has relatively low fixed costs of £10 which are inescapably associated with any positive level of use and a comparatively high unit variable cost of £2 up to the plant capacity of 15 units. By contrast, the larger of the plant sizes has the relatively high fixed cost of £30 but has only half the unit variable cost of the lower capacity plant. Note that this function exhibits *step discontinuity* against the y axis at $x = 0$ and $x = 15$.

A further example of the relevance of piecewise-defined functions is where a curvilinear expression is approximated by a series of linear segments as shown in Figure 4.3. Piecewise linear approximation represents a possible way of making more manageable problems that involve the optimization of nonlinear functions.

EXERCISES 4.1

1. In the absence of other restrictions, find the domain for the following functions:

   (i) $y = (x + 10)^{1/2}$
   (ii) $y = (40 - 2x)^{1/2}$

2. What is the range for the following functions:

   (i) $y = x^4$
   (ii) $y = 2x^2 - 10$

3 Given the specified domains, find the range for each of the following functions:
   (i) $y = 3x - 15$ Domain: $-5 \leqslant x \leqslant 20$
   (ii) $y = 100 - 5x$ Domain: $-10 \leqslant x \leqslant 100$
   (iii) $y = 250 - 10x$ Domain: $-200 \leqslant x \leqslant -100$

4 Do the following piecewise-defined functions exhibit step discontinuities against the y axis?

   (i) $f(x) = \begin{cases} x + 10 & \text{for } x < 30 \\ 2x - 20 & \text{for } x \geqslant 30 \end{cases}$

   (ii) $f(x) = \begin{cases} 5x + 5 & \text{for } 0 \leqslant x < 15 \\ 2x + 60 & \text{for } x \geqslant 15 \end{cases}$

4.2 QUADRATIC FUNCTIONS

If the dependent variable y is a quadratic function of x, this can be written as

$$y = ax^2 + bx + c \tag{4.2}$$

in which a, b and c are real constants and a, the *leading coefficient*, is non-zero. Where y is the dependent variable, the exponents of the terms in x will be 2, 1 or 0 (in the case of the constant term). Later we present the general equation of the second degree, but will concentrate on expressions of the form (4.2). The graph of a quadratic equation is called a *parabola*. Parabolas are U-shaped curves and, with y as the dependent variable, the parabola will make an upright or inverted U shape depending on the sign of the leading coefficient. Figure 4.4 shows the effect of varying the parameter a in the parabola $y = ax^2$. High and positive values of the leading coefficient produce a steeply sloping parabola which opens upwards. In this case, the extreme point of the parabola, the *vertex*, represents a minimum for the function. Negative values of a mean that the parabola opens downwards and the vertex represents a maximum for the function. The parabolas of Figure 4.4 are symmetric about the y axis, which is the *axis of symmetry* of the parabola. All parabolas given by the equation

$$y = ax^2 + bx + c$$

will have axes of symmetry parallel to the y axis. Figure 4.5 shows the effect of varying the value of the constant term c in the parabola

$$y = ax^2 + c$$

The value of c gives the y intercept of the parabola and the effect of varying c is to shift the whole parabola up if c increases or down if c is reduced.

202 Mathematics for business, finance and economics

$y = ax^2$

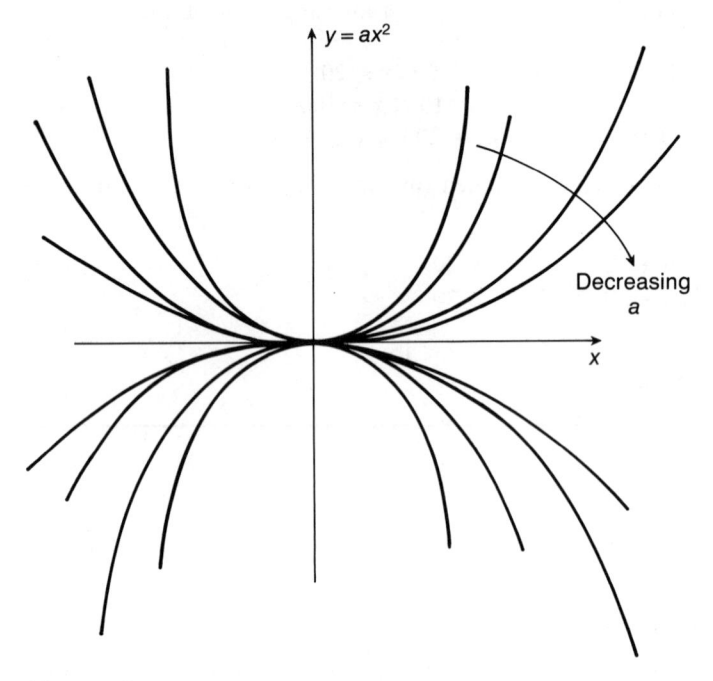

Decreasing a

Figure 4.4

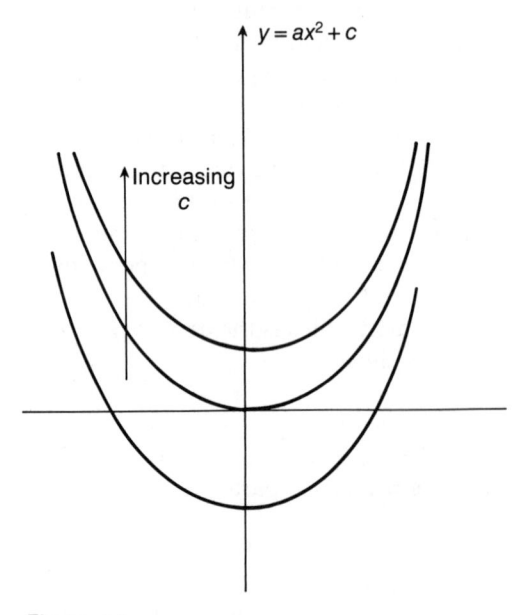

$y = ax^2 + c$

Increasing c

Figure 4.5

An important effect of such variations in the constant term is to change the points (if any) where the curve cuts the x axis. Note that the slope of the curve at any point is unaffected by the changes in c. Changes in the value of the x coefficient, b, in the quadratic equation

$$y = ax^2 + bx + c$$

move the vertex of the parabola relative to both of the axes, although the parabola remains 'upright' – the axis of symmetry remains parallel to the y axis. As b is decreased, the algebraic value of the slope is decreased. Thus if, at a point, the slope is positive to begin with, the curve becomes less steep. If the slope is negative, then the curve will decrease more rapidly at the given point. While variations in the value of c move the parabola relative to the vertical axis, and changes in b move the parabola relative to both axes, the curve can be shifted to the left or to the right alone by replacing x with $x - k$ (which will produce a shift to the right) or with $x + k$ (to produce a shift to the left) with $k > 0$ in both cases. Making such a change to the independent variable affects both b and c in such a way as to give a horizontal movement only. These shift effects can be confirmed with the parabola

$$y = ax^2 + bx + c$$

Substitute $x - k$ for x to give

$$y = a(x - k)^2 + b(x - k) + c$$

which must represent a rightwards shift, since any original value of x must now be increased by k in order to maintain the original value of y. Expansion of the brackets gives

$$y = ax^2 - 2akx + ak^2 + bx - bk + c$$
$$= ax^2 + (b - 2ak)x + c + (ak^2 - bk)$$

which form shows the changes to b and the constant term needed to effect a shift of the whole curve k units to the right. Consider the simplest parabola, $y = x^2$. A 1 unit shift to the right is represented by

$$y = (x - 1)^2$$
$$= x^2 - 2x + 1$$

Figures 4.6(a), 4.6(b) and 4.6(c) illustrate the shift as a two-part process. Figure 4.6(a) shows the original function, $y = x^2$. Figure 4.6(b) shows the effect of the subtraction of $2x$; this will move the parabola both across and down. In Figure 4.6(c) the constant, 1, has been added, with the original function indicated by the broken line. The overall effect is a lateral movement only.

204 Mathematics for business, finance and economics

(a)

(b)

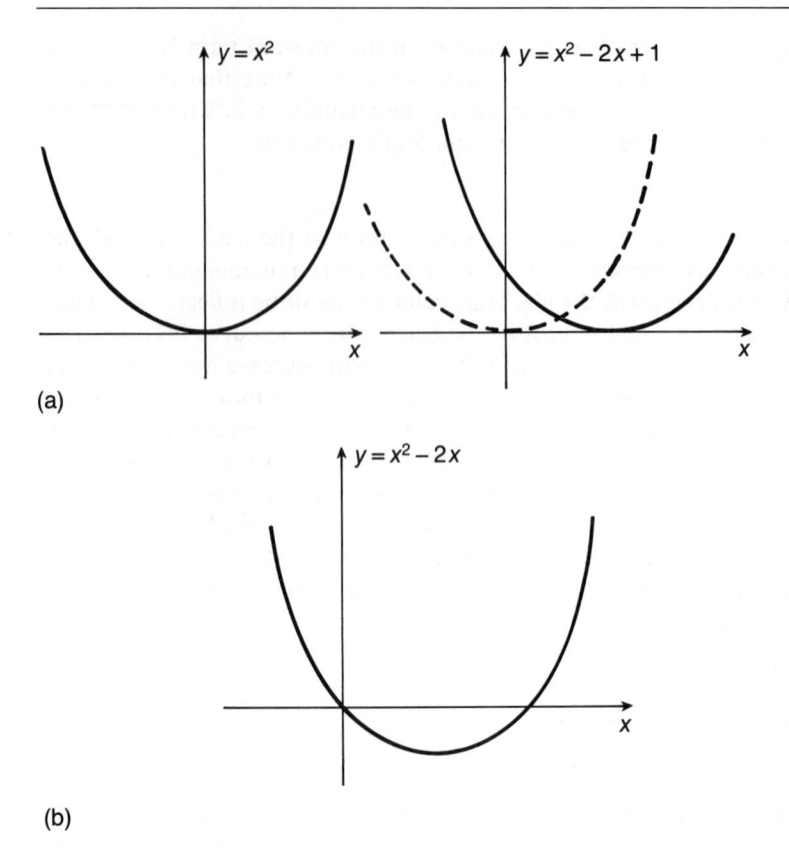

Figure 4.6

The effect on the slope of the function for any given value of x can be seen from Figure 4.6(c). For any value of x less than $\frac{1}{2}$, the function $y = x^2 - 2x + 1$ is further from its vertex than is $y = x^2$, and is steeper in absolute terms.[4] For any value of x greater than $\frac{1}{2}$, the function $y = x^2 - 2x + 1$ is nearer to its vertex than is $y = x^2$, and has a lesser slope. The x intercepts are of particular importance in most applications. These are the points where $y = 0$ and are the *zeros* of the function. The values of x which give rise to the zero values are called the *roots* of the equation

$$ax^2 + bx + c = 0$$

One way to obtain the roots of a quadratic equation is by *factoring*. This can be the best approach when the coefficients a, b and c assume convenient values. To factor a quadratic expression equated to zero is to express it as the product of two linear terms at least one of which must therefore be zero if the equation is to be satisfied. For example the equation

$$x^2 - 10x + 21 = 0$$

factors as

$$(x - 3)(x - 7) = 0$$

and so the roots of the equation are[5]

$$x = 3 \text{ and } x = 7$$

In other words, it is these values of the independent variable which produce the zeros of the function

$$y = x^2 - 10x + 21$$

Where a quadratic equation does not factor in a fairly obvious way, the *quadratic formula* is usually the best way to find the roots. The formula is

$$x = \frac{-b \pm (b^2 - 4ac)^{1/2}}{2a} \qquad (4.3)$$

The quadratic formula can always be used to solve for the exact values of the roots, and can be slightly rearranged as

$$x = \frac{-b}{2a} \pm \frac{(b^2 - 4ac)^{1/2}}{2a}$$

which shows more clearly that the two roots of the quadratic will be equal distances of

$$\frac{(b^2 - 4ac)^{1/2}}{2a}$$

on either side of the vertex. The vertex itself, as suggested by the fact that it lies on the axis of symmetry, is located centrally between the roots at $-b/2a$. The expression

$$b^2 - 4ac$$

is the *discriminant*. Three cases are distinguished in terms of the roots by the relation of the discriminant to zero. These are as follows.

Case 1 $b^2 - 4ac > 0$

Here the discriminant is positive and, as a result, the quadratic has *two distinct real roots*. In geometric terms the parabola crosses the x axis at two separate points. As an example consider the quadratic equation

$$y = x^2 - 3x - 10$$

On equating to zero, this factors as

$$(x+2)(x-5) = 0$$

so the roots are located at $x = -2$ and $x = 5$. The function is graphed in Figure 4.7.

Case 2 $b^2 - 4ac = 0$

A zero value for the discriminant means that the two roots are equal and, as can be seen by reference to the quadratic formula, will be equivalent to the vertex $-b/2a$. This is the case of *double roots*. Double roots have zero separation on either side of the vertex. In an equation of higher degree than a quadratic, where a root occurs three times, it is said to be a *triple root*. In general, a root occurring n times, where $n \geq 2$, is said to be a *multiple* or *repeated* root. In contrast, a single root is sometimes called a *simple root*. A case of a quadratic equation with double roots is illustrated in Figure 4.8. As will be seen from the figure, which is drawn for the equation

$$y = x^2 - 6x + 9$$

factoring and equating to zero gives

$$(x-3)(x-3) = 0$$

There is therefore a single zero of the function corresponding to the double root $x = 3$. The parabola just touches the horizontal axis, rather than intersecting it.

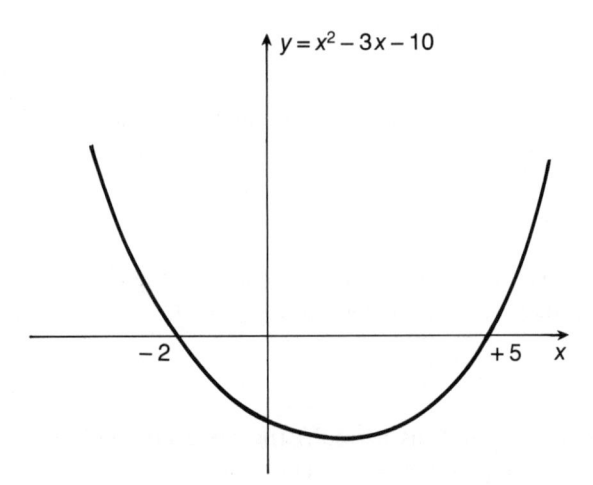

Figure 4.7

$$y = x^2 - 6x + 9$$

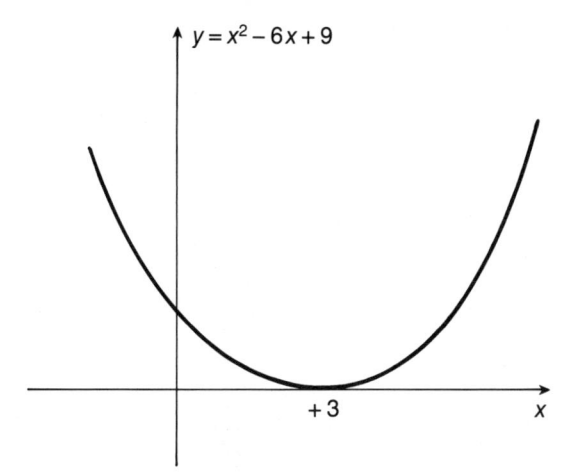

Figure 4.8

Case 3 $b^2 - 4ac < 0$

When the discriminant of the quadratic is negative, there is no solution to the equation in terms of the *real* numbers. The parabola has no point in common with the x axis. This case is illustrated in Figure 4.9, which is drawn for the equation

$$y = x^2 + 6x + 13$$

Application of the quadratic formula in this case produces the x values

$$x = \frac{-6 \pm \sqrt{(36 - 52)}}{2}$$

i.e.

$$x = -3 + 2i$$

and

$$x = -3 - 2i$$

where, as will be recalled from Chapter 1, i is the *imaginary* number $\sqrt{(-1)}$. So the roots of this quadratic, while distinct, are *complex numbers*. Note that the real component of the complex roots, which will be $-b/2a$ (-3 in this case), corresponds to the vertex of the parabola. The vertex is always located at $x = -b/2a$ whether the roots of the equation are real or complex. In practical work, it is often useful to make a sketch of the graph of a function as part of the process of analysis. For quadratic functions, points on the graph which are helpful for curve sketching can be produced by several criteria, of which we shall use five for the time being.

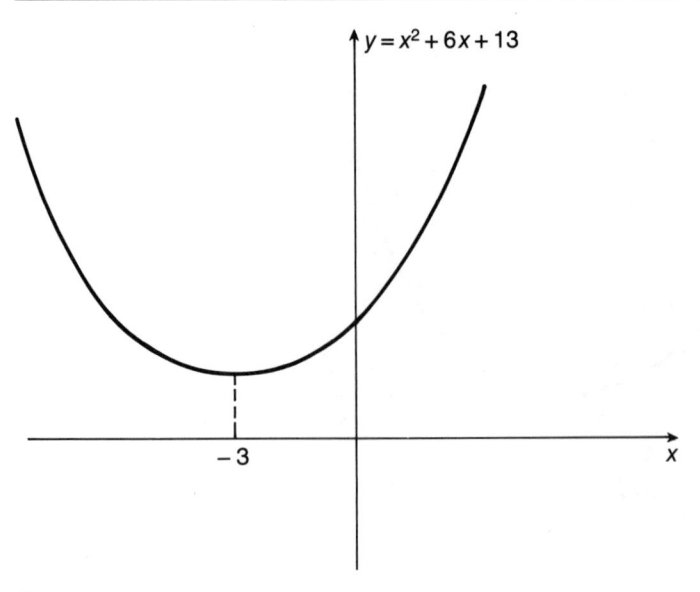

[Figure 4.9: graph of $y = x^2 + 6x + 13$, parabola with vertex at $x = -3$]

Figure 4.9

1 *The y intercept* is the point where the graph cuts the vertical axis, and is found by setting x equal to zero.
2 *The x intercepts* are the points, if any, at which the graph cuts the horizontal axis (so that y is zero).[6] These points may be found by use of the quadratic formula.
3 *The turning points*: the graph of a quadratic equation has a single turning point – the vertex of the parabola. This, as we have seen, is located where $x = -b/2a$. While there is this convenient result for quadratics, calculus methods will usually be required to locate the turning points of other functions.
4 *Extreme x values* are used to show the ultimate direction in which the function is heading. Appropriately large positive and negative values of x are inserted into the function.
5 *Other x values*: depending on the accuracy required, a suitable selection of x values not covered by 1–4 should be plotted.

An additional step of value in curve sketching (the identification of any *asymptotes*) will be mentioned in Section 4.5. Also, use can be made of any known properties of *symmetry*[7] possessed by the function. However, steps 1–5, appropriately employed, will usually be sufficient in most business and economic applications. Now consider the use of these five steps to sketch the graph of the function

$$y = x^2 - 4x - 21$$

1 When $x = 0$, $y = -21$, so that the y intercept lies below the x axis. The co-ordinates of this point are $(0, -21)$.
2 When $y = 0$, the quadratic formula (or factoring) results in

$$x = -3 \text{ and } x = +7$$

The co-ordinates of the points generated are: $(-3, 0)$ and $(+7, 0)$.
3 The x value at the vertex is

$$\frac{-(-4)}{2(1)} = +2$$

and the corresponding value of y is -25. The co-ordinates of the vertex are therefore $(2, -25)$.
4 When x is sufficiently large (positive *or* negative) for any parabola, the overwhelming influence will be effected by the term in x^2. This is called the *dominant term*, and in sketching any function it is useful to look for dominant terms.[8] Here x^2 has a positive coefficient and, since x^2 is positive for high x values at either end of the domain, the function as a whole must be large and positive in these regions. In terms of the sketch, this means that the curve will open upwards.
5 For sketching purposes, 'moderate' or obviously significant values of x are likely to be useful.[9] When the roots are not too close together, a selection of points for x values between the roots will usually suffice. In the present example, a possible selection of moderate x values and the corresponding values of y could be

x	y
-2	-9
-1	-16
$+3$	-24
$+5$	-16

which should be sufficient for sketching purposes. The appearance of the function is shown in Figure 4.10.

For parabolas in general much can be learned from the value of the leading coefficient.

(a) If $a > 0$ the parabola must open upwards, and the vertex will therefore represent a *minimum* of the function.
(b) $a < 0$ the parabola opens downwards, and the vertex corresponds to a *maximum*.

In historic terms, a quadratic equation of particular interest is the following:

$$x^2 - x - 870 = 0$$

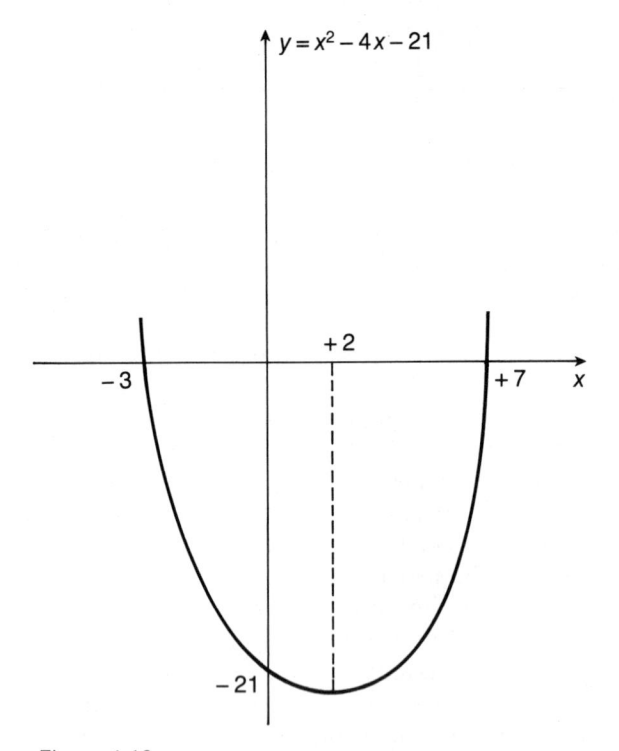

Figure 4.10

Using the quadratic formula, the x values produced are

$x = 30$ and $x = -29$

The interest of this equation is that its solution was first obtained no less than 3900 years ago in Mesopotamia. It arises directly from a problem stated on a Babylonian tablet dated 1900 BC. The problem was expressed as that of finding the side of a square the area of which exceeds the side by a factor of 870. The Babylonian algebra allowed the obtaining of the positive root or roots of a limited number of quadratic equations. This amazing civilization was also able to solve some equations of higher degree.[10] Now consider an example of quadratic equations and parabolas in the context of business management. Suppose that a firm's costs C are given by

$C = 4q + 450$

in which q represents both output level and sales. The product is sold at a price p under demand conditions given by

$p = 50 - 0.1q$

First consider the firm's *turnover*. Turnover, R, is total receipts from sales of the product and is also known as *total revenue, sales* or *income*. Turnover will therefore be

$$R = pq$$
$$= 50q - 0.1q^2$$

As a quadratic equation this is written out in full as

$$R = -0.1q^2 + 50q + 0 \tag{4.4}$$

in which form it is evident that

$$a = -0.1, \ b = 50, \ c = 0$$

Since the leading coefficient here is negative, the parabola will open downwards. The y intercept is at $(0, 0)$. The x intercepts are at $(0, 0)$ and $(500, 0)$ while the vertex has an x co-ordinate given by

$$-b/2a = -50/2(-0.1)$$
$$= 250$$

It is at this output that turnover is at a maximum. The value of R will then be

$$R = -0.1(250)^2 + 50(250)$$
$$= 6250$$

However, the firm's objective may well be to maximize the value of profits rather than sales. Profit, π, is the difference between income and costs and will therefore be

$$\pi = R - C$$
$$= 50q - 0.1q^2 - 4q - 450$$
$$= -0.1q^2 + 46q - 450 \tag{4.5}$$

Step 1 of the curve sketching procedure shows the y intercept at $(0, -450)$. Use of the quadratic formula shows that the zeros of the profit function occur at

$$q = 10 \text{ and } q = 450$$

These are the intercepts with the horizontal axis. The lower of these levels (with coordinates $(10, 0)$) represents the *breakeven* output level, while the other zero is a point of theoretical interest only, being well past the level at which a firm would choose to operate in practice. The vertex of the profit parabola is at $q = 230$ at which the maximum possible profit of $\pi = 4840$ is made. The revenue and profit parabolas are shown in Figure 4.11. The profit curve will always peak at a lower level of output than the revenue

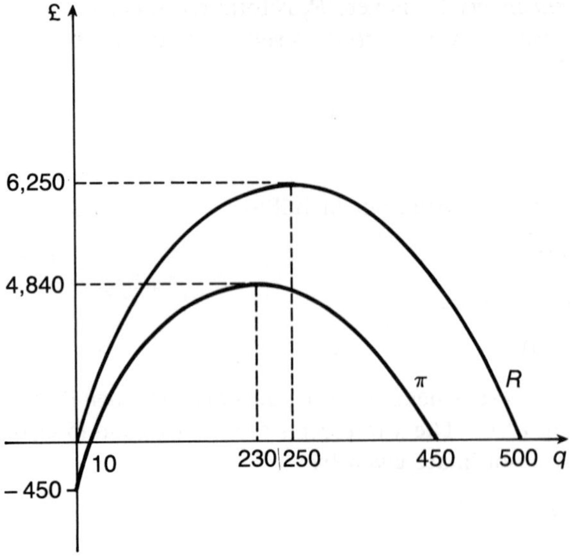
£
6,250
4,840
-450
10
230
250
450
500
q
π
R

Figure 4.11

curve (so long as costs always increase with output level). This will not be proved at this point, but it is a useful exercise to see how the subtraction of the cost expression affects the values of b and a (and hence $-b/2a$) in the profit equation compared with the turnover expression.

In Chapter 2, we considered the solution of systems of linear equations. Any system which contains even a single equation with a single non-linear term is a *non-linear system*. Non-linear systems of equations are much less tractable than linear systems and the statements relating the number of solutions to the numbers of independent equations and unknowns no longer apply. In practice, non-linear systems of equations are less widely used than linear systems and we shall confine the presentation here to reasonably simple cases.

One linear and one quadratic equation

Consider first of all the solution of a mixed system of one linear and one quadratic equation. For example suppose that

$$4x + 2y = 20$$

and that, simultaneously, it is required that

$$y = x^2 - 7x + 16$$

A solution procedure useful with mixed systems of this kind is first to solve for y in terms of x in the linear equation. Thus we obtain

$$y = 10 - 2x$$

The next step is to substitute for y in the quadratic equation. This produces

$$10 - 2x = x^2 - 7x + 16$$

which on rearrangement becomes

$$x^2 - 5x + 6 = 0$$

which is satisfied for $x = 2$ and $x = 3$. The corresponding values of y (obtained from either the quadratic or the linear expression) are $y = 6$ and $y = 4$. Thus the two solutions to the mixed system are

$$x = 2 \text{ and } y = 6$$

and

$$x = 3 \text{ and } y = 4$$

Now consider a further example of the solution of a mixed system of one quadratic and one linear equation. The result that will be obtained is not one which could occur with two linear equations. Suppose that

$$2x + y = 7$$

and

$$y = x^2 + 4x + 20$$

The result of substitution of $7 - 2x$ for y into the original quadratic equation produces

$$x^2 + 6x + 13 = 0$$

which has no solution in real terms. The solution values (as may be confirmed) are the complex numbers

$$x = -3 + 2i \text{ and } y = 13 - 4i$$

and

$$x = -3 - 2i \text{ and } y = 13 + 4i$$

Two quadratic equations

Systems that involve two simultaneous quadratic equations of the form

$$y = ax^2 + bx + c$$
$$y = dx^2 + ex + f$$

can also be solved, since they can be reduced to a single quadratic in x by subtraction. For example if

$$y = 3x^2 - 7x - 2$$

and

$$y = x^2 - 2x + 16$$

subtraction produces

$$0 = 2x^2 - 5x - 18$$

which solves in the normal way for

$$x = -2 \text{ and } x = 4.5$$

with the corresponding values of y (obtained from either of the quadratics) being

$$y = 24 \text{ and } y = 27.25$$

Two further applications of quadratic equations in economics can now be considered. The first example is of market equilibrium with a linear demand curve and a quadratic supply function. Suppose that demand conditions are given by

D: $p = 90 - 2q$

where p is price and q is output (sales) and that market supply is given by

S: $p = 20 + 0.5q^2$

The graph of this system is shown in Figure 4.12. Equating the value of price given by the supply and demand equations produces

$$p = 90 - 2q$$
$$= 20 + 0.5q^2$$

so that

$$0.5q^2 + 2q - 70 = 0$$

and so

$$q^2 + 4q - 140 = 0$$

which factors as

$$(q - 10)(q + 14) = 0$$

Therefore, market equilibrium is reached when

$$q = 10 \text{ and } p = 70$$

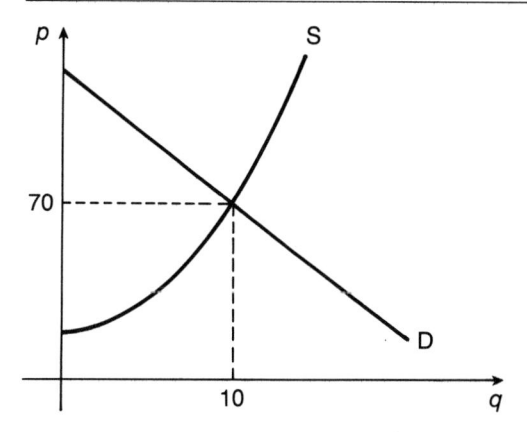

Figure 4.12

Note that the other solution to this mixed system (which does not correspond to a practically realizable situation) occurs at the values

$$q = -14 \text{ and } p = 118$$

One circumstance in which simultaneous quadratic equations can arise is in breakeven analysis where costs are a quadratic function of output level and where a straight line demand curve is faced. Suppose that demand is given by

D: $\quad p = 200 - 2.5q$

and that costs are given by

$$C = 1600 + 1.5q^2$$

Breakeven is the output level where sales revenue first covers costs. This is where

$$R = 200q - 2.5q^2$$
$$= 1600 + 1.5q^2$$
$$= C$$

Thus

$$4q^2 - 200q + 1600 = 0$$

i.e.

$$q^2 - 50q + 400 = 0$$

so that

$$(q - 10)(q - 40) = 0$$

giving

$$q = 10 \text{ and } q = 40$$

Thus the breakeven output level is 10. The situation is graphed in Figure 4.13.

In a *quadratic inequality*, there will be a solution set of values of x which satisfy the inequality. For example, consider the relationship

$$x^2 \geqslant 3x + 10$$

Rearrangement gives

$$x^2 - 3x - 10 \geqslant 0 \tag{4.6}$$

Taken as an equation, (4.6) solves for

$$x = 5 \text{ and } x = -2$$

The function

$$f(x) = x^2 - 3x - 10$$

is graphed in Figure 4.14. $f(x)$ is non-negative provided that $x \leqslant -5$ or $x \geqslant 2$ and so the solution set will be

$$-\infty < x \leqslant -5$$
$$2 \leqslant x < \infty$$

As a further illustration of quadratic inequalities, the relationship

$$4x + 35 \geqslant 14 + x^2$$

which rearranges to

$$-x^2 + 4x + 21 \geqslant 0$$

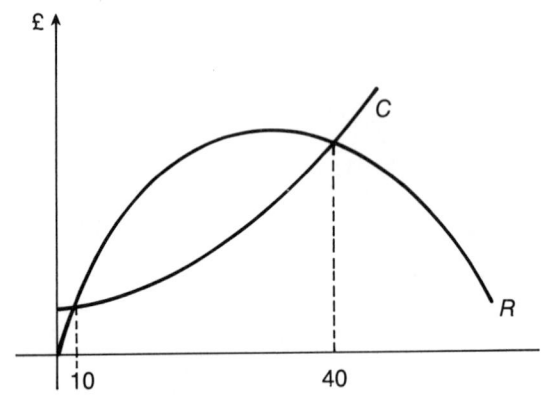

Figure 4.13

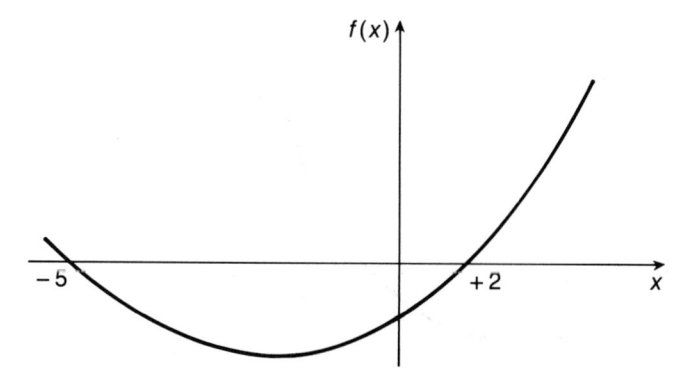

Figure 4.14

has the solution set

$$-3 \leqslant x \leqslant 7$$

Now consider an economic example in quadratic inequalities. Suppose that a firm's average costs of production, AC, are given by

$$AC = 0.05x^2 - 8x + 350$$

where x is output level. The firm can sell its product in a market where its price is set at $p = 210$. The management of the firm regards the product – one of several that it makes – in strategic rather than in necessarily profit making terms. However, it is not company policy to sell any product at a loss. Given these conditions, what range of output levels is available to management? The requirements mean that the range must be such that unit costs do not exceed unit revenue, i.e.

$$0.05x^2 - 8x + 350 \leqslant 210$$

so that

$$0.05x^2 - 8x + 140 \leqslant 0$$

which is more conveniently stated as

$$x^2 - 160x + 2800 \leqslant 0$$

As an equality, this factors to

$$(x - 20)(x - 140) = 0$$

so that the range of output levels available is

$$20 \leqslant x \leqslant 140$$

This example is shown in Figure 4.15.

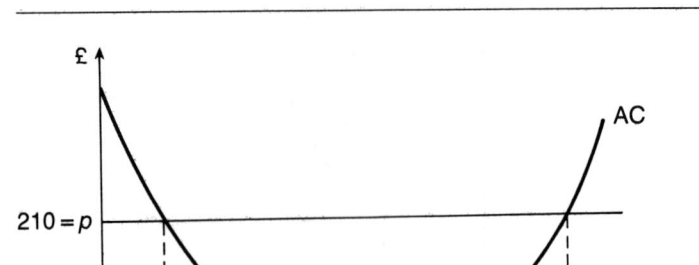

Figure 4.15

Just as the equation of a straight line could be expressed in general form, so there is a general form of the equation of a parabola. When the axis of the parabola is parallel to the y axis, the general form of the equation can be written as

$$Ax^2 + Dx + Ey + F = 0 \qquad (4.7)$$

in which $A \neq 0$, $E \neq 0$ and A, D, E and F are real numbers. The general form of the parabolic equation itself represents a special case of the *general quadratic equation*, which includes the possibility of terms in y^2 and xy:

$$Ax^2 + Bxy + Cy^2 + Dx + Ey + F = 0 \qquad (4.8)$$

in which not all of the coefficients A, B and C can be zero, and in which all of the coefficients are real numbers. The general form of a parabola with axis parallel to the y axis emerges when B and C are zero. Parabolas symmetric about axes not parallel to the y axis are generated when

$$B^2 - 4AC = 0$$

and where the values of A, B and C are not all zero. For example, if $A = 0$, $B = 0$ and $D = -1$, the resulting equation can be written as

$$x = Cy^2 + Ey + F$$

which describes parabolas with axes parallel to the x axis and opening to the right or to the left. The simplest such parabola is where

$E = F = 0$ and $C = 1$

in which case

$x = y^2$ or $y = \sqrt{x}$

The graph of $y = \sqrt{x}$ is shown in Figure 4.16. Recall that by limiting the range to the *positive root only*, the definition of a function will be satisfied, and the curve as a whole can be seen as a plot of the two functions:

$y = -\sqrt{x}$ and $y = +\sqrt{x}$

To complete this brief look at quadratic equations, note that in the general quadratic equation a *circle* will result if

$A = C$ and $B = 0$

An *ellipse* is produced if

$B^2 - 4AC < 0$

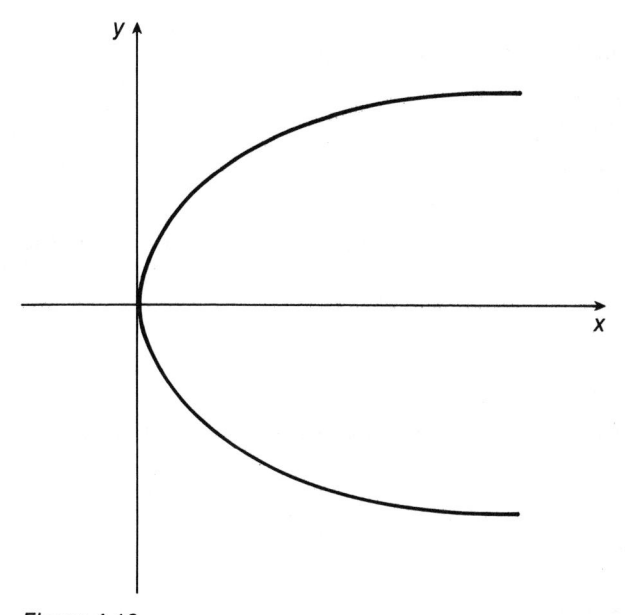

Figure 4.16

The outcome will be a *hyperbola* if

$$B^2 - 4AC > 0$$

Hyperbolas and their areas of application will be discussed later in the chapter under rational functions. There are also some very special cases which drop out of the general quadratic equation. For example, if the only non-zero coefficient is A, then the equation reduces to

$$Ax^2 = 0$$

and the y axis is produced.[11] In the event that all of A, B and C were zero, then equation (4.8) reduces to the general *linear* equation – which is therefore contained as a special case.

EXERCISES 4.2

1 Find the roots of the following quadratic equations:

(i) $x^2 - 7x + 12 = 0$
(ii) $x^2 + x - 30 = 0$
(iii) $x^2 + 9x + 14 = 0$
(iv) $x^2 - 16x + 64 = 0$
(v) $x^2 + 18x + 81 = 0$
(vi) $x^2 - 7x + 0 = 0$

2 Find the roots of the following quadratic equations:

(i) $2x^2 - 14x + 24 = 0$
(ii) $x^2 - 2.3x + 1.32 = 0$
(iii) $2x^2 - 3x - 9 = 0$
(iv) $x^2 + 99x - 100 = 0$

3 Find the roots of the following quadratic equations and comment on the results:

(i) $x^2 - 20x + 100 = 0$
(ii) $x^2 + 9 = 0$
(iii) $x^2 - 10x + 29 = 0$

4 Find the solutions to the following systems of one linear and one quadratic equation:

(i) $4x + y = 20$
    $y = x^2 - 25x + 118$
(ii) $6x + 2y = 100$
    $2x^2 - y - 68x + 550 = 0$

5 Find solutions to the following systems involving two quadratic equations:
   (i) $y = 4x^2 - 20x + 60$
       $y = 3x^2 - 5x + 10$
   (ii) $6x^2 - 10x + 20 - y = 0$
        $4x^2 - 5x + 18 - y = 0$

6 Find the solution sets to the following quadratic inequalities:
   (i) $x^2 - 2x - 24 \leq 0$
   (ii) $x^2 + 3x - 10 \geq 0$
   (iii) $3x^2 + 10x + 50 \leq 2x^2 + 21x + 26$

7 A firm faces the demand curve given by
   $$p = 700 - 5x$$
   where p is price and x is quantity produced and sold. Costs are given by
   $$C = 150 + 40x + x^2$$
   (i) What output level maximizes the firm's revenue from sales, and what is the maximum revenue level achievable?
   (ii) What output level maximizes profits, and what is the maximum profit so made?

8 If market demand is given by
   $$p = 720 - 4q$$
   while the market supply curve is given by
   $$p = 160 + 0.25q^2$$
   find the market clearing values of price and quantity.

9 A company's product sells at price $p = £10,500$. Its average costs are given by
   $$AC = 0.1x^2 - 80x + 17,500$$
   Within what range of output levels does revenue at least cover costs?

4.3 CUBICS AND QUARTICS

If the dependent variable y is a *cubic* function of x, this can be written as
$$y = ax^3 + bx^2 + cx + d \tag{4.9}$$
where in (4.9) the leading coefficient, a, is not zero, and all the coefficients

(a)

(b)

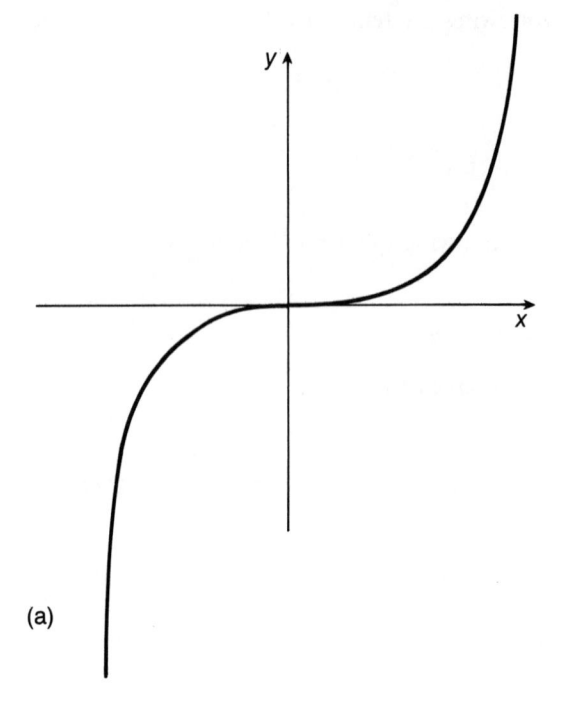

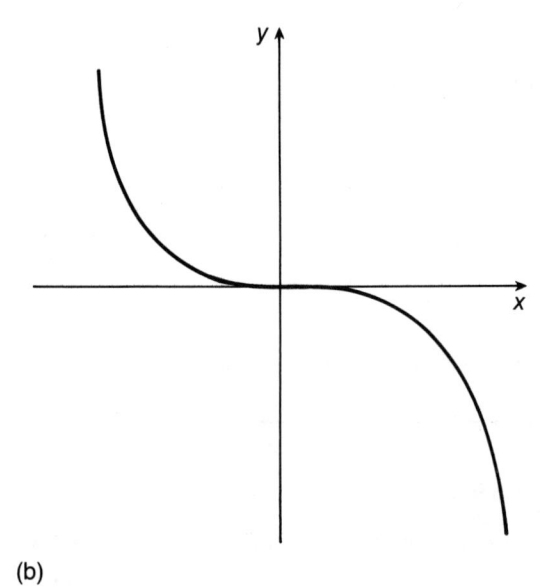

Figure 4.17

a, b, c and d are real numbers. The simplest form of cubic equation is

$$y = ax^3$$

the appearance of which is shown in Figure 4.17.

Figure 4.17(a) is drawn for $a > 0$, while Figure 4.17(b) shows the appearance when $a < 0$. If $a = 1$ then the function

$$y = x^3$$

results. This is sometimes called the *cubing function* for obvious reasons. If the reader plots the cubing function accurately on graph paper, it will be evident just how steeply the graph rises, and it is usually convenient to plot cubics using a different unit on the vertical axis. As is evident, the cubing function has a single zero at $x = 0$. At $x = 0$ the curvature of the function changes and at the same time the slope becomes zero. All cubics exhibit the change in 'curvature' at some point (they are said to *inflect*) and will have two turning points or none. A cubic cannot have a single turning point. The cubing function can be seen as a limiting case where two turning points have coalesced. Figure 4.18 shows the graph of a cubic with two turning points. The equation is

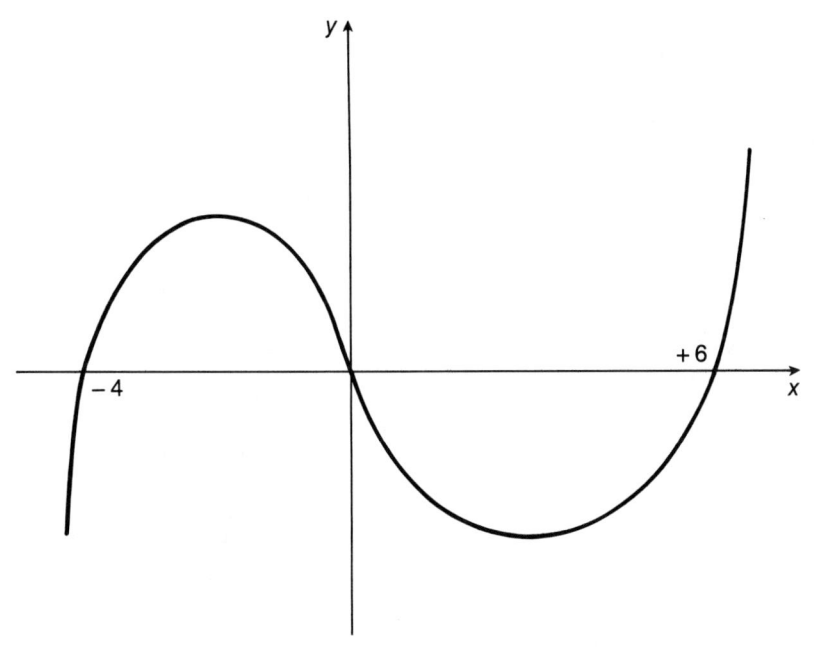

$$y = x^3 - 2x^2 - 24x$$

Figure 4.18

Note that with no constant term, the graph of this cubic will pass through the origin, which is one of the three zeros of the function.

While there is an analytical method for finding the roots of a cubic equation – a much more complicated version of the formula approach used for quadratics – it is not widely used and will not be detailed here.[12] In the following chapter, the turning points of cubics will be found using calculus. But there are special cases for which the roots of a cubic equation are easily obtained. For instance, if one of the roots is already known, then only a quadratic equation needs to be solved to find the other two roots. This is the case when a function passes through the origin, i.e. when the constant term is zero. For example in the function

$$y = x^3 - 2x^2 - 24x$$

because of the absence of the constant term, the right-hand side factors as

$$y = x(x^2 - 2x - 24)$$

the bracketed part of which also factors to produce

$$y = x(x+4)(x-6)$$

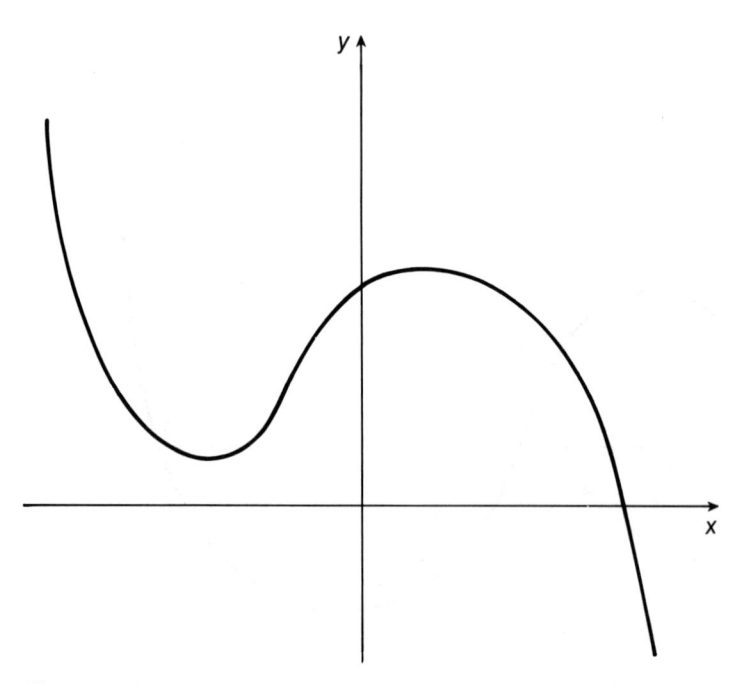

Figure 4.19

Thus if $y = 0$, it must be the case that

$$x = 0 \text{ or } x = -4 \text{ or } x = 6$$

The general appearance of this function, with the (local) high point before the low point, is a property of all cubics in which the leading coefficient is positive. If $a < 0$, the cubic with turning points will appear as in Figure 4.19, although of course the actual *positions* of the turning points need not be as shown. Note that the cubic illustrated in Figure 4.19 will have one real positive root, where the curve crosses the x axis, and two complex roots.

In economic theory cubics are frequently used to represent cost functions – relating total costs of production to output level. With appropriate coefficients, cubic cost functions have reasonable properties (the cost of producing a marginal unit falling at first and then rising later on). To illustrate, suppose that a firm's costs are given by

$$C = 0.2x^3 - 18x^2 + 800x + 100 \qquad (4.10)$$

That part of (4.10) lying in the positive quadrant is shown in Figure 4.20. This cubic inflects but does not have turning points. The presence of a local maximum would mean, implausibly, that total costs would actually go down as output expanded over a certain interval. The intercept with the vertical axis, 100, represents fixed costs.

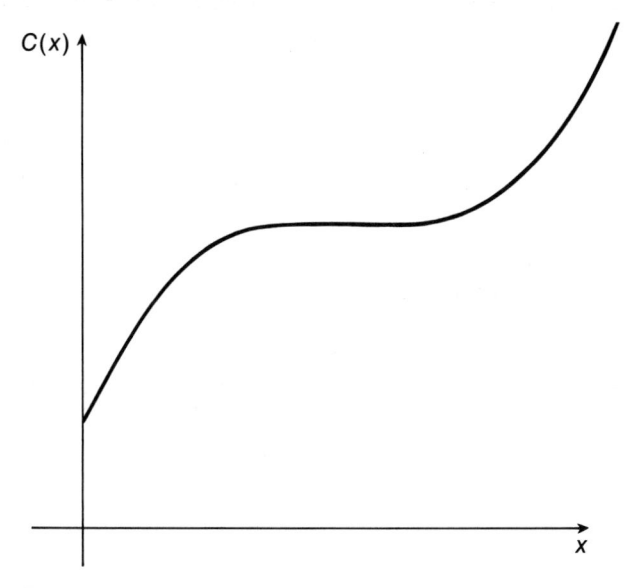

Figure 4.20

Quartics

Where the dependent variable is a *quartic* function of x, this can be written as

$$y = ax^4 + bx^3 + cx^2 + dx + e \tag{4.11}$$

in which the leading coefficient, a, is non-zero, and where a, b, c, d and e are real numbers. Quartic functions will have three turning points or one. Quartics cannot have two turning points, but turn they must. This is due to the fact that the leading term (by which the value of the function is dominated for values of x at either extreme) is an *even* power, so the function must eventually assume similar y values if x is large and negative as when x is large and positive. If the function turned twice, the values at the extremes would be of opposite sign. For a numerical example of a quartic equation, consider the function

$$y = x^4 - 125x^2 + 2500 \tag{4.12}$$

The function is sketched in Figure 4.21 (in which larger units on the vertical scale are implied). The function has zeros at

$$x = -10, \ x = -5, \ x = +5, \ x = +10$$

Equation (4.12) is an interesting quartic in two respects. First it is seen to be symmetric about the y axis. This will always be the case if there are no

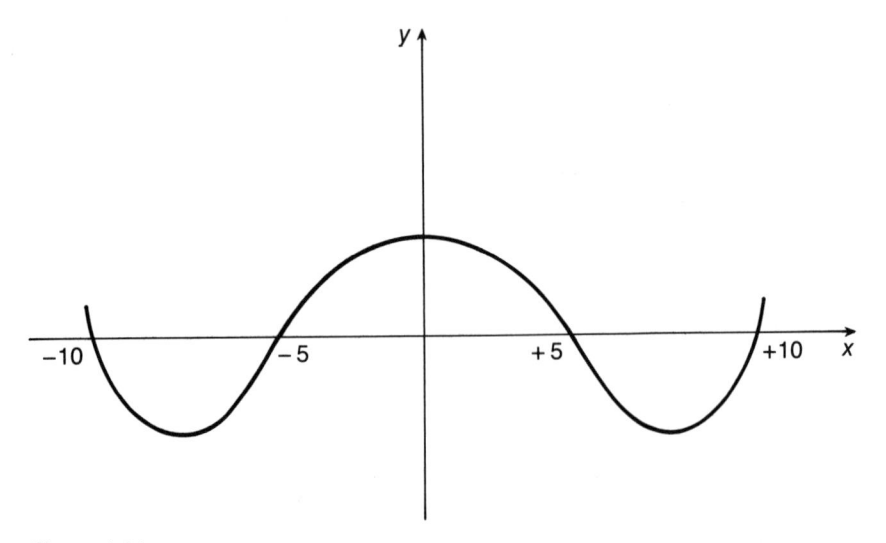

Figure 4.21

odd powers of x.[13] Second, although (4.12) is a quartic, it can be solved as a *quadratic*. First make the substitution

$$z = x^2$$

The original quartic equation in x then becomes the quadratic equation in z

$$z^2 - 125z + 2500 = 0$$

which conveniently factors as

$$(z - 100)(z - 25) = 0$$

so that

$$z = 100 \text{ or } z = 25$$

and therefore

$$x^2 = 100 \text{ or } x^2 = 25$$

thereby producing the stated results. The most notable special case of a quartic equation is one in which only even powers occur. This is known as a *biquadratic*. Biquadratic equations can be solved using the quadratic formula. The roots of quartics in general can be found by use of formulae, but the method is very long winded and if precise roots are needed it is preferable to use mathematical software such as *Derive* to obtain the results. For an example in an economic context, suppose that a company has fixed costs of £7.29m and an excess of revenue over variable costs of

$$9000x^2 - x^4$$

where x represents the level of output and sales. The firm's profit, π, will then be given by

$$\pi = 9000x^2 - x^4 - 7,290,000 \qquad (4.13)$$

for the domain $x \geqslant 0$. The profit function is as shown in Figure 4.22.

The reader may confirm by substitution that breakeven output is at $x = 30$. The other zero within the domain is at $x = 90$. The function (4.13) can be expressed as a quadratic in $z = x^2$:

$$-z^2 + 9000z - 7,290,000 = 0$$

which, as seen in the study of quadratics, produces a curve opening downwards, with the vertex representing a maximum. The vertex occurs when $z = -b/2a$, so the maximum of profit is found at

$$z = \frac{-9000}{-2}$$

$$= 4500$$

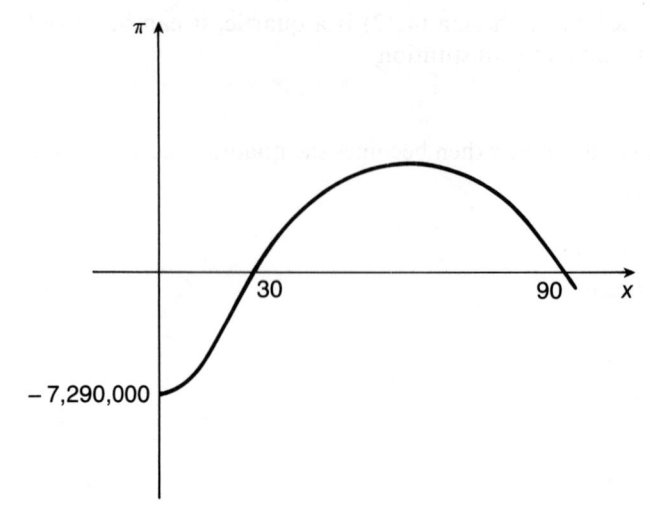

228 Mathematics for business, finance and economics

Figure 4.22

So in terms of the original independent variable the maximum occurs when

$$x = \sqrt{(4500)} \approx 67$$

at which output level the profit made is £12.96m.

EXERCISES 4.3

1 Find the x intercepts of the cubic functions:
   (i) $y = x^3 - 5x^2 - 50x$
   (ii) $y = x^3 - 2x^2 + x$

2 Find the x values which satisfy the quartic equations:
   (i) $x^4 - 13x^2 + 36 = 0$
   (ii) $x^4 - 37x^2 + 36 = 0$

3 A firm's total variable costs (VC) are given by
   $$VC = x^3 - 90x^2 + 4000x$$
   What is the level of total variable costs when average variable costs are at a minimum?

4 A company has fixed costs of £1,250,000, while its sales revenue exceeds variable costs by
   $$5000x^2 - x^4$$
   where x is the output level. At what level of output is the company's profit maximized, and what is the level of profit made?

4.4 POLYNOMIALS

Linear, quadratic, cubic and quartic functions are all examples of *polynomials*[14] in a single variable. In general, a polynomial expression in the variable x can be written as [15]

$$f(x) = a_n x^n + a_{n-1} x^{n-1} + a_{n-2} x^{n-2} + \cdots + a_3 x^3$$
$$+ a_2 x^2 + a_1 x + a_0 \tag{4.14}$$

where all of the coefficients a_i are real numbers and where n, a *non-negative integer*, is the *degree* of the polynomial so long as the leading coefficient, a_n, is non-zero. A cubic is a polynomial of the third degree, a polynomial of the fourth degree is a quartic, and a straight line is produced by a polynomial of the first degree.[16] In the extreme, the value of n could be zero; in other words, a constant could be said to be a polynomial of degree zero. In (4.14), a_0 is described as the *constant term*. Where the polynomial expression $f(x) = 0$, a polynomial *equation* results. The notation of (4.14) is very flexible, the subscript form for the coefficients allowing polynomials of any degree to be conveniently represented. For example, a polynomial of the fifth degree, a *quintic*, is written as

$$f(x) = a_5 x^5 + a_4 x^4 + a_3 x^3 + a_2 x^2 + a_1 x + a_0$$

in which the leading coefficient, a_5, must not be zero. The roots of quintics and polynomials of higher degree cannot usually be found by use of formulae.[17] The only exceptions are special cases where some of the terms have zero coefficients. For example the quintic equation

$$2x^5 - 486 = 0$$

conveniently solves for $x = 3$, which is the only *real* root of this quintic equation. As a further example, consider the sixth degree equation

$$x^6 - 9x^3 + 8 = 0$$

This is a quadratic in x^3 which solves for $x^3 = 1$ and $x^3 = 8$, so that the real roots are $x = 1$ and $x = 2$. Note that in (4.14) the exponents, i, of x must be non-negative integers. Therefore the function:

$$y = 4x^5 + 3x^4 - 2x^3 + x^2 - 7x + 3x^{-1}$$

is *not* a polynomial in view of the negative exponent in the final term. Neither is

$$y = 7x^2 + 3x^{1.5} + 2x + 6$$

because of the exponent 1.5. Nor is the following more extreme case a polynomial:

$$3x^\pi - 2x^{\sqrt{2}} + x^{\sqrt{3}} + 7$$

230   Mathematics for business, finance and economics

But note that, for an equation to be a polynomial, the *coefficients* of x^i are not limited to the integers or rational numbers. For example

$$y = \pi x^3 - \sqrt{2}x^2 + \sqrt{3}x + 7$$

is a valid polynomial.[18]

Often, the only roots of a polynomial that are of interest, or meaningful at all, are the positive roots. It is possible to get some idea of whether a polynomial has positive roots and an upper bound on how many positive roots there may be. Following the exercises we consider how this information can be obtained.

EXERCISES 4.4

1 Which of the following are polynomials?

 (i) $y = x^5 - 4x^2 + 2x + 1$
 (ii) $y = x^4 - \pi x^3 + 2x^2 + \sqrt{2}x$
 (iii) $y = x^5 - 5x^4 + 3x^{-2} + 2x + 1$
 (iv) $y = 4x - 1$
 (v) $y = x^2 + x^{1/2}$
 (vi) $y = x^{100}$

2 State the degrees of the following polynomials:

 (i) $f(x) = 4x^5 - 6x^3 + 2x$
 (ii) $f(x) = 1000 - 2x^3 - 0.001x^4$
 (iii) $y = 100 - x$
 (iv) $y = 7$

4.5 DESCARTES'S RULE OF SIGNS

From the *theory of equations*, Descartes's rule of signs tells us that the maximum number of positive roots for a polynomial is the *number of times the coefficients change sign* when the equation is written in the form (4.14). Descartes's rule states that the number of positive roots (counting repeated roots) is less than or equal to the number of sign changes in the coefficients and *equal in parity* to the number of sign changes. Parity means that if the number of sign changes is odd (even), then the number of positive roots must be odd (even). So an equation in which the coefficients change sign five times must have five, three or one positive root(s). Note the implication that a polynomial with an odd number of sign changes must have at least one positive root. If the number of sign changes had been four, then there may be four, two or possibly no positive root(s) to the equation.[19] For the first example, apply Descartes's rule of signs to the following cubic:

$$x^3 - 9x^2 + 26x - 24 = 0$$

The rule means that there could be up to three positive roots since, starting from the x^3 term, there are three changes in the signs of the coefficients (amongst which the constant term is included). It turns out that there are indeed three positive roots in this case:

$$x = 2 \quad x = 3 \quad x = 4$$

The quartic

$$x^4 - 125x^2 + 2500 = 0$$

has two changes of sign (recall that zero coefficients do not count as sign changes) and as we saw earlier it does indeed have two positive roots. The quintic

$$x^5 - 8x^4 - 49x^3 - 4x^2 + 180x + 144 = 0$$

has two sign changes, so the maximum number of positive roots is two. This even number of sign changes is consistent with there being no positive roots, so further investigation is needed to see if there are any at all. It turns out that there are two positive roots:

$$x = 12 \text{ and } x = 2$$

Recalling that Descartes's rule of signs states the *maximum* number of positive roots of an equation, consider the quadratic:

$$x^2 - 6x + 13 = 0$$

Here there are *two* changes of sign but in this case there are *no* positive real roots – the roots are the complex conjugate pair

$$3 + 2i \text{ and } 3 - 2i$$

Descartes's rule also states that the number of *negative* roots of a polynomial (including repeated roots) is no greater than the number of sign changes *in the polynomial in* $-x$. This polynomial is obtained by substituting $-x$ for x throughout the original equation. Thus, for example, the cubic equation

$$x^3 - 7x - 6 = 0$$

has one sign change, and because of the equal in parity requirement there must be one positive root. This is $x = 3$. The equation in $-x$ is

$$(-x)^3 - 7(-x) - 6 = 0$$

i.e.

$$-x^3 + 7x - 6 = 0$$

which shows *two* changes of sign. There may therefore be two negative roots or no negative roots. In fact there are two negative roots:

$x = -1$ and $x = -2$

The rule of signs provides us with a valuable practical tool. For example Descartes's rule can be used in financial calculations of the yield (internal rate of return) figure(s) for investment projects. The rule tells us whether multiple yields are possible and whether methods appropriate to that situation need to be applied.[20]

We have so far considered polynomials where a dependent variable, y, is expressed in terms of the powers of an independent variable, x. But a polynomial involving x and y may in general include powers of *both* x and y, as for example in the expression

$$3x^5 - 2x^3y^2 + xy - 2x^2y^3 + 3y^5 + 100 = 0 \tag{4.15}$$

In this context some symmetry properties of the graphs of polynomials are worth noting.

1. If the expression contains only even powers of x, then the graph of the function will be symmetrical about the y axis.
2. If only even powers of y are included, then the curve is symmetrical about the x axis.
3. If x and y can be interchanged in the expression without altering it, then the graph of the function will be symmetric about the line $y = x$.

As a result of (1) and (2) an expression containing only even powers of both variables will produce a curve symmetrical about both of the axes. The simplest example is the second degree equation

$$x^2 + y^2 - r^2 = 0$$

which is a circle of radius r centred at the origin. Note that condition 3 is also satisfied in this case. In respect of 3, note that equation (4.15) is symmetrical in y and x and the plot of this function is therefore symmetrical about the 45° line.

EXERCISES 4.5

1. Use Descartes's rule of signs to find the maximum number of positive roots to the following equations:

   (i) $4x^5 - 16x^4 - 49x^3 - 2x^2 + 45x + 18 = 0$
   (ii) $4x^4 - 125x^2 + 625 = 0$
   (iii) $0.5x^3 - 9x^2 + 52x - 96 = 0$

2 On the basis of Descartes's rule, what can be said about the numbers of positive and negative roots of the following equations?
   (i) $x^3 - 28x - 48 = 0$
   (ii) $4x^3 - 7x + 3 = 0$

4.6 RATIONAL FUNCTIONS

A rational function is the *ratio* of two polynomial functions.[21] Thus $f(x)$ is a rational function of x if it can be expressed in the following way:

$$f(x) = \frac{P(x)}{Q(x)} = \frac{a_n x^n + a_{n-1} x^{n-1} + a_{n-2} x^{n-2} + \cdots + a_2 x^2 + a_1 x + a_0}{b_m x^m + b_{m-1} x^{m-1} + b_{m-2} x^{m-2} + \cdots + b_2 x^2 + b_1 x + b_0}$$

where $P(x)$ and $Q(x)$ are the polynomials and $f(x)$ is defined for $Q(x) \neq 0$. So, for example, the following are rational functions:

(i) $f(x) = \dfrac{4x^5 - 8x^3 + x^2}{6x^4 - 7}$

(ii) $f(x) = \dfrac{x^2 - 1}{x - 1}$

(iii) $f(x) = \dfrac{1}{x}$

(iv) $f(x) = ax^2 + bx + c$

Example (i) illustrates a rational function in which $P(x)$ is a quintic (degree 5) and where $Q(x)$ is a quartic (degree 4). There is no requirement for P to be of greater degree than Q. In the case of example (ii), note that the numerator expression $x^2 - 1$ factors as

$(x + 1)(x - 1)$

so the rational expression apparently simplifies to $x + 1$, but there is a subtle difference in that for the rational function

$$f(x) = \frac{x^2 - 1}{x - 1}$$

$x = +1$ for which the denominator is zero is by definition *excluded from the domain*. Thus the graph of (ii) appears as the straight line $y = x + 1$ with one point excluded and is illustrated in Figure 4.23.[22]

Example (iii) is the *reciprocal function* and is graphed in Figure 4.24. The reciprocal function $y = 1/x$ is the simplest example of a rational function which is not a polynomial. Note that the curve separates into two *branches*, for $x > 0$ and $x < 0$. The origin, $x = 0$, is not included in the domain. The

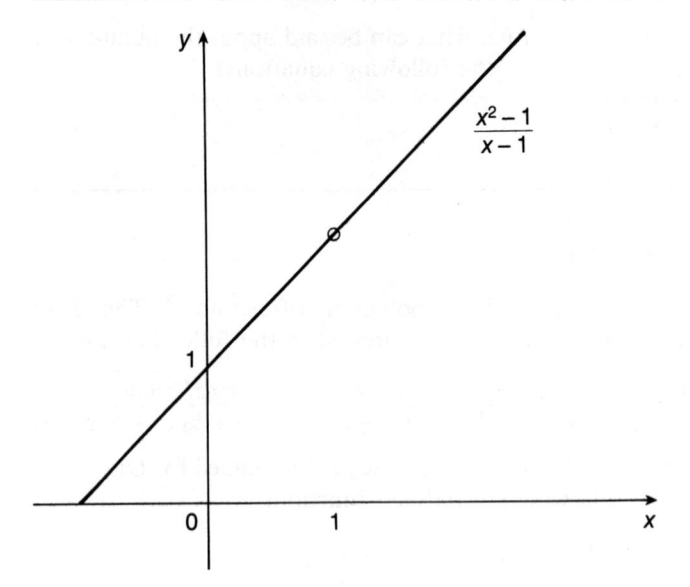

Figure 4.23

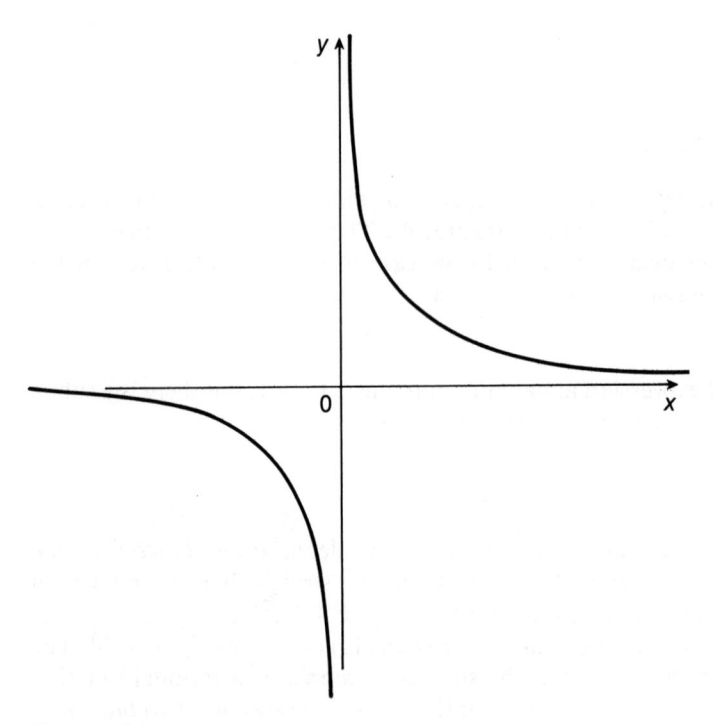

Figure 4.24

reciprocal function is an example of a *rectangular hyperbola*. For this function the *x* and *y* axes are *asymptotes* – straight lines which the function approaches but never quite reaches. Hyperbolas always plot in two sections with asymptotes, and in the case where the asymptotes are orthogonal the hyperbola is said to be *rectangular*. Note that the reciprocal function, by cross-multiplication, can be written as

$$xy = 1$$

or, apparently rather obscurely, as

$$(x + 0)(y + 0) - 1 = 0$$

The purpose of this manipulation is to point out that a rectangular hyperbola having asymptotes located at $x = -a$ and $y = -b$ can be written as

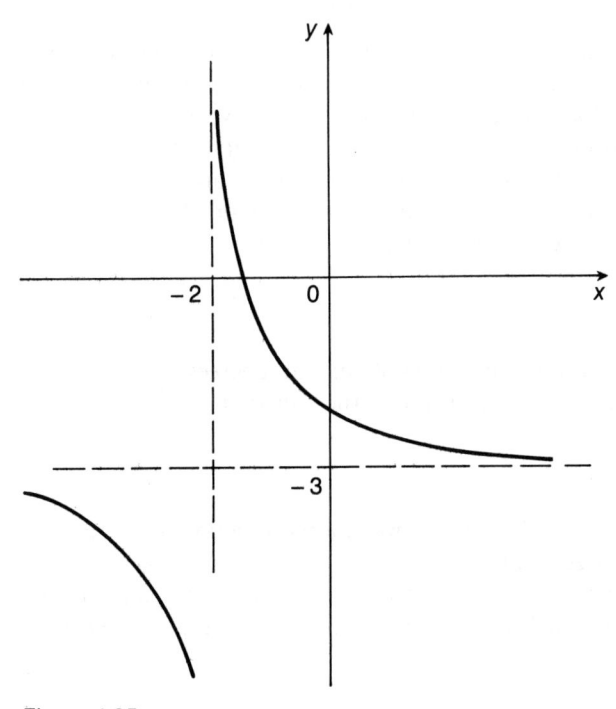

$$(x + a)(y + b) - c = 0 \tag{4.16}$$

Figure 4.25

So, for example, the rectangular hyperbola defined by the equation

$$(x+2)(y+3) - 1 = 0$$

which expands as

$$xy + 2y + 3x + 5 = 0$$

has asymptotes given by $x = -2$ and $y = -3$. This function is graphed in Figure 4.25.

A sixth step can now be added to the curve sketching procedure outlined in Section 4.2: the identification of any asymptotes of the function. The general equation of the hyperbola is given by the general quadratic equation (4.8) for values of A, B and C such that

$$B^2 > 4AC$$

The reciprocal function is a special case in which

$$A = C = D = E = 0 \qquad B = +1 \qquad F = -1$$

Example (iv) provides a quadratic illustration of the fact that any polynomial is also a rational function corresponding to the case where $Q(x)$ is of degree zero. Rational functions produce a wide variety of graphs. Indeed, rational functions going no further in complexity than the ratio of linear and quadratic expressions can have very varied appearances. Two examples are shown in Figures 4.26(a) and 4.26(b). Figure 4.26(a) is called *Newton's Serpentine*, the equation of which is given by

$$y = \frac{4x}{x^2 + 1}$$

and which represents a ratio of polynomials of degrees one and two respectively. Figure 4.26(b) is the graph of the function

$$y = \frac{8}{4 + x^2}$$

and is known as the *Witch of Agnesi*. This rational function is the ratio of polynomials of degrees zero and two.

Note, for completeness, that an expression which has both a rational part and an integral part is called a *mixed expression*, an example of which is

$$3x^2 + 5x + \frac{4}{x}$$

In a mixed expression, the term which dominates the value of the function[23] may vary depending on the value of x. For example, in the function

$$f(x) = ax + \frac{b}{x}$$

Functions and turning points 237

(a)

(b)

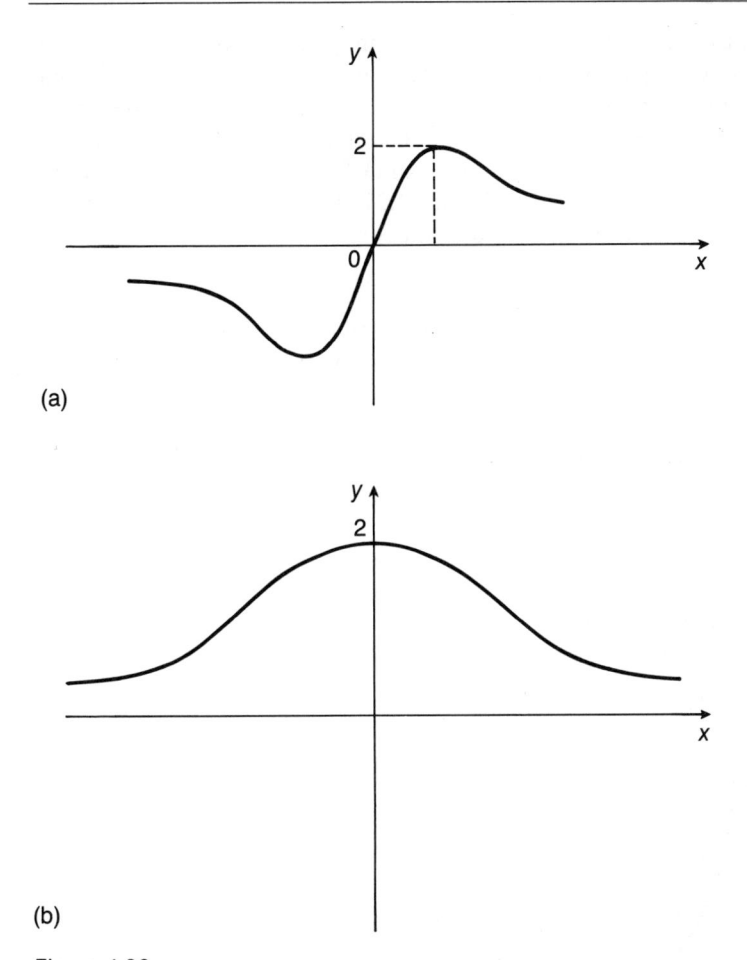

Figure 4.26

in which $a \neq 0$ and $b \neq 0$, when x is large the dominant term is ax, but when x is sufficiently close to zero the term b/x will dominate.

To illustrate the use of rational functions in economics consider *hyperbolic demand curves*. If the relationship between price and quantity is given by

$$p = \frac{k}{q} \qquad (4.17)$$

where k is a positive constant, the demand curve produced is a rectangular hyperbola with the axes as asymptotes. The curve has the special property that, since $pq = k$, the total expenditure of consumers is constant for all values of q in the domain ($q > 0$). This means that demand is *unit elastic* throughout the entire length of the curve. It may be argued that the demand curve (4.17), while plausible for moderate values of q, is unrealistic in that there is no finite limit to either price or quantity. These reasonable points can be met by use of a demand curve based on (4.16). This produces a hyperbolic curve which has definite maxima for both price and quantity. Such an example is displayed in Figure 4.27, for which $a > 0$ and $b > 0$.

The domain of the function is defined for

$$0 \leqslant q \leqslant q^0$$

where q^0 represents the *market saturation* level of demand if the good were free. With the equation linking price and quantity in this case being

$$(q + a)(p + b) = c$$

it will be seen that setting $p = 0$ produces

$$q^0 = \frac{c - ab}{b}$$

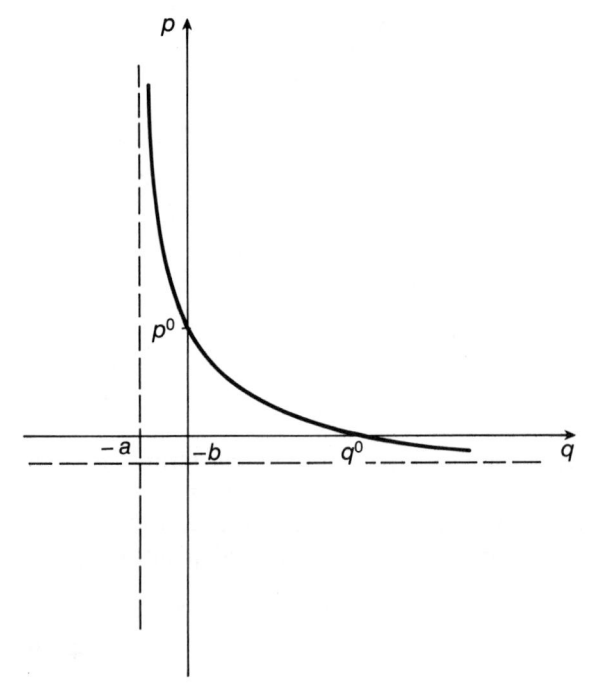

Figure 4.27

while the lowest price that will choke off all demand is p^0 where

$$p^0 = \frac{c - ab}{a}$$

The fact that consumer expenditure varies along this curve can be deduced from the fact that, while revenue is zero at p^0 and q^0, it is strictly positive at points in-between. As a numerical example, suppose that the demand curve is given by

$$(q + 5)(p + 8) = 240$$

Then the market saturation level of demand will be

$$q^0 = \frac{240 - 40}{8}$$
$$= 25$$

while the lowest price that eliminates all demand will be

$$p^0 = \frac{240 - 40}{5}$$
$$= 40$$

An example of a rectangular hyperbola in a business context is provided by the graph of overhead per unit output, *average fixed costs* – AFC. This concept is important in business, especially for firms which produce capital-intensive products. Similar concepts are also used as measures of efficiency or performance indicators. An example of the latter usage within the public sector is the Higher Education Funding Council's 1992 measure of a university's administrative costs per full-time-equivalent student. The function

$$\text{AFC} = \frac{1{,}000{,}000}{q}$$

might represent average fixed costs for an organization, and the graph of the branch of the hyperbola for $q > 0$ is shown in Figure 4.28.

Low levels of output mean that the price of the product must reflect a disproportionately large element of overhead, particularly if the variable costs are not great. A classic example of this situation is oil production and other examples are found in many public utilities such as railways. In these instances, increasing output level will mean that prices can be sharply reduced or high short-term profits can be made if the producer is a monopolistic provider.

AFC

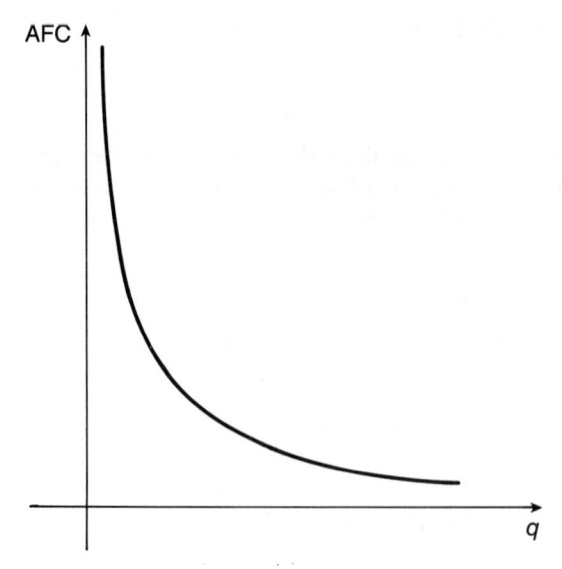

Figure 4.28

EXERCISES 4.6

1 Provide the extended function for the rational function

$$f(x) = \frac{x^2 - 4}{x - 2}$$

2 For the rectangular hyperbola

$$xy - y - 2x + 2 = 0$$

find the location of the asymptotes.

3 For the demand curve given by

$$(q + 10)(p + 20) = 5000$$

find

(i) the market saturation level of demand;
(ii) the lowest price that will choke off all demand.

4 If a firm's fixed costs are £500,000, what minimum level of output is needed to ensure that fixed costs per unit do not exceed £2.50?

4.7 FUNCTIONS: FURTHER CONSIDERATIONS

In this section we take up a number of topics concerning functions of which it is important to be aware, and to which we shall need to refer in later work. The subjects are as follows:

1 algebraic functions
2 composite functions
3 inverse functions
4 continuous and discontinuous functions
5 homogeneous functions
6 monotonic functions
7 other categories of function

4.7.1 Algebraic functions

Polynomials in a single independent variable, x, and rational functions of x are examples of *algebraic functions*. An algebraic function satisfies a polynomial equation $f(x, y) = 0$ in which both x and y may be raised to integral powers. One way of writing this is

$$P_n(x)y^n + P_{n-1}(x)y^{n-1} + P_{n-2}(x)y^{n-2} + \cdots + P_1(x)y + P_0(x) = 0$$

where $P_0(x), \ldots, P_n(x)$ are polynomials in x. For example, the function

$$y = (x^2 + 1)^{1/n}$$

is an algebraic function. Raising both sides to the nth power produces

$$y^n = x^2 + 1$$

and minor rearrangement gives the $f(x, y) = 0$ form:

$$y^n - x^2 - 1 = 0$$

Other examples of algebraic functions which are not rational functions or polynomials are

$$y = (x^4 - x^3 + x^2 + 10)^{1/2}$$

and

$$y = (x + 2x^2 + 3x^3)^{1.5}$$

Now consider the function

$$y = \frac{x}{(x-7)^{1/3}}$$

This is also algebraic, since rearrangement shows that it satisfies an equation of the required form. Proceed thus: first cube both sides. The result is

$$y^3 = \frac{x^3}{x-7}$$

from which cross-multiplication produces

$$y^3(x-7) = x^3$$

which in implicit form is

$$(x-7)y^3 - x^3 = 0$$

so that with $n = 3$ the polynomials are

$$P_3(x) = x - 7 \qquad P_2(x) = 0 \qquad P_1(x) = 0 \qquad P_0(x) = -x^3$$

and the equation is therefore an algebraic equation. Note, however, that functions such as $f(x) = a^x$, in which the independent variable is included in the exponent, are *non-algebraic* or *transcendental*.[24] Transcendental *numbers* were so named by Euler since they 'transcend' the power of algebraic processes. By this is meant that such numbers cannot appear as roots of an algebraic equation with rational coefficients. We stated in Chapter 1 that the irrationals π and e[25] were examples of transcendental numbers. By contrast, an irrational such as $\sqrt{2}$ is algebraic.

4.7.2 Composite functions

A *composite function* is a function of a *function*, rather than a function of an independent variable. For example if we begin with the relationships

$$y = 2x + 1 \qquad (4.19)$$

and

$$z = 3y - 5 \qquad (4.20)$$

then the variable z is a composite function of x. Note that z *ultimately* depends on an independent variable, and in the present example this relationship is easily worked out. Substituting for y in (4.20) from (4.19) yields the result

$$z = 3(2x + 1) - 5$$
$$= 6x - 2$$

As another example of a composite function, consider z as a function of x where

$$y = 5x + 4$$
$$z = 3y^2$$

Thus by substitution

$$z = 3(5x + 4)^2$$

This 'two-stage' relationship between z and x can be expressed in the alternative notational scheme as

$$z = f[g(x)]$$

where the function $g(x)$ is

$$g(x) = 5x + 4$$

and where the function f is 'squaring and multiplying by 3'. Note that it is possible to consider composite functions which involve more than one function of a function. Thus if

$$w = 5x + 4$$
$$y = 2w - 6$$
$$z = 0.5y + 7$$

it follows that

$$y = 10x + 2$$

and therefore that

$$z = 5x + 8$$

A check on this result is provided by taking a particular value of x, working the original equations through and then verifying this outcome by use of the composite function. So, if $x = 2$, then from the original equations $w = 14$ and hence $y = 22$ and $z = 18$, which result is confirmed by substitution of $x = 2$ directly into the composite function.

4.7.3 Inverse functions

The inverse of a function is the function which 'undoes' the operations of the original function. For example, if the operation of the original function is 'doubling' the inverse function will be 'halving'; if the original function is 'squaring', the inverse function is taking the square root (of appropriate sign). Suppose that the original function is doubling and adding 5. The inverse function is then halving and subtracting 2.5. These operations can be written as

dependent variable = 2(independent variable) + 5 (4.19a)

dependent variable = 0.5(independent variable) − 2.5 (4.19b)

Clearly (4.19a) and (4.19b) are *mutual* inverses. Inverse functions can be used in two ways. By far the most common usage in business and economics is to describe a given relationship between two variables with, as may best serve the purpose to hand, an exchange of roles between the dependent and the independent variables. Suppose that in (4.19a) the dependent variable is cost, C, and that the independent variable is output, q, so that the cost function is

$$C = 2q + 5 \qquad (4.20)$$

which is the most convenient form for ascertaining the cost level consequent upon a decision to operate at a specific output level. However, suppose that the relationship relates to the output at a particular plant. Suppose also that the plant manager is concerned that head office may set a budget which implies a particular level of output. The inverse function is more convenient for the purpose of knowing the output consequences of any given level of cost. This is

$$q = 0.5C - 2.5 \qquad (4.21)$$

Thus a budget of $C = 500$ would indicate an output rate of 247.5 per period time. Inverse functions are frequently used in the analysis of demand and supply. The *demand function* states quantity purchased as a function of price charged (and in general other factors too). Suppose that the demand function is

$$q = 1000 - 0.25p \qquad (4.22)$$

The *demand curve*, however, is conventionally written with price as the dependent variable. This will therefore be the inverse function obtained by rearranging (4.22) with p as the subject of the equation. Thus

$$p = 4000 - 4q \qquad (4.23)$$

In terms of notation, if the original function is

$$y = f(x) \qquad (4.24)$$

the relationship can be expressed with x as the subject by using the inverse function and writing

$$x = f^{-1}(y) \qquad (4.25)$$

where in (4.25) f^{-1} represents the inverse function. It is important to note that f^{-1} does *not* mean the reciprocal of the original function ($1/f(x)$). In equations (4.20) and (4.21), as in the case of (4.22) and (4.23), the same underlying relationship is being described. However, if the roles of dependent and independent variables are exchanged *and* the function is inverted as a separate step, the two variables retain their original roles and a different functional relationship results. For example, starting with

$$y = 2x + 5 \tag{4.26}$$

first substitute x for y and y for x giving

$$x = 2y + 5 \tag{4.27}$$

Now invert the function:

$$y = 0.5x - 2.5 \tag{4.28}$$

The function (4.28) involving, as it does, halving the dependent variable and subtracting 2.5, is the inverse of the function (4.26). In fact (4.28) will be the *mirror image* of (4.26) in the 45° line ($y = x$). This fact is illustrated in Figure 4.29, in which the straight lines intersect where $y = x = -5$.

Also shown in Figure 4.29 is the relationship between $y = x^2$ (for $x \geqslant 0$) and $y = +\sqrt{x}$. In the latter case, it will be recalled that $y = \sqrt{x}$ (with roots of both signs allowed) is not technically a function, but by restricting the domain of $y = x^2$ to non-negative values of x, we ensure that the inverse relationship meets the function criterion. In terms of the notation for inverse functions, if equation (4.26) represents a specific case of $y = f(x)$, then in (4.28) we are setting out $y = f^{-1}(x)$ as distinct from $x = f^{-1}(y)$. A test that two functions are inverses is provided by taking the composition of the two functions. For inverse functions

$$f^{-1}[f(x)] = x$$

i.e. if the inverse function is applied to the y value produced by a given x, then the original x value will be returned. If this is not the case then the two functions are not inverses. For the inverse function $f^{-1}(x)$ to exist, each element in the range of $f(x)$ must correspond to a unique element in the domain. This is in addition to the requirement for $f(x)$ to be a function that each x value in the domain gives a unique output from $f(x)$. Such functions are said to be *one-to-one*, a property possessed by always-increasing functions and always-decreasing functions, which are the only continuous functions for which inverses exist.[26]

4.7.4 Continuous and discontinuous functions

A *continuous function* of a single variable is one for which the graph of the function does not display any vertical or horizontal gaps. It is possible to draw the graph of a continuous function without lifting pencil from paper.

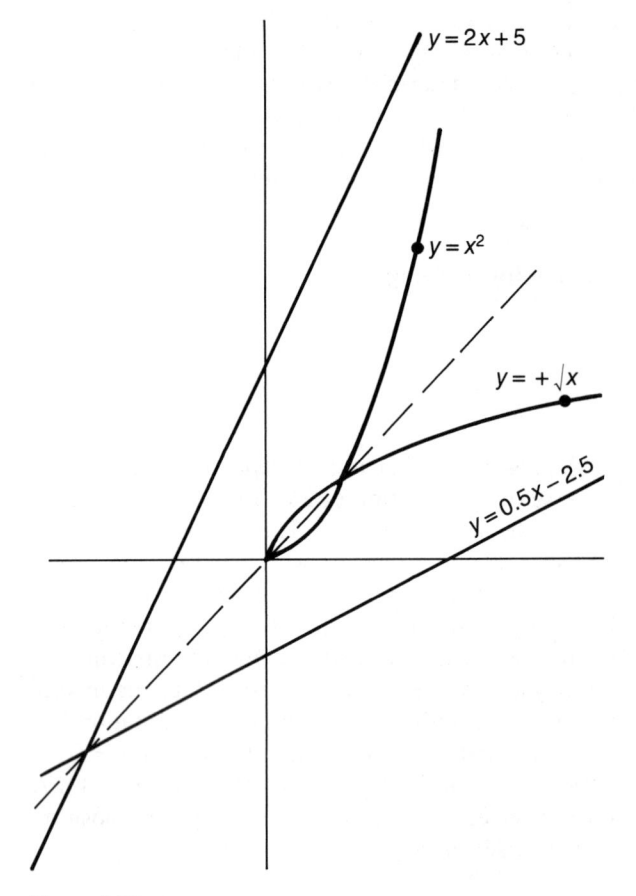

Figure 4.29

This means that the function must have no gaps when viewed from either axis. Continuous does not necessarily mean *smooth*, since a continuous function can have sharp corners or cusps with possible awkward consequences.[27]

Some important results follow if it is known that a function is continuous. Of special significance in the context of business and economic problems are the following:

1 If a function $f(x)$ is continuous at every point in a closed interval $[a, b]$, then the function *must* have an absolute maximum and an absolute minimum within the interval.
2 Where a and b are particular values of x, if $f(a)$ and $f(b)$ are of opposite sign, then the function *must* cross the x axis between a and b.

The import of consequence 1 is that the dependent variable within the interval must be contained within certain limits, while 2 means that there must be at least one real root between *a* and *b*. It can be shown that all polynomial functions are continuous at all points and rational functions are continuous at all of the points for which they are defined. It is also the case that sums, differences, products and quotients of continuous functions are also continuous. Indeed, a continuous function of a continuous function is continuous. Unless otherwise specified, we shall presume continuity (at least in the regions of interest) for functions used in the following chapters. This is not a sweeping assumption as many business and economic applications use polynomials. Functions which are not continuous at all points are often perfectly manageable for certain purposes. This is just as well, since many interesting and realistic situations in business and economics involve discontinuity. A *discontinuous function* is one in which at certain points there is an abrupt jump in the value of, or the absence of a value for, one variable as the other variable is continuously changed. Figure 4.30 illustrates a case of a *step discontinuity* against the *y* axis at $x = x^0$.

Many examples of step discontinuities occur in economics and business. For example, 'welfare traps' represent step discontinuities in the tax–benefit system. In any arrangement in which, at a threshold level, the rate of benefit or deduction changes and where the changed rate applies to *all*

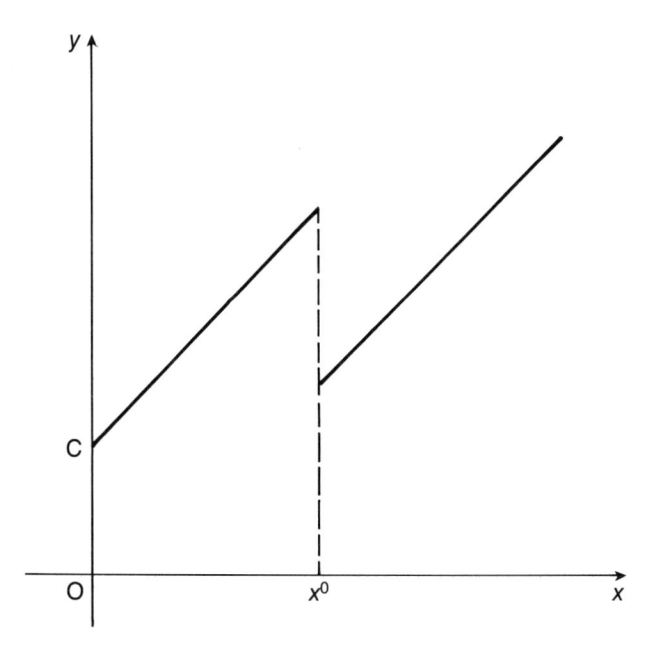

Figure 4.30

income (rather than that beyond the threshold level or between levels) there will be a built-in step discontinuity. Figure 4.30 could illustrate a particularly severe welfare trap. Let x represent wages and y represent total income inclusive of benefit. In the case shown, the full benefit (OC) is paid up to the wage level x^0 and then abruptly withdrawn at that point.[28] A business example of a step discontinuity is *price breaks*. A price break occurs when the unit cost of an item changes at a particular level of sales. In practice the new price usually applies to all items purchased rather than only those units beyond the threshold. The graph of total cost of purchases against quantity bought will then exhibit step discontinuities at the price break levels. The diagram that results has the appearance of Figure 4.31. In Figure 4.31, C represents cost and two price breaks are shown at sales levels q_0 and q_1.

A *step function* is one which takes different constant values for successive intervals in the domain. Such a pattern arises in the graph of marginal cost of production for a firm with a resource-constrained linear technology. At first, the cheapest possible production process is used, but as the required level of production is increased the first of the resource constraints is met. This will require the introduction of a less economical process which requires less of the resource at the limit of its availability. The marginal cost jumps at this point. A further jump occurs when the next resource

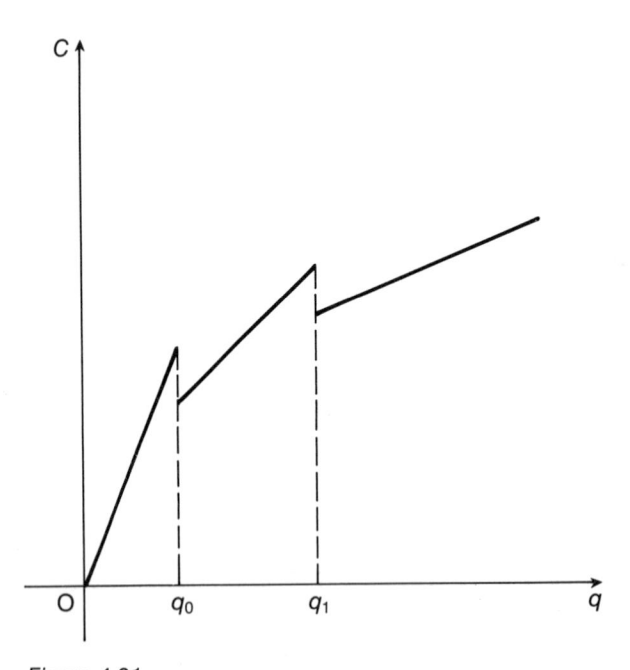

Figure 4.31

constraint becomes binding – calling for the introduction of a still less economical process. The resulting graph is that of a step function, and is illustrated in Figure 4.32.

A *discrete function* is a particular instance of a non-continuous function and is defined for discrete points in the domain.[29] Such a function is discontinuous everywhere. Discrete functions occur in business, economics and finance, but are often adequately approximated by continuous functions. Examples of discrete functions are shown in Figures 4.33(a), 4.33(b) and 4.33(c). Figure 4.33(a) might represent a product demand function for an item which, although sold in individual units, has sufficient volume to make a continuous approximation perfectly acceptable. This is often the case with consumer goods. In contrast, if the commodity was a capital good, a continuous approximation may well be inappropriate. An extreme example in this category is a power station. But there are low-volume consumer goods too. As an example, Figure 4.33(b) might show the costs of production of a highly specialized car in terms of the number of units produced per week. Even when the product is mass produced and costs only pence, the practice of packaging items for sale in quanta other than one unit will present the customer with a discrete unit cost function. For example,

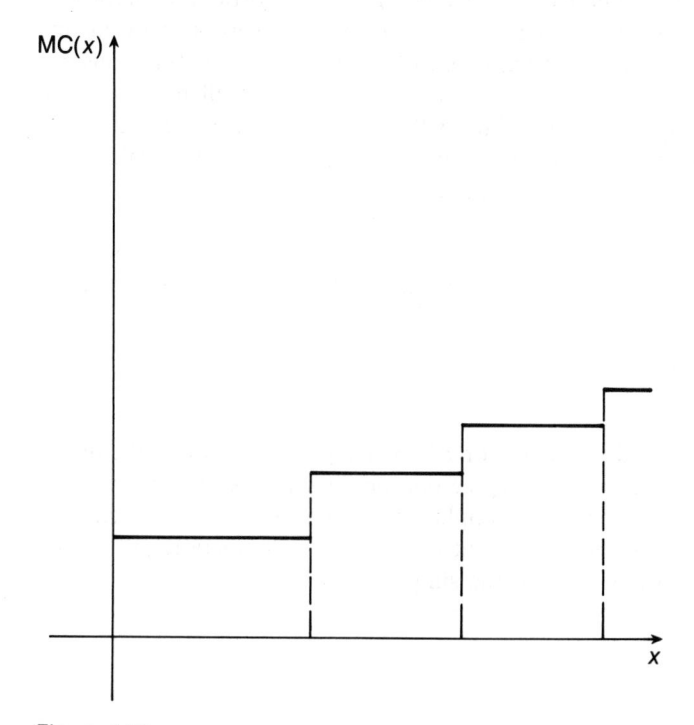

Figure 4.32

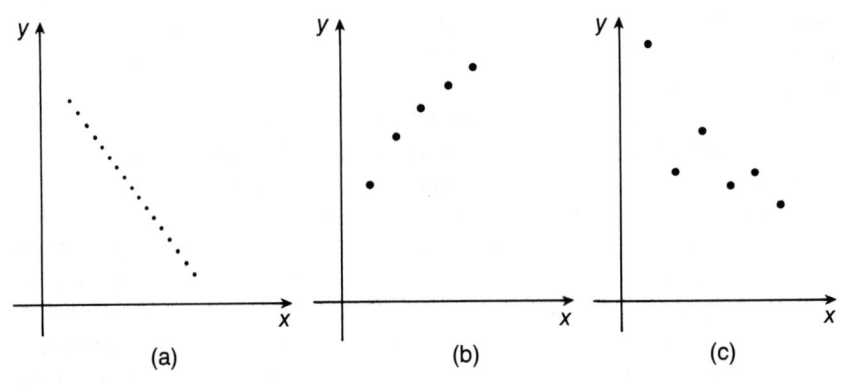

250 Mathematics for business, finance and economics

Figure 4.33

Figure 4.33(c) might represent the average cost to a consumer of items retailed in packets of two, four or six units.

4.7.5 Homogeneous functions

A *homogeneous function* is one in which a given proportionate alteration, p, in the value of the independent variable produces a proportionate change in the value of the dependent variable of p^n. Homogeneous functions are important in the theory of production and cost. We shall introduce the concept of homogeneity here, but will take up the topic again when functions of several variables have been introduced. Consider a straight line through the origin given by the equation

$$y = mx$$

If x is now multiplied by the factor p, the right-hand side of the equation becomes

$$m(px) = pmx$$
$$= py$$

so that y has increased by the factor p^1. A straight line through the origin is said to be homogeneous of degree one or *linearly homogeneous*. Note that the quality of homogeneity applies only to straight lines which pass through the origin. If the y intercept is not zero, the function will not be homogeneous to any degree. In the simplest quadratic equation

$$y = ax^2$$

if x is replaced by px the right-hand side becomes

$$a(px)^2 = ap^2x^2$$
$$= p^2ax^2$$
$$= p^2y$$

so the function $y = ax^2$ is homogeneous of *degree two*. In contrast, the function

$$y = x^2 - 2x$$

is not homogeneous. In this case, although there are specific values of y, p and n that will satisfy

$$(px)^2 - 2(px) = p^n y \tag{4.29}$$

for the function to be homogeneous, the relationship must hold for *arbitrary* positive p and x. The degree of homogeneity of a function need not be a whole number. For example, in the operation of taking a square root, a function results if only the roots of a specific sign are taken. Thus

$$y = +\sqrt{x}$$

is a function. Now multiplication of x by the factor p produces on the right-hand side

$$+\sqrt{px} = (px)^{1/2}$$
$$= p^{1/2}x^{1/2}$$
$$= p^{1/2}y$$

so, with the positive root of p also exclusively taken, the square root function is homogeneous of degree one-half. The degree of homogeneity need not be a positive number. For instance the reciprocal function $y = 1/x$ is homogeneous of degree -1. In the case of functions of several variables (as will be seen in a later chapter) degree zero homogeneity can be a convenient property. To illustrate, consider the breakeven formula

$$q = \frac{F}{p-b}$$

and let all of the parameters F, p and b be scaled by a factor t. The result is

$$\frac{tF}{tp - tb} = \frac{tF}{t(p-b)}$$
$$= \frac{F}{p-b}$$
$$= q$$
$$= qt^0$$

so the breakeven output level is unchanged. Thus, the breakeven output level is homogeneous of degree zero in fixed cost, unit cost and price when the cost function is linear and the demand curve is horizontal.

4.7.6 Monotonic functions

A function which is always increasing or always decreasing is *monotonic*. If the function always increases, the function is said to be *isotonic*. When a function is isotonic, for two values of x, x_2 and x_1, whenever

$$x_2 > x_1$$

it is always the case that

$$f(x_2) > f(x_1)$$

A distinction is drawn between *strictly isotonic* functions, for which the strict inequality holds as above, and *weakly isotonic* functions for which

$$f(x_2) \geq f(x_1) \text{ for all } x_2 > x_1$$

Functions for which

$$f(x_2) < f(x_1) \text{ whenever } x_2 > x_1$$

are said to be monotonic decreasing or *antitonic*. A similar distinction to that for isotonic functions may be drawn between weakly and strictly antitonic functions. Monotonicity can be defined over *parts* of the domain and different monotonicity may be exhibited in different parts of the domain. Correspondingly we can speak of functions which are *locally* or *globally* or *piecewise* monotonic. The possible monotonicity of a function can be checked by investigating the function describing the slope.[30] The slope function must not change its sign anywhere in the domain if the original function is to be monotonic. Monotonicity is an important property in a number of respects – e.g. in the context of transformations of variables in particular scales of measurement or in terms of the existence of the inverse function.[31]

4.7.7 Other categories of function

A function is said to be an *odd function* if it changes its sign but not its absolute value when the sign of the independent variable is reversed. Therefore, for any odd function,

$$f(-x) = -f(x)$$

An example of a function with this property is the cubic

$$y = 7x^3 + 6x$$

As with all odd functions, this cubic is symmetrical about the origin. Polynomials which have only odd powers of x are odd functions.[32] A function which is unchanged in value when x is reversed in sign is said to be an *even function*. Thus for even functions

$$f(-x) = f(x)$$

and the graph of an even function will be symmetrical about the y axis.[33] An example of an even function is the biquadratic

$$y = -5x^4 - 3x^2 - 10$$

This property is possessed by any polynomial for which all powers are even and where all powers of x have coefficients of the same sign.[34] A function is said to be a *periodic function* if, starting from any point, its value is unchanged when x is increased by a specific increment or any whole multiple of that increment. For a periodic function, therefore,

$$f(x) = f(x + na)$$

where n is any integer and where a is a constant which represents the period of the function. The trigonometric functions $y = \sin x$ and $y = \cos x$ are examples of periodic functions. It can be shown that all continuous and differentiable periodic functions can be expressed as algebraic combinations of sines and cosines. In economics, businesses such as travel agencies or agricultural producers will have strongly periodic elements within their cash flows. In management science, a good example of periodicity is provided by

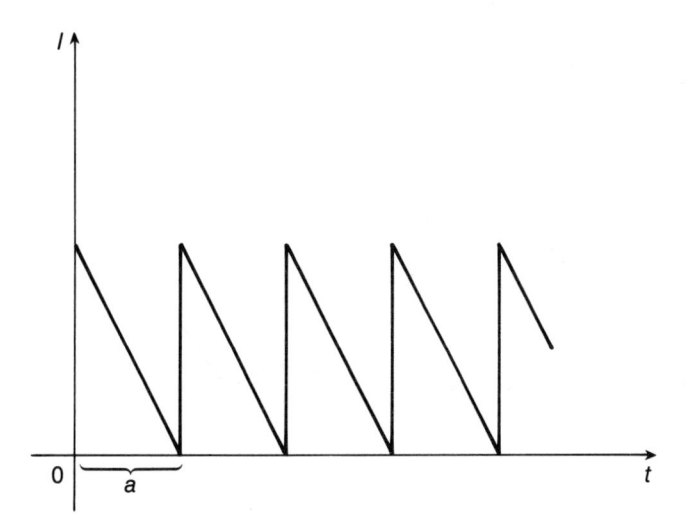

Figure 4.34

the track of inventory level against time in the classical static model of stock control as shown in Figure 4.34.

In Figure 4.34, the variable I represents the level of inventory, t is time and the period of the function, a, is the length of the inventory cycle. Note that the cycle may be taken from peak to peak, trough to trough, or from any given level of stock to that same level – which will re-occur a time periods later. Economic time series data also frequently contain a periodic or cyclical component. The concept of the absolute value (modulus) of numbers was introduced in Chapter 1. An *absolute value function* states the magnitude of an expression *regardless of sign*. The form of such functions is therefore obtained by taking the modulus of $f(x)$ and the function will then appear as

$$y = |f(x)|$$

Absolute value functions are plotted only in quadrants one and two and will change abruptly when the function would otherwise have extended below the x axis into quadrants three and four. Figure 4.35(a) shows the graph of $y = |x|$, while Figure 4.35(b) represents the function $y = |x^2 - c|$.

Absolute values are involved in calculations of distance and are an important concept in computer science. In statistics, absolute values are used in one approach to the problem of measuring the deviation of a variable from its average value. The statistic is called *mean absolute deviation* (MAD). Within management science, such measures of dispersion themselves find use in adaptive forecasting methods – as applied for example in some models in stock control.

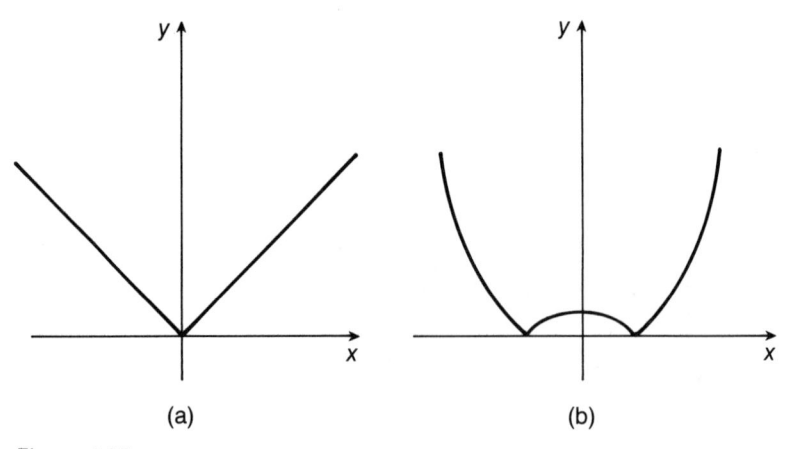

Figure 4.35

EXERCISES 4.7

1. Express z as a function of x where
   (i) $y = 5x - 6$
   $z = 3y + 4$
   (ii) $y = 100 - 4x$
   $z = 100 + 4y$

2. Express z as a function of x where
   (i) $w = 3x + 4$
   $y = 2w + 5$
   $z = 4y - 8$
   (ii) $w = 10 - 2x$
   $y = 3w - 50$
   $z = 100 - 4y$

3. Express z as a function of x where
   (i) $y = 2x + 5$
   $z = y^2 - 4y + 10$
   (ii) $y = x - 2$
   $z = y^3 + 6y^2 + 12y + 8$

4. Find the inverse function stating x as a function of y where
   (i) $y = 4x + 12$
   (ii) $y = 20x - 50$
   (iii) $y = 0.025x - 0.5$
   (iv) $y = +\sqrt{x}$
   (v) $y = 200 - \dfrac{10}{x}$

5. Establish whether or not the following functions are homogeneous, and, if so, state the degree of homogeneity in each case:
   (i) $y = 6x$
   (ii) $y = 6x + 5$
   (iii) $y = 3x^3$
   (iv) $y = x(3x + 4)$
   (v) $y = x^{1/2}$
   (vi) $y = \dfrac{1}{x^2}$

6 Establish whether the following are odd functions, even functions or neither:

   (i) $f(x) = 4x^5 + 2x^3 + 10x$
   (ii) $f(x) = x^4 + 3x^2 + 10$
   (iii) $f(x) = x^6 + x^3 + 10$

4.8 BASIC PROGRAM

In Section 4.2, we considered the simultaneous solution of a mixed system of one linear and one quadratic equation. The following program, *Linquad.bas*, provides the solution to such systems and allows for the possibility of complex roots. The program will run in QuickBasic or Q-Basic. The program listing is as follows.

Program 4.1

```
                        LINQUAD.BAS

DECLARE SUB discrim ()
DECLARE SUB realroot ()
DECLARE SUB compsoln ()
DECLARE SUB inpsub ()
COMMON SHARED a, b, c, d, j, g, h, l, m

CLS

PRINT 'This program finds the simultaneous solution to a'
PRINT 'quadratic equation in the form:'
PRINT
PRINT '                    y = ax² + lx + m'
PRINT
PRINT 'and a linear equation of the form:'
PRINT
PRINT '                         gx + hy = j'
PRINT

INPUT 'Coefficient of x squared '; a
PRINT
INPUT 'Coefficient of x in the quadratic equation '; l
PRINT
INPUT 'Constant term in the quadratic equation '; m
PRINT
INPUT 'Coefficient of x in the linear equation '; g
PRINT
```

INPUT 'Coefficient of y in the linear equation '; h
PRINT
INPUT 'Constant term in the linear equation '; j
PRINT

b = l + g/h

c = m – j/h

d = 0

e = 0

f = 0

realx = 0

imagx = 0

realy = 0

imagy = 0

CALL discrim

END

SUB compsoln

e = – d

f = SQR(e)

realx = – b/(2*a)

imagx = f/(2*a)

realy = (j – g*realx)/h

imagy = (g*imagx)/h

CLS

CALL inpsub

PRINT
PRINT 'The solutions are...'
PRINT

PRINT 'x = '; realx; ' + '; imagx; 'i',
PRINT 'and y = '; realy; ' – '; imagy; 'i'

PRINT

PRINT 'x = '; realx; ' – '; imagx; 'i',
PRINT 'and y = '; realy; ' + '; imagy; 'i'

END SUB

SUB discrim

d = b ^ 2 – 4*a*c

IF d >= 0 THEN

    CALL realroot

ELSE

    CALL compsoln

END IF

END SUB

SUB inpsub

PRINT 'The system you have input is:'
PRINT
PRINT '   y = '; a; 'x^2 + '; l; 'x + '; m
PRINT
PRINT 'and:'
PRINT
PRINT '   '; g; 'x + '; h; 'y = '; j
PRINT
PRINT '--'

END SUB

SUB realroot

f = SQR(d)

x1 = (– b + f)/(2*a)

x2 = (– b – f)/(2*a)

y2 = (j – g*x2)/h

y1 = (j – g*x1)/h

```
CLS

CALL inpsub

PRINT
PRINT 'The solutions are...'
PRINT

PRINT '                    x = '; x1, ' and y = '; y1
PRINT
PRINT '                    x = '; x2, ' and y = '; y2

END SUB
```

The program involves the 'calling' of subprograms, an arrangement which can make the program easier to understand and develop. Once again, at the expense of easy reading, the coding could be tightened up in a number of respects.

ADDITIONAL PROBLEMS

1 The demand curve for a company's product is given by

$$p = 1000 - 5q$$

where p is price and q is output level. The firm's costs are given by

$$C = 200q + 14{,}000$$

Use your knowledge of parabolas and quadratic equations to establish the following:

   (i) the output level that results in the maximum revenue from sales;
   (ii) the output level that produces the greatest possible profit;
   (iii) the output levels at which the company would break even.

2 A survey has revealed that the asbestos fibre content of the atmosphere in a factory is $x = 10$ fibres per unit volume. The cost of obtaining a lower level of contamination is given by the rational function

$$C = \frac{500x + 20{,}000}{x}$$

where C is cost.

   (i) Find the cost of achieving contamination levels of 8, 5, 4, 2, 1 and 0.5 fibres per unit volume.
   (ii) What contamination levels could be achieved for a budget of

   (a) £8500   (b) £80,500

REFERENCES AND FURTHER READING

1 Begg, D., Fischer, S. and Dornbush, R. (1987) *Economics* (Second Edition), McGraw-Hill.
2 Borowski, E. J. and Borwein, J. M. (1989) *Dictionary of Mathematics*, Collins.
3 Boyer, C. B. (1968) *A History of Mathematics*, Wiley.
4 Courant, R. (1937) *Differential and Integral Calculus* (Second Edition), Volume 1, Blackie.
5 Hart, W. L. (1966) *College Algebra* (Fifth Edition), Heath.
6 Hollingdale, S. (1989) *Makers of Mathematics*, Penguin.
7 Samuels, J. M., Wilkes, F. M. and Brayshaw, R. B. (1990) *Management of Company Finance* (Fifth Edition), Chapman and Hall.
8 Weber, J. E. (1982) *Mathematical Analysis: Business and Economic Applications* (Fourth Edition), Harper and Row.

SOLUTIONS TO EXERCISES

Exercises 4.1

1. (i) $x \geqslant -10$
   (ii) $x \leqslant 20$

2. (i) $y \geqslant 0$
   (ii) $y \geqslant -10$

3. (i) $-30 \leqslant y \leqslant 45$
   (ii) $-400 \leqslant y \leqslant 150$
   (iii) $1250 \leqslant y \leqslant 2250$

4. (i) No. While the *slope* of the function shows abrupt change at $x = 30$, there is no step discontinuity in $f(x)$ itself.
   (ii) Yes, the function $f(x)$ has a step discontinuity when $x = 15$.

Exercises 4.2

1. (i) $x = 3$, $x = 4$
   (ii) $x = -6$, $x = 5$
   (iii) $x = -2$, $x = -7$
   (iv) $x = 8$ (double root)
   (v) $x = -9$ (double root)
   (vi) $x = 0$, $x = 7$

   Note that the quadratic expression of (vi) is more conveniently written as

   $$x^2 - 7x = 0$$

   so that

   $$x(x - 7) = 0$$

   i.e.

   $$(x - 0)(x - 7) = 0$$

2. (i) $x = 3$, $x = 4$
   (ii) $x = 1.1$, $x = 1.2$
   (iii) $x = -1.5$, $x = 3$
   (iv) $x = 1$, $x = -100$

3. (i) $x = 10$ (repeated root)

   The discriminant here is zero. The curve touches the x axis at one point (from above, since the leading coefficient is positive).

   (ii) $x = +3i$, $x = -3i$

   The discriminant is negative in this case. Since the value of b in the quadratic formula is zero, the roots are the purely imaginary numbers $3\sqrt{(-1)}$ and $-3\sqrt{(-1)}$. The parabola never reaches the x axis.

   (iii) $x = 5 - 2i$, $x = 5 + 2i$

   Here the discriminant is negative and the roots are the complex conjugate pair, $5 \pm 2i$. The parabola never reaches the x axis, and it has its vertex at $x = 5$.

4. (i) $x = 7$, $y = -8$ and $x = 14$, $y = -36$
   (ii) $x = 20$, $y = -10$ and $x = 12.5$, $y = 12.5$

5. (i) $x = 5$, $y = 60$ and $x = 10$, $y = 260$
   (ii) $x = 0.5$, $y = 16.5$ and $x = 2$, $y = 24$

6. (i) $-4 \leqslant x \leqslant 6$
   (ii) $x \leqslant -5$ $x \geqslant 2$
   (iii) $3 \leqslant x \leqslant 8$

7. (i) The revenue parabola is

   $$R = 700x - 5x^2$$

   for which the vertex is at

   $$x = \frac{-700}{-10} = 70$$

   at which point revenue is

   $$R = 700(70) - 5(70)^2$$
   $$= 24,500$$

(ii) Profits, π, are given by

$$\pi = R - C$$
$$= 700x - 5x^2 - 150 - 40x - x^2$$
$$= 660x - 6x^2 - 150$$

which parabola has its vertex at

$$x = \frac{-660}{-12} = 55$$

at which point profit is

$$\pi = 36{,}300 - 18{,}150 - 150 = 18{,}000$$

8 Equating the value of price given by the supply and demand equations

$$720 - 4q = 160 + 0.25q^2$$

which reorganizes as the quadratic equation

$$560 - 4q - 0.25q^2 = 0$$

i.e.

$$q^2 + 16q - 2240 = 0$$

which factors as

$$(q - 40)(q + 56) = 0$$

Discarding the negative value of q leaves $q = 40$, at which point $p = 560$ from either the supply or the demand equations. These are the required market clearing values.

9 Equating unit cost and price gives

$$0.1x^2 - 80x + 17{,}500 = 10{,}500$$

which results in

$$x^2 - 800x + 70{,}000 = 0$$

which then factors as

$$(x - 100)(x - 700) = 0$$

so that, in order for costs to at least cover revenue, output must be in the interval

$$100 \leqslant x \leqslant 700$$

Functions and turning points 263

Exercises 4.3

1. (i) $x = -5$, $x = 0$, $x = 10$
   (ii) $x = 0$, $x = 1$

   Since there is no constant term in either case, $x = 0$ must be one root of each equation. In case (i), taking x out, the remaining quadratic

   $$x^2 - 5x - 50 = 0$$

   solves for

   $$x = -5 \text{ and } x = 10$$

   In case (ii), $x = 1$ is a double root.

2. (i) $x = \pm 2$, $x = \pm 3$
   (ii) $x = \pm 1$, $x = \pm 6$

   Note that these quartics are *biquadratics*, and can be solved as quadratics in x^2.

3. Average variable costs are

   $$\frac{VC}{x} = x^2 - 90x + 4000$$

   and the parabola for this equation opens upwards and has its vertex, a minimum, at

   $$\frac{-(-90)}{2} = 45$$

   at which point total variable costs are

   $$VC = (45)^3 - 90(45)^2 + 4000(45) = 88{,}875$$

4. Profit is given by the equation

   $$\pi = -x^4 + 5000x^2 - 1{,}250{,}000$$

   Since this equation is a biquadratic, letting $z = x^2$ produces

   $$\pi = -z^2 + 5000z - 1{,}250{,}000$$

   which is a parabola in z opening down so that the vertex represents a maximum. This is located at

   $$z = \frac{-5000}{-2} = 2500$$

   so that, discarding the negative root, $x = 50$, at which point

   $$\pi = -(50)^4 + 5000(50)^2 - 1{,}250{,}000 = 5{,}000{,}000$$

Exercises 4.4

1 (i), (ii), (iv) and (vi) are polynomials.

Note: Recall that only non-negative integers are allowed as exponents. So the term $3x^{-2}$ disqualifies (c) and the term $x^{1/2}$ disqualifies (e).

2 (i) The degree is 5; the polynomial is a quintic.
  (ii) The degree is 4; the polynomial is a quartic.
  (iii) The degree of the polynomial is 1.
  (iv) The degree of the polynomial is 0.

Exercises 4.5

1 (i) Maximum number of positive roots 3.
  (ii) Maximum number of positive roots 2.
  (iii) Maximum number of positive roots 3.

2 (i) One positive root, maximum of two negative roots.
  (ii) Maximum two positive roots, one negative root.

Note: In the case of part (i), the polynomial in x has one sign change and, because of the equal in parity requirement, must have precisely one positive root. The polynomial in $-x$ has two sign changes, so that there will be either two or no negative roots.

Exercises 4.6

1 The piecewise-defined extended function is

$$f(x) = \begin{cases} (x^2 - 4)/(x-2) & \text{for } x \neq 2 \\ 4 & \text{for } x = 2 \end{cases}$$

2 The given equation for the hyperbola

$$xy - y - 2x + 2 = 0$$

can be re-expressed as

$$(x-1)(y-2) = 0$$

from which the asymptotes are located at $x = 1$ and $y = 2$.

  (i) Market saturation occurs when $p = 0$, at which point

$$q^0 = \frac{5000 - 200}{20} = 240$$

  (ii) The lowest price that eliminates all demand will be

$$p^0 = \frac{5000 - 200}{10} = 480$$

Functions and turning points

4 Per unit fixed costs are given by the rational function

$$AC = \frac{500{,}000}{q}$$

so that an output level of at least $q = 200{,}000$ would be needed in order to ensure that per unit fixed costs do not exceed £2.5.

Exercises 4.7

1. (i) $z = 15x - 14$
   (ii) $z = 500 - 16x$

2. (i) $z = 24x + 44$
   (ii) $z = 24x + 180$

3. (i) $z = 4x^2 + 12x + 15$
   (ii) $z = x^3$

4. (i) $x = 0.25y - 3$
   (ii) $x = 0.05y + 2.5$
   (iii) $x = 4y + 20$
   (iv) $x = y^2$
   (v) $x = \dfrac{10}{200 - y}$

5. (i) Homogeneous of degree one.
   (ii) Not homogeneous.
   (iii) Homogeneous of degree three.
   (iv) Not homogeneous.
   (v) Homogeneous of degree one-half.
   (vi) Homogeneous of degree minus two.

6. (i) $f(x)$ is an odd function.
   (ii) $f(x)$ is an even function.
   (iii) $f(x)$ is neither an odd nor an even function.

Additional problems

1. (i) The revenue parabola is given by

   $$R = 1000q - 5q^2$$

   which has its vertex at

   $$q = \frac{-1000}{2(-5)} = 100$$

   The parabola opens down, the vertex giving a maximum of $R = 50{,}000$.

(ii) The equation for profit, π, is

$$\pi = \text{Revenue} - \text{cost}$$
$$= 1000q - 5q^2 - (200q + 14{,}000)$$
$$= -5q^2 + 800q - 14{,}000$$

which has its vertex at

$$q = \frac{-800}{2(-5)} = 80$$

at which point profit is maximized at $\pi = 18{,}000$.

(iii) At breakeven output, total revenue is equal to total cost, so

$$1000q - 5q^2 = 200q + 14{,}000$$

which condition gives the quadratic equation

$$5q^2 - 800q + 14{,}000 = 0$$

Applying the quadratic formula, the roots are

$$q = \frac{800 \pm \sqrt{[(-800)^2 - 4(5)14{,}000]}}{2(5)}$$

so the breakeven levels are

$$q = 20 \text{ and } q = 140$$

2 (i) The results are

x	C
8	3,000
5	4,500
4	5,500
2	10,500
1	20,500
0.5	40,500

(ii) (a) 2.5 fibres per unit volume.
(b) 0.25 fibres per unit volume.

NOTES

1 In the case of functions of several variables.
2 To meet the single outcome requirement, unless otherwise stated we shall assume that when a square root is taken, only the positive root is involved.
3 A *monotonic* function – see Section 4.7.6 below.

Functions and turning points 267

4 Concentrate on the negative values of x. For example, at $x = -4$, $y = x^2 - 2x + 1$ has a slope of -10 while $y = x^2$ has a slope of -8. Since $|-10| > |-8|$, $y = x^2 - 2x + 1$ is steeper in *absolute* terms at this point.

5 Another way of expressing the relationship of the factors to the equation itself is to say that $x^2 - 10x + 2$ is the *common multiple* of $x - 3$ and $x - 7$.

6 The x intercepts represent the zeros of the function, the roots of the quadratic equation.

7 Further significant points for curve sketching will arise in Chapter 5 following coverage of the calculus.

8 There may be more than one term which can be dominant depending on the value taken by x. See Section 4.6.

9 For example, outside of the present context of parabolas, in sketching the curve given by $1/(1-x)$ it is worth having a look at the curve for values of x close to 1.

10 For further details see Hollingdale (1989).

11 Which demonstrates that a particular conic section can produce a straight line rather than a curve. Such a section is said to be *degenerate*.

12 See Hart (1966). Methods for solving cubics and quartics were developed by the Italian mathematician and physician Cardano in the sixteenth century.

13 The same statement also applies to quadratics.

14 A polynomial expression can also be described as *multinomial* in that it contains more than one term, and is therefore distinguished from a *monomial*, or single-term expression, such as $4ax$.

15 In the general form of a polynomial equation, both x and y may be raised to integral powers. We shall concentrate here on polynomials in which only the dependent variable is raised to powers other than unity.

16 Where the slope of the line is finite and non-zero.

17 The general insolubility of quintic equations was proved by the Norwegian Niels Abel in 1821 at the age of 19.

18 Polynomials with complex coefficients can be considered, but these are beyond the scope of this book.

19 Note that a zero coefficient does not count as a sign change.

20 See Samuels, Wilkes and Brayshaw (1990).

21 Older terminology describes polynomials as rational *integral* functions and rational functions as rational *fractional* functions.

22 The function (ii) being undefined at $x = +1$ is not continuous throughout an interval that includes $x = +1$. To 'fill the gap' a *continuous extension* of the function can be defined in the form of the piecewise-defined *extended function*

$$f(x) = \begin{cases} (x^2 - 1)/(x - 1) & \text{if } x \neq 1 \\ 2 & \text{if } x = 1 \end{cases}$$

23 Which means that the value of the whole expression is approximately equal to the value of the dominant term alone.

24 The function $y = a^x$ is an *exponential* function and is covered in Chapter 9. Other examples of non-algebraic functions are the logarithm function $y = \log_a x$ (also covered in Chapter 9) and trigonometric functions such as $y = \sin x$.

25 The number e is introduced in Chapter 9.

26 See the discussion of monotonic functions below.

27 A *smooth function* is differentiable at every point. Both the function itself and the first derivative are continuous.

28 A further example of a step discontinuity in economic theory is the marginal revenue function in the kinked demand curve model of oligopoly.

29 A discrete function may be defined for a finite or for an infinite number of points (e.g. the integers).
30 This is the derivative. See Chapter 5.
31 And the corresponding applicability of the *inverse function rule* for differentiation. See Chapter 5.
32 Other examples of odd functions are the trigonometric functions sin x and tan x.
33 And all functions symmetrical about the y axis are even functions.
34 The trigonometric function $y = \cos x$ is a further example of an even function.

Chapter 5

Slopes, Derivatives and Turning Points

5.1	Introduction	270
5.2	Slope and turning points	270
5.3	An approach to the derivative	272
5.4	The power function rule	277
5.5	Differentiating polynomials	283
5.6	The product and quotient rules	290
5.7	The chain rule	295
5.8	The inverse function rule	302
5.9	Implicit differentiation	304
5.10	Higher order derivatives	307
5.11	Local maxima and minima	310
5.12	Global maxima and minima	316
5.13	Concavity, convexity and points of inflection	325
5.14	BASIC program	333
	Additional problems	337
	References and further reading	339
	Solutions to exercises	340

In this chapter, having seen the significance of turning points, you will be introduced to the concept of the derivative of a function — an expression from which the slope of the function can be found and which helps to identify any turning points. Commonly used rules for finding the derivative of a function of a single variable are then studied.

The derivative of a function, and higher order derivatives, are then applied to optimization problems — where the best value of functions is sought. Also considered in this chapter are points of inflection. The chapter concludes with a BASIC computer program to differentiate polynomials.

By the end of the chapter, you will be able to make use of the important rules for finding the derivatives of functions of a single variable; you will also have acquired the means by which to establish the nature of stationary values of polynomial and rational functions, and be able to find local and global optima in a variety of cases.

5.1 INTRODUCTION

In normative models in finance, economics and management science, the aim of the decision-maker is to find the values of the decision variables that produce the optimal value of a function that represents his or her objectives. There are many possible goals that may be relevant. For example, in the case of a manufacturer, it will be of value to know:

1 the output level at which the firm's unit costs of production are at a minimum;
2 how to combine processes so that a given output level is produced at minimum cost;
3 how best to apportion a scarce factor of production between possible products;
4 the volume of product sales at which company turnover peaks;
5 how to maximize the value of company profits or return on capital.

Addressing these problems in analytical terms is not simply a matter of finding solutions, but also involves finding the properties that characterize an optimum or equilibrium position. The analytical results make clear important concepts such as marginal revenue, marginal cost and elasticity of demand and it brings out their often complex interrelationships. Although decision-makers may seek a standard of performance that is merely satisfactory – such as the achievement of a predetermined target rate of return, or the sales growth or profit norm for the industry – there are few convincing reasons for not going for an optimum level of performance once all the relevant considerations have been taken into account. An optimum position usually corresponds to a maximum or minimum value of a function, but this covers a multitude of cases – e.g. the presence or absence of constraints – many of which we shall consider in later chapters. At present our task is to identify the properties possessed by maximum and minimum points and to use these properties to establish a powerful means of finding the extreme point themselves.

5.2 SLOPE AND TURNING POINTS

Consider the function shown in Figure 5.1. The expression, if any, that would produce a function of this general appearance would be of daunting complexity, but it is only the graph that is of interest to us here.

For values of x which are large in absolute terms, the function has the x axis as an asymptote and approaches this from above. The function has seven *turning points* at values of x from x_1 to x_7. It is clear that x_1, x_3, x_5 and x_7 produce values of the function which are the largest in their own neighbourhood – if not necessarily the largest overall. The values of the function at these points are said to be *local maxima*. The values of $x = x_2$,

Slopes, derivatives and turning points 271

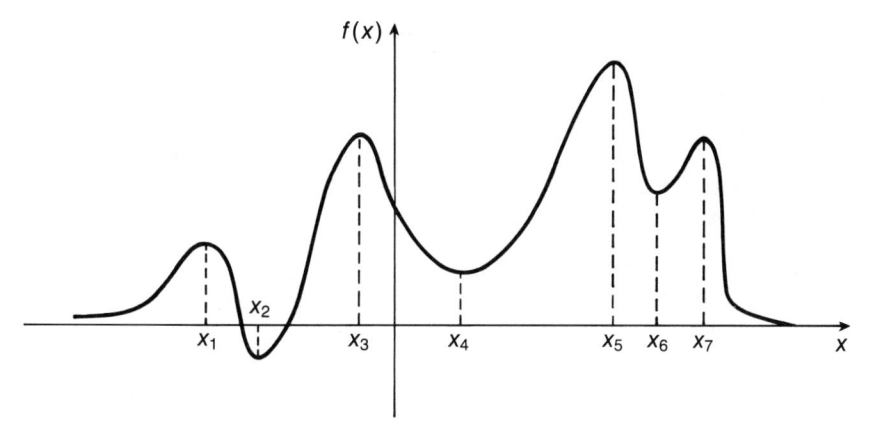

Figure 5.1

x_4 and x_6 produce the smallest value of the function in their immediate vicinity. The values of the function at these points are therefore *local minima*. Turning points can be either maxima or minima and are called *stationary points, extreme points* or sometimes *critical points* of the function.[1]

The term *stationary point* highlights the property common to all turning points – the fact that at the precise point itself the function is stationary, it has *zero slope*. If a function does not have zero slope at a point, then it must be increasing for a change in x value in one direction (either an increase in x or a decrease) and falling as x changes in the other direction. Consequently, the function cannot have either a maximum or a minimum value at the point itself. Zero slope is a *necessary condition* for a local maximum or minimum of a function. If a *maximum* is required, zero slope, while essential, is not enough on its own – it does not represent a *sufficient condition*. We will take up this point later on. All of the values $f(x_1), f(x_3)$, $f(x_5)$ and $f(x_7)$ are local maxima, but $f(x_5)$ alone is the *global* maximum. This is the highest value that is taken by the function at any point in the domain. Similarly $f(x_2)$ is the global minimum of the function. For a point to be the global maximum of a function with an unrestricted domain, the value of $f(x)$ at this point must be greater than at the other local maxima *and* be greater than $f(x)$ as x increases or decreases without limit;[2] that is, as

$x \to +\infty$ and as $x \to -\infty$

Where the domain of the function is restricted, it is necessary that the turning point that is the putative global maximum must not be less than the value of $f(x)$ at the endpoints of the domain.[3] The usual purpose of the exercise is to locate either the global maximum or the global minimum. In

272 Mathematics for business, finance and economics

the function of Figure 5.1, the global maximum is *unique*, but it may be the case that the overall highest value of the function occurs at *more* than one x value.[4] Such cases of *non-uniqueness* give the decision-maker an extra choice, and it may be clear that one of the points is more desirable on grounds that it was not appropriate to quantify in the function itself. On the basis of the discussion so far, a three-part strategy suggests itself for locating the global maximum of a function. Specifically:

1 locate all of the turning points;
2 distinguish between turning points representing local maxima and those which produce local minima;
3 find the global maximum from amongst the local maxima.

The first step requires a means by which to determine all values of x for which the function has zero slope. This, in turn, could be accomplished if

(a) an expression could be obtained which gives the slope of the function at each value of x, and
(b) this expression were set equal to zero and solved for x.

The values of the function corresponding to these x values include all the local maxima and minima produced by turning points.[5] So equipped, the procedures can be applied to business and economic models, e.g. finding the revenue or profits maximizing output levels for a firm, the level of use of plant that produces a minimum of unit cost or numerous other applications.

5.3 AN APPROACH TO THE DERIVATIVE

To move closer to finding an expression to state the slope of a function, consider the most convenient non-linear function, the squaring function:

$$y = x^2$$

Take any point P on this function. As shown in Figure 5.2, the slope of the function at point P is approximately equal to the slope of the straight line connecting P and the nearby point P^0 (a wide separation of P and P^0 is shown for illustrative purposes).[6]

Let the value of the independent variable at P^0 be $x + \Delta x$ where Δx, read as delta x, means a small change in x. The value of the function at P is x^2 and at P^0 it is $(x + \Delta x)^2$. Thus the change in the value of the function as x changes to $x + \Delta x$ will be Δy where

$$\Delta y = (x + \Delta x)^2 - x^2$$

Slopes, derivatives and turning points 273

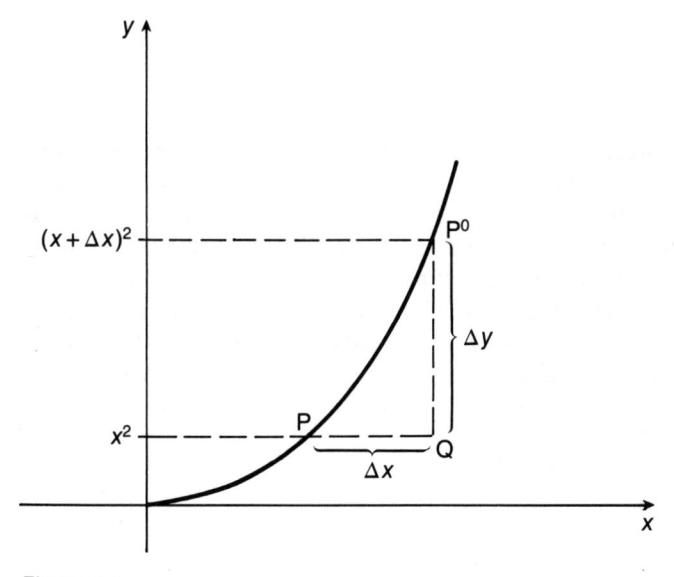

Figure 5.2

The slope of the line PP^0 (which approximates the slope of the function at P) is

$$\text{slope of } PP^0 = \frac{P^0Q}{PQ}$$

$$= \frac{\text{change in value of function}}{\text{change in value of } x}$$

$$= \frac{f(x + \Delta x) - f(x)}{\Delta x}$$

$$= \frac{\Delta y}{\Delta x}$$

This ratio will now be used to describe the slope of the line segment PP^0. From what we have already established we can write

$$\frac{\Delta y}{\Delta x} = \frac{(x + \Delta x)^2 - x^2}{\Delta x}$$

$$= \frac{x^2 + 2x \Delta x + (\Delta x)^2 - x^2}{\Delta x}$$

$$= \frac{2x \Delta x + (\Delta x)^2}{\Delta x}$$

so that

$$\frac{\Delta y}{\Delta x} = 2x + \Delta x \tag{5.1}$$

So $2x + \Delta x$ is the precise slope of the line PP⁰. Now as P⁰ gets progressively closer to P, so Δx becomes ever smaller and (5.1) becomes a better approximation to the slope of the curve itself at the point P. The value of (5.1) as Δx becomes smaller approaches $2x$ and no great leap of the imagination is required to see that the slope of the function itself at P is $2x$. Note that $2x$ is a *formula* which states the slope of the function $y = x^2$ at any point in the domain – this expression is the *derivative* of the function $y = x^2$. Thus when $x = 1$, the derivative gives the slope of $y = x^2$ as 2; when $x = -4$, the slope is -8, and when $x = +150$, the slope is $+300$ and so on. Figure 5.3

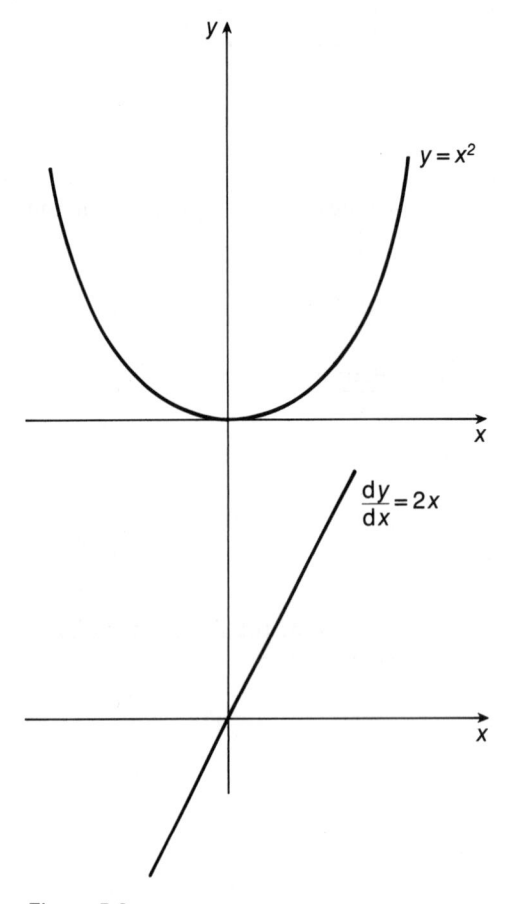

Figure 5.3

Slopes, derivatives and turning points 275

shows the derivative, $dy/dx = 2x$, on a graph set below that of the function $y = x^2$ itself. The same horizontal scale is employed in both cases. This is a useful device that enables significant points on a function to be related to the slope. Note that the *height* of the derivative graph gives the slope of the original function.

At this point we need to introduce some new notation to distinguish the derivative itself from the slope of the chord PP^0. The two most commonly used forms of notation for the derivative are shown in equation (5.2).[7]

$$\frac{dy}{dx} = 2x = f'(x) \tag{5.2}$$

The dy/dx notation for the derivative compares most directly with the *ratio* $\Delta y/\Delta x$, which gave the slope of the chord. This form of notation is due to Leibniz (1646–1716) who developed the calculus independently of Newton (1643–1727). The dy/dx notation is therefore called *Leibniz's notation*. In a similar vein, the derivative of $f(x)$ can also be indicated by

$$\frac{df}{dx} \quad \text{or} \quad \frac{df(x)}{dx}$$

The $f'(x)$ representation of the derivative is called *function notation*. The remaining notation that is commonly employed to indicate the derivative of a function is

$$Df(x) \quad \text{or} \quad D_x f(x)$$

Newton's original notation bears some resemblance to function notation but is now rarely used.[8] Both Leibniz's notation and function notation have their advantages and we shall use them both throughout this book. The approach used above to obtain the derivative of $y = x^2$ could be used, if somewhat laboriously, to obtain the derivatives of higher powers of x. For example, if $y = x^3$, then as before

$$\frac{\Delta y}{\Delta x} = \frac{\text{change in value of the function}}{\text{change in value of } x}$$

$$= \frac{(x + \Delta x)^3 - x^3}{\Delta x}$$

$$= \frac{x^3 + 3x^2 \Delta x + 3x(\Delta x)^2 + (\Delta x)^3 - x^3}{\Delta x}$$

so that

$$\frac{\Delta y}{\Delta x} = 3x^2 + 3x \Delta x + (\Delta x)^2 \tag{5.3}$$

276 Mathematics for business, finance and economics

The derivative dy/dx of $y = x^3$ is the value approached by (5.3) as Δx becomes ever smaller and approaches zero. As Δx approaches zero, $(\Delta x)^2$ approaches zero with greater rapidity. The term $3x\Delta x$ also approaches zero regardless of how large a value x itself takes. For any given value of x (say 1,000,000) a value of Δx can be selected that will make $3x\Delta x$ less than any pre-assigned quantity, however small. Thus if the pre-assigned quantity is, say, 0.00001, then if Δx is set at $\Delta x = 10^{-12}$

$$3x\Delta x = 0.000003 < 0.00001$$

We can therefore write for the derivative of $y = x^3$

$$\frac{dy}{dx} = 3x^2 = f'(x) \tag{5.4}$$

Using (5.4) we can see just how steep the cubing function rapidly becomes. If $x = 1$, the slope is $3(1)^2 = 3$, while if $x = 10$ the slope is $3(10)^2 = 300$. Figure 5.4 shows the graph of $dy/dx = 3x^2$ against the same horizontal scale as $y = x^3$.

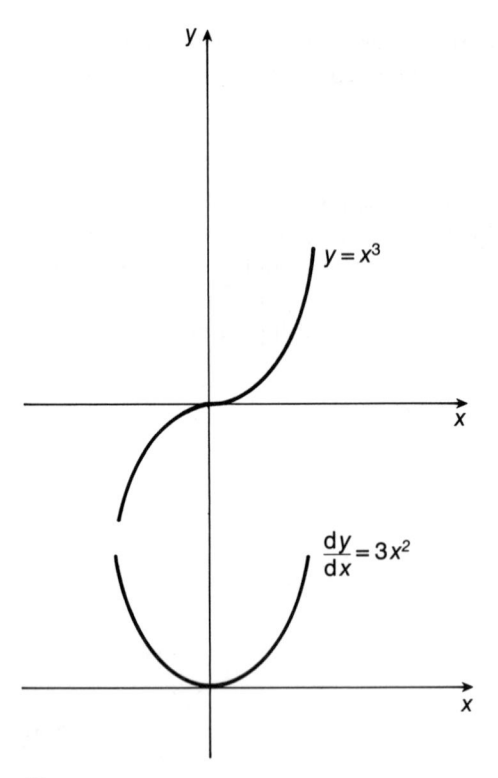

Figure 5.4

The slope of the cubing function is never negative but there is clearly an interesting situation at $x = 0$ which will be discussed later in the chapter. With sufficient patience, the approach can be used to establish that for

$$y = x^4 \quad dy/dx = 4x^3$$
$$y = x^5 \quad dy/dx = 5x^4$$

while moving in the other direction, for $y = x^1$ we obtain

$$\frac{\Delta y}{\Delta x} = \frac{(x + \Delta x) - x}{\Delta x}$$
$$= 1$$

and so dy/dx is also 1. Now suppose that y is a constant. The simplest case is where $y = 1$. Writing 1 as x raised to the power zero, we have

$$\frac{\Delta y}{\Delta x} = \frac{(x + \Delta x)^0 - x^0}{\Delta x}$$
$$= \frac{1 - 1}{\Delta x}$$
$$= 0$$

so that dy/dx is also zero. In the following section we bring these results together in more general considerations of the form of the derivative.

EXERCISES 5.3

1 Using the $\Delta y / \Delta x$ approach of Section 5.3, find the derivative of the function

$$y = 5x^2$$

2 Find the derivative of

$$y = x^4$$

5.4 THE POWER FUNCTION RULE

It is instructive to tabulate the results relating the form of the derivative to the original function that have been obtained so far. As will be seen from Table 5.1, a pattern begins to emerge.

Table 5.1 invites the general conclusion that, to obtain the derivative of

278 Mathematics for business, finance and economics

Table 5.1

Function (y =)	Derivative (dy/dx =)
x^0	0
x^1	x^0
x^2	$2x^1$
x^3	$3x^2$
x^4	$4x^3$
x^5	$5x^4$

x raised to any power, the power is reduced by one and the 'old' power becomes the coefficient of x in the derivative. That is, if

$$y = x^n$$

$$\frac{dy}{dx} = nx^{n-1} \tag{5.5}$$

This conclusion is correct, and (5.5) is an important formula which forms the basis of the *power function rule* for *differentiation*, the name given to the process of obtaining the derivative. We shall now consider some examples of the use of (5.5) for a variety of values of n.

When n is a *positive integer*:

(i) $\quad y = x^6$

$$\frac{dy}{dx} = 6x^5$$

For example, the slope of the function $y = x^6$ when $x = 3$ is 1458 and the slope when x is -2 is -192. Now consider another case when x is a positive integer, this time making use of function notation:

(ii) $\quad f(x) = x^{27}$

$\quad\quad f'(x) = 27x^{26}$

When n is a *negative integer*:

(iii) $\quad y = x^{-2}$

$$\frac{dy}{dx} = -2x^{-3}$$

What we have found here is the derivative of $1/x^2$, to which case it was not obvious that the rule applied. The rule *does* apply, and to make use of it the reciprocal should first be converted to the form where x is raised to a negative power. So by this means we have established that for

$$y = \frac{1}{x^2} \qquad \frac{dy}{dx} = \frac{-2}{x^3}$$

Now take another example:

(iv) $\quad f(x) = \dfrac{1}{x^5}$

$\qquad \quad = x^{-5}$

$\quad f'(x) = -5x^{-6}$

$\qquad \quad = \dfrac{-5}{x^6}$

When n is a *rational number*:

(v) $\quad y = x^{1/2}$

$\quad \dfrac{dy}{dx} = \dfrac{1}{2} x^{-1/2}$

In this case, note that what has been obtained is the derivative of $y = \sqrt{x}$. Again, the first step is to convert the expression into a form to which the rule can be directly applied.[9] In this way we have established that for

$$y = \sqrt{x}$$

$$\frac{dy}{dx} = \frac{1}{2\sqrt{x}}$$

which result would not otherwise have been evident. As with reciprocals, to obtain the derivative, terms involving radical signs should first be converted to power function form. Consider another example:

(vi) $\quad f(x) = \sqrt{x^3}$

Remembering that the radical sign is in effect an instruction to divide the exponent by 2, we can more conveniently express function (vi) as

$$f(x) = x^{3/2}$$

$$f'(x) = \frac{3x^{1/2}}{2}$$

which, in terms of the original notation, is $1.5\sqrt{x}$.

As a further example consider

(vii) $y = 1/\sqrt{x}$
$= x^{-1/2}$

$$\frac{dy}{dx} = \frac{-1x^{-3/2}}{2}$$

which, in terms of the original notation, is $-0.5/\sqrt{x^3}$.

When *n* is an *irrational number*:

(viii) $f(x) = x^{\sqrt{2}}$
$f'(x) = \sqrt{2}.x^{\sqrt{2}-1}$

from which it will be seen that the rule applies without variation to the case where *n* is an irrational number. As a further example,

(ix) $y = \dfrac{1}{x^{\sqrt{2}}}$

$= x^{-\sqrt{2}}$

$$\frac{dy}{dx} = -\sqrt{2}.x^{-\sqrt{2}-1}$$

$$= \frac{-\sqrt{2}}{x^{-\sqrt{2}-1}}$$

When *n* is a *transcendental number*:

(x) $y = x^{\pi}$

$$\frac{dy}{dx} = \pi x^{\pi-1}$$

Minor generalization of (5.5) produces the *power function rule*. A power function takes the form $y = cx^n$ where the *base*, x, is variable and the *exponent*, n, is a constant. The rule is

> **The power function rule** Where $y = cx^n$
>
> $$\frac{dy}{dx} = ncx^{n-1}$$

The power function rule is the first of several rules for differentiation that will be considered. It applies where c and n are any real valued constants. So, for example, if

(xi) $y = 4x^3$

$$\frac{dy}{dx} = 12x^2$$

and

(xii) $f(x) = -27.5x^{-4}$
$f'(x) = 110x^{-5}$

Now consider the special case of the power function rule where $n = 0$. For

$$y = cx^0$$

to which mechanical application of the power function rule produces a result of zero. This outcome is itself a named rule:

> The constant function rule Where $f(x) = c$
> $$f'(x) = 0$$

This gives rather a grand status to the obvious statement that the slope of a horizontal line is zero!

EXERCISES 5.4

1 Find the derivative for each of the following functions:

(i) $y = x^7$
(ii) $y = x^{100}$
(iii) $y = x^{-3}$
(iv) $y = \dfrac{1}{x^4}$
(v) $y = x^{1/3}$
(vi) $y = -x^{-2/3}$
(vii) $f(x) = x^{\sqrt{3}}$
(viii) $f(x) = \dfrac{-1}{x^{\sqrt{5}}}$
(ix) $f(x) = x^{2\pi}$
(x) $y = 5x^8$
(xi) $y = 0.25x^4$
(xii) $y = -\dfrac{0.2}{x^{10}}$
(xiii) $f(x) = 10x^0$

2 Find the derivatives for:

(i) $f(x) = x^9$
(ii) $f(x) = x^{-9}$
(iii) $f(x) = -x^{-8}$

(iv) $y = \dfrac{1}{x^{10}}$

(v) $y = \dfrac{1}{10x}$

(vi) $f(x) = x^{\sqrt{4}}$
(vii) $f(x) = -x^{-\sqrt{2}}$

(viii) $y = -\dfrac{10}{10x^{10}}$

(ix) $f(x) = x^{\sqrt{\pi}}$
(x) $f(x) = 100x^{100}$
(xi) $f(x) = -0.05x^{20}$

(xii) $y = \dfrac{0.1}{x^{0.1}}$

3 Find the derivatives of the following functions, in which a, b, m, n, p and π are constants:

(i) $f(x) = x$
(ii) $f(x) = x^m$
(iii) $f(x) = x^{-p}$
(iv) $f(x) = -x^{-(n-1)}$

(v) $y = -\dfrac{1}{x^n}$

(vi) $f(x) = ax^b$

(vii) $f(x) = \dfrac{x^m}{m}$

(viii) $y = x^{\sqrt{n}}$
(ix) $f(x) = x^{\pi^2}$

4 Find the derivatives for:

(i) $y = \dfrac{1}{x^1}$

(ii) $y = \dfrac{1}{x^0}$

(iii) $y = (\sqrt{2} - 1)x^{\sqrt{2}+1}$
(iv) $f(x) = -0.2x^{0.2}$

5.5 DIFFERENTIATING POLYNOMIALS

Most expressions contain more than one term and it would be a cramped calculus that could not address multi-term expressions. The *sum–difference rule* allows the power function rule to be extended to all polynomials and any function that consists of differentiable terms separated by plus and minus signs. The statement of the rule is as follows.

> *The sum–difference rule*: Where $y = f(x) \pm g(x)$
>
> $$\frac{dy}{dx} = f'(x) \pm g'(x) \qquad (5.6)$$

In (5.6) the separated components are shown as *functions*. They may be individual power function terms, or themselves comprise several terms. In either case the expression as a whole consists of a number of terms each of which can be differentiated individually. This is the nub of a powerful rule. No matter how many functions are linked by the plus and minus signs, it always boils down to a number of distinct terms which can be tackled one at a time. As the first example of the use of the rule we shall apply it to the sum of two terms:

(xiii) $\quad y = 8x^4 + x^3$

$$\frac{dy}{dx} = 32x^3 + 3x^2$$

Now apply the rule to the difference between two terms:

(xiv) $\quad y = 7x^5 - 9x^2$

$$\frac{dy}{dx} = 35x^4 - 18x$$

As an example involving both sums of and differences between terms consider

(xv) $\quad y = x^7 - 2x^6 + 3x^5 - 7x^3 + 4x^2 + 25$

$$\frac{dy}{dx} = 7x^6 - 12x^5 + 15x^4 - 21x^2 + 8x$$

The rule can also be applied to the general nth degree polynomial:

(xvi) $\quad y = a_n x^n + a_{n-1} x^{n-1} + \cdots + a_2 x^2 + a_1 x + a_0$

$$\frac{dy}{dx} = na_n x^{n-1} + (n-1)a_{n-1} x^{n-2} + \cdots + 2a_2 x + a_1$$

Now use the rule to differentiate a non-polynomial expression with terms linked by addition and subtraction:

(xvii) $\quad y = x^5 - 4x^{4.5} + \sqrt{2}.x^{\sqrt{2}} - x^{-2}$

$$\frac{dy}{dx} = 5x^4 - 18x^{3.5} + 2x^{\sqrt{2}-1} + 2x^{-3}$$

The expression for y in example (xvii) is not a polynomial as not all of the exponents are positive integers. The sum–difference rule can be applied when the variable base (x) is raised to any real power. Now consider the application of the sum–difference rule to the marginal analysis of the firm. In Chapter 4 we took the example of a firm with quadratic revenue and profit functions and analysed the case using the known properties of parabolas. The same data can now be used to demonstrate the effectiveness of derivatives in this context. The revenue function was

$$R = 50x - 0.1x^2 \qquad (5.7)$$

where R is total sales revenue and x represents physical sales volume (equals output in this model). We know that equation (5.7) produces a parabola which opens down, so that the turning point (the vertex) is a maximum. Later on we shall distinguish maxima from minima using derivatives only. Applying the power function and sum–difference rules to (5.7), the derivative of the revenue function is

$$\frac{dR}{dx} = 50 - 0.2x \qquad (5.8)$$

The derivative of revenue with respect to output shows the rate of change of income as output (equals sales in this model) is varied. This is an important economic concept and is called *marginal revenue*, MR. Defined in this way as a derivative, marginal revenue is a *rate of change at a point*, but it is often helpful to think in terms of a practical quantum of variation – usually one unit of sales. This *finite approximation* to MR is defined as the extra revenue from one more unit of output or, as convenience of use dictates, the loss of revenue that would result from the sale of one *less* unit. To illustrate these distinctions in the present example, consider an output level of a hundred units. Setting $x = 100$ in (5.8) produces

$$\frac{dR}{dx} = 50 - 0.2(100)$$

$$= 30 = \text{MR}$$

So the rate of change of revenue at the $x = 100$ point on the function is 30. Now consider the finite approximations. We can do this by using (5.7) to calculate revenue levels at $x = 100$, $x = 101$ and $x = 99$. The results are as shown in Table 5.2(a). The finite approximations to marginal revenue are within a third of 1 per cent of the rate of change figure. In absolute terms the finite change in revenue itself goes down by 0.2 for each unit increase in x. This constant rate of change (-0.2) is the slope of the marginal revenue line. The value -0.2 is the *slope of the slope* of the revenue function. Now repeat the calculations for a level of output where the unit change is a greater proportion of the initial output level. At $x = 2$, (5.8) gives the marginal revenue as 49.6. Table 5.2(b) shows the finite change calculations around the initial $x = 2$ position. The finite measures in this case represent an even better approximation accurate to a fifth of a per cent of the rate of change figure. The significance of this precision (which would be ample for practical purposes) is that where, as here, the derivative is easy to calculate it can be used to assess the consequences of finite changes. Where the derivative is hard to find, finite measures may approximate well enough to the rate of change at a point. As we have already seen, in order to produce a stationary value of the function, the derivative must be zero. In the present example, the one stationary value is a maximum taken for a value of x satisfying

$$\frac{dR}{dx} = 50 - 0.2x$$
$$= 0$$

so that

$$x = 250$$

Sales of 250 units per period time will produce revenue of 6250 (by substitution of $x = 250$ in (5.7)). Figure 5.5 graphs the problem.

Note that in Figure 5.5 marginal revenue is initially 50 and is a straight line function, being the derivative of a quadratic. Under normal circumstances, a firm would not choose to operate at a level of output at which marginal revenue was negative. A revenue minded firm would set output level so that marginal revenue was zero while, as we shall later see, the *profit*

Table 5.2

	x	TR	TR(x) – TR(100)
(a)	100	4000	0
	101	4029.9	+29.9
	99	3969.9	−30.1
(b)	2	99.6	0
	3	149.1	+49.5
	1	49.9	−49.7

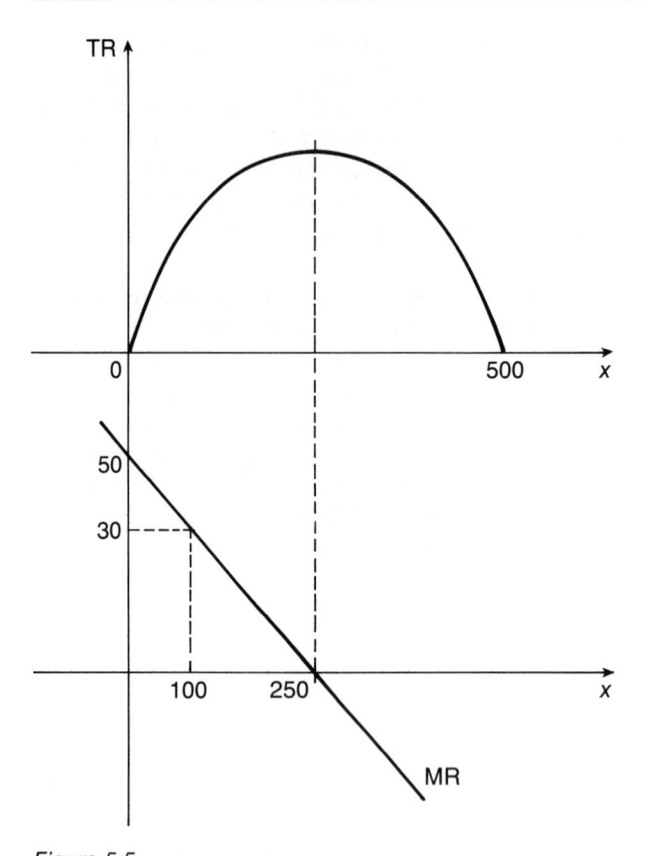

Figure 5.5

minded firm will choose an output level at which marginal revenue is *positive*. The *equilibrium condition* for maximum revenue is MR = 0. The equilibrium condition in this case has a common-sense appeal, but this is not always so. If a condition seems to be counter intuitive, it can be of help in understanding the validity of the condition to confirm that an optimum would *not* be achieved if the condition did *not* hold. In the present case, if marginal revenue was positive, revenue could be increased by increasing output level, while if marginal revenue was negative, revenue could be increased by *reducing* output (sales) level. Consequently, it is essential that marginal revenue be zero for a maximum of total revenue. Now consider profits again. Given the cost function $C = 4x + 450$ profits are

$$\pi = 46x - 0.1x^2 - 450 \tag{5.9}$$

Using the knowledge that the profits parabola opens down, a maximum of profit will occur when the derivative of (5.9) is zero. That is,

$$\frac{d\pi}{dx} = 46 - 0.2x \qquad (5.10)$$
$$= 0$$

from which

$$x = 230$$

which confirms the result obtained in Chapter 4 using the properties of parabolas. The derivative $d\pi/dx$ is *marginal profit*, which must be zero at a maximum. However, the equilibrium condition for maximum profit is usually stated in terms of the constituent parts of profit—marginal revenue and *marginal cost*, dC/dx. Since by definition

$$\pi = R - C$$

$$\frac{d\pi}{dx} = \frac{dR}{dx} - \frac{dC}{dx} = 0 \qquad (5.11)$$

and so

$$\frac{dR}{dx} = \frac{dC}{dx} \qquad (5.12)$$

Equation (5.12) states that, for a maximum of profit, marginal revenue should equal marginal cost:

$$MR = MC$$

To confirm this in the present example:

$$MR = 50 - 0.2(230)$$
$$= 4$$

$$MC = \frac{dC}{dx} = 4$$

Appreciation of this fundamental economic principle can be firmed up by supposing that the firm was operating at a point where $MR \neq MC$. In particular, suppose that the output level was such that

$$MR = 2 \qquad MC = 4$$

In this case, the firm could increase profit by reducing output, as the rate of loss of revenue (2) is less than the rate of reduction of cost – so that the excess of revenue over cost must increase. The alternative description of marginal cost as *escapable cost* comes to mind in this context. Again if an output level (225 units in fact) was set such that

MR = 5

and

MC = 4

the firm could increase profits by stepping up production as the rise in costs would be more than offset by the increase in revenue. Now consider an economic example involving a function to be *minimized*. Suppose that the average total cost (unit cost) for the manufacture of a product is

$$y = x^2 - 90x + 4000 + 25{,}000x^{-1} \qquad (5.13)$$

where, in (5.13), y is average cost and x is output level. The domain of this rational function is $x > 0$, and the graph is illustrated in Figure 5.6. The function is asymptotic to the y axis, and for high values of x it approaches unit *variable* costs as given by the parabola

$$x^2 - 90x + 4000$$

as unit *fixed* costs described by the rational term

$$\frac{25{,}000}{x}$$

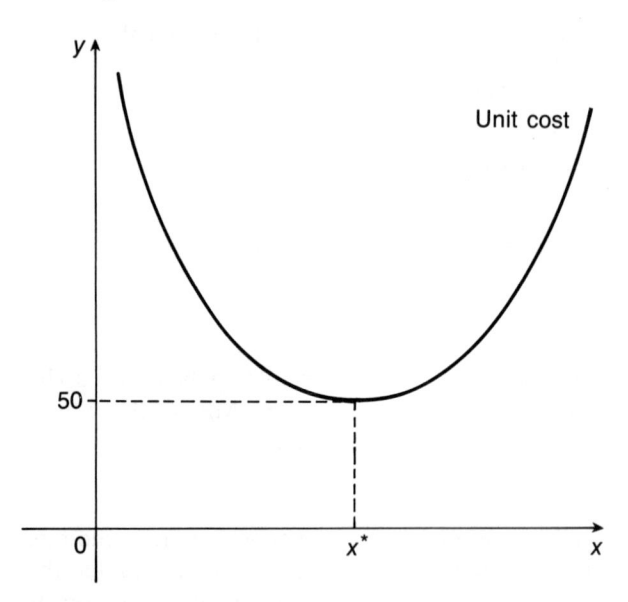

Figure 5.6

become ever smaller. The firm wishes to know the level of output, x^*, for which unit cost is lowest. This will be the output level at which its most competitive price can be charged without sustaining a loss. While such an output level is not likely to be the most profitable point for the firm in the short term, it can be strategically important in competitive situations. In theoretical terms, a minimum unit cost position such as this should in any case result under conditions of pure competition in the long run. Knowing the general shape of the curve, the minimum unit cost position is found by setting the derivative of (5.13) to zero. That is:

$$\frac{dy}{dx} = 2x - 90 - 25{,}000 x^{-2} = 0 \qquad (5.14)$$

Multiplying both sides of (5.14) by $x^2 \neq 0$ produces the cubic equation

$$2x^3 - 90x^2 - 25{,}000 = 0$$

This equation has just one real solution at $x^* = 50$, the other two roots being complex numbers.[10] An output level of 50 units minimizes unit cost at £2500. It is a useful exercise to compare this value of $x = 50$ with the output level at which unit *variable* cost reaches a minimum. Given the shape of the unit fixed cost hyperbola, unit variable cost will always reach its minimum at a *lower* output level than unit cost as a whole. In the present case unit variable cost reaches its minimum value at $x = 45$.

EXERCISES 5.5

1 Use the sum–difference rule to find the derivative of each of the following expressions:

(i) $y = x^4 + x^3$
(ii) $y = x^5 - x^3$
(iii) $y = x^6 + x^5 - x^4$
(iv) $f(x) = 5x^4 + 3x^3 - 7x^2$
(v) $f(x) = 9x^5 - 7x^4 + 5x^3 - 8x^2 + 14x - 100$

2 Find the derivative for each of the following expressions in which a, b, c, d, k, m and n are constants.

(i) $f(x) = ax^m + bx^n$
(ii) $f(x) = cx^{m+1} - dx^{n-1}$
(iii) $f(x) = ax^3 + bx^2 + cx + d$
(iv) $y = \dfrac{x^{n+1}}{n+1} + \dfrac{x^{m+1}}{m+1} + k$
(v) $y = a_0 x^5 + a_1 x^4 + a_2 x^3 + a_3 x^2 + a_4 x + a_5$

3 Find the derivative of each of the following functions, and the points at which the functions have stationary values.

(i) $y = 6x^2 - 156x + 100$
(ii) $y = 5x^2 + 100x - 20$
(iii) $y = 2x^3 - 9x^2 - 60x + 100$

5.6 THE PRODUCT AND QUOTIENT RULES

It is often necessary to find the derivative of expressions which can be regarded as either the ratio or the product of two or more functions. The 'component parts' may be much more convenient to differentiate than the original expression as a whole. Therefore, if the derivative of the whole can be set out in terms of the derivatives of its parts, a more convenient means of obtaining the derivative will result. In this section we consider the two rules for differentiation involved here. These are:

1 the product rule
2 the quotient rule

The product rule

For a function which is itself the product of two functions, the *product rule* for differentiation states that if

$$y = f(x)g(x)$$

then the derivative is given by

$$\frac{dy}{dx} = f(x)g'(x) + g(x)f'(x) \tag{5.15}$$

The right-hand side of (5.15) is usually written in the more economical form

$$fg' + gf'$$

In considering examples of the use of the product rule, we shall begin with an expression for which the result is already known. For example, the cubing function $y = x^3$ can be represented as the product of two functions:

$$f(x) = x^2 \quad \text{and} \quad g(x) = x$$

Expressed in this way, our task is to find the derivative of

(xviii) $y = x^2 x$

Slopes, derivatives and turning points 291

The application of the product rule as stated by (5.15) then produces

$$\frac{dy}{dx} = x^2.1 + x.2x$$
$$= x^2 + 2x^2$$
$$= 3x^2$$

which result confirms the outcome obtained by the direct differentiation of $y = x^3$. Now consider obtaining the derivative of

(xiv) $y = 4x^2(x^3 + 2x^5)$

This expression can be seen as the product of the two functions

$$f(x) = 4x^2$$
$$g(x) = (x^3 + 2x^5)$$

Application of the product rule produces

$$\frac{dy}{dx} = 4x^2(3x^2 + 10x^4) + (x^3 + 2x^5)8x$$
$$= 12x^4 + 40x^6 + 8x^4 + 16x^6$$
$$= 20x^4 + 56x^6$$

This result could be checked by multiplying out the bracket in y and differentiating the result using the sum–difference rule. As may be confirmed, the slope of the function when $x = 3$ is 42,444. The more complicated $f(x)$ and $g(x)$ are, the greater is the value of the product rule. For example, consider the function

(xx) $y = (x^3 + 2x^2 - 5x)(x^5 - x^4 + 4x^3 - 7x^2 + x)$

Applying the rule:

$$\frac{dy}{dx} = (x^3 + 2x^2 - 5x)(5x^4 - 4x^3 + 12x^2 - 14x + 1)$$
$$+ (3x^2 + 4x - 5)(x^5 - x^4 + 4x^3 - 7x^2 + x)$$

which, if desired, could be simplified to

$$x(8x^6 + 7x^5 - 18x^4 + 30x^3 - 132x^2 + 111x - 10)$$

One of the convenient features of the product rule is that the unsimplified form of the derivative may well suffice. For example the objective may simply be to find the numerical value of the slope and it is usually quicker to obtain the slope from the unsimplified form. The value of the product rule is further enhanced in cases which require the use of other rules (such as the *chain rule*, shortly to be discussed). Like the sum–difference rule, the product rule also applies to expressions that involve the product of more than two functions. For example, in the case where the dependent variable is the product of three functions,

$$y = f(x)g(x)h(x)$$

$$\frac{dy}{dx} = fgh' + fg'h + f'gh$$

Thus for

(xxi) $y = 2x(3x - 1)(4x + 2)$

$$\frac{dy}{dx} = 2x(3x - 1)4 + 2x.3(4x + 2) + 2(3x - 1)(4x + 2)$$

which simplifies to

$$72x^2 + 8x - 4$$

This result can be confirmed by expansion of y and use of the sum–difference rule. The original function becomes

$$y = 24x^3 + 4x^2 - 4x$$

so that

$$\frac{dy}{dx} = 72x^2 + 8x - 4$$

Now consider the second of the rules.

The quotient rule

As its name suggests, the *quotient rule* applies to the differentiation of the ratio of two functions. For

$$y = \frac{f(x)}{g(x)}$$

the derivative is given by

$$\frac{dy}{dx} = \frac{gf' - fg'}{g^2} \qquad (5.16)$$

First apply the rule to a familiar case. Consider the cubing function, but where this is expressed as the rational function:

$$y = \frac{x^5}{x^2} = x^3$$

Applying the quotient rule:

$$\frac{dy}{dx} = \frac{x^2 \cdot 5x^4 - x^5 \cdot 2x}{x^4}$$

$$= \frac{5x^6 - 2x^6}{x^4}$$

$$= \frac{3x^6}{x^4}$$

$$= 3x^2$$

The purpose of the rule is in its application to more involved expressions such as

(xxiii) $\quad y = \dfrac{x^3 + x}{3 + x^2}$

in which case, use of the quotient rule produces

$$\frac{dy}{dx} = \frac{(3 + x^2)(3x^2 + 1) - (x^3 + x)2x}{(3 + x^2)^2}$$

which could, if required, be simplified to

$$\frac{dy}{dx} = \frac{x^4 + 8x^2 + 3}{(3 + x^2)^2}$$

Using this expression, for example, when $x = -2$ the slope of the function y is $dy/dx = 51/49 \approx 1.04$. As a further example consider

$$y = \frac{x^5 - 2x^3}{4 - x^7}$$

Here the derivative is

$$\frac{dy}{dx} = \frac{(4 - x^7)(5x^4 - 6x^2) + 7x^6(x^5 - 2x^3)}{(4 - x^7)^2}$$

which could be re-expressed in the simpler form

$$\frac{2x^2(x^9 - 4x^7 + 10x^2 - 12)}{(x^7 - 4)^2}$$

Now apply the quotient rule to an expression first encountered in Chapter 4:

$$f(x) = \frac{x^2 - 1}{x + 1}$$

so that

$$f'(x) = \frac{(x+1)2x - (x^2-1)1}{(x+1)^2}$$

$$= \frac{x^2 + 2x + 1}{x^2 + 2x + 1} = 1$$

A special case of the quotient rule arises when the numerator of the ratio is unity. The result is the *reciprocal rule* which gives the derivative of the reciprocal of a function and states that if

$$y = \frac{1}{f(x)}$$

then

$$\frac{dy}{dx} = \frac{-f'(x)}{[f(x)]^2}$$

For example, if $f(x) = x^2$, then by the reciprocal rule

$$y = \frac{1}{x^2} \quad \text{and} \quad \frac{dy}{dx} = \frac{-2x}{x^4} = \frac{-2}{x^3}$$

which confirms the result already obtained by other means. The quotient rule can be obtained from the product rule. Indeed, the quotient rule is the product rule applied to the case of

$$f(x)[g(x)]^{-1}$$

Our original example of

$$y = x^5 (x^2)^{-1}$$

could be seen in this guise. A demonstration of the result involves the chain rule – to which, following the exercises, we now turn.

EXERCISES 5.6

1 Use the product rule to find the derivative for:

(i) $y = (x^2 + x)(x^3 - x^2)$
(ii) $y = (5x^4 - 4x^2)(10x^3 + 7x^5)$
(iii) $y = (4x^3 + 3x^2 + 8x)(2x + x^2 + x^4)$
(iv) $y = (x + 1)(x - 1)$

2 Use the quotient rule to find the derivative for:

(i) $y = \dfrac{x^3 + x^2}{x^5 - x^4}$

(ii) $y = \dfrac{5x^4 + 2x^7}{3x^4 - 2x^3}$

(iii) $f(x) = \dfrac{x}{7x^3 + 5}$

(iv) $f(x) = \dfrac{4x^2 - 5x + 3}{x^3}$

3 (i) Use the quotient rule to find the derivative of

$$f(x) = (x^2 - 1)x^{-1}$$

Can your result be confirmed by proceeding in a different way?

(ii) Find the derivative of

$$f(x) = (x^3 - 2x^2)(x^2 + 2x)(3x - 5)^{-1}$$

and evaluate the slope of the function $f(x)$ at

(a) $x = 1$ (b) $x = 2$

5.7 THE CHAIN RULE

The *chain rule* is applied to expressions which are a *function of a function*. This is an alternative name for this rule which is also known as the *composite function rule*.[11] Consider the function

$$y = (x^2 + x)^3 \qquad (5.17)$$

At present, our only means of finding the derivative for (5.17) is by first expanding the bracket. While expansion is practical here, one has only to suppose that the exponent of the bracket had been 53 instead of 3; it is then clear that a more efficient technique is required. In (5.17), the expression within the bracket is itself a function – which we call $g(x)$. That is:

$$x^2 + x = g(x)$$

To obtain y, $g(x)$ must have a further functional operation applied to it, in this case cubing. So with particular reference to (5.17), we write here

$$y = [g(x)]^3$$

The original expression (5.17) is therefore a function (cubing) of a function (adding its square to the original value), and so is a composite function. In general, the second operation could be referred to as $f(x)$, and the chain rule can be stated as follows:

> The chain rule: For $y = f[g(x)]$
> $$\frac{dy}{dx} = \frac{dy}{dg}\frac{dg}{dx} \qquad (5.18)$$

Begin by applying the chain rule to a case for which the outcome is known. The function

$$y = x^6$$

could be expressed as

$$y = (x^2)^3$$

for which, in terms of the chain rule,

$$g(x) = x^2 \quad \text{and} \quad y = f[g(x)] = [g(x)]^3$$

Now in applying the chain rule

$$\frac{dg}{dx} = 2x \quad \text{and} \quad \frac{dy}{dg} = 3[g(x)]^2$$

so that from (5.18)

$$\frac{dy}{dx} = 3[x^2]^2 2x$$
$$= 3x^4 2x$$
$$= 6x^5$$

Now applying the rule to the original example where $y = (x^2 + x)^3$, in this case

$$\frac{dy}{dg} = 3(x^2 + x)^2$$

and

$$\frac{dg}{dx} = 2x + 1$$

so that from the chain rule (5.18)

$$\frac{dy}{dx} = 3(x^2 + x)^2 (2x + 1)$$

In fixing an intuitive grasp of (5.18), it may help to envisage dy/dg and dg/dx as *ratios*, in which case the dg would cancel. The following equally informal interpretation may also help to anchor the rule. Consider the character x as the juxtaposition of a closing bracket) and an opening bracket (. Now simply reverse the order of the brackets to obtain a new symbol, (), and consider

$$y = (\ )^3$$

Clearly

$$\frac{dy}{d(\ )} = 3(\ )^2$$

But suppose that

$$(\ ) = g(x)$$

$d(\ )/dx$ could be found in the normal way, and the product

$$\frac{dy}{d(\ )} \frac{d(\ )}{dx}$$

could be formed which, with the $d(\ )$ 'cancelled', is dy/dx. In terms of our illustration with $y = x^6$

(xxv) $\qquad y = x^6 = (x^2)^3$

and so since

$$\frac{dy}{dx} = \frac{dy}{d(\ )} \frac{d(\ )}{dx}$$

$$\frac{dy}{dx} = 3(\ )^2 2x$$

$$= 3(x^2)^2 2x$$
$$= 3x^4 2x$$
$$= 6x^5$$

as previously obtained. Now consider the chain rule as applied to the function

(xxvi) $\qquad y = 4(x^3 - 7x^2 + 8x)^5$

In this case the derivative is

$$\frac{dy}{dx} = 20(x^3 - 7x^2 + 8x)^4 (3x^2 - 14x + 8)$$

The chain rule can be used with the sum–difference rule so that it is applied at a number of points within an expression made up of several functions of functions. As an example consider

(xxvii) $\quad y = (x^3 + 4x)^3 - (x^5 + x^2)^2$

$$\frac{dy}{dx} = 3(x^3 + 4x)^2(3x^2 + 4) - 2(x^5 + x^2)(5x^4 + 2x)$$

from which we can find, for example, the slope of the function at $x = 2$. This is 6240. As is evident, the chain rule is applied separately to each of the terms of y. If each of several functions can be differentiated individually, then at least in principle sums, differences, products, quotients and powers of the functions are also differentiable. The chain rule and the product rule can also be used together. As an example consider

(xxviii) $\quad y = (x^2 - x)^7 (x + x^2)^4$

$$\frac{dy}{dx} = (x^2 - x)^7 [4(x + x^2)^3 (1 + 2x)]$$

$$+ (x + x^2)^4 [7(x^2 - x)^6 (2x - 1)]$$

in which, to obtain the derivative, the chain rule is applied within the product rule. The result could be simplified, if required, to

$$x^{10}(x - 1)^6 (x + 1)^3 (22x^2 + 3x - 11)$$

It may also be necessary to use the chain and quotient rules together. As an example consider

(xxix) $\quad y = \dfrac{(2x - 1)^3}{7x^2}$

for which the derivative

$$\frac{dy}{dx} = \frac{7x^2 [3(2x - 1)^2 2] - (2x - 1)^3 14x}{49x^4}$$

This expression eventually simplifies to

$$\frac{dy}{dx} = \frac{2(x + 1)(2x - 1)^2}{7x^3}$$

We have applied the rule to polynomials and ratios of polynomials, but the chain rule also applies to functions in which bracketed terms are raised to non-integral powers. As an example consider

(xxx) $\quad y = (2x + 1)^{1/2}$

Application of the chain rule produces

$$\frac{dy}{dx} = \frac{1}{2}(2x+1)^{-1/2} \cdot 2$$

$$= \frac{1}{(2x+1)^{1/2}}$$

Now consider the chain rule applied to the following case:

(xxxi) $y = 3z^2$ and $z = 2x^2$

We have, as usual,

$$\frac{dy}{dx} = \frac{dy}{dz}\frac{dz}{dx}$$

where

$$\frac{dy}{dz} = 6z$$

and

$$\frac{dz}{dx} = 4x$$

so that by the chain rule

$$\frac{dy}{dx} = 6z \cdot 4x$$

$$= 24zx$$

into which, substitution for z gives

$$\frac{dy}{dx} = 24(2x^2)x$$

$$= 48x^3$$

It will be noted that this result could be confirmed by substitution for z in the original function:

$$y = 3(2x^2)^2$$
$$= 12x^4$$

so that

$$\frac{dy}{dx} = 48x^3$$

as previously obtained. Now consider a further example[12] of the use of the chain rule where the relationship between y and x is mediated through another variable z:

(xxxii) $\quad y = z^2 + z$ and $z = x^2 - x$

so that, applying the rule,

$$\frac{dy}{dx} = (2z + 1)(2x - 1)$$
$$= [2(x^2 - x) + 1](2x - 1)$$
$$= 4x^3 - 6x^2 + 4x - 1$$

The chain rule can also be applied to find the derivative dy/dx where there are several 'intermediate' variables. For example, if

$$y = f(z) \quad \text{and} \quad z = g(w) \quad \text{and} \quad w = h(x)$$

then, by the chain rule,

$$\frac{dy}{dx} = \frac{dy}{dz}\frac{dz}{dw}\frac{dw}{dx}$$

Thus, given the particular functions

$$y = z^2 \quad z = w^2 + w \quad w = 5x^2 + 1$$

the derivative of y with respect to x is

$$\frac{dy}{dx} = 2z(2w + 1)(10x)$$

which, after substitution for z in terms of w, gives

$$\frac{dy}{dx} = 2(w^2 + w)(2w + 1)(10x)$$

which, after substitution for w in terms of x, gives

$$\frac{dy}{dx} = 2[(5x^2 + 1)^2 + (5x^2 + 1)][2(5x^2 + 1) + 1](10x)$$

which simplifies to

$$20x(5x^2 + 1)(50x^4 + 35x^2 + 6)$$

which in expanded form is

$$5000x^7 + 4500x^5 + 1300x^3 + 120x$$

so that, for example, when $x = 1$, $dy/dx = 10{,}920$. As progressively more functions are interposed between the variables y and x, the more work will be involved in finding the derivative dy/dx even with the use of the chain

Slopes, derivatives and turning points 301

rule and the more complicated will be the eventual expression. However, the principle of the chain rule still stands good.

In summary, the *chain rule*, the *power function rule*, the *sum–difference rule*, the *product rule* and the *quotient rule* constitute a useful tool-kit of techniques that allow derivatives to be obtained for a wide range of functions in business and economic problems. But the functions with which we must deal do not always present themselves in the most convenient manner, and the next two sections will address such situations.

EXERCISES 5.7

1 Use the chain rule to find the derivative for:
   (i) $f(x) = (x^3 + 1)^4$
   (ii) $f(x) = (x^4 + x^2)^5$
   (iii) $f(x) = (4x^5 + 3x^4 - 7x^2 + 2x)^7$
   (iv) $f(x) = 5(1 - x^5)^5$
   (v) $y = (x^2 + x)^3 + (x^3 + 2x)^2$
   (vi) $y = (6x^7 - 5x^4 + 6x^2)^6 - (x^8 - x^7)^9$

2 Use the chain rule in conjunction with the product rule and/or the quotient rule to find the derivatives for:
   (i) $y = (x^3 + x^4)^5 (x^5 - x^3)^6$
   (ii) $y = \dfrac{4x^3 + 6x}{(x^2 - 1)^3}$
   (iii) $y = \dfrac{(3x^2 - 2x)^3}{8x^4}$

3 Use the chain rule to differentiate:
   (i) $f(x) = (5x^3 - 7x^2)^{1/2}$
   (ii) $f(x) = (4x^3 + 8x)^{0.25}$

4 Find the derivatives of:
   (i) $f(x) = (ax^2 + bx + c)^n$
   (ii) $f(x) = \dfrac{(m-1)[(m+1)x + m]^{m+1}}{m+1}$

5 Use the chain rule to find the derivative dy/dx for:
   (i) $y = 5z^3$ and $z = 4x^4$
   (ii) $y = z^2 - z + 2$ and $z = 2x^2 - 2x$

6 Given the functions:

$$y = z^2 + 2z \qquad z = 2w \qquad w = x^3 - x$$

(i) Find the derivative dy/dx.
(ii) Evaluate the derivative at $x = 1$.

5.8 THE INVERSE FUNCTION RULE

Suppose that the variables x and y are related by the following equation:

$$x = y^3 + 9y^2 + 39y + 100 \qquad (5.19)$$

With y as the independent variable, what is dy/dx and how can this derivative be found? To answer the questions we begin by recalling the definition of a function. The relationship $y = f(x)$ is a function if each value of x in the domain yields a *unique* corresponding value of y. As we saw in Chapter 4, if it is also the case that each value of y results from a unique corresponding value of x, then the inverse function exists. The inverse function rule is as follows.

> *Inverse function rule*: Where $y = f(x)$
> and $x = f^{-1}(y)$
>
> $$\frac{dy}{dx} = \frac{1}{dx/dy} \qquad (5.20)$$
>
> where $dx/dy \neq 0$.

Take an example that can be checked by other means:

$$x = 0.25y - 3 \qquad (5.21)$$

As the inverse function always exists for straight lines (so long as the slope is not zero or undefined), we can first find the derivative $dx/dy = 0.25$ and then apply the rule to produce the result:

$$\frac{dy}{dx} = \frac{1}{dx/dy} = \frac{1}{0.25} = 4$$

This result can be confirmed by obtaining the inverse function for (5.21) and obtaining dy/dx directly. The inverse function will be

$$y = 4x + 12$$

for which

$$\frac{dy}{dx} = 4$$

The one-to-one relationship between x and y needed for the inverse function rule can be expressed in terms of slope. The inverse function exists if the slope of $x = g(y)$ does not change sign within the domain – in other words if the function is *monotonic*. If the function does not change slope then for any value of y (x) there is only one real value of x (y).[13] An example where this does *not* apply is $y = x^2$ where because of the change in slope there are two values of x for any (non-zero) y in the range. So to use the inverse function rule in the case of the function

$$x = y^2$$

the domain should be restricted to either $y > 0$ or $y < 0$. This done, y expressed in terms of x will satisfy the requirements for a function. Therefore, the inverse function rule will apply, and we can write

$$\frac{dy}{dx} = \frac{1}{dx/dy}$$

$$= \frac{1}{2y}$$

$$= \frac{1}{2\sqrt{x}} = 2x^{-1/2}$$

The inverse function rule comes into its own with more complicated expressions where the inverse function cannot be explicitly stated. A good example is provided by an otherwise intractable case such as (5.19). Given

$$x = y^3 + 9y^2 + 39y + 100$$

for which

$$\frac{dx}{dy} = 3y^2 + 18y + 39$$

application of the inverse function rule produces

$$\frac{dy}{dx} = \frac{1}{3y^2 + 18y + 39}$$

The use of this expression to find the slope of the function for given x would require the solution of the cubic (5.19) and the use of the resulting real root.

Hence, when $x = 222$ (to choose what turns out to be a relatively convenient value), $y = 2$ and

$$\frac{dy}{dx} = \frac{1}{3(2)^2 + 18(2) + 39}$$

$$= \frac{1}{87} \approx 0.01149$$

Before taking the result of the application of the inverse function rule for granted, we should be satisfied that the rule does, in fact, apply. In this case, it can be confirmed that the slope of (5.19) is always positive. This can be deduced from the fact that the roots of the derivative dx/dy are complex – meaning that the graph does not cross the axis and that the derivative (the slope of the original function) always stays the same sign.

EXERCISES 5.8

1 Use the inverse function rule to find the derivative dy/dx for the following functions:

  (i) $x = 4 + 0.2y$
  (ii) $0.25x = 0.125y - 2.5$
  (iii) $x = y^5$
  (iv) $x = 2y^7 - 3y^5 + 6y - 100$

2 Find the derivative dy/dx for:

  (i) $x = (2y^3 + y)(y^4 + y^2)$
  (ii) $x = (5y^3 + 12y - 10)^3$

3 Find the derivative dy/dx for the function

$$y = x^5 + 2x^3 + 5x$$

and evaluate the result for $x = 1$.

5.9 IMPLICIT DIFFERENTIATION

Much work using functions of a single variable involves the functional form where one variable is the *subject of the equation*, as is y in

$$y = f(x) \tag{5.22}$$

Another way of describing the relationship (5.22) is to say that y is stated as an *explicit function* of x. If (5.22) is marginally rearranged as

$$y - f(x) = 0$$

or, in general, as

$$f(x, y) = 0$$

then the relationship between x and y is stated as an *implicit* function, in which all of the terms are on the same side of the equation. An implicit function may for some purposes be more convenient because it is not asymmetric in the way that the variables are treated or because the relationship is presented in this form. Since we must deal with implicit functions, our purpose is to show how dy/dx can be found in these cases. Consider

$$f(x, y) = xy - 5x - 6y + 4 = 0 \qquad (5.23)$$

The procedure is to differentiate each term in (5.23) with respect to x and, where a term includes y, to treat y as an unknown function of x. Consider the first term, xy. To differentiate this term with respect to x, the product rule must be deployed. Thus, bearing in mind the way that we intend to treat y, use of the product rule produces

$$\frac{d(xy)}{dx} = x\frac{dy}{dx} + y\frac{dx}{dx}$$

$$= x\frac{dy}{dx} + y$$

The term $-5x$ differentiates with respect to x as -5 and the term $+4$ has zero as its derivative. The derivative with respect to x of the term $-6y$ is

$$-6\frac{dy}{dx}$$

Thus, the derivative with respect to x of (5.23) as a whole, obtained in this process of implicit differentiation, is

$$x\frac{dy}{dx} + y - 5 - 6\frac{dy}{dx} = 0 \qquad (5.24)$$

Note that (5.24) is an implicit function involving dy/dx which can be rearranged to state dy/dx *explicitly* as

$$\frac{dy}{dx} = \frac{5 - y}{x - 6} \qquad (5.25)$$

Note in (5.25) that the derivative dy/dx is expressed in terms of both x and y. In (5.23) we chose an example that *could* be rearranged to state y as an

306 Mathematics for business, finance and economics

explicit function of x. We shall now make that rearrangement and differentiate the results to confirm (5.25). Equation (5.23) can be written as

$$y(x-6) - 5x + 4 = 0$$

from which

$$y = \frac{5x-4}{x-6} \tag{5.26}$$

from which dy/dx can be obtained by use of the quotient rule:

$$\frac{dy}{dx} = \frac{(x-6)5 - (5x-4)}{(x-6)^2}$$

$$= \frac{-26}{(x-6)^2} \tag{5.27}$$

At first glance, (5.27) does not look a great deal like equation (5.25), but substitution for y from (5.26) into (5.25) results in

$$\frac{dy}{dx} = \frac{5 - \frac{5x-4}{x-6}}{x-6}$$

$$= \frac{-26}{(x-6)^2}$$

which confirms our original result. Now try implicit differentiation on a function of higher degree:

$$f(x, y) = x^2 y^2 + 3y - 4x = 0 \tag{5.28}$$

Differentiation of the first term of (5.28) with respect to x produces

$$x^2 2y \frac{dy}{dx} + y^2 2x \tag{5.29}$$

where in (5.29) the chain rule is employed in producing the derivative of y^2 with respect to x as

$$2y \frac{dy}{dx}$$

The derivatives of the other terms follow and the overall result is

$$2x^2 y \frac{dy}{dx} + 2xy^2 + 3 \frac{dy}{dx} - 4 = 0$$

which rearranges as

$$\frac{dy}{dx} = \frac{4 - 2xy^2}{2x^2y + 3}$$

Implicit differentiation completes our presentation of the rules and procedures for differentiation of functions of a single variable for the time being. Later on, some further results will be presented to cover exponential and logarithmic functions.

EXERCISES 5.9

1 Find the derivative, dy/dx for the following implicit functions:

   (i) $4x - 5y + 10 = 0$
   (ii) $10x + 5xy - 2y - 100 = 0$

2 In the case of the function 1(ii) above, rearrange the expression to state y as an explicit function of x, and obtain an expression for dy/dx by use of the quotient rule. Confirm the equivalence of the result so obtained to that of 1(ii) by evaluating the expressions at the points

   (i) $x = 10$ $y = 0$
   (ii) $x = 1$ $y = 30$

3 Use the implicit function rule to find the derivative dy/dx for the following functions:

   (i) $f(x, y) = 5y - 2x + xy^2 = 0$
   (ii) $f(x, y) = 2x^3y^2 - 3x^2 + xy - 7y = 0$.

5.10 HIGHER ORDER DERIVATIVES

The derivative of a function of x is in general also a function of x. This function can be differentiated to obtain the derivative of the derivative – the *second derivative*. In turn, the second derivative, also typically a function of x, can be differentiated to obtain the *third derivative*. This process can be continued to find still higher order derivatives where these exist. Just as the first derivative is the expression giving the slope of the original function, so the second derivative gives the *slope of the slope*. This is an important consideration in distinguishing maxima from minima. To illustrate second and higher order derivatives, consider the quartic function

$$y = 2x^4 - 7x^3 + 4x^2 + 8x + 20 \tag{5.30}$$

A complete listing of the derivatives of this function is as follows.

First derivative:

$$\frac{dy}{dx} = 8x^3 - 21x^2 + 8x + 8$$

Second derivative:

$$\frac{d^2y}{dx^2} = 24x^2 - 42x + 8$$

Third derivative:

$$\frac{d^3y}{dx^3} = 48x - 42$$

Fourth derivative:

$$\frac{d^4y}{dx^4} = 48$$

Fifth and all higher order derivatives:

$$\frac{d^n y}{dy^n} = 0 \quad \text{for } n \geq 5$$

Note that the second derivative is written as

$$\frac{d^2y}{dx^2}$$

rather than

$$\frac{dy^2}{dx^2}$$

The reason for this notation is that the second derivative is the derivative with respect to x of dy/dx. This could be stated as

$$\frac{d^2y}{dx^2} = \frac{d}{dx} \quad \text{of} \quad \frac{dy}{dx}$$

so that it is the 'd' on the top line that appears twice rather than y. Similar reasoning applies to the notation for the higher order derivatives. In terms of function notation, the second derivative of $f(x)$ with respect to x can be represented as

$$f''(x)$$

while the third derivative would be shown as

$$f'''(x)$$

Although higher order derivatives can be represented in this manner, the notation soon becomes cumbersome. A more economical usage is to write

$$f^{(n)}(x)$$

for the nth derivative of $f(x)$ with respect to x. For an example, consider

$$f(x) = x^{10} - 2x^8 + 4x^5 + x^4$$

for which the derivatives down to the fourth are

$$f'(x) = 10x^9 - 16x^7 + 20x^4 + 4x^3$$
$$f''(x) = 90x^8 - 112x^6 + 80x^3 + 12x^2$$
$$f'''(x) = 720x^7 - 672x^5 + 240x^2 + 24x$$
$$f^{iv}(x) = 5040x^6 - 3360x^4 + 480x + 24$$

A polynomial of the fourth degree such as (5.30) will have non-zero derivatives down to the fourth derivative. Similarly, for an nth degree polynomial, all derivatives beyond order n will be zero. For other classes of function – such as rationals – there may be no limit to the order of derivative that can be obtained. For example, with

$$y = x^{-1}$$

the second derivative is

$$\frac{d^2 y}{dx^2} = 2x^{-3}$$

and, for example, the sixth derivative is

$$\frac{d^6 y}{dy^6} = 720 y^{-7}$$

Or consider

$$y = x^\pi$$

the fourth derivative of which is

$$\frac{d^4 y}{dx^4} = \pi(\pi - 1)(\pi - 2)(\pi - 3)x^{\pi - 4}$$

Of the higher order derivatives, the second derivative is the most useful on the majority of occasions. Applications in business and economics rarely go beyond the second derivative – although later we shall need the third derivative on one occasion and will state one result in terms of the nth derivative. So we shall concentrate here on the properties and uses of the second derivative.

EXERCISES 5.10

1 Find the second derivative for the following functions:

   (i) $f(x) = x^5$
   (ii) $f(x) = 7x^6$
   (iii) $f(x) = 5x^4 + 3x^2 + 2x$
   (iv) $f(x) = 8x^5 - 9x^4 + 7x^3 - 8x^2$

2 Find the third and fourth derivatives for:

   (i) $y = x^6$
   (ii) $y = 2x^5 - 3x^4$
   (iii) $y = 0.01x^{10} - 0.005x^{20}$

3 Find the second derivative for:

   (i) $f(x) = x^{-1}$
   (ii) $f(x) = \dfrac{1}{x^2}$
   (iii) $f(x) = x^{0.5}$
   (iv) $f(x) = (x^2 + 5)^7$

4 Find the second derivatives for the following functions:

   (i) $y = ax^4 + bx^3 + cx^2 + dx + e$
   (ii) $y = -mx^{-m} + nx^{-n}$

5.11 LOCAL MAXIMA AND MINIMA

One of the most important uses of the second derivative is in distinguishing local (relative) maxima from local minima. For stationary points of a specific character, the slope will change in a particular way. Through a maximum, we saw that the slope of the function *diminished* – so the slope of the slope would be negative. In other words, the second derivative is *negative*. Through a local minimum, the slope of the function *increased* as x increased – the second derivative being *positive*.[14] These considerations produce the *second derivative test* to identify the nature of stationary points in functions of a single variable:

1 If, for the x value that produces $f'(x) = 0$, it is also the case that $f''(x) < 0$, then the x value produces a local *maximum* of the function $f(x)$.

2 If, for the x value that produces $f'(x) = 0$, it is also the case that $f''(x) > 0$, then the x value produces a local *minimum* of the function $f(x)$.

Slopes, derivatives and turning points 311

3 If, for the x value giving $f'(x) = 0$, it is also true that $f''(x) = 0$, further investigation is needed to determine the nature of the stationary point.

We shall consider situation 3 later on. The point could be a maximum, a minimum, or a point at which the function inflects. But in the majority of cases, 1 or 2 will apply, and the second derivative test will be decisive. We now look at the way the test operates in a number of examples. Consider the quadratic function

$$f(x) = 50x - 0.1x^2$$

The first derivative is

$$f'(x) = 50 - 0.2x$$

which is zero when $x = 250$. To establish whether the value $x = 250$ produces a maximum or a minimum, consider the second derivative. This is

$$f''(x) = -0.2$$

Since the second derivative is negative for *all* values of x, any turning point must represent a maximum of the function. Figure 5.7 adds to Figure 5.5 a third level showing the graph of the second derivative (a constant function at the level -0.2).

Now consider two further examples involving the maximization of quadratic functions. Suppose that a maximum is sought for

$$y = 500 - x^2 - 20x$$

Setting the first derivative to zero gives

$$\frac{dy}{dx} = -2x - 20 = 0$$

which solves for $x = -10$. The second derivative is

$$\frac{d^2y}{dx^2} = -2 < 0$$

meaning that $x = -10$ produces a maximum, at which point $y = 600$. This example illustrates the fact that *there is no guarantee that the maximum producing value of x will be positive*. Now find the maximum of

$$y = 40x - x^2 - 500$$

Setting the first derivative to zero gives

$$\frac{dy}{dx} = 40 - 2x = 0$$

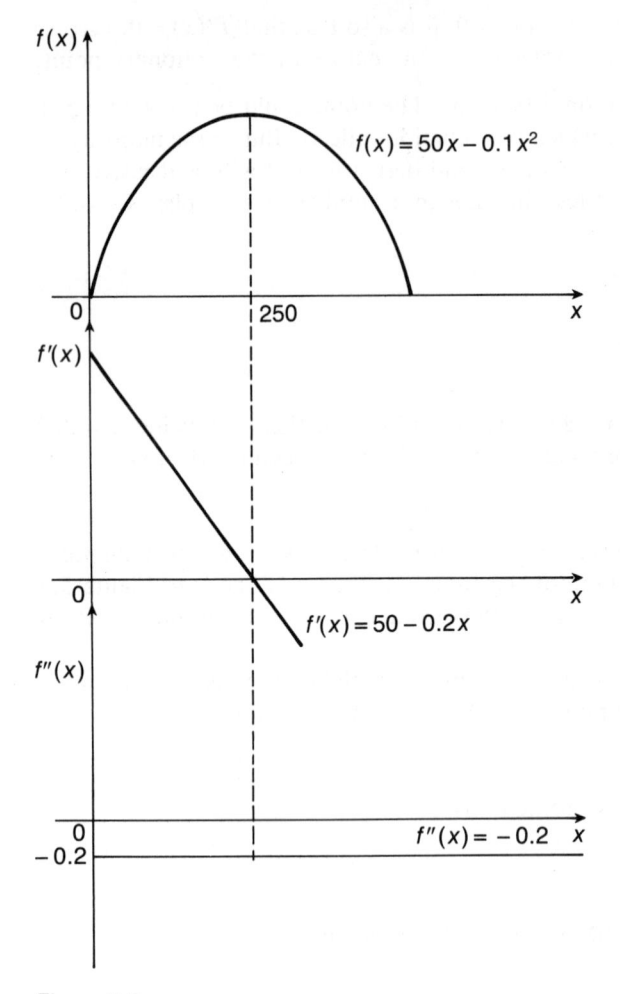

Figure 5.7

which solves for $x = 20$. The second derivative is

$$\frac{d^2y}{dx^2} = -2 < 0$$

meaning that $x = 20$ produces a maximum. At this point $y = -100$. This example illustrates the fact that *there is no guarantee that the maximum value of the function will be positive*. Now consider the cubic

$$y = x^3 + 3x^2 - 24x + 20 \tag{5.31}$$

Slopes, derivatives and turning points 313

The first derivative is

$$\frac{dy}{dx} = 3x^2 + 6x - 24 \qquad (5.32)$$

and the second derivative is

$$\frac{d^2y}{dx^2} = 6x + 6 \qquad (5.33)$$

Now setting (5.32) to zero in order to identify the stationary values of the function, we have

$$3x^2 + 6x - 24 = 0$$

so with division by 3 throughout

$$x^2 + 2x - 8 = 0$$

and therefore

$$(x+4)(x-2) = 0$$

so the two turning points of the function are at $x = -4$ and $x = +2$. As with all cubics, the second derivative is a linear function, into which the values of $x = -4$ and $x = 2$ are inserted. Thus, at $x = -4$,

$$\frac{d^2y}{dx^2} = -18$$

which by the second derivative test indicates a maximum. However, when $x = 2$

$$\frac{d^2y}{dx^2} = +18$$

so $x = 2$ produces a minimum. The function is graphed in Figure 5.8.

Points to note in this case are that it happens to be the smaller value of x which produces the local maximum.[15] The sign of x at a turning point is of considerable significance in practical work, and we shall take up this subject in a later chapter. For a second cubic example, consider the function

$$y = 500 - x^3 + 7.5x^2 + 150x$$

for which the first derivative is

$$\frac{dy}{dx} = -3x^2 + 15x + 150$$

which has the roots $x = -5$ and $x = 10$. The second derivative is

$$\frac{d^2y}{dx^2} = -6x + 15$$

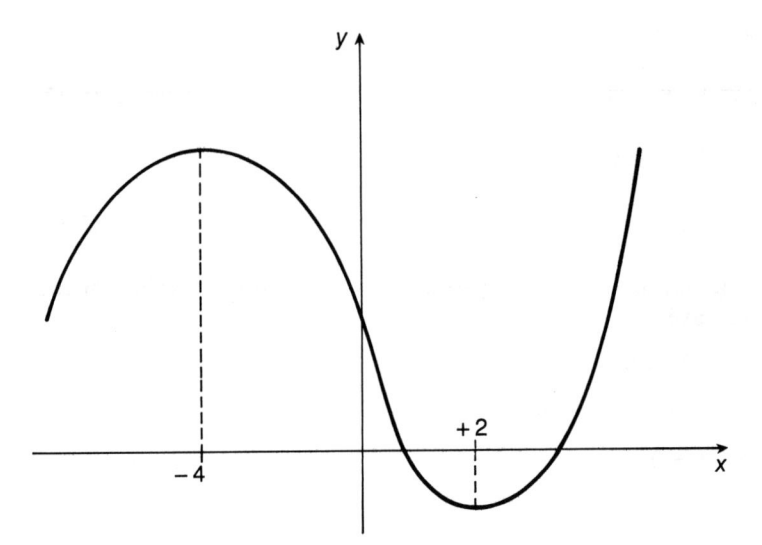

Figure 5.8

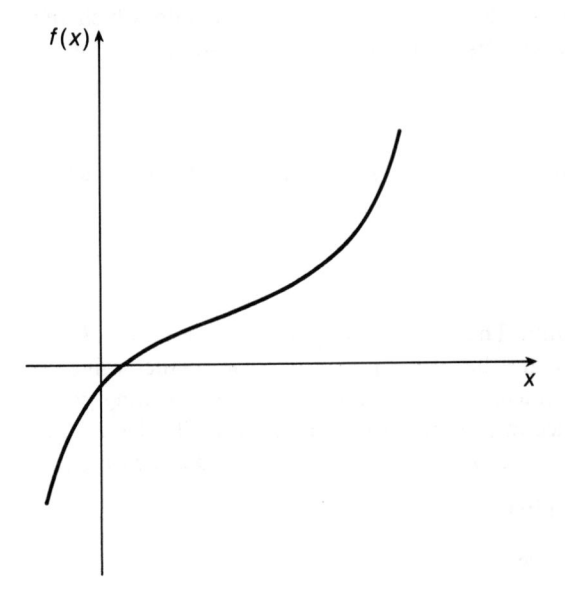

Figure 5.9

which is positive at $x = -5$ and negative when $x = 10$. Therefore, $x = -5$ corresponds to a local minimum and $x = 10$ produces the local maximum, at which point $y = 1750$.[16] Now consider

$$f(x) = x^3 - 15x^2 + 87x - 100$$

for which

$$f'(x) = 3x^2 - 30x + 87$$

but setting $f'(x)$ to zero produces only the *complex* roots

$$x = 5 + 2i \quad \text{and} \quad x = 5 - 2i$$

The absence of a solution in real numbers means that this cubic has *no* turning points. It appears as graphed in Figure 5.9.

Having examined the use of the second derivative test in discriminating between *local* high and low points, following the end of section exercises we take a broader look at turning points.

EXERCISES 5.11

1 Find the stationary point of the following functions and use the second derivative test to establish whether the point represents a maximum or a minimum:

  (i) $y = 100x - 2.5x^2$
  (ii) $y = 4x^2 - 72x + 100$

2 Find the stationary points of the following functions and establish their character:

  (i) $f(x) = 0.5x^3 - 2.25x^2 - 60x$
  (ii) $f(x) = 1000 - 2x^3 + 6x^2 + 48x$
  (iii) $f(x) = x^4 - 12x^3 + 28x^2$
  (iv) $f(x) = \dfrac{x^5}{5} - \dfrac{5x^3}{3} + 4x + 2$

3 (i) The stationary points of the function:

$$y = x^4 - 8x^3 + 22x^2 - 24x + 100$$

  occur at

  $x = 1 \quad x = 2 \quad x = 3$

  Use the second derivative test to establish the nature of each stationary point.

(ii) The stationary points of the function

$$y = 6x^5 - 67.5x^4 - 320x^3 + 2700x^2 + 12{,}000x + 500$$

occur at

$$x = -4 \qquad x = -2 \qquad x = 5 \qquad x = 10$$

Use the second derivative test to establish the nature of each stationary point.

4 Do the following functions have turning points? Provide as much information as possible in each case.

(i) $f(x) = 3x^3 - 18x^2 - 1053x + 1000$
(ii) $f(x) = x^3 - 3x^2 + 15x + 100$
(iii) $f(x) = x^4 - 12x^3 + 80x^2 - 400x + 20$

5.12 GLOBAL MAXIMA AND MINIMA

When applied to those x values where the first derivative is zero, the second derivative classifies the points as maxima or minima in their immediate vicinities. But from amongst this list of points the test says nothing about which of the maxima gives the greatest value overall. So unless there is only one local maximum, further work will be required to identify the *global* or *absolute* maximum.[17] This may take the form of enumeration – calculating the value of the function at each of the local maxima. This may be an efficient way to proceed with straightforward expressions. In business or economic modelling, meaningless outcomes need to be excluded – e.g. negative output levels or those beyond the maximum capacity of plant. The decision-maker expects the model to provide sensible values of the control variables – and ideally those values which produce an optimum value of the objective function. Irrelevant values of the decision variables can be excluded in a variety of ways. For example:

1 the selection of a function for which the extrema occur only for realistic values of the decision variables;
2 the placing of specific limits on the domain of the function;
3 by discarding the extrema associated with unrealistic values of the decision variables.

A risk involved with approach 3 is the implicit presumption that amongst the list of extrema producing x values there will be *some* within the reasonable range. If this is not the case, then either approach 2 must be used or the function must be redefined. In approach 1, from amongst polynomials, quadratics and quartics with leading coefficients negative for maximization

Slopes, derivatives and turning points 317

problems and positive for minimization problems will produce finite maxima and minima respectively. With appropriately selected coefficients, the extrema can also be confined to the positive quadrant. For example, a quadratic profits function will follow from a linear demand curve and a linear or quadratic cost function. Suppose that the monthly profit function for a small firm assembling a personal computer is

$$\pi = -1000 + 200x - x^2$$

where the -1000 term represents fixed costs and where the first-order condition for a maximum

$$\frac{d\pi}{dx} = 200 - 2x = 0$$

is satisfied within the positive quadrant for $x = 100$. The second derivative test

$$\frac{d^2\pi}{dx^2} = -2$$

confirms that the stationary value is a maximum. As the one 'candidate' local maximum this will also be the global maximum. We presume that the level of profit achieved, $\pi = £9000$ (representing a margin of £90 per unit output) is reasonable and that the output level is within the productive capacity of the company. Thus in terms of a reasonable value, it happens that things have worked out well in this case. Now consider a quartic case. The objective function to be maximized is

$$f(x) = -0.75x^4 + 7x^3 - 21x^2 + 24x$$

for which the first derivative is

$$f'(x) = -3x^3 + 21x^2 - 42x + 24$$

and the second derivative is

$$f''(x) = -9x^2 + 42x - 42$$

When the first derivative is set equal to zero, the solutions obtained are

$$x = 1 \quad x = 2 \quad x = 4$$

Insertion of these values into the second derivative shows that the values $x = 1$ and $x = 4$ produce local maxima while $x = 2$ gives a local minimum. The global maximum is determined by evaluating $f(x)$ at $x = 1$ and $x = 4$. The superior point is $x = 4$ which produces $f(x) = 16$, in comparison with $f(x) = 9.25$ at $x = 1$. This function is graphed in Figure 5.10.

The function itself can be used to discriminate between the local maxima and in this case involved the evaluation of two possibilities. But a function may produce values greater than that for any stationary value. An obvious

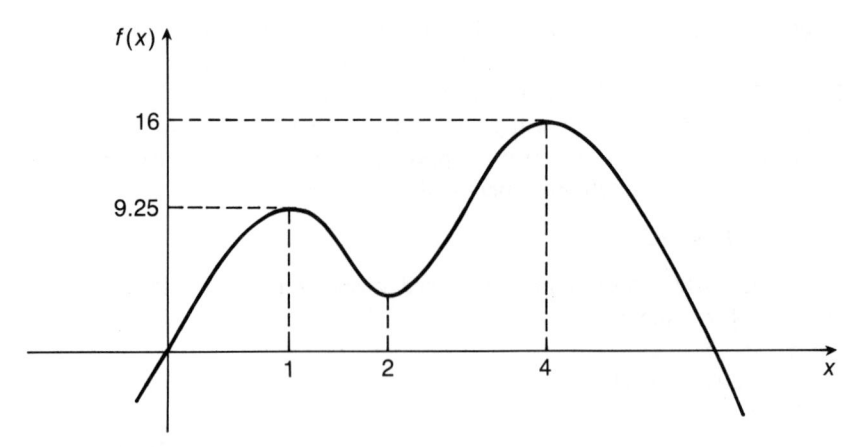

318 Mathematics for business, finance and economics

Figure 5.10

case is where the function increases without limit as x moves without restriction in one or other direction. This is so for all cubic functions. For example, with the function

$$y = 2x^3 - 45x^2 + 300x$$

the first derivative is

$$\frac{dy}{dx} = 6x^2 - 90x + 300$$

which, when equated to zero, solves for $x = 5$ and $x = 10$. At the second order, the derivative is

$$\frac{d^2y}{dx^2} = 12x - 90$$

which is negative at $x = 5$ and positive at $x = 10$, so there is a local maximum at $x = 5$ producing $y = 625$ while the local minimum at $x = 10$ produces $y = 500$. The function appears as in Figure 5.11.

However, the value of $x = 5$ does not produce the global maximum of the function, and $x = 10$ does not correspond to the overall minimum of y, as the function is unbounded in either direction. Other functions may take unlimited values at particular points within the domain. This is the case for rational functions where the denominator is zero for specific values of x. Consider

$$y = \frac{1}{x - 10} \tag{5.34}$$

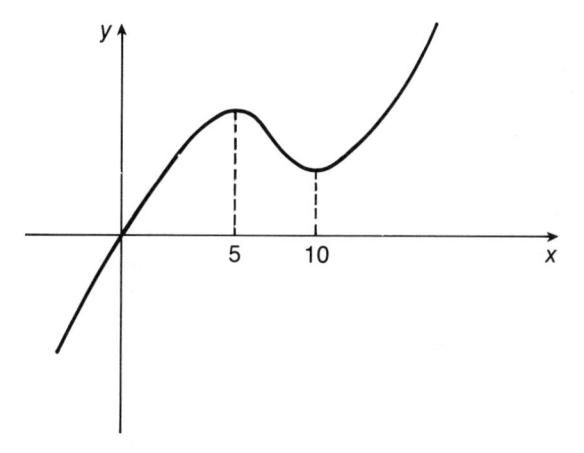

Figure 5.11

the derivative of which is

$$\frac{dy}{dx} = \frac{-1}{(x-10)^2}$$

An attempt to set this derivative to zero cannot succeed for finite x. The function has no stationary value. The function is graphed in Figure 5.12 from which it is evident that the function becomes arbitrarily large and positive as x approaches 10 from above, and arbitrarily large and negative as x approaches 10 from below.

In order to take cases such as (5.34) into account, an exhaustive process for identifying the global extremes of a function would need to include a search for vertical asymptotes. There are more esoteric examples of functions for which the value of the expression oscillates between ever larger extremes – both positive and negative – as some crucial value of x is approached. In addition, *piecewise-defined* functions can be given exceptional properties at individual points. However, functions with such characteristics will not be our primary concern here. Rather, we now consider the principles of approach 2, where explicit account is taken of the function's domain. Suppose that a maximum value is sought for the cubic

$$y = 500 + 22.5x^2 - 150x - x^3$$

for which

$$\frac{dy}{dx} = 45x - 150 - 3x^2 = 0$$

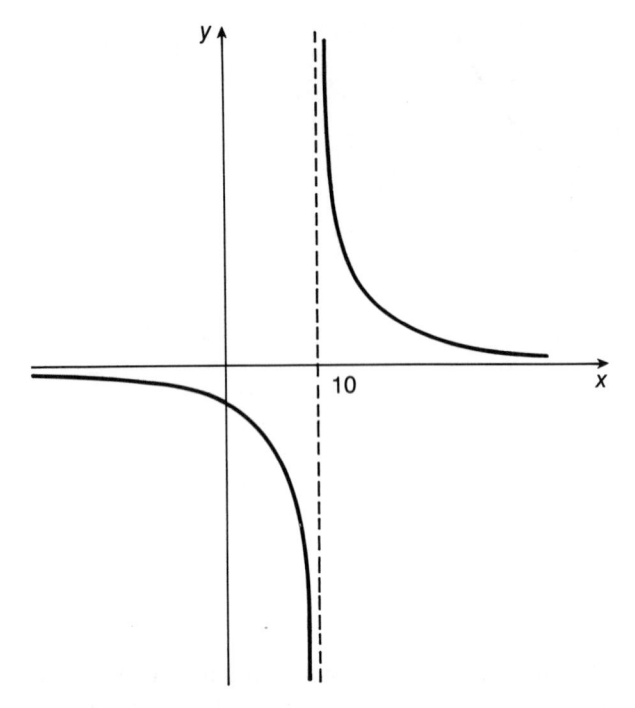

Figure 5.12

which solves for $x = 5$ and $x = 10$ and

$$\frac{d^2y}{dx^2} = 45 - 6x$$

which is positive for $x = 5$ and negative for $x = 10$. Thus $x = 5$ produces a local minimum while $x = 10$ gives a local maximum. The function is graphed in Figure 5.13.

Now suppose that the function represents a desirable factor such as net income, y, against the decision variable x. For example, x may be the output level of a possible new commodity to be added to the existing product range, so that negative values of x would have no meaning. What is the optimal output level for the new product? In the context of the model, the optimal decision is the value of x *within the domain* $x \geqslant 0$ that produces the highest value of y. Inspection of Figure 5.13 reveals this to be at $x = 0$ (at which value $y = 500$) rather than at the local maximum where $x = 10$ (when $y = 250$). This is an example of cases where the global optimum occurs at an endpoint of the domain – which here means that the new product should not be produced. A maximum at an endpoint is a possibility that must be investigated whenever there is a restricted domain. The

Slopes, derivatives and turning points 321

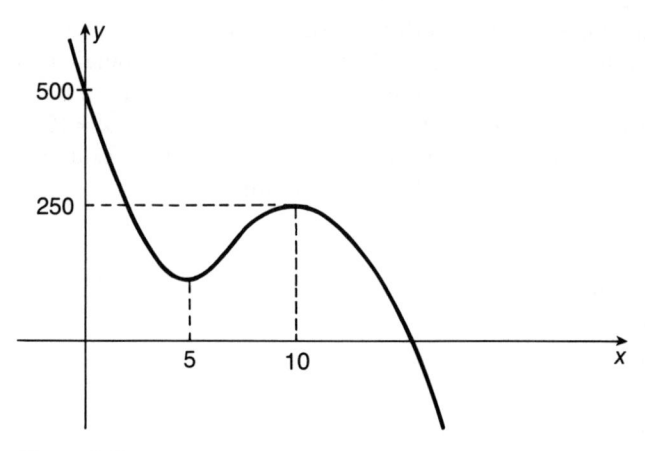

Figure 5.13

function should be evaluated at the endpoints as well as at the turning points within the domain.

Consider the quartic illustrated in Figure 5.14. There is a fixed lower endpoint, L, for the domain but there are three places where the upper endpoint, U, might be set: U_1, U_2 and U_3. The function itself increases without limit for values of x below L or above U_3. The decision-maker seeks the global maximum within a domain in which L is the lower endpoint and where the upper endpoint is given by one of the U values. It is clear from Figure 5.14 that with the upper endpoint set at

U_1 the overall maximum is at the lower endpoint $x = L$
U_2 the overall maximum is at the turning point where $x = x^0$
U_3 the overall maximum is at the upper endpoint $x = U_3$

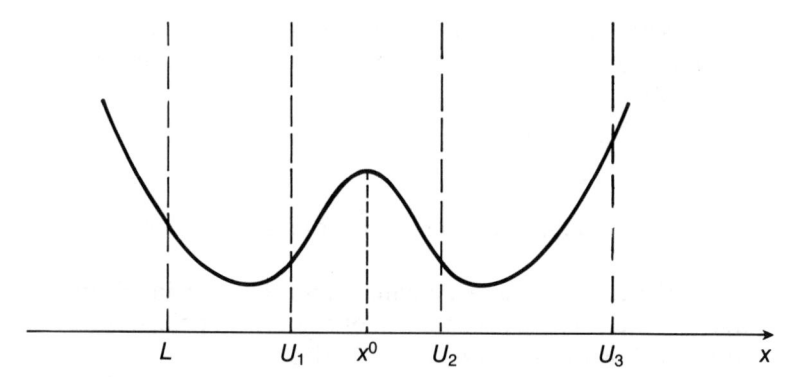

Figure 5.14

When seeking the global maximum with a restricted domain an exhaustive procedure evaluates the function at all local maxima within the domain and at the extremities of the domain. In addition, if the function is not a polynomial, any points at which the function is not defined (such as $x = 10$ in Figure 5.12) will need to be taken into account. As a numerical example, consider the problem of finding the global maximum of

$$y = 3x^4 - 124x^3 + 1380x^2 - 2400x + 2000$$

within the domain restricted to

(i) $x = 0$ to $x = 2$
(ii) $x = 0$ to $x = 4$
(iii) $x = 0$ to $x = 12$
(iv) $x = 0$ to $x = 30$
(v) $x = -10$ to $x = 30$

First, the turning points of the function will need to be found in order to locate the local maxima. Thus

$$\frac{dy}{dx} = 12x^3 - 372x^2 + 2760x - 2400 = 0$$

This cubic solves for

$x = 1 \qquad x = 10 \qquad x = 20$

The second-order condition is

$$\frac{d^2y}{dx^2} = 36x^2 - 744x + 2760 < 0$$

The second derivative is positive at $x = 1$ and $x = 20$ and negative at $x = 10$ which is therefore the only local maximum. Now consider the position of the global maximum in the specified domains.

(i) Since there is no local maximum in this interval, the overall maximum *must* occur at one of the endpoints. Checking the value of the function at these points,

at $x = 0$ $\qquad y = 2000$
at $x = 2$ $\qquad y = 1776$

so the global maximum of the function within this domain is 2000 at $x = 0$.

(ii) Here again there is no local maximum within the domain. Therefore the value of y must be at a maximum at one of the endpoints. At $x = 4$, $y = 7312$, which is the global high within this domain.

(iii) In this case the local maximum falls within the domain, so that there are now three points at which the function should be evaluated. These are

$$x = 0 \qquad x = 10 \qquad x = 12$$

At $x = 10$, $y = 22,000$, while at $x = 12$, $y = 19,856$, so that this time the overall maximum occurs at the turning point.

(iv) Here the upper endpoint of case (iii) is raised, and the value of the function at $x = 30$ must now be compared with the figure 22,000 at $x = 10$. At $x = 30$, $y = 254,000$, so this is definitely the global maximum.

(v) In this case, the overall maximum occurs when $x = -10$, where $y = 318,000$. Note how rapidly the function increases for negative x values since all of the terms will now be positive.

We have seen that a restricted domain can arise because of the nature of the independent variable. A case in point is where the decision variable (x) is *probability* where the domain is

$$0 \leq x \leq 1$$

The domain may be narrower if there are other considerations. To illustrate: in a two-player *zero-sum game* with two courses of action open to each player, player A must attach a probability $0 \leq x \leq 1$ to the selection of the first course of action. The opponent, B, has an equivalent choice and, given A's probability value, can determine which of two expected *pay-offs* accrues to A: E_1 or E_2, as shown in Figure 5.15. Player B, whose losses are

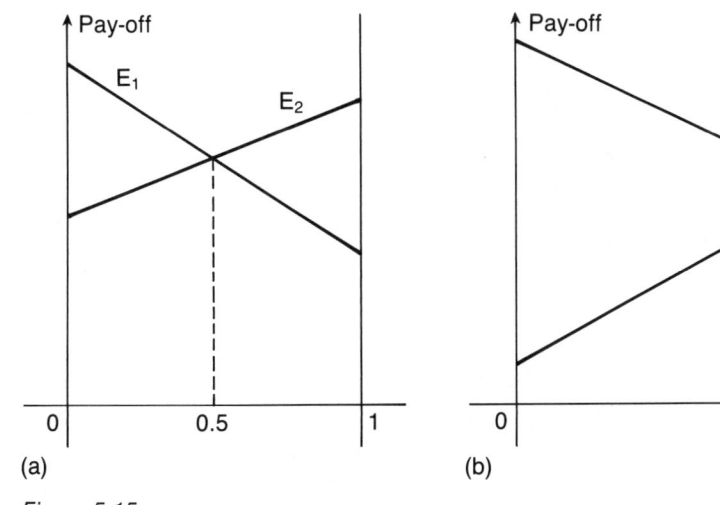

Figure 5.15

324 Mathematics for business, finance and economics

the gains of A, acts so as to produce the lower of the values E_1 and E_2. Thus only the *lower envelope* of the two lines is relevant to A who will set the probability x so that the lower envelope is at its highest point. Therefore, given the lines E_1 and E_2 of Figure 5.15(a), player A will attach probability $x = 0.5$ to the selection of the first course of action. In Figure 5.15(b), the maximum of the minimum values of E_1 and E_2 reaches its highest point when $x = 1$, meaning that player A should use the first course of action with certainty. Figure 5.16 shows an example from inventory control. In this model, in which demand for the stocked product is uncertain, the decision-maker must decide the level of service provided to customers. The service level is measured by the probability of *not* being out of stock.

In Figure 5.16, two components of cost are shown as functions of the no-stockout probability. These are *buffer stock* holding costs (BSHC) and *stockout* costs. Stockout costs (SOC) are those costs attributable to not being able to supply customers directly from stock. These costs will decline (typically linearly) with the service level. BSHC are the costs of holding a reserve or *buffer* of stock against higher than expected demand or a longer than expected re-supply interval. These costs rise with increasing steepness and ultimately without limit as 100 per cent service is approached. The decision-maker's objective is to minimize the sum of these costs, T, where

$$T = \text{BSHC} + \text{SOC}$$

In Figure 5.16, the service level in the best interests of the company is 97 per cent (corresponding to a 0.97 supply-from-stock probability).

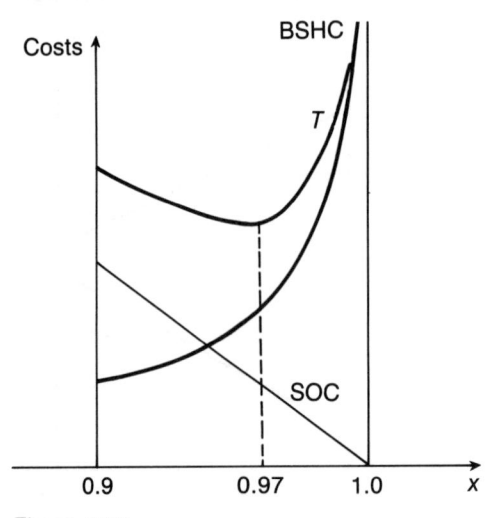

Figure 5.16

EXERCISES 5.12

1 With the specified turning points, identify the global maximum for each of the following functions:

   (i) $f(x) = 1000 - 3x^4 + 56x^3 - 336x^2 + 768x$
   (turning points at $x = 2$, $x = 4$, $x = 8$)
   (ii) $f(x) = 250 + 12x - 10.5x^2 + 7x^3 - 1.5x^4$
   (turning points at $x = 0.5$, $x = 1$, $x = 2$)

2 The following function has turning points at

   $x = 3 \quad x = 6 \quad x = 12$

   Find the global minimum of $f(x)$:

   $f(x) = x^4 - 28x^3 + 252x^2 - 864x + 5000$

3 Which of the following functions have finite global maxima?

   (i) $f(x) = 100{,}000 + 2500x - 0.0005x^2$
   (ii) $f(x) = 0.01x^2 - 5000x - 10{,}000{,}000$
   (iii) $f(x) = 0.01x^3 - 1{,}000{,}000x^2$
   (iv) $f(x) = 75 + 20x + 30x^2 - 0.1x^3$
   (v) $f(x) = x^{-1} - x^{10}$
   (vi) $f(x) = 4000x - x^4$

4 Find the global maximum of the function $f(x)$ in each of the following cases:

   (i) $f(x) = 1500 - x^3 + 39x^2 - 360x$
   (with domain restricted to $5 \leqslant x \leqslant 25$)
   (ii) $f(x) = 1500 - x^3 + 33x^2 - 360x$
   (with domain restricted to $2 \leqslant x \leqslant 15$)
   (iii) $f(x) = 1500 - x^3 - 18x^2 - 60x$
   (with domain restricted to $0 \leqslant x \leqslant 3$)
   (iv) $f(x) = x^3 - 45x^2 + 600x$
   (with domain restricted to $5 \leqslant x \leqslant 30$).

5.13 CONCAVITY, CONVEXITY AND POINTS OF INFLECTION

In the second derivative test, we saw how the value taken by the second derivative allowed local maxima to be distinguished from local minima. The test can be regarded as distinguishing those stationary values at which the function's curvature[18] makes it locally 'hollow from below', or *concave*, from those points where the function is locally 'hollow from above', or *convex*.[19] Local concavity means that a stationary value will be a local maximum of the function, while local convexity implies that the turning point produces a local minimum. Figure 5.17 illustrates this distinction for a cubic.

A function is concave over an interval of x if for *any* two points in the interval the function lies on or above the straight line connecting the corresponding points on the graph. If the function is always above the straight line, it is *strictly* concave over the interval. The function is convex over the interval if the function lies on or *below* a straight line connecting the points on the graph corresponding to any two points in the interval. The function is *strictly* convex over the interval if it is always below the straight line. Figure 5.17 gives examples of concavity or convexity over an interval. The function is concave between points a and c because, no matter how closely the endpoints of the interval are approached, the function lies above the straight line connecting those points. However, care is required. For instance, the function of Figure 5.17 cannot be said to be concave over the entire interval a–e even though the function lies above the straight line connecting a and e. The reason is that points can be selected near to e where this is not the case. The second derivative being negative ($f''(x) < 0$) is sufficient[20] for strict concavity. This corresponds to the slope of the function diminishing and the graph curving down. The second derivative being positive ($f''(x) > 0$) is sufficient for strict convexity. This corresponds to the slope of the function increasing and the graph bending up. Note that in a concave or convex section the slope of the curve itself could be either negative or positive.

Some functions (such as quadratics) have the same curvature throughout, while others (such as cubics) change in curvature[21] as x varies through the domain. A function for which $f''(x) \leq 0$ is globally concave,[22] and is said to be a *concave function*. A function for which $f''(x) \geq 0$ is globally convex,[23] and is said to be a *convex function*. A linear function satisfies

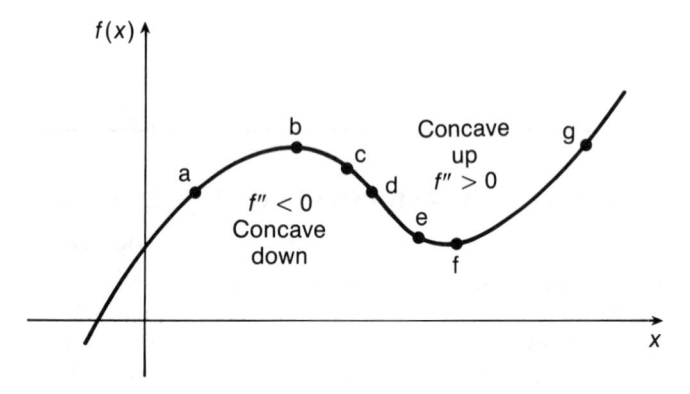

Figure 5.17

both of these conditions (since $f''(x) = 0$) and is therefore *both* a concave function and a convex function. However, a function such as that illustrated in Figure 5.17 is *neither* a concave function nor a convex function. In maximization problems it is very useful to know if a function is a concave function. If this is so, a local maximum, if any, will also be a global maximum. More than this, if a function is known to be globally concave, then all that is required for a global maximum at $x = x^0$ is the first-order condition

$$f'(x) = 0 \text{ at } x = x^0$$

which is both necessary and sufficient for a global maximum of a concave function. If the function is strictly concave throughout, then x^0 produces the *unique* global maximum. In minimization problems it is of similar utility to know that a function is *convex* over the whole domain. In such a case, a necessary and sufficient condition for a global minimum of the function at $x = x^0$ is

$$f'(x) = 0 \quad \text{at} \quad x = x^0$$

Here again, if the function is strictly convex throughout, then x^0 produces the *unique* global minimum. Three further points concerning concave and convex functions are worth noting.

1 If $f(x)$ is a concave function, then $g(x) = -f(x)$ is a convex function.
2 The sum of convex (concave) functions is a convex (concave) function.
3 Points lying on or below a concave function form a convex set. Points lying on or above a convex function form a convex set.

Point 2 means that in a function with several terms, if the terms each produce a concave function, then the overall function is also concave. Point 3 is illustrated in Figures 5.18(a) and 5.18(b).

Points of inflection

Where a function changes from strict concavity to strict convexity (or vice versa) over an interval, there will be a point in the interval where the second derivative is zero.[24] Such a point is an example of a *point of inflection*.[25] In Figure 5.17, point d is an inflection point. We shall concentrate on those cases where the second derivative of the function exists where the point of inflection arises. But note that points at which the second derivative is *not defined* may also be points of inflection. For example, $y = x^{1/3}$ inflects at $x = 0$, which is a critical point in the sense that the tangent is vertical. In such cases, the convexity or concavity of the function must be examined on either side of the point in question.

A distinction is drawn between inflection points such as d, where the function is itself changing in value, and those points – *stationary* points of

f(x) ↑   f(x) ↑

          (a)   (b)

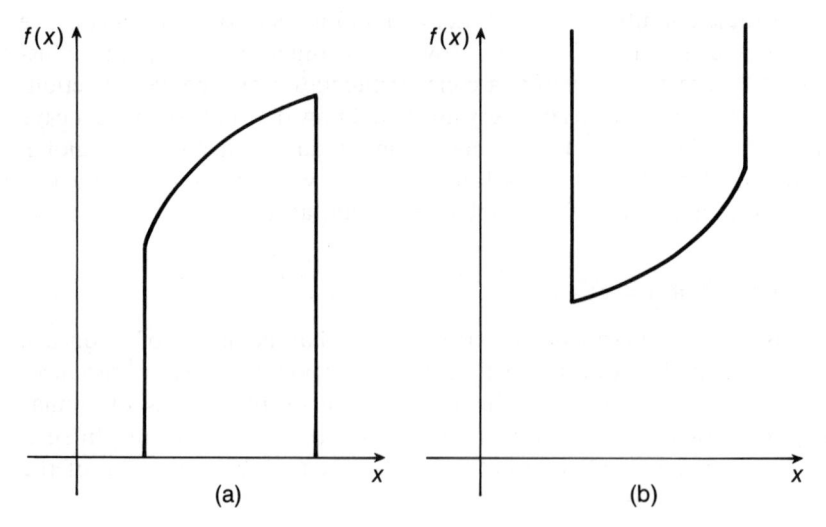

Figure 5.18

inflection – where the function is static and the first derivative is also zero. Point d is therefore an example of a *non-stationary* point of inflection. The usual way in which a stationary point of inflection is discovered is if, in the search for a maximum or minimum of a function, a value of x is found which produces $f'(x) = 0$ and for which $f''(x)$ is also zero. However, it should be noted that not all points at which $f''(x) = 0$ are points of inflection.[26] Consider some examples. First find the point of inflection of the function

$$f(x) = -0.5x^3 + 15x^2 + 20x + 100 \tag{5.35}$$

Here we have the advantage of knowing that all cubics have a point of inflection. We shall now locate this point. The first derivative is

$$f'(x) = -1.5x^2 + 30x + 20$$

Obtaining the second derivative and setting this equal to zero produces

$$f''(x) = -3x + 30 = 0$$

which solves for $x = 10$. This is a *possible* inflection point. It *will* be an inflection point if the second derivative also *changes sign* as it passes through $x = 10$. This possibility can be checked by finding the value of $f''(x)$ on either side of $x = 10$ and close to this value. For example, taking a one unit variation,

$$x = 9 \text{ gives } f''(x) = 3$$

while

$x = 11$ gives $f''(x) = -3$

So in this case the function does change sign through $x = 10$, which therefore produces a point of inflection. Finally, substitution of $x = 10$ into $f'(x)$ produces a non-zero result (170) so that $x = 10$ in fact gives a *non-stationary* point of inflection. The requirement that $f''(x)$ changes sign through the x value for which $f''(x)$ is zero is seen to be necessary when it is recalled that on one side of the point of inflection the function is concave ($f''(x) < 0$) while on the other side it is convex ($f''(x) > 0$). If the *third* derivative is non-zero at the candidate point of inflection, this is sufficient to ensure that the second derivative changes sign. Such is the case for (5.35) for which

$f'''(x) = -3$

The sign of $f'''(x)$ is immaterial – the important fact is that it is non-zero. As we shall see, however, the third derivative could also be zero, and yet the point could *still* be a point of inflection. Where $f'''(x)$ is easily calculated, this is a quicker check than working out $f''(x)$ either side of the point in question. Now consider another example. Let

$$f(x) = x^3 - 60x^2 + 1200x + 10{,}000 \qquad (5.36)$$

The first derivative is

$f'(x) = 3x^2 - 120x + 1200$

and the second derivative is

$f''(x) = 6x - 120$

From equating the second derivative to zero, the value of $x = 20$ emerges. The fact that the third derivative

$f'''(x) = 6 \neq 0$

confirms that $x = 20$ produces a point of inflection. In this case at $x = 20$ the value of $f'(x)$ is also zero so that a *stationary* point of inflection is produced. It is possible for a function to inflect at a point even if the third derivative is zero. This is confirmed by examination of the quintic

$$f(x) = x^5 \qquad (5.37)$$

which inflects at $x = 0$ (a stationary point of inflection) at which point all derivatives down to the fifth are zero. These discussions will lead us to a comprehensive rule for sorting out the character of the turning and inflecting points of functions, but first consider one further example. Does the quartic

$$f(x) = 0.25x^4 - 10x^3 + 150x^2 - 1000x \qquad (5.38)$$

have a point of inflection? The first derivative of this function is

$$f'(x) = x^3 - 30x^2 + 300x - 1000$$

and the second derivative is

$$f''(x) = 3x^2 - 60x + 300$$

Setting $f''(x) = 0$ gives the equation

$$3x^2 - 60x + 300 = 0$$

i.e.

$$x^2 - 20x + 100$$

which factors as

$$(x - 10)^2 = 0$$

so there is a double root, and the second derivative is zero only at $x = 10$. Checking the third derivative at this point,

$$f'''(x) = 6x - 60$$
$$= 0 \text{ at } x = 10$$

which does not resolve the issue. Trying $f''(x)$ for one unit departures from $x = 10$, setting $x = 9$ gives

$$f''(x) = 3(81) - 60(9) + 300$$
$$= +3$$

and when $x = 11$

$$f''(x) = 3(121) - 60(11) + 300$$
$$= +3$$

These results show that the second derivative does not change sign as x passes through the value 10. It follows that $x = 10$ does *not* produce a point of inflection for the function. Note in this case that

$$f'(x) = 0 \quad \text{at} \quad x = 10$$

so what kind of a stationary value is $x = 10$? In fact it produces a *minimum* of (5.38). This fact can be deduced from the following general rule for establishing the nature of stationary values:

> If all derivatives as far as $f^n(x)$ are zero at $x = x^0$ but $f^n(x) \neq 0$, then if n is even, $f^n(x) < 0$ means that the point is a local maximum, while $f^n(x) > 0$ implies a local minimum. If n is an *odd* number then $x = x^0$ produces a point of inflection.

Using this rule in the case of (5.38), the first non-zero derivative is $f^{iv} = 6$. Thus since the fourth (even) derivative is positive, the value $x = 10$ produces a minimum. Using the rule on the quintic (5.37), the first non-zero derivative at $x = 0$ is $f^v = 120$. Since it is the fifth (odd) derivative that is the first to be non-zero, we conclude that there must be a point of inflection at $x = 0$. Another, sometimes more convenient, way of determining the character of a point where several higher derivatives are zero is to find out if $f'(x)$ changes sign as x increases through the point in question. If it does, the point will be a maximum if the change in sign is from positive to negative or a minimum for a negative to positive change. If $f'(x)$ does not change in sign, a stationary point of inflection is identified.

In microeconomics, in the theory of production and cost, a cubic cost function is often employed for its plausible economic properties. For example, suppose that the daily costs of a car manufacturer are given by

$$C(x) = 0.1x^3 - 30x^2 + 3100x + 10{,}000 \tag{5.39}$$

where $C(x)$ is the total cost of producing x cars per day. *Marginal cost* is the rate of change of total cost, $C'(x)$, and is given by

$$C'(x) = 0.3x^2 - 60x + 3100$$

The point of inflection of the total cost function (5.39) is the daily output level that corresponds to the minimum of marginal cost. This is given by

$$C''(x) = 0.6x - 60 = 0$$

which solves for $x = 100$. Beyond this output marginal costs begin to rise – even though average costs continue to fall for some time. The example illustrates the fact that to find a point of inflection is to find the maximum or minimum of the first derivative.

EXERCISES 5.13

1 Find the point of inflection for:

   (i) $f(x) = 1000 - 2.5x^3 + 60x^2 + 300x$
   (ii) $f(x) = 4x^3 - 48x^2 + 25x + 100$
   (iii) $f(x) = 0.25x^4 - 9x^3 + 108x^2 + 20x$
   (iv) $f(x) = 0.15x^5 - 0.75x^4 + x^3 + 15x + 300$

2 Establish whether the points of inflection for each of the following functions are stationary or non-stationary:

   (i) $f(x) = x^3 - 18x^2 + 108x + 100$
   (ii) $f(x) = 3x^3 - 27x^2 + 100x - 50$
   (iii) $f(x) = x^4 - 14x^3 + 60x^2 - 50x + 100$

3 Do the following functions have points of inflection?
   (i) $f(x) = 0.25x^4 - 5x^3 + 37.5x^2 - 10x + 50$
   (ii) $f(x) = 0.5x^4 - 3x^3 - 30x^2 + 5x + 20$
   (iii) $f(x) = 3x^5 - 100x^4 + 1000x^3$

5.14 BASIC PROGRAM

In this section we list a program to find the derivatives of a polynomial function. The program, *Polydiff.bas*, will provide any requested derivative of the polynomial, and can be of value in checking results, particularly if higher order derivatives are required. The program will run in QuickBasic or Q-Basic. The listing is as follows.

Program 5.4

```
'                       POLYDIFF.BAS

DECLARE SUB differentiate ()

COMMON SHARED a(), a$(), product(), power(), n%, m%

CLS

PRINT 'Program differentiates polynomials.'
PRINT 'Please input coefficients as requested.'
PRINT '-------------------------------------'
PRINT

INPUT 'Degree of polynomial'; n%

PRINT

DIM a(n%)
DIM a$(n%)
DIM product(n%)
DIM power(n%)

FOR i% = n% TO 2 STEP −1

      PRINT 'Coefficient of x^ '; i%

      INPUT a(i%)

      PRINT

NEXT i%

PRINT 'Coefficient of x'
INPUT a(1)

PRINT

PRINT 'Constant term'
INPUT a(0)
```

```
CLS
PRINT 'The polynomial is...'
PRINT
PRINT 'y = '; a(n%); 'x^ '; n%;
IF a(n%) < 0 THEN a$(n%) = ' – '
FOR i% = n% –1 TO 2 STEP  –1
    IF a(i%) > 0 THEN
        a$(i%) = ' + '
        PRINT a$(i%); a(i%); 'x^ '; i%;
    ELSEIF a(i%) < 0 THEN
        a$(i%) = ' – '
        PRINT a$(i%); ABS(a(i%)); 'x^ '; i%;
    END IF
NEXT i%
IF a(1) > 0 THEN
    a$(1) = ' + '
    PRINT a$(1); a(1); 'x';
ELSEIF a(1) < 0 THEN
    a$(1) = ' – '
    PRINT a$(1); ABS(a(1)); 'x';
END IF
IF a(0) > 0 THEN
    a$(0) = ' + '
    PRINT a$(0); a(0)
ELSEIF a(0) < 0 THEN
    a$(0) = ' – '
    PRINT a$(0); ABS(a(0))
```

END IF

PRINT
PRINT '--'
PRINT

DO

    CALL differentiate

    PRINT

    INPUT 'Find another derivative y/n'; r$
    PRINT

LOOP WHILE r$ = 'y' OR r$ = 'Y'

END

SUB differentiate

startup:

PRINT 'Order of derivative required'
INPUT 'First = 1, second = 2, third = 3 etc.'; order
PRINT

IF order < 1 OR INT(order) < order THEN

    PRINT 'Order must be positive integer, try again.'
    PRINT

    GOTO startup

END IF

m% = order

IF m% > n% THEN

    PRINT 'All derivatives of order > '; n%; 'are zero';
    PRINT

    EXIT SUB

END IF

    FOR i% = n% TO m% STEP −1

```
        IF a(i%) < > 0 THEN

            product(i%) = a(i%)

            FOR k% = 1 TO m%

                product(i%) = product(i%)*(i% +1 – k%)

            NEXT k%

        END IF

    NEXT i%

    FOR i% = n% – 1 TO m% STEP –1

        power(i%) = i% – m%

    NEXT i%

    SELECT CASE m%

        CASE 1

            PRINT ' The first derivative is...'
            PRINT

        CASE 2

            PRINT 'The second derivative is...'
            PRINT

        CASE 3

            PRINT 'The third derivative is...'
            PRINT

        CASE ELSE

            PRINT 'The'; m%; 'th derivative is...'
            PRINT

    END SELECT

    IF n% – m% > 1 THEN

        IF product(n%) < > 0 THEN

            PRINT product(n%); 'x^ '; n% – m%;

        END IF
```

```
END IF

FOR i% = n%−1 TO m%+2 STEP −1

    IF product(i%) < > 0 THEN

        PRINT a$(i%); ABS(product(i%)); 'x^ '; power(i%);

    END IF

NEXT i%

IF m% < n% THEN

    IF product(m%+1) < > 0 THEN

        PRINT a$(m%+1); ABS(product(m%+1)); 'x';

    END IF

END IF

IF m% < n%+1 THEN

    IF product(m%) < > 0 THEN

        PRINT a$(m%); ABS(product(m%))

    END IF

END IF

END SUB
```

ADDITIONAL PROBLEMS

1 The demand curve for a company's product is given by

   $p = 200 - 4q$

where p is price and q is quantity. The firm's cost function is

   $C = 100 + 20q + q^2$

   (i) What output leads to the highest level of revenue from sales?
   (ii) Find the *marginal revenue* at $q = 10$.
   (iii) Find the value of *marginal cost* at $q = 15$.
   (iv) For what output level is marginal revenue the same as marginal cost?
   (v) What output level gives the highest profit?
   (vi) Will the profit parabola always peak at a lower output level than the sales revenue parabola?

2 A commodity has market demand and supply curves given by

   Demand: $p = 500 - x$
   Supply: $p = 100 + 3x$

   (i) Find the equilibrium (market clearing) quantity.

   Now suppose that the government imposes an excise duty in the form of a tax of £t per unit supplied, so that the supply curve becomes

   $p = 100 + 3x + t$

   (ii) Find the equilibrium quantity if $t = 40$. What would be the 'tax take' in this position?
   (iii) What would be the cost to the government of a *subsidy* of 40 per unit?
   (iv) What levels of the per unit tax would produce revenue of £7500 for the government?
   (v) What level of t maximizes the government's tax receipts?

3 The numbers of kilometres travelled per litre of fuel consumed for two types of car are given by

   Model A $y = 14 - 0.008x^2 + 0.64x$
   Model B $y = 18 - 0.016x^2 + 0.96x$

   where y is kilometres per litre and x is speed in kilometres per hour.

   (i) Find the most economical speed for each model and the fuel consumption figures achieved.
   (ii) For what range of speeds does model B have superior fuel consumption to model A?

4 For a particular firm, the relationship between its level of profit, π, and the selling price of its product, p, is given by

   $\pi = 500p - 12.5p^2 - 950$

   (i) For what range of values of p is profit an increasing function of selling price?
   (ii) What is the profits maximizing price and the best level of profits?
   (iii) What is the highest price that the company can charge and still not make a loss?

5 The total cost function for a firm is

$$C = x^3 - 18x^2 + 120x + 100$$

where C is total cost and x is output level.

(i) Find the level of output that gives rise to a point of inflection in the cost function.
(ii) At what point does marginal cost reach its minimum?

6 *Price elasticity of demand*, e, is a dimensionless measure of the sensitivity of the demand for a product at a point to price changes, and is given by

$$e = -\frac{p}{q}\frac{dq}{dp}$$

where q is quantity demanded and p is price. In the case where

$$p = 1000 - 2q$$

find the elasticity of demand when $q = 5$, $q = 50$ and $q = 500$.

7 A *consumption function* relates aggregate consumption expenditure (C) to the level of national income (Y). In a particular case, suppose that the consumption function is

$$C = 570 - \frac{160{,}000}{Y + 200}$$

The *marginal propensity to consume* (MPC) shows the rate of change of consumption expenditure as income varies. For the above case, find C and MPC for $Y = 300$, $Y = 600$ and $Y = 800$.

REFERENCES AND FURTHER READING

1 Borowski, E. J. and Borwein, J. M. (1989) *Dictionary of Mathematics*, Collins.
2 Chiang, A. C. (1984) *Fundamental Methods of Mathematical Economics* (Third Edition), McGraw-Hill.
3 Hart, W. L. (1966) *College Algebra* (Fifth Edition), Heath.
4 Mizrahi, A. and Sullivan, M. (1988) *Mathematics for Business and Social Sciences* (Fourth Edition), Wiley.
5 Weber, J. E. (1982) *Mathematical Analysis: Business and Economic Applications* (Fourth Edition), Harper and Row.

SOLUTIONS TO EXERCISES

Exercises 5.3

1 The slope of the straight line segment connecting two points on $y = 5x^2$ for which the x co-ordinate differs by Δx is given by

$$\frac{\Delta y}{\Delta x} = \frac{5(x + \Delta x)^2 - 5x^2}{\Delta x}$$

$$= \frac{5[x^2 + 2x\,\Delta x + (\Delta x)^2] - 5x^2}{\Delta x}$$

$$= \frac{5x^2 + 10x\,\Delta x + 5(\Delta x)^2 - 5x^2}{\Delta x}$$

$$= \frac{10x\,\Delta x + 5(\Delta x)^2}{\Delta x}$$

$$= 10x + 5\,\Delta x$$

but as Δx approaches zero, so does $5\,\Delta x$, so that the derivative of $y = 5x^2$ is

$$\frac{dy}{dx} = 10x$$

2 The slope of the straight line segment connecting two points on $y = x^4$ for which the x co-ordinate differs by Δx is given by

$$\frac{\Delta y}{\Delta x} = \frac{(x + \Delta x)^4 - x^4}{\Delta x}$$

so we shall need to expand $(x + \Delta x)^4$. This is

$$x^4 + 4x^3\,\Delta x + 6x^2(\Delta x)^2 + 4x(\Delta x)^3 + (\Delta x)^4$$

from which subtraction of x^4 and division through by Δx gives

$$\frac{\Delta y}{\Delta x} = 4x^3 + 6x^2\,\Delta x + 4x(\Delta x)^2 + (\Delta x)^3$$

and as $\Delta x \to 0$, so do all of the above terms that include Δx, with the result that

$$\frac{dy}{dx} = 4x^3$$

Exercises 5.4

1 (i) $\dfrac{dy}{dx} = 7x^6$

(ii) $\dfrac{dy}{dx} = 100x^{99}$

(iii) $\dfrac{dy}{dx} = -3x^{-4}$, i.e. $\dfrac{-3}{x^4}$

(iv) $\dfrac{dy}{dx} = -4x^{-5}$, i.e. $\dfrac{-4}{x^5}$

(v) $\dfrac{dy}{dx} = (1/3)x^{-2/3}$, i.e. $\dfrac{1}{3x^{2/3}}$

(vi) $\dfrac{dy}{dx} = (2/3)x^{-5/3}$, i.e. $\dfrac{2}{3x^{5/3}}$

(vii) $f'(x) = \sqrt{3}x^{\sqrt{3}-1}$

(viii) $f'(x) = \sqrt{5}x^{-\sqrt{5}-1}$, i.e. $\dfrac{\sqrt{5}}{x^{\sqrt{5}+1}}$

(ix) $f'(x) = 2\pi x^{2\pi - 1}$

(x) $\dfrac{dy}{dx} = 40x^7$

(xi) $\dfrac{dy}{dx} = x^3$

(xii) $\dfrac{dy}{dx} = 2x^{-11}$, i.e. $\dfrac{2}{x^{11}}$

(xiii) $f'(x) = 0$

2 (i) $f'(x) = 9x^8$

(ii) $f'(x) = -9x^{-10}$, i.e. $\dfrac{-9}{x^{10}}$

(iii) $f'(x) = 8x^{-9}$, i.e. $\dfrac{8}{x^9}$

(iv) $\dfrac{dy}{dx} = \dfrac{-10}{x^{11}}$, i.e. $-10x^{-11}$

(v) $\dfrac{dy}{dx} = \dfrac{-1}{10x^2}$, i.e. $-0.1x^{-2}$

(vi) $f'(x) = \sqrt{4}x^{\sqrt{4}-1}$

342 Mathematics for business, finance and economics

(vii) $f'(x) = \sqrt{2} x^{-\sqrt{2}-1}$, i.e. $\dfrac{\sqrt{2}}{x^{\sqrt{2}+1}}$

(viii) $\dfrac{dy}{dx} = \dfrac{10}{x^{11}}$, i.e. $10x^{-11}$

(ix) $f'(x) = \sqrt{\pi} x^{\sqrt{\pi}-1}$
(x) $f'(x) = 10{,}000 x^{99}$
(xi) $f'(x) = -x^{19}$

(x) $\dfrac{dy}{dx} = -0.01 x^{-1.1}$, i.e. $-\dfrac{0.01}{x^{1.1}}$

3 (i) $f'(x) = 1$
(ii) $f'(x) = mx^{m-1}$
(iii) $f'(x) = -px^{-p-1}$
(iv) $f'(x) = (n-1)x^{-n}$

(v) $\dfrac{dy}{dx} = \dfrac{n}{x^{n+1}}$, i.e. nx^{-n-1}

(vi) $f'(x) = bax^{b-1}$
(vii) $f'(x) = x^{m-1}$

(viii) $\dfrac{dy}{dx} = \sqrt{n} x^{\sqrt{n}-1}$

(ix) $f'(x) = \pi^2 x^{\pi^2-1}$

4 (i) $\dfrac{dy}{dx} = -\dfrac{1}{x^2}$, i.e. $-x^{-2}$

(ii) $\dfrac{dy}{dx} = 0$

(iii) $\dfrac{dy}{dx} = x^{\sqrt{2}}$

(iv) $f'(x) = -0.04 x^{-0.8}$

Note that, in part (iii), $(\sqrt{2}-1)(\sqrt{2}+1) = 2 - 1 = 1$.

Exercises 5.5

1 (i) $\dfrac{dy}{dx} = 4x^3 + 3x^2$

(ii) $\dfrac{dy}{dx} = 5x^4 - 3x^2$

Slopes, derivatives and turning points 343

(iii) $\dfrac{dy}{dx} = 6x^5 + 5x^4 - 4x^3$

(iv) $f'(x) = 20x^3 + 9x^2 - 14x$

(v) $f'(x) = 45x^4 - 28x^3 + 15x^2 - 16x + 14$

2 (i) $f'(x) = max^{m-1} + nbx^{n-1}$

(ii) $f'(x) = (m+1)cx^m - (n-1)dx^{n-2}$

(iii) $f'(x) = 3ax^2 + 2bx + c$

(iv) $\dfrac{dy}{dx} = x^n + x^m$

(v) $\dfrac{dy}{dx} = 5a_0x^4 + 4a_1x^3 + 3a_2x^2 + 2a_3x + a_4$

3 (i) $\dfrac{dy}{dx} = 12x - 156$

So a stationary value occurs at $x = 13$ where the derivative is zero.

(ii) $\dfrac{dy}{dx} = 10x + 100$

So a stationary value occurs at $x = -10$.

(iii) $\dfrac{dy}{dx} = 6x^2 - 18x - 60$

So stationary values occur at $x = -2$ and $x = 5$.

Exercises 5.6

1 (i) $\dfrac{dy}{dx} = (x^2 + x)(3x^2 - 2x) + (x^3 - x^2)(2x + 1)$

(ii) $\dfrac{dy}{dx} = (5x^4 - 4x^2)(30x^2 + 35x^4)$

$\quad + (10x^3 + 7x^5)(20x^3 - 8x)$

(iii) $\dfrac{dy}{dx} = (4x^3 + 3x^2 + 8x)(2 + 2x + 4x^3)$

$\quad + (2x + x^2 + x^4)(12x^2 + 6x + 8)$

(iv) $\dfrac{dy}{dx} = (x+1) + (x-1) = 2x$

2 (i) $\dfrac{dy}{dx} = \dfrac{(x^5 - x^4)(3x^2 + 2x) - (x^3 + x^2)(5x^4 - 4x^3)}{(x^5 - x^4)^2}$

(ii) $\dfrac{dy}{dx} = \dfrac{(3x^4 - 2x^3)(20x^3 + 14x^6) - (5x^4 + 2x^7)(12x^3 - 6x^2)}{(3x^4 - 2x^3)^2}$

(iii) $f'(x) = \dfrac{(7x^3 + 5) - x(21x^2)}{(7x^3 + 5)^2}$

(iv) $f'(x) = \dfrac{x^3(8x - 5) - (4x^2 - 5x + 3)3x^2}{x^6}$

3 (i) The quotient rule is relevant, since the function is

$$f(x) = \dfrac{x^2 - 1}{x}$$

Using the rule:

$$f'(x) = \dfrac{x(2x) - (x^2 - 1)}{x^2}$$

$$= \dfrac{2x^2 - x^2 + 1}{x^2}$$

$$= \dfrac{x^2 + 1}{x^2}$$

$$= 1 + \dfrac{1}{x^2}$$

This result could have been obtained without the use of the quotient rule, since the function $f(x)$ simplifies to

$$f(x) = x - \dfrac{1}{x}$$

(ii) Using both the product and the quotient rules, with the first two brackets representing the numerator, the derivative is

$$f'(x) = [(3x - 5)[(x^3 - 2x^2)(2x + 2) + (x^2 + 2x)(3x^2 - 4x)] - (x^3 - 2x^2)(x^2 + 2x)3] \div (3x - 5)^2$$

i.e.

$$f'(x) = \dfrac{x^2(12x^3 - 25x^2 - 24x + 60)}{(3x - 5)^2}$$

for which, at $x = 1$,

$$f'(x) = \dfrac{23}{4}$$

and at $x = 2$, $f'(x)$ evaluates as 32.

Slopes, derivatives and turning points 345

Exercises 5.7

1. (i) $f'(x) = 4(x^3 + 1)^3(3x^2)$
   (ii) $f'(x) = 5(x^4 + x^2)^4(4x^3 + 2x)$
   (iii) $f'(x) = 7(4x^5 + 3x^4 - 7x^2 + 2x)^6(20x^4 + 12x^3 - 14x + 2)$
   (iv) $f'(x) = -25(1 - x^5)^4(5x^4) = -125x^4(1 - x^5)^4$
   (v) $\dfrac{dy}{dx} = 3(x^2 + x)^2(2x + 1) + 2(x^3 + 2x)(3x^2 + 2)$
   (vi) $\dfrac{dy}{dx} = 6(6x^7 - 5x^4 + 6x^2)^5(42x^6 - 20x^3 + 12x)$
   $\qquad - 9(x^8 - x^7)^8(8x^7 - 7x^6)$

2. (i) $\dfrac{dy}{dx} = (x^3 + x^4)^5 6(x^5 - x^3)^5(5x^6 - 3x^2)$
   $\qquad + (x^5 - x^3)^6 5(x^3 + x^4)^4(3x^2 + 4x^3)$
   (ii) $\dfrac{dy}{dx} = \dfrac{(x^2 - 1)^3(12x^2 + 6) - (4x^3 + 6x)3(x^2 - 1)^2 2x}{(x^2 - 1)^6}$
   (iii) $\dfrac{dy}{dx} = \dfrac{8x^4 3(3x^2 - 2x)^2(6x - 2) - 32x^3(3x^2 - 2x)^3}{64x^8}$

3. (i) $f'(x) = \tfrac{1}{2}(5x^3 - 7x^2)^{-1/2}(15x^2 - 14x)$
   (ii) $f'(x) = 0.25(4x^3 + 8x)^{-0.75}(12x^2 + 8)$

4. (i) $f'(x) = n(ax^2 + bx + c)^{n-1}(2ax + b)$
   (ii) $f'(x) = (m^2 - 1)[(m + 1)x + m]^m$

5. (i) $\dfrac{dy}{dx} = 15z^2(16x^3)$
   $\qquad = 240z^2 x^3$
   $\qquad = 240(4x^4)^2 x^3$
   $\qquad = 240(16)(x^8)x^3$
   $\qquad = 3840 x^{11}$

   (ii) $\dfrac{dy}{dx} = (2z - 1)(4x - 2)$
   $\qquad = (4x^2 - 4x - 1)(4x - 2)$
   $\qquad = 16x^3 - 24x^2 + 4x + 2$

6 (i) $\dfrac{dy}{dx} = (2z+z)(2)(3x^2-1)$

$= (4w+2)(2)(3x^2-1)$
$= (8w+4)(3x^2-1)$
$= 4(3x^2-1)(2x^3-2x+1)$

which expands as

$24x^5 - 32x^3 + 12x^2 + 8x - 4$

(ii) When $x = 1$, the derivative dy/dx evaluates as 8.

Exercises 5.8

1 (i) $\dfrac{dx}{dy} = 0.2$ so that $\dfrac{dy}{dx} = \dfrac{1}{0.2} = 5$

(ii) After multiplication by 4 throughout the equation:

$\dfrac{dx}{dy} = 0.5$ so that $\dfrac{dy}{dx} = \dfrac{1}{0.5} = 2$

(iii) $\dfrac{dx}{dy} = 5y^4$ so that $\dfrac{dy}{dx} = \dfrac{1}{5y^4}$

(iv) $\dfrac{dx}{dy} = 8y^3 - 9y^2 + 8y$ so that

$\dfrac{dy}{dx} = \dfrac{1}{8y^3 - 9y^2 + 8y}$

2 (i) $\dfrac{dy}{dx} = \dfrac{1}{(y^4+y^2)(6y^2+1)+(2y^3+y)(4y^3+2y)}$

(ii) $\dfrac{dy}{dx} = \dfrac{1}{3(5y^3+12y-10)^2(15y^2+12)}$

3 $\dfrac{dy}{dx} = 5x^4 + 6x^2 + 5$

so that, by the inverse function rule,

$\dfrac{dx}{dy} = \dfrac{1}{5x^4 + 6x^2 + 5}$

and, when $x = 1$,

$\dfrac{dx}{dy} = \dfrac{1}{16} = 0.0625$

Slopes, derivatives and turning points 347

Exercises 5.9

1 (i) $\dfrac{dy}{dx} = \dfrac{4}{5}$

(ii) $\dfrac{dy}{dx} = \dfrac{5y + 10}{2 - 5x}$

2 Having first expressed y as an explicit function of x, the derivative dy/dx is then

$$\dfrac{dy}{dx} = \dfrac{-480}{(2 - 5x)^2}$$

which evaluates as follows:

(i) at $x = 10$, $y = 0$, $\dfrac{dy}{dx} = \dfrac{-480}{2304} = -0.20833$

(ii) at $x = 1$, $y = 30$, $\dfrac{dy}{dx} = \dfrac{-480}{9} = -53.33$

and in terms of the original derivatives obtained in 1(i) and 1(ii), the corresponding evaluations are

(i) at $x = 10$, $y = 0$, $\dfrac{dy}{dx} = \dfrac{10}{-48} = -0.20833$

(ii) at $x = 1$, $y = 30$, $\dfrac{dy}{dx} = \dfrac{160}{-3} = -53.33$

which confirms the equivalence of the two results.

3 (i) $\dfrac{dy}{dx} = \dfrac{2 - y^2}{2xy + 5}$

(ii) $\dfrac{dy}{dx} = \dfrac{6x - y - 6x^2 y^2}{4x^3 y + x - 7}$

Exercises 5.10

1 (i) $f''(x) = 20x^3$
 (ii) $f''(x) = 210x^4$
 (iii) $f''(x) = 60x^2 + 6$
 (iv) $f''(x) = 160x^3 - 108x^2 + 42x - 16$

2 (i) $\dfrac{d^3 y}{dx^3} = 120x^3$ $\dfrac{d^4 y}{dx^4} = 360x^2$

(ii) $\dfrac{d^3y}{dx^3} = 120x^2 - 72x$ $\dfrac{d^4y}{dx^4} = 240x - 72$

(iii) $\dfrac{d^3y}{dx^3} = 7.2x^7 - 34.2x^{17}$ $\dfrac{d^4y}{dx^4} = 50.4x^6 - 581.4x^{16}$

3 (i) $f''(x) = 2x^{-3}$

(ii) $f''(x) = \dfrac{-6}{x^4}$, i.e. $-6x^{-4}$

(iii) $f''(x) = -0.25x^{-1.5}$

(iv) Using the chain rule:

$$f'(x) = 7(x^2 + 5)^6 2x = 14x(x^2 + 5)^6$$

To obtain the second derivative, the product and chain rules should be used. The result is

$$f''(x) = 14x[6(x^2 + 5)^5 2x] + (x^2 + 5)^6 14$$
$$= 168x^2(x^2 + 5)^5 + 14(x^2 + 5)^6$$

4 (i) $\dfrac{d^2y}{dx^2} = 12ax^2 + 6bx + 2c$

(ii) $\dfrac{d^2y}{dx^2} = -m^2(m+1)x^{-m-2} + n^2(n+1)x^{-n-2}$

Exercises 5.11

1 (i) For a stationary value of the function:

$$\dfrac{dy}{dx} = 100 - 5x = 0$$

so that $x = 100/5 = 20$ is a turning point. At the second order:

$$\dfrac{d^2y}{dx^2} = -5 < 0$$

so the turning point represents a maximum.

(ii) $\dfrac{dy}{dx} = 8x - 72 = 0$

so that $x = 9$ is a turning point. At the second order:

$$\dfrac{d^2y}{dx^2} = +8$$

so that a *minimum* of the function occurs at $x = 9$.

2 (i) $f'(x) = 1.5x^2 - 4.5x - 60 = 0$

solves for $x = -5$, $x = 8$. At the second order:

$$f''(x) = 3x - 4.5$$

which is negative for $x = -5$ and positive for $x = 8$, so that

$x = -5$ is a maximum

$x = 8$ is a minimum

(ii) $f'(x) = -6x^2 + 12x + 48 = 0$

solves for $x = -2$, $x = 4$. At the second order:

$$f''(x) = -12x + 12$$

which is negative for $x = 4$ and positive for $x = -2$, so that

$x = 4$ is a maximum

$x = -2$ is a minimum

(iii) $f'(x) = 4x^3 - 36x^2 + 56x = 0$

in which, since there is no constant term, one root must be $x = 0$. We can therefore write $f'(x)$ as

$$(x - 0)(4x^2 - 36x + 56) = 0$$

Now taking 4 out of the large bracket and factoring gives

$$4x(x - 2)(x - 7) = 0$$

so that the roots are $x = 0$, $x = 2$ and $x = 7$. The second derivative is

$$f''(x) = 12x^2 - 72x + 56$$

and

when $x = 0$, $f''(x) = 56 > 0$

so that $x = 0$ corresponds to a *minimum* of $f(x)$;

when $x = 2$, $f''(x) = -40 < 0$

so that $x = 2$ corresponds to a *maximum* of $f(x)$. Finally,

when $x = 7$, $f''(x) = 140 > 0$

so that $x = 7$ corresponds to a *minimum* of $f(x)$.

(iv) $f'(x) = x^4 - 5x^2 + 4 = 0$

Note that the function $f'(x)$ is a quadratic in x^2 which factors as

$$f'(x) = (x^2 - 1)(x^2 - 4) = 0$$

in which each bracket in turn factors to produce

$$(x + 1)(x - 1)(x + 2)(x - 2) = 0$$

so that the stationary values of the function are $x = -1$, $x = 1$, $x = -2$ and $x = 2$. The second derivative is

$$f''(x) = 4x^3 - 10x$$

for which

at $x = -1$, $f''(x) = 6$ so that $x = -1$ is a minimum;
at $x = 1$, $f''(x) = -6$ so that $x = 1$ is a maximum;
at $x = -2$, $f''(x) = -12$ so that $x = -2$ is a maximum;
at $x = 2$, $f''(x) = 12$ so that $x = 2$ is a minimum.

3 (i) $\dfrac{dy}{dx} = 4x^3 - 24x^2 + 44x - 24$

$\dfrac{d^2y}{dx^2} = 12x^2 - 48x + 44$

At $x = 1$, $\dfrac{d^2y}{dx^2} = 8$ so that $x = 1$ produces a minimum

At $x = 2$, $\dfrac{d^2y}{dx^2} = -4$ so that $x = 2$ produces a maximum

At $x = 3$, $\dfrac{d^2y}{dx^2} = 8$ so that $x = 3$ produces a minimum

(ii) $\dfrac{dy}{dx} = 30x^4 - 270x^3 - 960x^2 + 5400x + 12{,}000$

$\dfrac{d^2y}{dx^2} = 120x^3 - 810x^2 - 1920x + 5400$

At $x = -4$, $\dfrac{d^2y}{dx^2} = -7560$ so that $x = -4$ produces a maximum

At $x = -2$, $\dfrac{d^2y}{dx^2} = 5040$ so that $x = -2$ produces a minimum

At $x = 5$, $\dfrac{d^2y}{dx^2} = -9450$ so that $x = 5$ produces a maximum

At $x = 10$, $\dfrac{d^2 y}{dx^2} = 25{,}200$ so that $x = 10$ produces a minimum

4   (i) $f'(x) = 9x^2 - 36x - 1053 = 0$

for which the roots are $x = -9$ and $x = 13$. There are therefore two turning points: -9 (maximum) and 13 (minimum).

(ii) $f'(x) = 3x^2 - 6x + 15 = 0$

the roots for which are the complex pair

$x = 1 + 2i$ and $x = 1 - 2i$

so that $f(x)$ has no turning points.

(iii) $f'(x) = 4x^3 - 36x^2 + 160x - 400 = 0$

From Descartes's rule of signs, the *maximum* number of positive roots of $f'(x)$ is given by the number of sign changes (three) in the coefficients. But the number of positive roots is equal in parity with the number of sign changes. Therefore there must be three positive roots or one. In either case there is a real root, and the original quintic must have a turning point. In fact there is one real root, and the roots of $f'(x)$ are

$x = 5$ $x = 2 + 4i$ $x = 2 - 4i$

Exercises 5.12

1   (i) $f''(x) = -36x^2 + 336x - 672$

so that

when $x = 2$, $f''(x) = -144$ (so this is a local maximum)
when $x = 4$, $f''(x) = 96$ (local minimum)
when $x = 8$, $f''(x) = -288$ (local maximum)

Since the leading coefficient of $f(x)$ is negative and the power of the leading term is even, the function will decrease without limit at either extreme, and the global maximum will be at either $x = 2$ or $x = 8$. Evaluating $f(x)$ gives

$x = 2$ $f(x) = 1592$
$x = 8$ $f(x) = 2024$

so that the global maximum will be at $x = 8$.

(ii) $f''(x) = -21 + 42x - 18x^2$

so that

when $x = 0.5$, $f''(x) = -4.5$ (local maximum)
when $x = 1$, $f''(x) = 3$ (local minimum)
when $x = 2$, $f''(x) = -9$ (local maximum)

Evaluating $f(x)$ at the local maxima gives

$x = 0.5$, $f(x) = 254.15625$
$x = 2$, $f(x) = 264$

so that $x = 2$ is the global maximum.

2 $f''(x) = 12x^2 - 168x + 504$

so that

when $x = 3$, $f''(x) = 108$ (minimum)
when $x = 6$, $f''(x) = -72$ (maximum)
when $x = 12$, $f''(x) = 216$ (minimum)

Since the leading term is positive and increases without limit for x either positive or negative, then either $x = 3$ or $x = 12$ will be the global minimum.

When $x = 3$, $f(x) = 4001$
When $x = 12$, $f(x) = 3272$

Thus $x = 12$ constitutes the global minimum where $f(x) = 3272$.

3 (i) Since the leading coefficient of the quadratic is negative, there is a finite maximum.
   (ii) No finite maximum – the leading coefficient of the quadratic is positive.
   (iii) No finite maximum.
   (iv) No finite maximum ($f(x)$ increases without limit for negative x).
   (v) No finite maximum. The function increases without limit as x approaches zero.
   (vi) There is a finite maximum. $F(x)$ is unbounded from below at either extreme. Note that for

   $f'(x) = 4000 - 4x^3 = 0$

   $x = 10$ is a multiple root, and since

   $f''(x) = -12x^2 < 0$

   is satisfied for all x, then $x = 10$ is the single local maximum and the global maximum.

4 (i) First establish where the turning points are, and whether a local maximum lies within the permitted domain:

$$f'(x) = -3x^2 + 78x - 360 = 0$$

which solves for $x = 6$ and $x = 20$. The second derivative

$$f''(x) = -6x + 66$$

is negative at $x = 20$, showing that the local maximum lies within the restricted domain. The procedure described therefore calls for the function to be evaluated for the local maximum and the endpoints of the domain. The results are

$x = 5$ $f(x) = 550$
$x = 20$ $f(x) = 1900$
$x = 25$ $f(x) = 1250$

So that the global maximum over the restricted domain is 1900 at $x = 20$.

(ii) Establishing whether a local maximum lies within the permitted domain,

$$f'(x) = -3x^2 + 66x - 360 = 0$$

which solves for $x = 10$ and $x = 12$. The second derivative

$$f''(x) = -6x + 66$$

is negative at $x = 12$, showing that the local maximum lies within the domain. Evaluating the function for the local maximum and the endpoints of the domain, the results are

$x = 2$ $f(x) = 904$
$x = 12$ $f(x) = 204$
$x = 25$ $f(x) = 150$

so that the global maximum over the restricted domain is 904 at the endpoint $x = 2$.

(iii) The first derivative

$$f'(x) = -3x^2 - 36x - 60 = 0$$

solves for $x = -10$ and $x = -2$, both of which points lie outside the permitted domain. The function is then evaluated at the endpoints only giving

$x = 0$ $f(x) = 1500$
$x = 3$ $f(x) = 1131$

so that the global optimum in the restricted domain is at the endpoint $x = 0$.

(iv) The first derivative is

$$f'(x) = 3x^2 - 90x + 600 = 0$$

which solves for

$x = 10$ and $x = 20$

which are both within the domain. The second derivative

$$f''(x) = 6x - 90$$

identifies $x = 10$ as the local maximum, so, evaluating the function at this point and the endpoints,

$x = 5$ produces $f(x) = 2000$
$x = 10$ produces $f(x) = 2500$
$x = 30$ produces $f(x) = 4500$

so the endpoint $x = 30$ gives the global maximum of $f(x)$ in the domain.

Exercises 5.13

1 (i) The second derivative

$$f''(x) = -15x + 120$$

is zero for $x = 8$, and since the third derivative

$$f'''(x) = -15$$

is non-zero, $x = 8$ is confirmed as producing a point of inflection of $f(x)$.

(ii) The second derivative

$$f''(x) = 24x - 90$$

is zero for $x = 4$. The third derivative

$$f'''(x) = 24$$

is non-zero, so that, at $x = 4$, $f(x)$ has a point of inflection.

(iii) The second derivative

$$f''(x) = 3x^2 - 54x + 216$$

is zero for $x = 6$ and $x = 12$. The third derivative

$$f'''(x) = 6x - 54$$

is non-zero at both $x = 6$ and $x = 12$ which are confirmed as points of inflection.

Slopes, derivatives and turning points

(iv) The second derivative requirement

$$f''(x) = 3x^3 - 9x^2 + 6x = 0$$

has $x = 0$ as a solution, and with $3x$ taken out

$$3x(x^2 - 3x + 2) = 0$$

and so

$$3x(x-1)(x-2) = 0$$

from which it is clear that the other two roots are $x = 1$ and $x = 2$. The third derivative

$$f'''(x) = 9x^2 - 18x + 6$$

is non-zero at all three values, which are therefore confirmed as points of inflection.

2 (i) The first derivative is

$$f'(x) = 3x^2 - 36x + 108$$

The second derivative condition is

$$f''(x) = 6x - 36 = 0$$

which solves for $x = 6$ which represents an inflection point since $f'''(x) = 6$. When $x = 6$, the first derivative is zero, so that $x = 6$ produces a *stationary* point of inflection.

(ii) The first derivative is

$$f'(x) = 9x^2 - 54x + 100$$

The second derivative condition

$$f''(x) = 18x - 54 = 0$$

solves for $x = 3$ (which is an inflection point since $f'''(x) = 18$). At $x = 3$, the first derivative takes the value 19, so that $x = 3$ produces a *non*-stationary point of inflection.

(iii) The three relevant derivatives are

$$f'(x) = 4x^3 - 42x^2 + 120x - 50$$
$$f''(x) = 12x^2 - 84x + 120$$
$$f'''(x) = 24x - 84$$

From the second and third derivatives, $x = 2$ and $x = 5$ are confirmed as points of inflection. When $x = 2$, the first derivative is 104, and when $x = 5$, $f'(x) = 0$. So $x = 2$ is a non-stationary point of inflection, while $x = 5$ is a stationary inflection point.

3 (i) The second derivative condition
$$f''(x) = 3x^2 - 30x + 75 = 0$$
is satisfied only for $x = 5$, at which point the third derivative
$$f'''(x) = 6x - 30$$
is zero. As may be confirmed, the second derivative does not change sign through the point $x = 5$, so that the function does not inflect.

(ii) The second-order condition
$$f''(x) = 6x^2 - 18x - 60 = 0$$
solves for $x = 5$ and $x = -2$, which reference to the third derivative
$$f'''(x) = 12x - 18$$
confirms as an inflection point in both cases.

(iii) The second-order condition
$$f''(x) = 60x^3 - 1200x^2 + 6000x = 0$$
solves for $x = 0$ and $x = 10$ (double root). The third derivative
$$f'''(x) = 180x^2 - 2400x + 6000$$
is non-zero for $x = 0$, confirming this point as one of inflection. But at $x = 10$, $f'''(x) = 0$, and as may be borne out by checking the sign of $f''(x)$ at (say) $x = 9$ and $x = 11$ (positive in both cases), $x = 10$ does *not* produce an inflection point.

Additional Problems

1 (i) Revenue $(R) = pq$
$$= 200 - 4q^2$$
At a maximum of revenue
$$\text{MR} = \frac{dR}{dq} = 200 - 8q = 0$$
so the optimal value of q (for revenue maximization) is $q = 25$.

(ii) When $q = 10$
$$\text{MR} = 200 - 8(10)$$
$$= 120$$

(iii) $\text{MC} = \dfrac{dC}{dq} = 20 + 2q$

so that, at $q = 15$, marginal cost is 50.

(iv) The equation
$$200 - 8q = 20 + 2q$$
solves for $q = 18$.
(v) Profit is maximized where marginal revenue equals marginal cost, i.e. $q = 18$.
(vi) Yes, so long as marginal cost is positive, as can be seen from (iv).

2 (i) Equilibrium occurs where
$$500 - x = 100 + 3x$$
which solves for $x = 100$.
(ii) Here the equilibrium condition is
$$500 - x = 100 + 3x + 40$$
which solves for $x = 90$. The tax take is
$$T = tx = 40(90) = 3600$$
(iii) In this case $t = -40$, so the market clears when
$$500 - x = 100 + 3x - 40$$
which solves for $x = 110$, so that the cost of the subsidy is $110(40) = £4400$.
(iv) We need first to establish a formula for the tax yield as a function of t. Since
$$500 - x = 100 + 3x + t$$
then
$$x = 100 - 0.25t$$
so that the tax revenue is
$$T = tx = 100t - 0.25t^2$$
and the quadratic equation
$$100t - 0.25t^2 = 7500$$
solves for $t = 100$ and $t = 300$.
(v) For a maximum,
$$\frac{dT}{dt} = 100 - 0.5t = 0$$
so that $t = 200$ for maximum tax receipts.

3 (i) For model A the best position is where

$$\frac{dy}{dx} = -0.016x + 0.64 = 0$$

which solves for $x = 40$ km/h. The second derivative is negative for all values of x; this point is a maximum, and the fuel consumption achieved is

$$y = 14 - 0.008(40)^2 + 0.64(40)$$
$$= 26.8 \text{ km/litre}$$

For model B the optimum must satisfy

$$\frac{dy}{dx} = -0.032x + 0.96 = 0$$

which solves for $x = 30$ km/h, at which speed

$$y = 18 - 0.016(30)^2 + 0.96(30)$$
$$= 32.4 \text{ km/litre}$$

(ii) For x such that

$$18 - 0.016x^2 + 0.96x > 14 - 0.008x^2 + 0.64x$$

i.e.

$$0.008x^2 - 0.32x - 4 < 0$$

i.e.

$$x^2 - 40x - 500 < 0$$

the solution set is

$$-10 < x < 50$$

so model B has superior fuel consumption at speeds up to 50 km/h.

4 (i) The condition is that

$$\frac{d\pi}{dp} = 500 - 25p > 0$$

so therefore $p < 20$.

(ii) $p = 20$, at which point $\pi = 4050$.

(iii) The higher value of p satisfying $\pi = 0$. This is $p = 38$.

5 (i) The second derivative condition is

$$\frac{d^2C}{dx^2} = 6x - 36 = 0$$

which is satisfied for $x = 6$. This is a point of inflection as the derivative changes sign around the value $x = 6$.

(ii) Marginal cost is the first derivative:

$$\frac{dC}{dx} = 3x^2 - 36x + 120$$

This reaches its minimum where its derivative (the second derivative of the original function) is zero. This is the same point as the point of inflection, $x = 6$.

6 $dp/dq = -2$ and, by the inverse function rule, $dq/dp = -0.5$, so that

$$e = -\frac{p}{q}(-0.5)$$

$$= \frac{500 - q}{q}$$

so when $q = 5$, $e = 99$, when $q = 50$, $e = 9$, and when $q = 500$, $e = 0$.

7 $\text{MPC} = \dfrac{dC}{dY} = \dfrac{160{,}000}{(Y + 200)^2}$

so the requisite values are

Y	C	MPC
300	250	0.64
600	370	0.25
800	410	0.16

NOTES

1 The interpretation of the term *critical point* as a synonym for *stationary point* is mainly US usage. The term *critical point* can also be understood to mean a point where the tangent is vertical (and where the slope is therefore undefined).
2 As we shall see, there may be other points to consider. Things are rather simpler if the function is *globally concave*. See Section 5.13 below.
3 In terms of sufficiency we shall need to assume that there are no points within the domain where the function is unbounded, as for example at the point $x = a$ in a rational function with denominator $x - a$.
4 As in the case of the biquadratic illustrated in a later section.
5 Other points may be included – as we shall see later. In problems with restricted domain, endpoints must also be considered.
6 The line segment PP^0 is referred to as a *chord*, while a line intersecting a curve (usually at more than one point) is known as a *secant* to the curve.
7 The expression for dy/dx is strictly referred to as the *derived function*.
8 Newton's term for the rate of change of a function was *fluxion*. His notation for the derivative was a central superscript dot, as in $\dot{z}$. This notation is still occasionally used in physical applications such as acceleration.
9 This is an instance of the principle of re-expressing a problem, without distortion, to fit a means for its solution.

10 The solution technique for the general cubic equation in one variable is described in Hart (1966). Software packages such as *Derive* are an efficient way to solve polynomial equations.
11 A special case of the chain rule applied to the differentiation of a function raised to a given power, as in $y = [f(x)]^n$, is called the *power rule*. Although many of the examples here are of this form we shall avoid use of the term 'power rule' to avoid possible confusion with the power *function* rule.
12 The result from which could also be verified by substitution for z in y.
13 Care is needed when a monotonic function is not *strictly* monotonic. For example, $y = x^3$ is monotonic and the inverse function exists, but the inverse function is not differentiable at $x = 0$.
14 Strictly speaking, we can only say that at an interior maximum the slope of a function will not be increasing and at an interior minimum the slope will not be diminishing. Cases where the second derivative is zero are considered below.
15 Note that the local maximum does not produce the highest value of the function, which has no upper limit if x is allowed to become arbitrarily large and positive. This point will be considered in a later section.
16 Note that in this case the function has no finite overall maximum if x is allowed to become arbitrarily large and negative.
17 As we shall later see, there may be other possibilities for the global maximum.
18 The term *curvature* is used here in its common sense. For a formal definition of curvature see Borowski and Borwein (1989).
19 *Concave up* is sometimes used in place of convex.
20 But not necessary.
21 In the informal sense used earlier.
22 The curve always bends *downwards* – where it bends at all.
23 The curve always bends *upwards* – where it bends at all.
24 Assuming that the second derivative is defined at the point in question.
25 Convexity/concavity may change *without* an inflection point – as between the two branches of the hyperbola $y = 1/x$.
26 A point of inflection may also be defined in geometric terms as a point where a function crosses its tangent.

Chapter 6

Functions of More Than One Variable

6.1	Introduction	362
6.2	Linear functions of several variables	364
6.3	Quadratic functions	369
6.4	Slopes and first-order derivatives	373
6.5	Higher order partial derivatives	382
6.6	Local maxima and minima	387
6.7	Saddle points	392
6.8	Stationary values: résumé	398
6.9	Introduction to constrained optimization	402
6.10	The method of substitution	408
6.11	Substitution: concluding remarks	419
6.12	BASIC program	420
	Additional problems	427
	References and further reading	428
	Solutions to exercises	428

This chapter extends to problems involving two or more independent variables the concepts of functions and the differential calculus introduced in Chapters 4 and 5. You will encounter the idea of a partial derivative, and see how partial derivatives can be used to identify stationary points of functions of several variables. A basic program is provided that will carry out this identification for quadratic functions.

The chapter also provides a simple introduction to problems in which the decision-maker has limited room for manoeuvre. These are problems in which the choice of values of the decision variables is restricted, perhaps by resource limitations. We begin examination of approaches to the solution of such problems and the economic and financial insight that can be gained from them.

By the end of this chapter, you should feel at home with the principles of partial derivatives, be able to differentiate polynomial functions in two or more variables, understand the concept of a constrained optimum, employ the method of substitution, and appreciate the significance of this work in a business and economic context.

6.1 INTRODUCTION

Essential aspects of economic problems can often be discussed in terms of the relationship between two key variables. The classical Marshallian demand curve is an example of an outstandingly useful two-variable relationship. But there are limitations to the ability to draw generally applicable conclusions from two-variable models. As they originally present themselves, most quantitative problems in business and economics involve relationships between more than two variables. For example, consider demand again. The quantity demanded, x, of a product per period time might be expressed as

$$x = f(p, p_c, p_s, Y, i, \ldots) \tag{6.1}$$

where equation (6.1) is the *demand function*. Very many factors influence the demand for a product. Those identified in (6.1) are

p	the product's own price
p_c	the price of a complementary good
p_s	the price of a substitute good
Y	disposable income
i	a measure of interest rates

In (6.1), x is shown as a function of five variables. The demand curve relationship between quantity demanded and the product's own price would be a special case of (6.1) – one demand curve being defined for each set of values of p_c, p_s, Y, i and any other influencing factors involved. As another example of a function of several variables from microeconomics, the *production function* states output per period time – the dependent variable – as a function of the amounts used of the necessary factors of production – the independent variables or *arguments* of the function. At a high level of aggregation we can write

$$q = f(L, K, M) \tag{6.2}$$

where q is product output, L is labour input, K is capital applied and M is materials used. The form of the production function, the way in which quantities of input translate into output, represents the technology of the manufacturer. In the course of this and subsequent chapters we shall consider a number of forms that the production function might take. For the most part we shall concentrate on the relationship between one dependent variable and two independent variables. This will illustrate most of the important and relevant properties of functions of several variables. We shall use the notation

$$z = f(x, y) \tag{6.3}$$

In (6.3), x and y are sometimes referred to as *input variables* as well as the more familiar terminology of independent or decision variables. In this vein, the dependent variable z is an *output variable*. The domain is the set of x and y combinations to which the function is restricted or for which it is defined. The range of the function is the corresponding set of output values of z. For a function of a single variable, the domain will be the real line or those sections of it for which the function operation can be performed or to which it is otherwise restricted. Domains for functions of two variables, if not the whole of the x, y plane, are most commonly *areas* on the plane such as triangles, rectangles, other polygons, disks or irregular regions. But the domain might also be defined by a line, curve, or parts thereof or discrete points. A common domain in applied work is the first quadrant in which both of the independent variables are non-negative. This may be further restricted with one or more functions of x and y (often linear in practice) giving bounds to the domain. Where we deal with restricted domains, the feasible region will generally be taken to be a convex subset of the non-negative orthant.

In geometric terms, the function (6.3) produces a *surface*, rather than a solid, in three-dimensional space. The space is *Euclidean three space*, E^3, the space of everyday experience. Functions of a single variable generate diagrams in Euclidean *two* space, E^2. We shall not be concerned with other, non-Euclidean, spaces such as the curved two-dimensional space formed by the surface of a sphere, although these spaces can be important in particular contexts. It may be helpful to examine a particular instance of equation (6.3). Consider the function

$$z = 5x^2 + 3xy - y^2 - 2x + 10y + 100$$

Note that this function is a quadratic in x and y, all of the terms being of degree two, one or zero. The constant term means that the origin is not a point on the surface although $(0, 0, 100)$ clearly is. Insertion of any other pair of values of x and y into the function will give the corresponding value of z. For example, when $x = 2$ and $y = 3$, then

$$z = 20 + 18 - 9 - 4 + 30 + 100$$
$$= 155$$

Manual curve sketching in three dimensions, while possible, calls for considerable skills of draughting. While the general principles of curve sketching outlined in Chapter 4 still apply, it is rarely worth attempting more than an outline by hand. Several computer software packages are available that will produce wire-frame diagrams of functions of two variables. Such software can be of great assistance in understanding the properties of a function, and several diagrams in this book are generated using *Derive*.[1] But by no means all work with functions of two variables requires the assistance of computer software. The simplest form of (6.3) is that of a *linear* function of two independent variables, and it is with linear functions that we shall begin.

6.2 LINEAR FUNCTIONS OF SEVERAL VARIABLES

Linear functions of two or more variables are important in much practical work. For example, the objective function in linear programming takes such a form and we shall consider an example in this context shortly. First, however, examine the function

$$z = 40 - x - 0.5y \qquad (6.4)$$

Linear functions in two independent variables produce plane (flat) surfaces in three dimensions. That part of the plane corresponding to (6.4) which lies in the non-negative orthant is shown in Figure 6.1.

Three rectangular co-ordinate axes are required. By convention, the dependent variable, z, is measured against the vertical axis, while the independent variables, x and y, are at right angles (and to z) in the horizontal plane. The axes can be envisaged as the bottom left corner of a cube set on a desk-top. The x, y and z axes are the three *fundamental directions* which it may be helpful to visualize as North (y), East (x) and Up (z). Points on

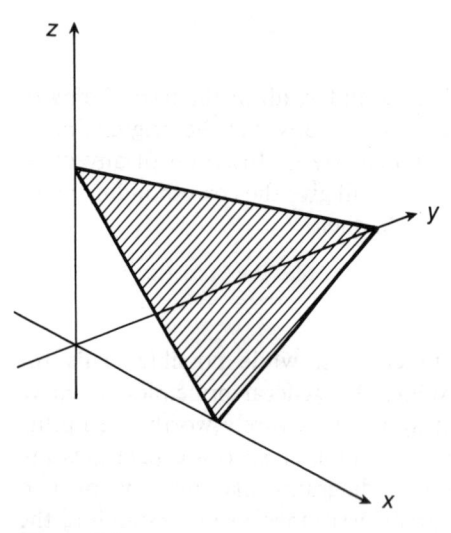

Figure 6.1

Functions of more than one variable 365

the plane are identified by *co-ordinate triples* of x, y and z values satisfying (6.4). For example, the points

$P = (20, 30, 5)$ and $Q = (10, -30, 45)$

lie on the plane (although of course Q would be outside the region shown in Figure 6.1). In contrast, the point

$R = (20, 20, 35)$

lies above the plane while

$S = (-20, 40, 35)$

lies below it. *Contours* of the surface – lines on the x, y plane above which the surface is at a constant height – may be obtained by 'slicing through' the surface for a given value of z in a plane parallel to the horizontal x, y plane. The resulting edge is then projected vertically down to the x, y plane to form the contour. This process is illustrated for $z = 15$ in Figure 6.2. The figure shows the part of the contour lying in the positive x, y quadrant.

In Figure 6.2, ABC represents the original plane, which is intersected at height $H = 15$ by a plane parallel to the x, y axes. With attention confined

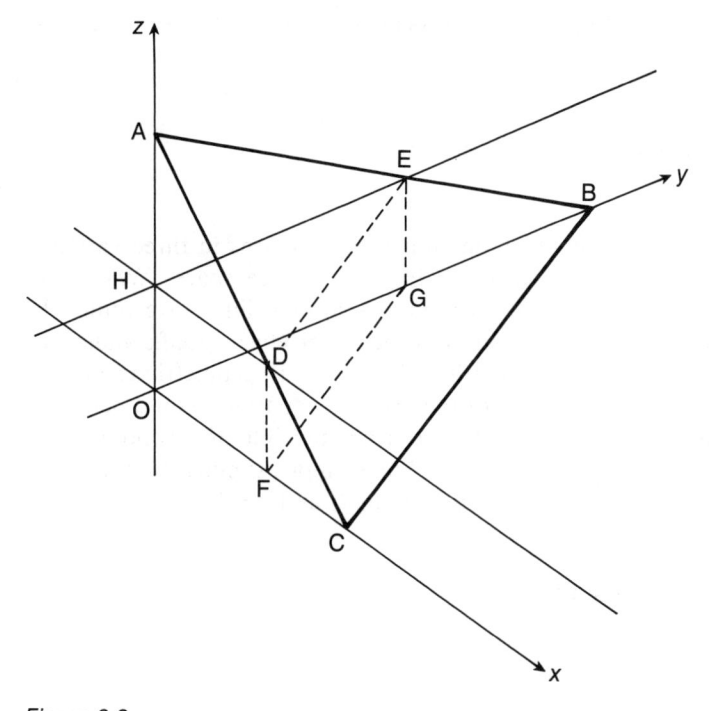

Figure 6.2

to the non-negative orthant, the line DE represents the 'edge' where the two planes given by (6.4) and $z = 15$ meet. The contour itself is FG. Note that the contour for $z = 0$ is the line BC (which is contour and edge simultaneously, as the plane for the given value of z is the x, y plane itself). Note also that the contour for $z = 40$ is the origin. This is because the only point in the non-negative orthant having a point in common with the plane at $z = 40$ is A, which is directly above the origin, O. The equation of any of the contours above can be found by the solution of the simultaneous equations for the planes that intersect in each case. Thus where $z = 15$ the equations are

$$z = 40 - x - 0.5y$$
$$z = 15$$

so that

$$0 = 25 - x - 0.5y$$

By rearrangement, the equation of this contour could be expressed explicitly as

$$y = 50 - 2x$$

The reader may verify by a similar process that the equation of the $z = 0$ contour is

$$y = 80 - 2x$$

and that the equation of the $z = 40$ contour is

$$y = -2x$$

We have seen that consistent simultaneous equations in three variables correspond to intersecting planes in Euclidean three space. Where two equations are specified, the intersection is a straight line — we found the equations of the line in three cases. In order to produce specific values for the variables, a third equation needs to be added. In geometric terms, the line representing the intersection of the first two equations would penetrate the third plane at a single point. The co-ordinate triple of this point is the solution of the simultaneous equations. For example, suppose that the third equation added to the system comprising (6.4) and $z = 15$ is

$$z = 55 - x - 2y$$

and therefore, since $z = 15$,

$$15 = 55 - x - 2y$$

so that

$$x + 2y = 40$$

and this equation and the equation of the contour already obtained

$$y = 50 - 2x$$

can be solved simultaneously as

$$2x + y = 50$$
$$x + 2y = 40$$

which by elimination or substitution produce the solution

$$x = 20 \quad \text{and} \quad y = 10$$

Thus the co-ordinate triple of the one point that lies on all three planes is (20, 10, 15). Now consider the straight line distance between any two points (x_1, y_1, z_1) and (x_2, y_2, z_2) in three space. This is found by use of Pythagoras's theorem as d, where

$$d = [(x_2 - x_1)^2 + (y_2 - y_1)^2 + (z_2 - z_1)^2]^{1/2} \tag{6.5}$$

Applying (6.5) to find the distance between points $P = (10, 20, 50)$ and $Q = (15, 8, 70)$ the workings are

$$d = [(15 - 10)^2 + (8 - 20)^2 + (70 - 50)^2]^{1/2}$$
$$= [25 + 144 + 400]^{1/2}$$
$$= \sqrt{(569)} \approx 23.85$$

Notice that in the use of (6.5), because of the squaring, it does not matter which way round the corresponding co-ordinates are subtracted. The Pythagorean formula also applies to higher dimensional Euclidean spaces. For example in E^4, the distance between the points $P = (8, 3, -2, 4)$ and $Q = (6, 7, -7, -2)$ is

$$d = [(8 - 6)^2 + (3 - 7)^2 + (-2 - (-7))^2 + (4 - (-2))^2]^{1/2}$$
$$= [4 + 16 + 25 + 36]^{1/2}$$
$$= \sqrt{(81)} = 9$$

The Pythagorean formula cannot be applied to determine minimum distances between points in curved spaces – e.g. between points substantially apart on the Earth's surface. As map makers know only too well it is impossible to preserve exactly the distances between locations when a projection of a spherical surface is made onto a plane. The final concept introduced in this section is that of a *simplex*. This arises in management science – for instance in linear programming (where the principal method of computation is the simplex method) and in decision theory where the idea of a simplex is useful in the theory of games. A simplex may have any number of dimensions. In geometric terms, a two-dimensional simplex is the smallest region *without indentations* (i.e. a convex figure) that contains three points which do not lie on the same straight line. So a two-dimensional simplex is a triangle. A one-dimensional simplex is the smallest convex region that will contain two points which are not superimposed: a straight line segment. A zero-dimensional simplex is a single point. A three-dimensional simplex is the smallest non re-entrant solid that will contain four points which do not lie in the same plane.[2] This will be a tetrahedron. A simplex is said to be *regular* if its vertices (which are the points) are evenly spaced. Simplices of up to three dimensions are illustrated in Figure 6.3. An *n*-dimensional simplex is defined as the smallest convex region containing $n + 1$ points in independent locations. *Independent locations* means that the points are not all contained within $n - 1$ dimensions. In the case of three points not all must lie on the same straight line (a one-dimensional construct).

(a)

(b)

(c)

(d)

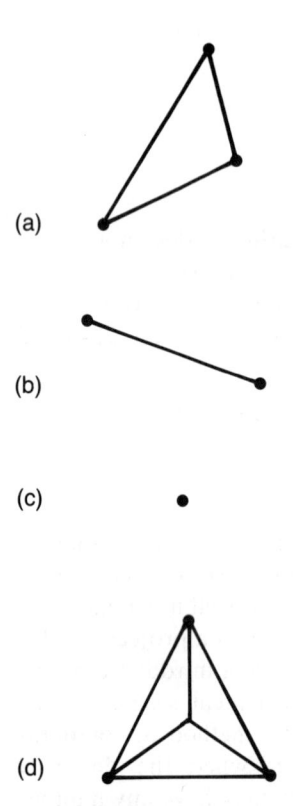

Figure 6.3

EXERCISES 6.2

1 Given the plane defined by the function

$$z = 200 - 5x - 2y$$

do the points below lie on, above or below the plane?

(i) (20, 50, 0)
(ii) (−10, 25, 210)
(iii) (− 20, −10, 330)
(iv) (40, 125, − 240)
(v) (− 80, 500, − 440)
(vi) (60, − 60, 20)

2 Find the straight line distance between the following points in three space:

(i) P: (10, 20, 30)
(ii) Q: (2, 16, 38)
(iii) P: (2, 18, − 5)
(iv) Q: (− 2, 6, 1)

3 Find the equations of the contours of the function

$$z = 3x + 0.5y - 20$$

for

(i) $z = 100$ (ii) $z = -20$ (iii) $z = 50$

4 Find the point in common between the three planes defined by

(i) $z = 9 - 4x - y$
$x + y + z - 6 = 0$
$3y + 4z - x = 17$

(ii) $5x - 2y + z = 140$
$3x + y - 0.5z = 40$
$0.25x + y + z = 30$

6.3 QUADRATIC FUNCTIONS

A quadratic function in two independent variables is one in which the degree of every term (obtained by adding the powers to which each variable in the term is raised) is two, one or zero, there being at least one term of the second degree. Consider the function

$$z = 100x + 60y - 2xy - 2x^2 - y^2 - 300 \tag{6.6}$$

370 Mathematics for business, finance and economics

This is a quadratic expression in the two independent variables x and y. A function of this form could well arise where a firm makes and sells two products in amounts x and y, has a quadratic cost function and faces linear demand equations for the two products. The value of z would represent profit. Later in the chapter, we shall show how (6.6) might be arrived at from basic demand and cost information and we shall also find the best overall profit position. For the time being, consider the general features of the function, the appearance of which is shown in Figure 6.4.

The function has a single turning point which is a local and the global maximum. This peak occurs in the positive orthant – an important feature if x and y are measures of physical entities such as output levels. An elevation of the function is shown in Figure 6.5.

The figure shows a vertical section, profile or elevation of the surface when one of the variables is set at a specific value. The vertical section corresponding to $y = 5$ is illustrated. As may be verified by substitution into equation (6.6), the equation of the elevation curve is

$$z_{(y=5)} = 90x - 2x^2 - 25$$

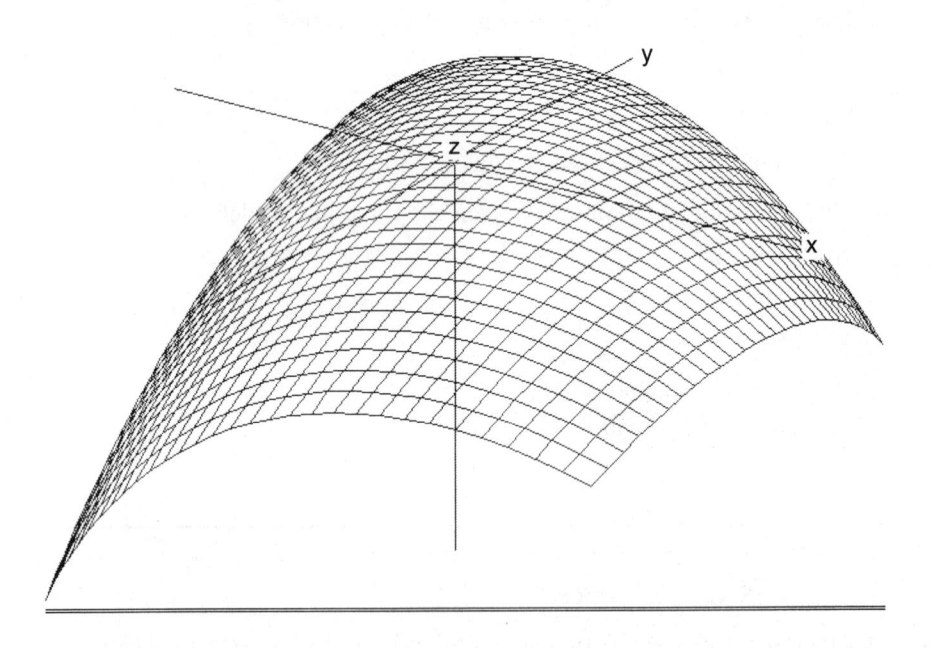

Center x: 0   y: 0   Length x: 200   y: 200   Derive 3D-plot

Figure 6.4

Functions of more than one variable 371

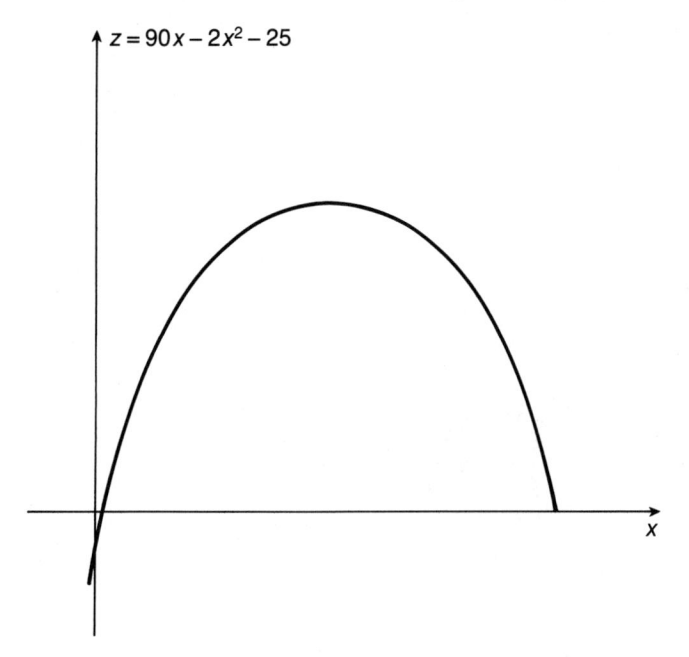

$z = 90x - 2x^2 - 25$

Figure 6.5

Similarly, if x is set at a specific value, then a vertical section of the surface results for which z is a function of y alone. For example, with $x = 10$

$$z_{(x=10)} = 500 + 40y - y^2$$

All elevations of this function will be quadratics with the same general appearance but with different intersection points with the axes and corresponding to different values of z at their highest points. In microeconomic theory, when the function is a production function (showing output level as a function of the input levels) a vertical section of the function shows output graphed against variations in the level of one input only. Such a curve is called a *total product curve* or *productivity curve*. The invariant input(s) determine the particular total product curve obtained. Horizontal sections through the surface z projected to the x, y plane generate contours of the surface as in the linear case. Here, the contours resemble those on an ordnance survey map – indeed 'contour' is to be interpreted in just the same way as on a map. The contours of the function for $z = -300$ and $z = +500$ are shown in Figure 6.6.

All of the contours of (6.6) will have the same general shape. The equation of any one contour will be given by

$$100x + 60y - 2xy - 2x^2 - y^2 - 300 = \text{constant}$$

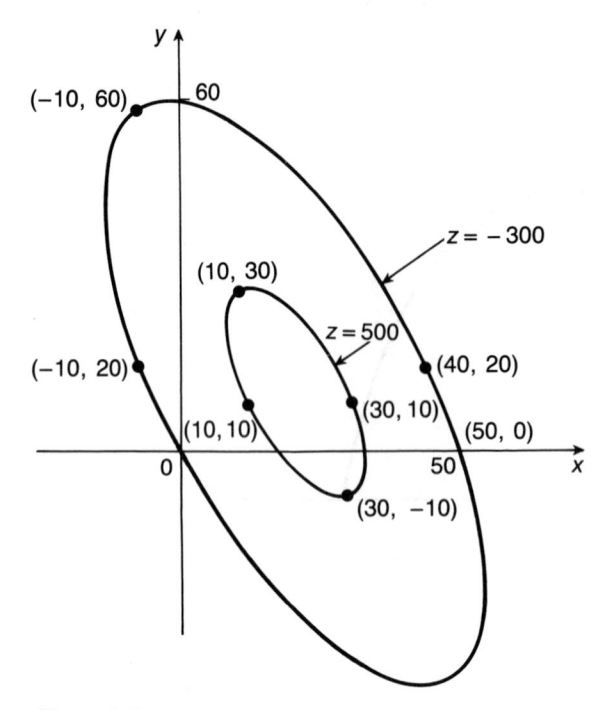

372 Mathematics for business, finance and economics

Figure 6.6

So, for example, in the case of the $z = 800$ contour, the equation is

$$100x + 60y - 2xy - 2x^2 - y^2 - 1100 = 0$$

The contour map of a function will be useful when considering the geometry of constrained optimization. In microeconomics, contours of functions often represent important economic concepts. For example, where the function under consideration is a production function, the contours show input combinations that produce the same level of output and are given the name *isoquands*. In the context of the example we have been considering, the contours are described as *iso-profit lines*, the prefix 'iso' meaning equal. We have studied a particular function of two independent variables in some detail. But the number of independent variables in a quadratic function can exceed two. As we have seen, there are many independent influences on the demand for a product, and there are numerous variable inputs in a typical production process. The function

$$z = 3xy + 2y^2 - wx + w^2 + 5yw + x + y + w - 20$$

is of degree two in the three independent variables x, y and w. Note that if we take only the terms of degree two in the function above, and therefore write

$$z = 3xy + 2y^2 - wx + w^2 + 5yw$$

then the result is a function *homogeneous* of degree two. The homogeneity means that if *all* of the independent variables are scaled by a factor p, then z will be scaled by a factor of p^2. Homogeneous functions play a role in the theory of production.

EXERCISES 6.3

1 Given the production function

$$z = 120x + 50y - 2xy - 4x^2 - 0.5y^2$$

find the total product curves for

(i) $y = 10$
(ii) $x = 25$

2 Find the equation of the contours of the function

$$z = 10y - 3xy - 100 + 6x - 9x^2$$

for

(i) $z = 900$
(ii) $z = 0$

3 Establish whether the following functions are homogeneous:

(i) $z = y^2 + 2xw - 3wy + w^2$
(ii) $z = 3x^2 + 2y^2 + xyw - w^2$

(iii) $z = wx - xy + \dfrac{x^2 y}{w}$

(iv) $z = wx + \dfrac{y^3}{x} + (x-1)^2$

6.4 SLOPES AND FIRST-ORDER DERIVATIVES

In the context of functions of several variables, the concept of slope is just as important as in the case of functions of a single variable. Having found expressions for the slope of various forms of function, we shall then

374 Mathematics for business, finance and economics

establish conditions for maxima and minima expressed in terms of slope. First consider a *linear* function of two independent variables:

$$z = 0.6x + 0.2y + 20$$

The function is graphed in Figure 6.7 and an arbitrary point P is indicated.

What is the slope of the surface at the point P? The first observation marks an important distinction between functions of several variables and functions of a single independent variable:

> The slope of a function of several variables at any point depends upon the direction of movement.

In the linear case, wherever the initial location on the surface is, a one unit increase in x alone increases z by 0.6, while a one unit increase in y alone increases z by 0.2. These values 0.6 and 0.2 are the slopes of the surface in the fundamental x and y directions respectively. Movement in other directions changes z by an amount which can be easily determined in the linear case. For example, movement in a direction at $45°$ to the x and y axes caused by a unit increase in both x and y would increase z by

$$0.6 + 0.2 = 0.8$$

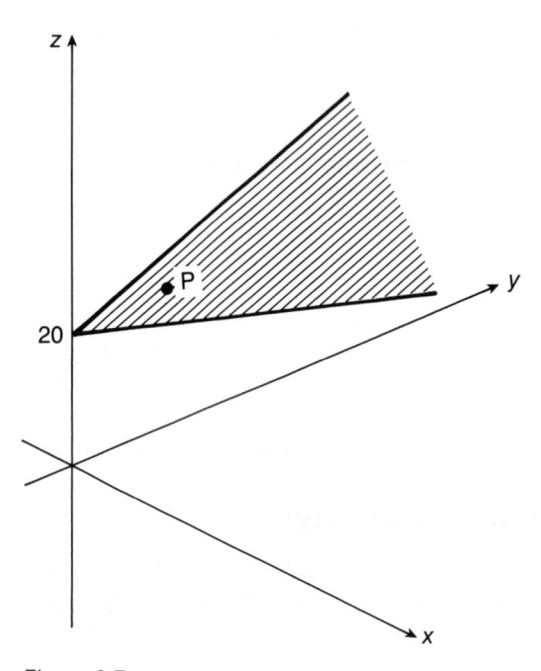

Figure 6.7

Since the horizontal distance away from the original point is $\sqrt{2}$ units, the gradient in the 'north easterly' direction is

$$0.8/\sqrt{2} = 0.5657$$

But as we shall see, the pivotal role in terms of optimization is played by the slopes in the fundamental directions. We shall therefore concentrate on these slopes which are given by the *partial derivatives* of the function. A partial derivative is an expression for the rate of change of a function when *one variable only* changes. The partial derivative of a function z with respect to x is written as

$$\frac{\partial z}{\partial x} \text{ or } f_x$$

As with functions of a single variable there are two forms of notation. The symbol

$$\frac{\partial z}{\partial x}$$

is read as 'partial dee z dee x' and is obtained by differentiating the function in the ordinary way except that *all 'non-x' terms and parts of terms are treated as constants*. So, for the linear function

$$z = 0.6x + 0.2y$$

when partially differentiating with respect to x, the term $0.2y$ is treated as a constant and will therefore disappear. The result is

$$\frac{\partial z}{\partial x} = 0.6 = f_x$$

Similarly:

$$\frac{\partial z}{\partial y} = 0.2 = f_y$$

As for functions of a single variable, both notations will be used. Now consider the non-linear function:

(i) $z = 2xy$

When differentiating partially with respect to x, the 'non-x' part of the term (i.e. $2y$) is treated as a constant. In this context, $2y$ is simply regarded as the coefficient of x. Thus with respect to x

$$\frac{\partial z}{\partial x} = 2y = f_x$$

and similarly with respect to y

$$\frac{\partial z}{\partial y} = 2x = f_y$$

These expressions give the slope of the function z in the two fundamental directions. Thus at the point $(3, 5, 30)$

$$f_x = 2(5) = 10$$
$$f_y = 2(3) = 6$$

so this function slopes more steeply in the x direction at this point. Now consider some further examples of non-linear functions:

(ii) $\quad f(x, y) = xy^2$

for which the partial derivative with respect to x is

$$f_x = y^2$$

while the partial derivative with respect to y is

$$f_y = x2y$$
$$= 2xy$$

In the case of the partial derivative with respect to x, the x part of the term is linear with coefficient y^2 which therefore 'comes down' into the derivative as would any constant. In differentiating partially with respect to y, the 'non-y' component, x, comes down into the derivative unaltered to multiply the derivative of y^2 with respect to y. Now consider the function

(iii) $\quad z = x^3 y^2$

When differentiating this function partially with respect to x, the 'non-x' component, y^2, is treated as a constant and multiplies the derivative of x^3 with respect to x (i.e. $3x^2$). Thus

$$\frac{\partial z}{\partial x} = 3x^2 y^2$$

The derivative with respect to y has the x^3 component treated as a constant and so

$$\frac{\partial z}{\partial y} = x^3 2y$$
$$= 2x^3 y$$

Now consider the partial derivatives of a function that contains more than one term:

(iv) $\quad z = 2xy + xy^2 - x^3 y^2$

The sum–difference rule for differentiation applies equally well to functions of several variables, and dealing with each term in turn produces

$$\frac{\partial z}{\partial x} = 2y + y^2 - 3x^2 y^2$$

and the partial derivative with respect to y is

$$\frac{\partial z}{\partial y} = 2x + 2xy - 2x^3 y$$

Consider again the profits function of Section 6.3. This was

$$z = 100x + 60y - 2xy - 2x^2 - y^2 - 300$$

and the partial derivatives will be

$$\frac{\partial z}{\partial x} = 100 - 2y - 4x \tag{6.7a}$$

and

$$\frac{\partial z}{\partial y} = 60 - 2x - 2y \tag{6.7b}$$

As we shall see, partial derivatives are vitally important in establishing equilibrium conditions – statements that need to be true – for an optimum to obtain. Furthermore, the derivatives usually have economic significance in themselves and an understanding of this import gives useful insight into economic problems and a vitally important feel for the meaning of the symbolic manipulation. But there is also an immediate practical use of the derivatives. The slope of the function in the fundamental directions gives the *marginal profit* from expansion of x or y production alone. For example, if the firm is operating at initial output levels of $x = 12$ and $y = 8$, the rates of change of profit at the margin can be found by substitution into (6.7a) and (6.7b) respectively. The results are

$$\frac{\partial z}{\partial x} = 36 \quad \text{and} \quad \frac{\partial z}{\partial y} = 20$$

These values for marginal profit are instantaneous rates of change at the point given by the original output levels. Since the partial derivatives are themselves functions of x and y, any change in x or y will bring about a change in the value of the derivative. The values of the partial derivatives are therefore useful in obtaining an *approximate* value of the actual change in z for finite increments in x or y. Consider an example. Suppose that the output of x is increased by one unit to 13; what will be the *actual* change in profit? This can be found (somewhat longwindedly) by substitution of

$x = 13$ and $y = 8$ into z and subtraction of the value of z at $x = 12$, $y = 8$. At the new output levels the value of profit is

$$z = 100(13) + 60(8) - 2(13)(8) - 2(13)^2 - (8)^2 - 300$$
$$= 870$$

while when x was 12, overall profits were

$$z = 100(12) + 60(8) - 2(12)(8) - 2(12)^2 - (8)^2 - 300$$
$$= 836$$

So the actual change in profit level following a one unit increase in x is

$$870 - 836 = 34$$

Consequently, the value of the partial derivative, 36, would in most practical situations be a good enough approximation to the actual change in profit. As an illustration of a practical situation suppose that a way could be found to increase output of x by one unit but at a cost (over and above the costs implicit in z) of £100. The proposal should be rejected and the figure given by the derivative is sufficient to do this. It is possible to be more specific in many cases about what the values of $\partial z/\partial x$ and $\partial z/\partial y$ represent in terms of finite increments in x or y. For example, in the present context, $\partial z/\partial x$ represents an *upper bound* on the change in z consequent on a unit increase in x. For an increase of n units in x, an upper bound would be given by $n(\partial z/\partial x)$. The upper bound property arises because the function z is concave in the x direction. This property can be inferred in the present case without explicit reference to the second derivative, because the linearity of (6.7a) and the negative coefficient for x make it clear that the derivative is a diminishing function of x: as x goes up by any amount, $\partial z/\partial x$, the *instantaneous* rate of change, will always go down. So we can conclude that the derivative will overstate, to some extent, the change in z for any finite increase in x: the smaller the increase in x the less will be the extent of the overstatement. The same upper bound property is true of the derivative with respect to y. If y is incremented by one unit to 9 while x remains at 12, the actual change in z is 19 compared with the value of 20 for $\partial z/\partial y$ at $x = 12$, $y = 8$. Now consider another example in partial differentiation. Let

$$z = 3x^4 y^5 - 2x^3 + 4y^6 = f(x, y) \tag{6.8}$$

In finding the partial derivatives in this case, we shall highlight the constancy under partial differentiation of those parts of terms that are not expressions in the variable with respect to which the differentiation is being carried out. For partial differentiation with respect to x, rearrange the terms in z as follows:

$$z = x^4 [3y^5] - [2] x^3 + [4y^6]$$

Those terms/parts of terms in square brackets are either constants or are treated as constants in the process of differentiation with respect to x. So in $\partial z/\partial x$, $[3y^5]$ will multiply the result of differentiation of x^4. The term $[4y^6]$ will disappear under differentiation with respect to x as it is not the coefficient of an expression in x. Thus

$$\frac{\partial z}{\partial x} = 4x^3[3y^5] - [2]3x^2$$
$$= 12x^3y^5 - 6x^2$$

Approaching the derivative with respect to y in the same way, z can be written as

$$z = [3x^4]y^5 - [2x^3] + [4]y^6$$

and the partial derivative with respect to y will be

$$\frac{\partial z}{\partial y} = [3x^4]5y^4 + [4]6y^5$$
$$= 15x^4y^4 + 24y^5$$

Now consider an example involving a function with more than two independent variables. Let

$$f(x, y, w) = 7x^2y^3w^4 - 8x^3y^2 + 9w^5y^3 + 6xw^2$$

The three first-order partial derivatives are obtained in just the same way as for functions with two independent variables. The derivatives are

$$f_x = 14xy^3w^4 - 24x^2y^2 + 6w^2$$
$$f_y = 21x^2y^2w^4 - 16x^3y + 27w^5y^2$$
$$f_w = 28x^2y^3w^3 + 45w^4y^3 + 12xw$$

The product, quotient and chain rules introduced in Chapter 5 in the context of functions of a single independent variable can also be used for finding the partial derivatives of functions of several variables. For example, consider the function

$$f(x, y) = (2 + 3x + 4y)^2$$

Using the chain rule, the derivative with respect to x is

$$f_x = 2(2 + 3x + 4y)(3)$$
$$= 12 + 18x + 24y$$

and the derivative with respect to y is

$$f_y = 2(2 + 3x + 4y)(4)$$
$$= 16 + 24x + 32y$$

These results can be checked by expanding the original function. The function is

$$f(x,y) = 4 + 12x + 16y + 9x^2 + 24xy + 16y^2$$

so

$$f_x = 12 + 18x + 24y$$

and

$$f_y = 16 + 24x + 32y$$

The *product rule* could also be used here, since the function can be written as

$$f(x,y) = (2 + 3x + 4y)(2 + 3x + 4y)$$

so that by the product rule

$$f_x = (2 + 3x + 4y)(3) + (2 + 3x + 4y)(3)$$
$$= 12 + 18x + 24y$$

and

$$f_y = (2 + 3x + 4y)(4) + (2 + 3x + 4y)(4)$$
$$= 16 + 24x + 32y$$

Now consider another example using the chain rule. If

$$f(x,y) = (3x^3 - 7y^4 + 5xy^2 - 9x^2y)^4$$

then

$$f_x = 4(3x^3 - 7y^4 + 5xy^2 - 9x^2y)^3(9x^2 + 5y^2 - 18xy)$$

and

$$f_y = 4(3x^3 - 7y^4 + 5xy^2 - 9x^2y)^3(-28y^3 + 10xy - 9x^2)$$

A further example of the use of the product rule is given by

$$f(x,y) = (4x^2y^2 + 6x^3y^4)(3x^3y^2 - 7xy)$$

for which

$$f_x = (3x^3y^2 - 7xy)(8xy^2 + 18x^2y^4)$$
$$+ (4x^2y^2 + 6x^3y^4)(9x^2y^2 - 7y)$$

and

$$f_y = (3x^3y^2 - 7xy)(8x^2y + 24x^3y^3)$$
$$+ (4x^2y^2 + 6x^3y^4)(6x^3y - 7x)$$

Now consider the use of the quotient rule in partial differentiation. Given

$$f(x, y) = \frac{9xy + 2x^2y^3}{x^3 + y^4}$$

then

$$f_x = \frac{(x^3 + y^4)(9y + 4xy^3) - (9xy + 2x^2y^3)3x^2}{(x^3 + y^4)^2}$$

and

$$f_y = \frac{(x^3 + y^4)(9x + 6x^2y^2) - (9xy + 2x^2y^3)4y^3}{(x^3 + y^4)^2}$$

The rules are applicable to functions with more than two independent variables. As an example, consider the use of the chain rule to find the partial derivative f_w, where

$$f(x, y, w) = (x^2y^2w^2 + x^3w^4)^7$$

Making use of the rule:

$$f_w = 7(x^2y^2w^2 + x^3w^4)^6(2x^2y^2w + 4x^3w^3)$$

EXERCISES 6.4

1. What is the slope of the plane

   $$z = 1000 - 2x + 0.5y$$

   in the fundamental directions?

2. (i) Find the first-order partial derivatives of the function

   $$z = x^2 - 4xy + y$$

   (ii) What is the slope of the function in the fundamental directions at the following points:

   (a) (10, 2, 22)
   (b) (−2, 7, 67)

3. For the function

   $$f(x, y) = x^2y^3$$

   find the partial derivatives and the slope of the function in the fundamental directions at the point (5, 4, 1600).

4. Find the first-order partial derivatives of the following functions:

   (i) $f(x, y) = x^7 + xy^7 + x^7y^7$
   (ii) $f(x, y) = 100x^{99}y^{101}$
   (iii) $f(x, y) = 4x^4 + 2y^5 - 3x^3y^4 + 4x^4y^3$

(iv) $q = l^{0.7}k^{0.4}$
(v) $q = l^a k^b$ (for a and b as constants)
(vi) $z = nx^m y^{n-1}$ (for m and n as constants)

(vii) $z = \dfrac{x}{y}$

(viii) $z = x^5 y + yx^{-2}$

5 Find the first-order partial derivatives for

   (i) $f(x, y, w) = 10x^3 y^4 w^2 - 2x^4 y^3 - 5x^2 w^4 + 4y^3 w^2$
   (ii) $Q = AL^r K^s M^t$

   where A, r, s and t are constants.

6 Given the profits function

   $$z = 500x + 400y + 4xy - 3x^2 - 2y^2 - 1000$$

   where z represents profits and x and y are output levels, find expressions for the marginal profits on the two products.

7 Find the partial derivatives for

   (i) $f(x, y) = (6x^2 y^3 + 2xy^2)(3x^2 y - 8x^3 y^3)$
   (ii) $f(x, y) = (1 + Axy)(1 + Bxy)$

8 Find the partial derivatives for

   (i) $f(x, y) = \dfrac{xy}{x - y}$

   (ii) $f(x, y) = (2x^2 y^2 - x^3)(y^2 - 5xy)^{-1}$

9 Find the partial derivatives for

   (i) $f(x, y) = (6x^4 y^5 - 3x^4 y^3)^6$
   (ii) $f(x, y) = (ax^2 + bxy + cy^2)^d$

10 Find the partial derivatives for the function

   $$f(x, y, w) = (x^3 - x^2 y^2 w^2)(y^3 - w^4 x^2)$$

6.5 HIGHER ORDER PARTIAL DERIVATIVES

The partial derivatives of a function will, in general, be functions of the independent variables and can themselves usually be differentiated. Take for example the function (6.8) with which we concluded the previous section. This was

$$z = 3x^4 y^5 - 2x^3 + 4y^6$$

It had been established that

$$\frac{\partial z}{\partial x} = 12x^3y^5 - 6x^2 \qquad (6.9)$$

If now (6.9) is itself differentiated partially with respect to x, the result is written as

$$\frac{\partial^2 z}{\partial x^2} = 36x^2y^5 - 12x = f_{xx} \qquad (6.10)$$

The process of obtaining this derivative is no different from that by which the first derivative of z with respect to x was found. Note the two forms of notation. Expression (6.10) is said to be the second *direct partial derivative* of z with respect to x. The use of the term 'direct' will be clarified shortly. First consider the second direct derivative with respect to y. At the first order we had

$$\frac{\partial z}{\partial y} = 15x^4y^4 + 24y^5 \qquad (6.11)$$

and differentiation of (6.11) partially with respect to y produces

$$\frac{\partial^2 z}{\partial y^2} = 60x^4y^3 + 120y^4 = f_{yy} \qquad (6.12)$$

It will have occurred to the reader that it should be possible to partially differentiate a function first with respect to one variable and then with respect to another. For example, there is no reason why the first derivative with respect to x, equation (6.9), should not then be differentiated with respect to y. The result is written as

$$\frac{\partial^2 z}{\partial y \, \partial x} = 60x^3y^4 = f_{xy} \qquad (6.13)$$

The process by which $60x^3y^4$ was arrived at should now be familiar, and the function notation for this derivative, f_{xy}, is that which common sense would have suggested. The result, (6.13), is known as a second-order *cross partial derivative* or *mixed derivative*. The somewhat counter intuitive notation

$$\frac{\partial^2 z}{\partial y \, \partial x}$$

may appear to have the denominator the wrong way round. But when viewed as the derivative with respect to y of the derivative with respect to x or

$$\frac{\partial}{\partial y} \frac{\partial z}{\partial x}$$

then the ordering of ∂y and ∂x in the denominator looks more reasonable. Now find the cross partial derivative of (6.8) first with respect to y and then with respect to x. At the first order we had found

$$f_y = 15x^4y^4 + 24y^5$$

and partial differentiation with respect to x produces

$$f_{yx} = 60x^3y^4 = \frac{\partial^2 z}{\partial x \, \partial y} \tag{6.14}$$

It will be seen that, notation aside, the outcome (6.14) is identical to (6.13). This is no mere coincidence, and it can be shown that, for a very wide range of functions of two independent variables,

Second-order cross partial derivatives are equal.

Exceptions to this property tend to be rather special expressions of mainly theoretical interest, and they will not concern us here. The equality of second-order mixed derivatives means that there is no need to fuss over the order in which ∂x and ∂y should appear in the denominator. This is more than just an academic point. There are usually interesting interpretative insights to be gained in the context of a particular application. Also, in obtaining the mixed derivatives, it may be far easier to proceed by one sequence of derivations than another. For example, given the function

$$z = \frac{y^4 + 4}{y^3 - 7} - 4xy^3$$

and the requirement to obtain the second-order cross partial derivative

$$\frac{\partial^2 z}{\partial x \, \partial y}$$

it is far more convenient to proceed by taking advantage of the property of the equality of mixed derivatives by first differentiating with respect to x rather than y. The benefit of this lies in eliminating the awkward quotient at the first stage. The effort of differentiating the quotient, which contains only y, proves to be wasted when the subsequent derivative is taken with respect to x. Thus, proceeding in the most efficient manner,

$$\frac{\partial z}{\partial x} = -4y^3$$

and so

$$\frac{\partial^2 z}{\partial y \, \partial x} = -12y^2 = \frac{\partial^2 z}{\partial x \, \partial y}$$

As further practice in partial differentiation consider the expression
$$z = 4x^3 - 5xy^2 + 4x^2y - 2y^3$$
with a view to finding the first-order partial derivatives, the slopes of the function in the two fundamental directions at the point (2, 3, −64) and the second-order direct and cross partial derivatives. At the first order
$$\frac{\partial z}{\partial x} = 12x^2 - 5y^2 + 8xy$$
and
$$\frac{\partial z}{\partial y} = -10xy + 4x^2 - 6y^2$$
So the slope in the x direction at (2, 3, −64) found by substitution in $\partial z/\partial x$ will be
$$\frac{\partial z}{\partial x} = 12(2)^2 - 5(3)^2 + 8(2)(3)$$
$$= 51$$
and the slope in the fundamental y direction at this point will be
$$\frac{\partial z}{\partial y} = -10(2)(3) + 4(2)^2 - 6(3)^2$$
$$= -98$$
The direct second-order partial derivatives are
$$\frac{\partial^2 z}{\partial x^2} = 24x + 8y$$
$$\frac{\partial^2 z}{\partial y^2} = -10x - 12y$$
while the second-order cross partial derivatives of the function work out as
$$\frac{\partial^2 z}{\partial y \, \partial x} = -10y + 8x = \frac{\partial^2 z}{\partial x \, \partial y}$$
The cross partial derivative can be found by differentiating $\partial z/\partial x$ partially with respect to y and can be confirmed by differentiation of $\partial z/\partial y$ partially with respect to x. As with functions of a single variable, higher order partial derivatives can be taken. For example, the third-order cross partial derivative
$$\frac{\partial^3 z}{\partial^2 x \, \partial y}$$

involves partially differentiating z twice with respect to x and once with respect to y. The result in the present example would be $+8$. As with second-order mixed derivatives, it does not matter in what sequence the differentiation is carried out. Similarly, the fourth-order mixed derivative

$$\frac{\partial^4 z}{\partial^2 x \, \partial^2 y}$$

requires partial differentiation twice with respect to both x and y. In the case of the current example this derivative is zero. While all this is possible, partial derivatives of higher order than the second are rarely required, and our focus here will remain firmly on the first- and second-order derivatives. We now turn to the most important context in which the first- and second-order derivatives find their principal role — optimization.

EXERCISES 6.5

1 Find the second-order direct derivative $\partial^2 z/\partial x^2$ for the following functions:

   (i) $z = x^5 y^2$
   (ii) $z = 7y^9 - 8x^5 y^3 - 2x^7 y$
   (iii) $z = 5x^3 y^3 - 6x^2 y^2 w^2 - xyw$

2 For the functions above, find the second-order direct derivative with respect to y.

3 For the functions 1(i), 1(ii) and 1(iii) above find the second-order cross partial derivatives.

4 For the function

$$z = 2x^3 + y^3 - 4xy^2 + 5x^2 y + x^2 - 3y^2 + 7xy - 3x + 8y + 9$$

   (i) find the slope in the x direction when $x = 1$ and $y = 2$;
   (ii) find the slope in the y direction when $x = 1$ and $y = 2$;
   (iii) confirm that $\partial^2 z/\partial y \, \partial x = \partial^2 z/\partial x \, \partial y$ and evaluate this term at $x = 1$, $y = 2$;
   (iv) evaluate $\partial^2 z/\partial x^2$ at $x = 1$, $y = 2$;
   (v) show that the slope in the y direction falls for increasing y at $x = 1$, $y = 2$;
   (vi) find $\partial^3 z/\partial x^3$ and $\partial^3 z/\partial y^3$.

5 (i) Given the function

$$f(x, y) = 2x^6 y^7 - 3x^5 y^2 + 2x^2 y^4$$

   find all of the second-order partial derivatives.

(ii) For the function
$$f(x, y, w) = x^5 y^4 w^6$$
find the partial derivatives:

(a) f_{xy} (b) f_{xw} (c) f_{yw}

6 For the function
$$z = x^5 y^6 - 4x^3 y^2 + 6x^2 y^4$$
find

(i) $\dfrac{\partial^3 z}{\partial x^3}$

(ii) $\dfrac{\partial^3 z}{\partial x \, \partial y^2}$

(iii) $\dfrac{\partial^4 z}{\partial x^2 \, \partial y^2}$

(iv) $\dfrac{\partial^5 z}{\partial x \, \partial y^4}$

7 For the function
$$f(x, y, w) = 3x^4 y^4 w - 4x^5 w^5 - 5y^6 w^6$$
find

(i) f_w (ii) f_{yw} (iii) f_{xyw} (iv) f_{www}

6.6 LOCAL MAXIMA AND MINIMA

At the first order, for a local maximum or a local minimum of the function
$$z = f(x, y)$$
the function must have zero slope in the fundamental directions. That is
$$\frac{\partial z}{\partial x} = 0 \quad \frac{\partial z}{\partial y} = 0 \tag{6.15}$$

The *first-order conditions* (6.15) assert that the function must not be in process of changing in the fundamental directions at the putative maximum. The conditions (6.15) do not discriminate between maxima and minima (or, in fact, another possibility which will be considered in the next section). Any point which satisfies (6.15) is said to be a *stationary point* of the function. The term *critical value* is also sometimes employed.[3] It is at the second order that the character of the stationary value is identified, and the conditions here fall into two parts. The first part is what would be expected from the single-variable case:

$$\frac{\partial^2 z}{\partial x^2} \leqslant 0 \qquad \frac{\partial^2 z}{\partial y^2} \leqslant 0$$

i.e. the second-order direct partial derivatives must be non-positive for a maximum. However, we will work with the strict inequalities

$$\frac{\partial^2 z}{\partial x^2} < 0 \qquad \frac{\partial^2 z}{\partial y^2} < 0 \qquad (6.16a)$$

and will not consider at this point cases where the second derivatives are zero.[4] But there is another condition that must be fulfilled and this involves the cross partial derivatives. The condition (again expressed as a strict inequality) is

$$\frac{\partial^2 z}{\partial x^2} \frac{\partial^2 z}{\partial y^2} - \left(\frac{\partial^2 z}{\partial x \, \partial y}\right)^2 > 0 \qquad (6.16b)$$

The left-hand side of (6.16b) is called the *discriminant* of the function.[5] Later on, we shall see that if (6.16a) is satisfied but (6.16b) is not, then a different kind of stationary value is produced. The second-order conditions as a whole (6.16a) and (6.16b), if satisfied, mean that the function z is locally concave. If the second-order conditions hold *for all* values of x and y, then $z = f(x, y)$ is a *concave function*. Consider a maximization example. For

$$z = 35 + 10x - 8y + 2xy - x^2 - 2y^2 \qquad (6.17)$$

the task is to establish whether or not the stationary value is a maximum and to identify its location. The first-order conditions are

$$\frac{\partial z}{\partial x} = 10 + 2y - 2x = 0$$

$$\frac{\partial z}{\partial y} = -8 + 2x - 4y = 0$$

The second-order direct derivatives are

$$\frac{\partial^2 z}{\partial x^2} = -2 \quad \text{and} \quad \frac{\partial^2 z}{\partial y^2} = -4$$

while the mixed derivatives work out as

$$\frac{\partial^2 z}{\partial x \, \partial y} = +2$$

The negativity of the direct second-order derivatives meets the requirement (6.16a). In general, the second-order derivatives will be functions of x and y and must be evaluated for the values of these variables that satisfy the first-order conditions. But when, as here, z is a quadratic the second derivatives will be constants. Condition (6.16b) is also satisfied since

$$(-2)(-4) - (+2)^2 > 0$$

So the discriminant is of the appropriate sign and therefore the stationary point is a maximum. The location of the point is determined by the first-order conditions which can be solved as simultaneous equations. It may be noted that adding the two equations in this case will eliminate x leaving

$$2 - 2y = 0$$

so that $y = 1$ and substitution into either equation produces $x = 6$. Note that there is no other solution to the first-order conditions so the function has a global maximum of $z = 61$ at $x = 6$, $y = 1$. Now consider a further example. Show that the function

$$z = 1152x - 84y + 2xy^2 + x^2y - 4x^3 - 8y^3$$

has a maximum at the point $x = 10$, $y = 2$. The first-order conditions are

$$\frac{\partial z}{\partial x} = 1152 + 2y^2 + 2xy - 12x^2 = 0$$

$$\frac{\partial z}{\partial y} = -84 + 4xy + x^2 - 24y^2 = 0$$

Insertion of $x = 10$ and $y = 2$ shows that the equations are satisfied and the point is a turning point. At the second order, the requirements are that

$$\frac{\partial^2 z}{\partial x^2} = 2y - 24x < 0$$

$$\frac{\partial^2 z}{\partial y^2} = 4x - 48y < 0$$

Note also that

$$\frac{\partial^2 z}{\partial x \, \partial y} = 4y + 2x$$

and the fact that condition (6.16b) must also be satisfied. In this example, with the second derivatives being functions of x and y, the conditions must be evaluated at the point in question. Thus

$$\frac{\partial^2 z}{\partial x^2} = -236 < 0$$

$$\frac{\partial^2 z}{\partial y^2} = -56 < 0$$

so that condition (6.16a) is satisfied. Furthermore, since

$$\frac{\partial^2 z}{\partial x \, \partial y} = +28$$

condition (6.16b) will produce

$$(-236)(-56) - (28)^2 > 0$$

and, as all the conditions are fulfilled, the point (10, 2, 3968) is a maximum.

For functions of more than two independent variables, the first-order conditions for a maximum are that all first-order partial derivatives should be zero. Second-order conditions are not convenient to express without the use of *determinants*. Discussion of second-order conditions for both maxima and minima of functions of more than two variables is deferred until the Appendix.

Conditions for a minimum

Now consider the conditions for a local minimum of a function of two variables. At the first order, the conditions are the same as for a local maximum, i.e.

$$\frac{\partial z}{\partial x} = 0 \quad \text{and} \quad \frac{\partial z}{\partial y} = 0$$

At the second order, condition (6.16a) is replaced, in strict inequality form, by

$$\frac{\partial^2 z}{\partial x^2} > 0 \quad \frac{\partial^2 z}{\partial y^2} > 0 \qquad (6.18a)$$

Condition (6.18a) is the intuitive presumption – that the point produces a minimum for the function in the fundamental directions. In this case, too, there is the further condition on the discriminant that

$$\frac{\partial^2 z}{\partial x^2} \frac{\partial^2 z}{\partial y^2} - \left(\frac{\partial^2 z}{\partial x \, \partial y}\right)^2 > 0 \qquad (6.18b)$$

This is the same condition as (6.16b).[6] The only difference between the conditions for a maximum and for a minimum are the signs of the direct second-order derivatives. Consider an example. Find the minimum value of the function

$$z = 2x - 3y - 0.5xy + 0.5x^2 + 0.25y^2$$

The first-order conditions for a minimum are

$$\frac{\partial z}{\partial x} = 2 - 0.5y + x = 0$$

$$\frac{\partial z}{\partial y} = -3 - 0.5x + 0.5y = 0$$

These conditions solve uniquely for $x = 2$, $y = 8$ which gives the only stationary value. Is this a minimum? The second-order derivatives are

$$\frac{\partial^2 z}{\partial x^2} = 1 \qquad \frac{\partial^2 z}{\partial y^2} = 0.5 \qquad \frac{\partial^2 z}{\partial x \, \partial y} = -0.5$$

The direct derivatives are positive as required by (6.18a), and condition (6.18b) is also satisfied since

$$(1)(0.5) - (-0.5)^2 = 0.25 > 0$$

so that the point $x = 2$, $y = 8$ *is* a minimum, giving a value of the function of $z = -10$. Now consider a further numerical example. Show that the function

$$z = 2x^3 - x^2 y + 21x - xy^2 - 576y + 8y^3$$

has a minimum at $x = 2$, $y = 5$. Equating the first derivatives to zero,

$$\frac{\partial z}{\partial x} = 6x^2 - 2xy + 21 - y^2 = 0$$

$$\frac{\partial z}{\partial y} = -x^2 - 2xy - 576 + 24y^2 = 0$$

substitution of the given values into the derivatives shows that the first-order conditions are satisfied. The conditions on the direct second-order derivatives are

$$\frac{\partial^2 z}{\partial x^2} = 12x - 2y > 0$$

$$\frac{\partial^2 z}{\partial y^2} = -2x + 48y > 0$$

which are again fulfilled at $x = 2$, $y = 5$. The second-order cross partial derivative is

$$\frac{\partial^2 z}{\partial x \, \partial y} = -2x - 2y$$

$$= -14 \quad \text{at} \quad x = 2, \ y = 5$$

so that the sign requirement for the discriminant

$$\frac{\partial^2 z}{\partial x^2}\frac{\partial^2 z}{\partial y^2} - \left(\frac{\partial^2 z}{\partial x\,\partial y}\right)^2 > 0$$

works out as

$$(14)(236) - (-14)^2 > 0$$

Therefore the point satisfies all the criteria and is a minimum of the function.

EXERCISES 6.6

1. For the following functions, establish whether or not the stationary value satisfies the conditions for a local maximum.

   (i) $f(x, y) = 20x + 64y + 8xy - x^2 - 32y^2$
   (ii) $f(x, y) = 2x^2 + y^2 - 16x - 20y - 2xy$
   (iii) $f(x, y) = 5x - 4y + xy - 0.25x^2 - 0.5y^2$

2. Find the maximum value of the following functions:

   (i) $f(x, y) = 100 - 4x + 5y + xy - x^2 - 0.5y^2$
   (ii) $f(x, y) = 100 + 5x - 16y + 2xy - 0.25x^2 - 8y^2$

3. Confirm that the function

   $$z = 1000 - 21x + 576y + x^2 y + xy^2 - 2x^3 - 8y^3$$

   has a local maximum when $x = 2$ and $y = 5$.

4. Find the minimum value of the following functions:

   (i) $f(x, y) = 100 - 14x + 4y - 2.5xy + 1.5x^2 + 2y^2$
   (ii) $f(x, y) = 126 + 6x - 54y - 3xy + 0.5x^2 + 9y^2$

5. Show that the function

   $$f(x, y) = x^3 - 288x - 0.5x^2 y + 42y - 2xy^2 + 16y^3$$

   has a local minimum value when $x = 10$ and $y = 1$.

6.7 SADDLE POINTS

In considering the stationary points of functions of two independent variables, we have so far examined maxima and minima. There is another possible form for a stationary point here, a possibility that did not exist in the case of functions of a single independent variable. This is the case of a *saddle point*. A saddle point has the property of being a relative *maximum* in one direction and a relative *minimum* in the orthogonal direction. Saddle points are an important concept in the theory of constrained optimization and the theory of games.[7] The first-order conditions for a saddle point are the same as for any other stationary value, namely

$$\frac{\partial z}{\partial x} = 0 \quad \text{and} \quad \frac{\partial z}{\partial y} = 0$$

But at the second order, a sufficient condition for a saddle point to exist is that the discriminant is negative.[8] That is,

$$\frac{\partial^2 z}{\partial x^2} \frac{\partial^2 z}{\partial y^2} - \left(\frac{\partial^2 z}{\partial x \partial y}\right)^2 < 0 \tag{6.18c}$$

These points are so named since they have the appearance of a saddle. Another helpful analogy is that of a *pass* through a range of mountains, with the route reaching its highest point at a local low between the hills. The appearance of a saddle point is illustrated in Figure 6.8.

It might be thought that for there to be a relative maximum in one direction and a relative minimum at right angles, then in addition to (16.8c), one of the direct second-order derivatives ought to be negative (thereby corresponding to the maximum) while the other should be positive (implying a minimum). This is not required, but does arise when the relative maximum

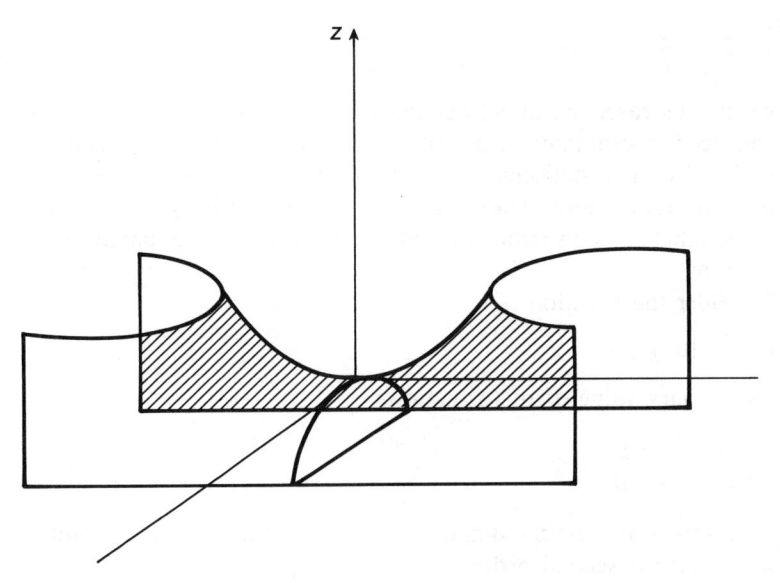

Figure 6.8

and minimum of the function are in the *fundamental* directions. Imagine such a case. Now rotate the surface through 45° in the horizontal plane. This has not changed the fact that a saddle point exists, but it *will* have altered the directions relative to which the relative extrema exist, and what happens in the fundamental directions. Unlike the case of, for example, a relative maximum in which movement in *all* directions must produce a decrease in z, what happens to z as movement occurs away from a saddle point depends on the direction of movement. At a saddle point the direct second-order derivatives may have the same sign or opposite signs as the numerical illustrations will show. First, we shall confirm that the one stationary value of the function

$$z = 25 + 12x - 30y + 6xy + 1.5x^2 - 3y^2$$

is a saddle point. As for any stationary value, the first-order conditions are that

$$\frac{\partial z}{\partial x} = 12 + 6y + 3x = 0$$

and

$$\frac{\partial z}{\partial y} = -30 + 6x - 6y = 0$$

which conditions solve uniquely for the values $x = 2$ and $y = -3$. At the second order

$$\frac{\partial^2 z}{\partial x^2} = 3 \qquad \frac{\partial^2 z}{\partial y^2} = -6 \qquad \frac{\partial^2 z}{\partial x\, \partial y} = 6$$

But since, in this case, the direct partial derivatives are of opposite sign, there is no need to evaluate the discriminant as condition (6.18c) is automatically fulfilled. The stationary value of the function z at $(2, -3)$ must therefore be a saddle point. The function is graphed in Figure 6.9, where it will be seen that the minimum is in the x direction and the maximum in the y direction.

Now consider the function

$$f(x, y) = 10xy + x^2 + y^2$$

For any stationary value of this function

$$f_x = 10y + 2x = 0$$
$$f_y = 10x + 2y = 0$$

which conditions solve for $x = 0$ and $y = 0$, so that the stationary point is at the origin. At the second order

$$f_{xx} = 2 \qquad f_{yy} = 2 \qquad f_{xy} = 10$$

Center x: 0   y: 0   Length x: 10   y: 10   Derive 3D-plot

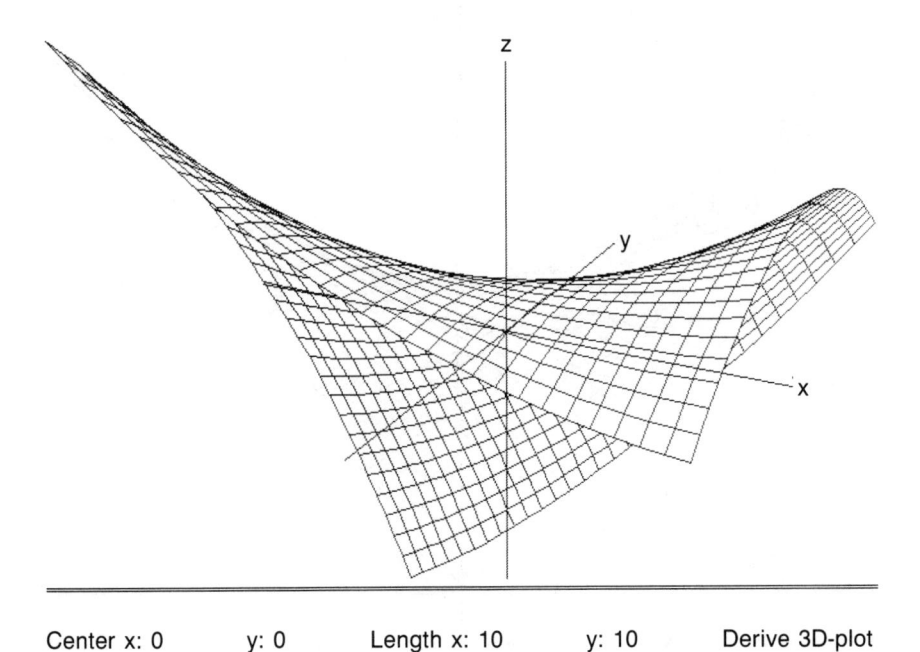

Figure 6.9

so that the discriminant is

$$f_{xx}f_{yy} - (f_{xy})^2 = (2)(2) - (10)^2$$
$$= -96 < 0$$

and a saddle point occurs, as shown in Figure 6.10.

As can be seen from Figure 6.10, in this case the 'saddle' is oriented at 45° to the x, y axes. In terms of the fundamental directions themselves, the fact that both of the direct second-order partial derivatives are positive in this case means that the function, while zero at the stationary value, is in process of increasing in both of the fundamental directions. Now consider the function

$$f(x, y) = 5xy - 0.25x^2 - 0.25y^2$$

396 Mathematics for business, finance and economics

Center x: 0   y: 0   Length x: 10   y: 10   Derive 3D-plot

Figure 6.10

At the first order

$$f_x = 5y - 0.5x = 0$$
$$f_y = 5x - 0.5y = 0$$

and at the second order

$$f_{xx} = -0.5 \qquad f_{yy} = -0.5 \qquad f_{xy} = 5$$

so that again the discriminant is negative, ensuring a saddle point. The function is graphed in Figure 6.11.

The overall appearance of the function is similar to that of Figure 6.10.[9] In this case, however, the function while stationary at (0, 0, 0) is in process of decreasing in both of the fundamental directions. The two cases demonstrate that the signs of f_{xx} and f_{yy} are not material to the existence of a saddle point.

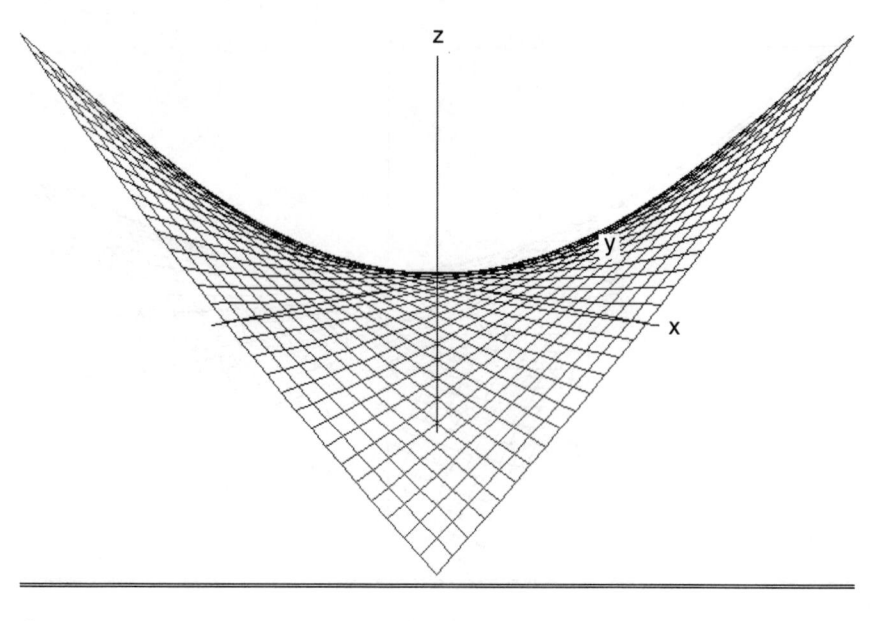

Center x: 0   y: 0   Length x: 10   y: 10   Derive 3D-plot

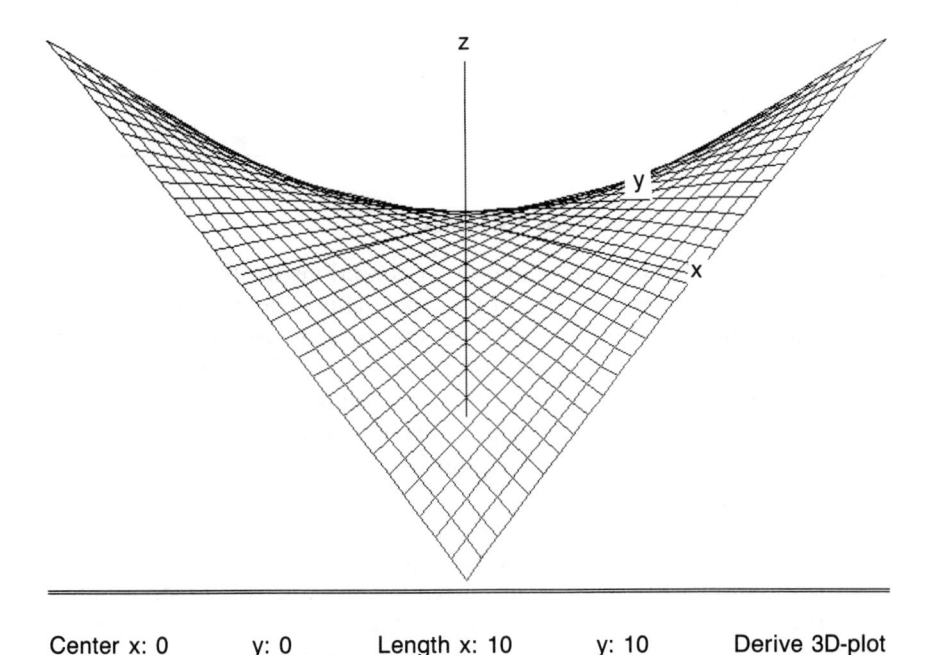

Figure 6.11

EXERCISES 6.7

1. Identify the stationary point of each of the following functions and confirm that the points correspond to saddle points:

    (i) $f(x, y) = 8x - 5y + 2xy - 2x^2 - 0.25y^2$
    (ii) $f(x, y) = 15x + 12y - 3xy - 0.75x^2 + 1.5y^2$
    (iii) $f(x, y) = 2x^2 + 0.5y^2 + 8xy - 200$

2. Find the stationary point of the functions below, and establish whether the point is a maximum, a minimum or a saddle point.

    (i) $f(x, y) = 1000 - 220x + 310y - 14xy + 5x^2 + 10y^2$
    (ii) $f(x, y) = 15x + 17y - 7xy - 3x^2 - 4y^2$
    (iii) $f(x, y) = 34x - 10y - xy - 2.5x^2 + 2y^2$

3. Locate and establish the nature of the stationary points of the following functions:

    (i) $z = 20x + 68y + 2xy + 4x^2 - 5y^2$
    (ii) $z = 17x - 15y + 7xy - 4x^2 - 3y^2$

(iii) $z = 8x - 12y - 2xy + 2x^2 + y^2$
(iv) $z = 8x - 20y + 4xy + x^2 - 2y^2$
(v) $z = 500 - 2.5x^2 + 7xy - 5y^2$
(vi) $z = 4x^2 - 0.5y^2 + 4xy + 32x - 20y$.

6.8 STATIONARY VALUES: RÉSUMÉ

It may be helpful to put the second-order conditions for the various types of stationary point into the form of a table. This is done in Table 6.1.

In Table 6.1 the bottom left box corresponds to an unattainable admixture. With the direct second-order derivatives of opposite sign, it is not then possible for condition (6.18b) to be fulfilled. Two further matters are worth raising at this point. First, as previously mentioned, it should be noted that cases where the discriminant is zero, i.e. where

$$f_{xx}f_{yy} - (f_{xy})^2 = 0$$

can still correspond to stationary values of $f(x, y)$. The further analysis that is required in these instances is considered in the Appendix. Second, in the numerical examples we have concentrated on quadratic functions. Furthermore, in the cases that were illustrated a stationary value of one kind or another was always found. There are other possible outcomes. For example, when the function z is a polynomial of higher degree than two, or is not a polynomial at all, the number of stationary values may be a finite number greater than one. And even in the case where z is a quadratic function of x and y, there are possible outcomes other than that of a single stationary value. The outcomes can be categorized as follows:

1 a single stationary value;
2 no stationary value;
3 a continuum of stationary values.

One outcome that cannot arise in the quadratic case is that of a finite

Table 6.1

		Discriminant > 0	Discriminant < 0
Direct derivatives	Negative	Maximum	Saddle point
	Positive	Minimum	Saddle point
	Opposite sign	*	Saddle point

number of stationary values greater than one. The reason for this is straightforward, and arises from the fact that, where z is a quadratic expression, the first-order conditions represent a system of simultaneous linear equations.[10] However, as we saw when studying this topic in Chapter 2, such a linear system can have one solution, no solutions or an infinity of solutions. This is what gives rise to cases 1 to 3 above. We have already seen numerous examples under category 1, so first let us look at a function of two variables that has no stationary values. Consider

$$z = 6x^2 + 1.5y^2 + 6xy - 18x$$

for which the first-order conditions for a stationary point are

$$\frac{\partial z}{\partial x} = 12x + 6y - 18 = 0$$

$$\frac{\partial z}{\partial y} = 3y + 6x = 0$$

Inspection of these conditions reveals that the simultaneous equations represented by the first-order conditions are *inconsistent*. They can be written as

$$12x + 6y = 18$$
$$6x + 3y = 0$$

for which multiplication of the lower equation by 2, and subtraction from the upper equation produces the unfulfillable condition

$$0 = 18$$

Therefore there are *no* values of x and y which can satisfy both necessary conditions and consequently there is no point at which the function is stationary. In this case, the graphs of $\partial z/\partial x = 0$ and $\partial z/\partial y = 0$ produce parallel lines in the x, y plane as shown in Figure 6.12(a).

A minor modification to the function z above results in a case of infinitely many solutions to the first-order conditions. If, for example,

$$z = 6x^2 + 1.5y^2 + 6xy - 18x - 9y + 13.5$$

then[11]

$$\frac{\partial z}{\partial x} = 12x + 6y - 18 = 0$$

$$\frac{\partial z}{\partial y} = 3y + 6x - 9 = 0$$

400 Mathematics for business, finance and economics

(a)   (b)

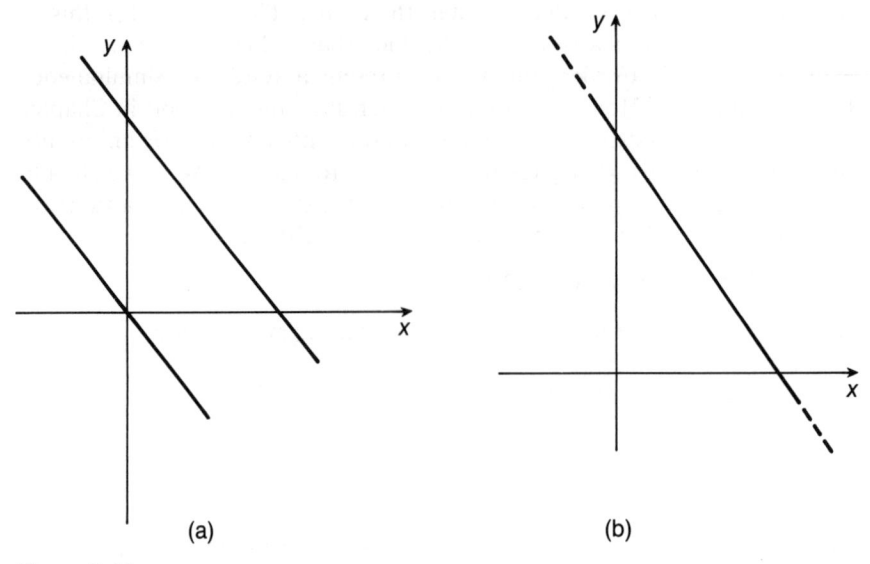

Figure 6.12

Each of these conditions can be simply rearranged as the explicit linear equation

$$y = 3 - 2x$$

This situation corresponds to the single line locus of $\partial z/\partial x = 0$ or $\partial z/\partial y = 0$ shown in Figure 6.12(b). The graph of z is shown in Figure 6.13.

The surface produced by this function resembles a bent sheet of paper which touches the x, y plane along a straight line with the equation $y = 3 - 2x$. Note that at the second order in this case

$$\frac{\partial^2 z}{\partial x^2} = 12 \quad \frac{\partial^2 z}{\partial y^2} = 3 \quad \frac{\partial^2 z}{\partial x \partial y} = 6$$

so that the direct second-order derivatives satisfy the conditions for a minimum (and are also consistent with the existence of a saddle point). But the discriminant test to distinguish these possibilities fails in this case, as the discriminant itself vanishes:

$$(12)(3) - (6)^2 = 0$$

However, as is evident from Figure 6.13, the points for which the co-ordinates are related by $y = 3 - 2x$ in fact constitute local minima. This can be confirmed at any point on the line by calculation of z for variation in x or y alone.

The ground covered so far is sufficient, at least in principle, to enable local maxima and minima of polynomial expressions to be identified. This

Functions of more than one variable 401

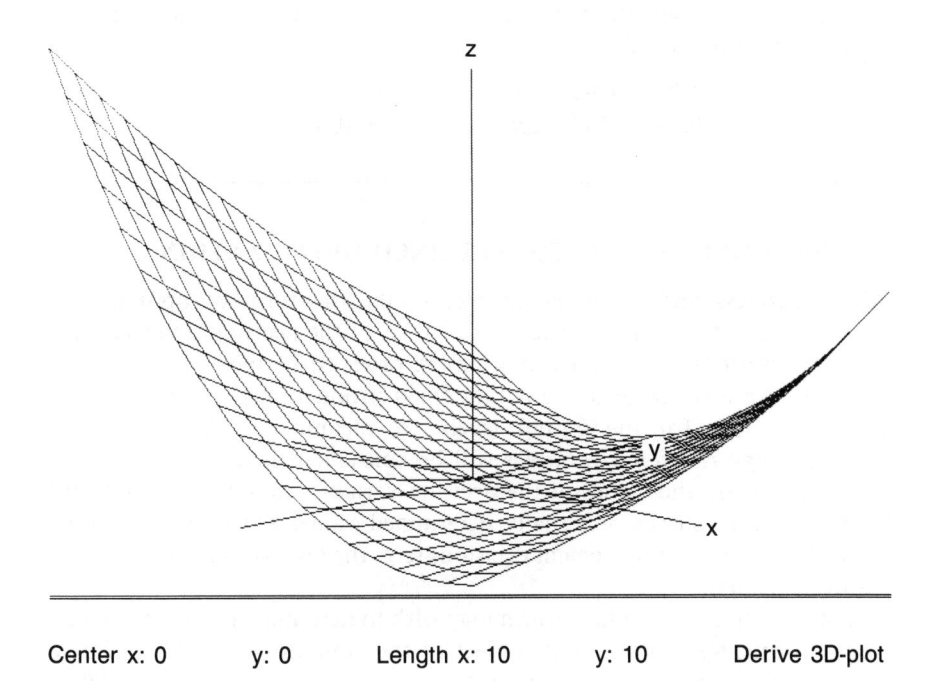

Center x: 0   y: 0   Length x: 10   y: 10   Derive 3D-plot

Figure 6.13

is so provided that there are no effective restrictions on the values that may be taken by the independent variables – it being assumed that whatever values emerge from the calculations will make economic or commercial sense. All this is the realm of *unconstrained* optimization in which the values of the variables may be freely chosen. Many such cases do occur in practice, particularly in problems involving rates of change. But, following the end of section exercises, we shall consider the other, very important, class of problems in which there are explicit limits to the decision-maker's freedom of choice.

EXERCISES 6.8

1 For the following functions, establish whether or not a stationary value exists. Where a stationary value does exist, establish its nature.

 (i) $f(x, y) = 12x^2 + 12xy + 3y^2 + 8x - 10y$
 (ii) $f(x, y) = 60x + 64.5y - 2.5x^2 + 3.5y^2 + 0.5xy$
 (iii) $f(x, y) = 1000 - 20x^2 - 5y^2 + 20xy - 50x + 30y$

2 Establish whether there are many or no stationary values for the following functions:

(i) $f(x, y) = 2.25x^2 + 9y^2 - 9xy - 9x + 18y + 100$
(ii) $f(x, y) = 0.5x^2 + 2y^2 + 2xy - 5x - 8y + 100$

6.9 INTRODUCTION TO CONSTRAINED OPTIMIZATION

Many business and economic problems take the form of optimization subject to constraint. In such cases the values of the decision (independent) variables must fulfil certain conditions in order to be meaningful or permissible. An obvious example is that for a production problem in which negative values for the output levels are meaningless. Such cases of explicitly *sign restricted variables* will be considered in Chapter 7. Other requirements on the values of the decision variables may relate to scarcity of available resources, externally determined values (such as the size of orders) or internal management decisions (e.g. the apportionment of output between plants).

Consider some examples. A firm may wish to determine production levels so as to maximize profits but subject to the condition that the available supplies of materials for the manufacturing process are not exceeded. Alternatively, the firm may wish to minimize the cost of producing an order of given size (i.e. cost minimization subject to an output constraint). In terms of service provision, management may find that the queuing times of its customers are unacceptable beyond a given level, and may wish to minimize the cost of arranging its services so as to meet a waiting time condition. To begin to see how this sort of problem can be approached we start with a particular function and seek the maximum *without* constraint. We then see what difference is made by the introduction of a single, joint constraint on the values that can be taken by the independent variables. The function is

$$z = 4x + 4y + 2xy - x^2 - 1.5y^2 \tag{6.19}$$

With no restrictions on the choice of values of x and y, for a maximum it will be required that

$$\frac{\partial z}{\partial x} = 4 + 2y - 2x = 0$$

and

$$\frac{\partial z}{\partial y} = 4 + 2x - 3y = 0$$

These first-order conditions solve for $x = 10$ and $y = 8$. At the second order

$$\frac{\partial^2 z}{\partial x^2} = -2 \qquad \frac{\partial^2 z}{\partial y^2} = -3 \qquad \frac{\partial^2 z}{\partial x \, \partial y} = +2$$

So all of the conditions are fulfilled, because the direct second-order derivatives are negative and (6.18b) is satisfied too, since

$$(-2)(-3) - (2)^2 = 2 > 0$$

Thus the point with x and y co-ordinates (10, 8) is a maximum, and by substitution into the function this maximal value of z is 36. The point (10, 8, 36) is the single turning point of the function (there being no other solutions to the first-order conditions in this quadratic case) and is therefore the global, as well as a local, maximum. The point represents the unconstrained or *free maximum* of the function, which is illustrated in Figure 6.14.

Now suppose that the decision-maker does not have unrestricted freedom in setting the values of the independent variables. Specifically, suppose that the selection of values of x and y has to satisfy the requirement

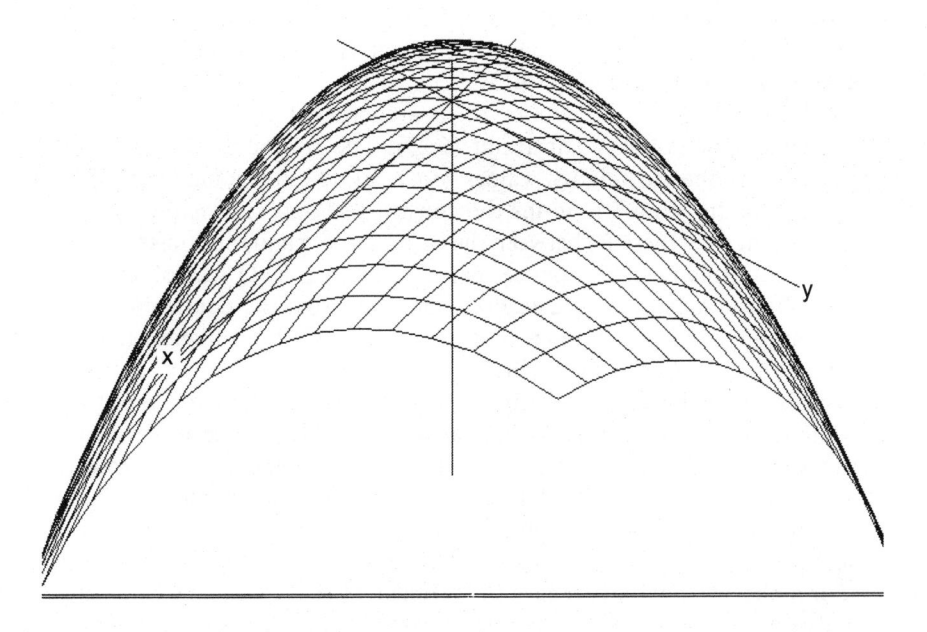

$$2x + y = 20 \tag{6.20}$$

Center x: 0   y: 0   Length x: 10   y: 10   Derive 3D-plot

Figure 6.14

404 Mathematics for business, finance and economics

The condition (6.20) might represent a budget constraint with x and y representing the amounts of two raw materials used in a production process, the prices of which are £2 and £1 per unit respectively, and with £20 as the overall budget limit. The function (6.19) would represent profits made from production expressed in terms of the materials used. The aim here would be to secure a maximum of (6.19) for values of x and y which can be purchased within the given budget and which therefore must satisfy (6.20). The reader will no doubt recognize a structural similarity to linear programming problems discussed in Chapter 3, except that in this case the objective function, (6.19), is non-linear. The task that faces us is indeed a programming problem,[12] which we will address here solely in terms of the calculus. There are several ways of approaching constrained optimization problems. Amongst these are mathematical programming (of which linear programming is a special case) and numerical methods. While due reference will be made to programming techniques, this book concentrates on three methods for the solution of constrained optimization problems. These are:

1 the tangency method;
2 the method of substitution;
3 Lagrange multiplier methods.

The tangency approach is outlined in the present section, and the method of substitution will be described in Section 6.10. Lagrange multiplier methods are the principal subject of Chapter 7. To illustrate the tangency method, first consider Figure 6.15.

In Figure 6.15, the contours of z are crossed by the straight line representing the equality constraint (6.20). The contours crossed, in this case, correspond to higher values of z as the free maximum P is approached. In geometric terms the constrained optimum is given by the point of tangency (here R) between the contours of z and the constraint line. Why should this be so? Suppose that a point such as Q was obtained. Q cannot be the maximum because the contours cross the constraint line at this point. This in turn implies that contours inside (and higher) than the contour through Q can be reached by movement away from Q along the constraint − in this case towards R − with x increasing at the expense of y. Similarly, if the original point had been Q′, higher contours can be reached by substituting y for x in movement across higher contours towards R. It is only at point R itself that movement in *either* possible direction along the constraint line would necessarily mean movement onto a lower contour.

For a physical analogy, regard the problem as that of gaining the highest point reached by a road (the constraint line) on an Ordnance Survey map with the contours being of the topographical variety. The fundamental y direction then represents movement north while the x direction corresponds to east. Movement towards R from above represents motion in a direction

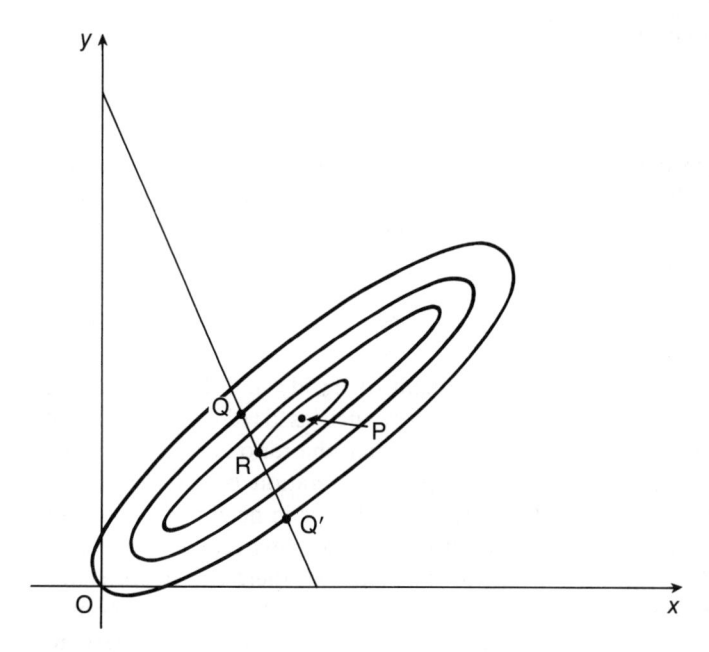

Figure 6.15

approximately $26\frac{1}{2}°$ west of south! The highest point of the road will not usually, as here, be on the brow of the hill, which roads normally avoid, but at the location where it forms a tangent to a contour of the hill.

Although feasible in principle, the purely geometric approach of drawing contours and finding the tangency point on a diagram is long-winded and inaccurate as a method of solution. However, it does give a number of useful insights. At a tangency point slopes are equal and equality of slope means equality of *rates of change* or *rates of substitution*. Such equivalence usually yields meaningful interpretation. In the present case the optimum is reached when the rate at which x can be substituted for y *maintaining cost* is the same as the rate at which the substitution can be made *maintaining profit*; advantage can be taken of any disjunction either to increase profit or to reduce cost or both. The tangency property of the optimum – spelt out in terms of the expressions for slope – can be used to *solve* for the optimal values of x and y. This is superior to an attempt to measure these values off the axes. The slope of the constraint line is easily obtained by writing the constraint with y as an explicit function of x,

$$y = 20 - 2x$$

so that the slope of the line is

$$\frac{dy}{dx} = -2$$

But how is the slope of a contour of z to be obtained? The answer requires the concept of the *differential*. The differential of z, dz, is the approximate *increment* in z following increments dx and dy in the independent variables. The differential of z is

$$dz = \frac{\partial z}{\partial x} dx + \frac{\partial z}{\partial y} dy \qquad (6.21)$$

In equation (6.21), the singular treatment of dz, dy and dx is at variance with the use, up to this point, of a term such as dy/dx as a composite *symbol*, representing the value towards which the ratio $\Delta y/\Delta x$ tends. But there are occasions when the separate interpretation of dx, dy and dz proves to be valuable.[13] In (6.21) dz is the change in height of the tangent plane to the surface at a point, and this is approximately equal to the actual change in z. Now with a contour of the function, there can be *no* change in z so long as any movement from the original point is along the contour through the point. Thus for movement along a contour dz must be zero and we can write

$$dz = \frac{\partial z}{\partial x} dx + \frac{\partial z}{\partial y} dy = 0 \qquad (6.22)$$

Rearrangement of (6.22) yields

$$\frac{\partial z}{\partial y} dy = -\frac{\partial z}{\partial x} dx$$

which in turn can be re-expressed as

$$\frac{dy}{dx} = -\frac{\partial z/\partial x}{\partial z/\partial y} \qquad (6.23)$$

where (6.23) is known as the *implicit function rule*. The usefulness of the implicit function rule in the present context is in finding the slope of the contours of a function of two independent variables. Given the requirement that tangency must occur between a contour and the constraint line at the constrained optimum, the implicit function rule can be used to find the precise co-ordinates of the optimal point. We have already established that the slope of the constraint line is -2. Now in terms of the function, z, since

$$\frac{\partial z}{\partial x} = 4 + 2y - 2x$$

and

$$\frac{\partial z}{\partial y} = 4 + 2x - 3y$$

the slope of the contours at any point can be obtained by use of (6.23). The expression that results is

$$\frac{dy}{dx} = \frac{-(4 + 2y - 2x)}{4 + 2x - 3y}$$

Tangency requires the equation of slopes. Thus

$$-2 = \frac{-(4 + 2y - 2x)}{4 + 2x - 3y}$$

and so

$$2(4 + 2x - 3y) = 4 + 2y - 2x$$
$$8 + 4x - 6y = 4 + 2y - 2x$$

so therefore

$$6x - 8y = -4 \qquad (6.24)$$

where (6.24) is a condition that must be fulfilled at *any* point of tangency. Our *particular* point of tangency is located on the constraint line given by $2x + y = 20$, and the co-ordinates of the constrained optimum are obtained by solving (6.24) simultaneously with the constraint line expression. Using the elimination procedure and multiplying the constraint line equation by 3 gives the system

$$6x + 3y = 60$$
$$6x - 8y = -4$$

so that

$$11y = 64$$

Thus

$$y = 64/11 \approx 5.82$$

and, by substitution into either equation,

$$x = 78/11 \approx 7.09$$

With these values, the constrained maximum of z (obtained by substitution of the optimal values of x and y into (6.19)) emerges as

$$z = 364/11 \approx 33.09$$

This procedure – establishing and applying the conditions for tangency –

is used in the microeconomic theory of the firm to find *expansion paths* for output combinations as the level of available resource increases. It is also used in portfolio theory to find the *critical line* – which shows the ideal balance between investments that ensures minimum risk for each value of return. In these and other instances, the interpretative insight obtained usually compensates to some extent for the frequently laborious nature of the procedure itself.

We have been taking it for granted that the point of tangency – which represents the solution of first-order conditions – produces a maximum (as is required in this case) rather than a minimum. While the nature of the function, the context of the application – or both – may make it obvious that an appropriate extremity is produced by tangency, in general the question of second-order conditions needs to be considered. We shall begin to address this matter in the following section and in rather more detail in Chapter 7, with further material on second-order conditions in constrained optimization problems in the Appendix.

EXERCISES 6.9

1 Use the tangency approach to find the maximum value of

$$f(x, y) = 22x + 33y + 3xy - 1.5x^2 - 2.25y^2$$

subject to the requirement that

$$x + 3y = 55$$

2 For the function

$$f(x, y) = 20x + 16y - 0.1x^2 - 0.05y^2$$

in comparison with the unconstrained optimum, by how much is the maximum value of $f(x,y)$ reduced when x and y must satisfy the constraint

$$5x + 2y = 325$$

6.10 THE METHOD OF SUBSTITUTION

We begin by considering the *method of substitution* in the case of a function of two variables to be maximized subject to a single linear constraint. Later, we shall look at examples going beyond the two-variable, single linear constraint case. For comparative purposes we shall take the problem of Section 6.9 as the first numerical example. The substitution method starts by expressing the constraint equation as an explicit function of one of the variables. That is, (6.20) is written either as

$$y = 20 - 2x$$

or as

$$x = 10 - 0.5y$$

One of these forms is selected on the basis of convenience, and the dependent variable therein is then replaced in the function z by the appropriate expression. For example, substituting for y gives

$$z^* = 4x + 4(20 - 2x) + 2x(20 - 2x) - x^2 - 1.5(20 - 2x)^2$$
$$= 156x - 11x^2 - 520 \qquad (6.25)$$

There are two important points to note about (6.25). First, because of the substitution, the problem has been reduced to one involving a function of a *single* variable, to which the known procedure for unconstrained maximization can be applied. Second, we have labelled the function z^* rather than z to remind us that what we shall be optimizing is the value of z *provided that the constraint is satisfied*. A different z^* expression would obviously have resulted if a different constraint had been used. Proceeding with

$$z^* = 156x - 11x^2 - 520$$

a maximum requires that

$$\frac{dz^*}{dx} = 156 - 22x = 0$$

and, at the second order,

$$\frac{d^2z^*}{dx^2} = -22 < 0$$

Solving the first-order condition gives

$$x = 156/22 = 78/11$$

as previously obtained via the tangency approach. The corresponding values of y and z can be confirmed by substitution of the optimal x value into the expressions for z and y. As an exercise the reader should rework the calculation making the substitution $10 - 0.5y$ for x. The plot of z^* against x is shown in Figure 6.16. The figure shows that the z^* parabola has its vertex above $x = 7.09$. If x had been the variable initially substituted out instead of y, the outcome would have been the same. To confirm, beginning with

$$z = 4x + 4y + 2xy - x^2 - 1.5y^2$$

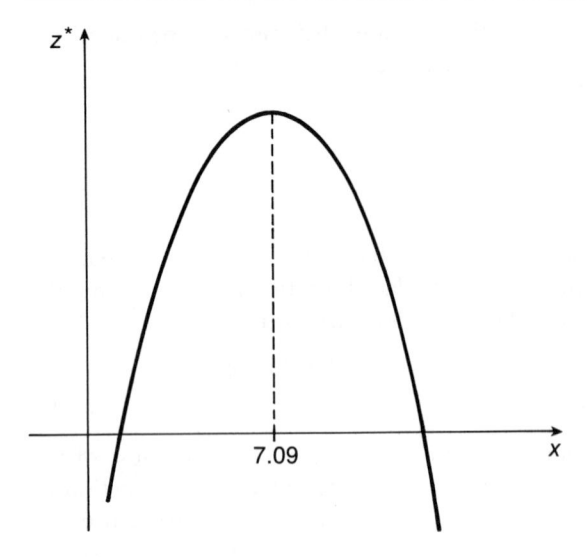

410 Mathematics for business, finance and economics

Figure 6.16

using the substitution
$$x = 10 - 0.5y$$
we obtain
$$z^* = 4(10 - 0.5y) + 4y + 2(10 - 0.5y)y - (10 - 0.5y)^2 - 1.5y^2$$
$$= -60 + 32y - 2.75y^2$$
for which the requirements on the derivatives are
$$\frac{dz^*}{dy} = 32 - 5.5y = 0$$
and, at the second order,
$$\frac{d^2z^*}{dy^2} = -5.5 < 0$$
From the first-order condition
$$y = \frac{32}{5.5} = 5.\overline{81}$$
and hence
$$x = 7.\overline{09} \quad \text{and} \quad z = 33.\overline{09}$$
So substitution for either variable produces the same result. The result is a *local* maximum of z^* and therefore a *local* maximum of z subject to the

constraint. In the present example, with quadratic z^*, that local maximum will also be the *global* maximum. If the function z^* is cubic or higher, either because of a cubic z or a non-linear constraint, then despite the constraint there may be no finite maximum or minimum of z. For example, the problem

maximize $\quad z = -180y + x^2 y$
subject to $\quad 2x + y = 12$

on substitution for y produces

$$z^* = -2160 + 360x + 12x^2 - 2x^3$$

and while the first-order condition solves for $x = -6$ (minimum) and $x = 10$ (maximum) these values give only *local* maxima and minima. Without further restriction, z^* is unbounded. As may be seen from the sign of the dominant cubic term, it becomes arbitrarily large and positive for negative x and arbitrarily large and negative for positive x. In practice, of course, there *will* be further restrictions on the values that the decision variables can take. These restrictions often take the form of *sign requirements* on the variables, which may also be individually bounded from above. While allowance for sign requirements can be tacked on to the substitution approach, the Lagrange multiplier framework of Chapter 7 is more appropriate for this purpose. Within the present context, we shall simply remain on informal alert for unbounded cases. The use of the method of substitution illustrates a very valuable problem solving principle. When there appears to be no means of solving a problem it is worth asking the following question:

> Can the problem be converted – without altering its essential character – into a form for which techniques of solution are available?

In applying this principle, it is important to note the specific proviso, '*without altering its essential character*', the problem solving art being to discern a modification that achieves this end. Now consider a further numerical example in the use of the method of substitution in the case of a single linear constraint. The example will be used to introduce an additional aspect of constrained optimization problems: the useful economic and financial information that can usually be extracted from them. The problem is this:

maximize $\quad z = xy$
subject to $\quad x + y = 12$

Note in passing that, in the absence of an effective constraint, this function would have no finite maximum. To solve the problem, the constraint is first rearranged to express y in terms of x. This gives

$$y = 12 - x$$

Now substitute for y in the objective function

$$z^* = x(12 - x)$$
$$= 12x - x^2$$

so at the first order

$$\frac{dz^*}{dx} = 12 - 2x = 0$$

The second order condition will be satisfied since

$$\frac{d^2z^*}{dx^2} = -2$$

From the first derivative, therefore, $x = 6$ implying $y = 6$ and $z = 36$ is the maximum value achievable given the constraint. There are two points to note here. First, the problem is *symmetrical* in x and y. If x and y were exchanged in the objective function and constraint the problem would be unchanged. This being the case, and since the optimum is unique, the optimal values of x and y *must* be the same.[14] Note that symmetry within the objective function or constraint alone is not enough to achieve this result. The second point is that this is a further case in which the constrained optimum of z is finite. As is clear from the sign of the leading term in the quadratic z^*, the value $x = 6$ produces the *global* maximum under constraint. Note, however, that in the absence of constraints there is no finite maximum of z.

There is a further aspect of constrained optimization problems that leads to some important economic and business considerations. What would happen if the right-hand side of the constraint was increased by one unit? You can be sure that, if the problem models a real business situation, it will be a managerial priority to *relax* any constraints that restrict the achievement of company objectives. In the present case it is clear that an increase in the right-hand side value of 12 will enable an improved position to be obtained and therefore an increase can be seen as a relaxation of the constraint. It should be noted, however, that in general the direction which represents easement is not always clear in the case of equality constraints. Suppose that the value on the right-hand side of the constraint had been 13. Reworking the problem:

$$y = 13 - x$$
$$z^* = x(13 - x) = 13x - x^2$$
$$\frac{dz^*}{dx} = 13 - 2x = 0$$

so that

$$x = 6.5 \qquad y = 6.5 \qquad z = 42.25$$

The increase in z of

$$42.25 - 36 = 6.25$$

is directly attributable to the extra unit on the right-hand side, and 6.25 is a *marginal valuation* of the scarce resource (or whatever is represented by the right-hand side). In programming terms, the figure of 6.25 is the *shadow price* or *dual value* of the resource represented by the right-hand side value, and is the *maximum* amount worth paying in order to secure the extra unit. This is management information of particular importance in that it allows judgement to be made on costed proposals to relax the constraint. The most common type of constraint used in business and economic modelling is the linear form, and the method of substitution is well suited to the case of a single linear equality constraint. But it is also possible to use the method with a non-linear constraint if the constraint can be expressed as an explicit function. For example, suppose that it is required to find the maximum of the function

$$z = 30x + 5y$$

subject to the constraint

$$4x^2 + 2y - 20 = 0$$

First, the constraint is written in explicit form. In this case, it is more convenient to state y as a function of x. That is,

$$y = 10 - 2x^2$$

Substitution into the objective function produces

$$z^* = 30x + 50 - 10x^2$$

for which the first- and second-order derivatives are

$$\frac{dz^*}{dx} = 30 - 20x = 0$$

and

$$\frac{d^2z^*}{dx^2} = -20 < 0$$

so that the solution to the first-order condition produces a maximum. The resulting values are

$$x = 1.5 \qquad y = 5.5 \qquad z = 72.5$$

Some care may be required, as the next example illustrates. The objective is to maximize

$$z = xy$$

subject to the constraint

$$0.5y - x^2 + 21x = 120 \qquad (6.26)$$

From the constraint, we obtain y as a function of x as

$$y = 2x^2 - 42x + 240$$

which, when substituted into the objective function, produces the result

$$z^* = 2x^3 - 42x^2 + 240x$$

for which the derivatives are

$$\frac{dz^*}{dx} = 6x^2 - 84x + 240$$

and at the second order

$$\frac{d^2z^*}{dx^2} = 12x - 84 < 0$$

The first-order condition can be divided through by 6 to give the more convenient quadratic equation

$$x^2 - 14x + 40 = 0$$

which factors as

$$(x - 4)(x - 10) = 0$$

so that the stationary points are where $x = 4$ and $x = 10$. At the second order, for $x = 4$,

$$\frac{d^2z^*}{dx^2} = 12(4) - 84 < 0$$

so $x = 4$ corresponds to a maximum, where $y = 104$ and $z = 416$. But the point which has been produced is only a *local* maximum of z^*, and therefore only a local maximum of z with respect to the constraint. In fact, as the positive coefficient of x^3 in z^* indicates, there is no finite maximum to

z^* and therefore to z subject to (6.26), witness the following extract from a program printout:

x	y	z
0	240	0
1	200	200
2	164	328
3	132	396
4	104	416
5	80	400
6	60	400
7	44	308
8	32	256
9	24	216
10	20	200
11	20	220
12	24	288
13	32	416
14	44	616
15	60	900
16	80	1280
17	104	1768
18	132	2376
19	164	3116
20	200	4000

The method of substitution can be used in some cases where there is more than one constraint. The effect of 'using up' the information in any one constraint by making a substitution in the objective function is to reduce the number of freely determinable variables by one. Thus, when we were concerned with a function of two variables to be maximized with respect to a single constraint, the z^* function was of a single variable. In the case where a function of two variables is to be maximized with respect to *two* linear constraints, there will be $2 - 2 = 0$ variables over which the decision-maker has control.[15] That is, the decision-maker is restricted to the particular values of the variables that satisfy the simultaneous equations given by the constraints. In this case, what is faced is not so much a problem as a *fact*. If it was required to maximize

$$z = xy$$

subject to

$$2x + y = 16$$

and

$$x + 3y = 23$$

416   Mathematics for business, finance and economics

since the constraints solve uniquely for $x = 5$ and $y = 6$, the value of z is, perforce, 30. However, if z is a function of *more* than two variables, then the presence of two constraints will still leave one or more variables to be freely chosen. Consider the problem

$$\text{maximize} \quad z = wx + wy + xy + 6x$$

subject to the constraints

$$2x + y + w = 16$$
$$x + 3y - 2w = 23$$

The first constraint will be used to express w in terms of x and y. The right-hand side of the result

$$w = 16 - 2x - y \tag{6.27}$$

is now substituted for w in the second constraint. The outcome is

$$x + 3y - 2(16 - 2x - y) = 23$$
$$5x + 5y - 32 = 23$$
$$x + y = 11$$

i.e.

$$y = 11 - x \tag{6.28a}$$

which, when substituted into (6.27), gives

$$w = 5 - x \tag{6.28b}$$

As ever, when there is one more variable than there are equations, we are able to solve for all but one of the unknowns in terms of the remaining variable, the numeraire, which in this case is x. Now substituting (6.28a) and (6.28b) into z gives

$$z^* = (5 - x)x + (5 - x)(11 - x) + x(11 - x) + 6x$$
$$= 6x - x^2 + 55$$

so that, at the first order, it is required that

$$\frac{dz^*}{dx} = 6 - 2x = 0$$

while at the second order the condition for a maximum is automatically fulfilled since

$$\frac{d^2z^*}{dz^{*2}} = -2 < 0$$

Therefore, solution of the first-order condition and substitution of the result into (6.28a) and (6.27) gives the outcome

$$x = 3 \qquad y = 8 \qquad w = 2 \qquad z^* = 64$$

and in this case, since z^* is a quadratic, we can be confident that the solution obtained represents a global maximum of z subject to the constraint. In the case of a function of three variables to be optimized subject to a single constraint, the use of the method of substitution reduces the problem to that of unconstrained optimization of a function of two variables. Consider the problem

$$\text{maximize} \qquad z = 6wx - w^2 + xy$$
$$\text{subject to} \qquad 5x + 2w + y = 20$$

The most convenient explicit statement of the constraint is

$$y = 20 - 5x - 2w$$

and substitution for y in z produces

$$z^* = 6wx - w^2 + x(20 - 5x - 2w)$$
$$= 4wx + 20x - 5x^2 - w^2$$

for which

$$\frac{\partial z^*}{\partial w} = 4x - 2w = 0$$

and

$$\frac{\partial z^*}{\partial x} = 4w + 20 - 10x = 0$$

while at the second order

$$\frac{\partial^2 z^*}{\partial w^2} = -2 \qquad \frac{\partial^2 z^*}{\partial x^2} = -10 \qquad \frac{\partial^2 z^*}{\partial x \, \partial w} = 4$$

so that the second-order conditions for a maximum will be fulfilled. Solving the first-order conditions produces

$$x = 10 \qquad w = 20 \qquad y = -70 \qquad z = 100$$

which represents the global optimum as well as a local one.

EXERCISES 6.10

1. Find the maximum value of
$$z = 100x + 50y + 2.5xy$$
subject to the constraint that
$$8x + 5y = 200$$

2. Given the condition
$$-2x + y = 100$$
find the maximum value of the function
$$z = 20y - 0.1x^2$$

3. Use the method of substitution to find the maximum value of the function
$$z = 1650x^2 - 80x^3 - 4x^2y^2 + 1000$$
subject to the constraint
$$0.5x + y = 20$$

4. Find the maximum value that can be taken by z where
$$z = wx + 4xy + 10wy + 30w$$
subject to the constraints that
$$x + 10y + 5w = 80$$
$$-2x + 30y + 2.5w = 115$$

5. Find the maximum value of the function
$$f(w, x, y) = 1000 - 5wx + 6xy - 3y^2$$
subject to the single constraint that
$$5x + 6y - 15w = 30$$

6. For the following problems, find the value at the margin of an extra unit of the resource represented by the right-hand side:
   (i) Maximize $f(x, y) = xy$
      subject to $0.25x + 0.25y = 120$
   (ii) Maximize $f(x, y) = 10x + 0.25xy + 20y$
      subject to $2.5x + 4y = 100$
   (iii) Maximize $f(x, y) = 100 + 200x - y^2$
      subject to $2x - y = 100$
   (iv) Maximize $f(x, y) = 100x^2 - 50xy + 200y^2$
      subject to $2x + y = 600.$

6.11 SUBSTITUTION: CONCLUDING REMARKS

A typical business problem involves more than one constraint. For example, when looking at an investment plan over the medium term, a company will usually be faced with limited investment funds in each year. The manufacturer will usually face, at least in the short term, limitations on capital equipment and floorspace as well as limited availability of some materials and components. While, as we have seen, some such multi-constraint problems can be addressed by the method of substitution, this would not, as a rule, be the technique of choice in the majority of cases. The Lagrange multiplier method provides a better framework in which to address multi-constraint problems, and would usually be preferred. Substitution is a very useful technique with the advantage of being a simple concept but it has a number of limitations. In particular, it can become impractical or unusable if

(a) there are many constraints;
(b) there is a constraint which cannot be stated as an explicit function;
(c) the constraints are inequalities rather than equations;
(d) there are sign restrictions on the decision variables.

What is required is a general purpose method that can accommodate these possibilities, at least so far as the formulation of the problem and the characterization of the solution are concerned. It is to such an approach, the method of Lagrange multipliers, that the next chapter is devoted. But first a tailpiece.

We have looked at the method of substitution as a process by which an optimization problem in n variables and m constraints is reduced to a problem in $n - m$ variables to which the methods of unconstrained optimization can be applied. Take the two-variable, one-constraint case. An intriguing question presents itself: was the problem *really* one in two variables in the first place? It certainly was not a problem in *two free choices*, because the determination of the value of one of the variables automatically (through the constraint) implied a specific value of the other variable. Maximization-subject-to-constraint is the form in which problems are likely to arise, but they could equally well be viewed as problems in a *single* decision variable masquerading as problems with two. Indeed, an unlimited number of apparently different numerical problems of maximization subject to constraint reduce, after substitution, to the same z^* function. Consider for example the following problems.

Problem A:

maximize $z = 50x - 45y - 3xy$

subject to $-2x + 6y = 20$

Problem B:

maximize $z = 0.4xy - 2.5x - 3y$

subject to $5x + 2y = 100$

For problem A, stating y in terms of x from the constraint gives

$$y = \frac{10}{3} + \frac{1}{3}x$$

which when substituted into z produces

$$z^* = 25x - x^2 - 150$$

for which the conditions for a maximum are

$$\frac{dz^*}{dx} = 25 - 2x = 0$$

and

$$\frac{d^2z^*}{dx^2} = -2 < 0$$

The first-order condition solves for $x = 12.5$ and therefore $y = 7.5$ and $z = 6.25$. For problem B, the constraint can be rearranged as

$$y = 50 - 2.5x$$

which when substituted into z gives the result

$$z^* = 25x - x^2 - 150$$

exactly as before. The value of x is therefore 12.5 at the optimum, but in this case $y = 18.75$. The value of z is 6.25 as before. Could the unconstrained problem in z^* be seen as the underlying 'real' structure with problems A and B as two particular manifestations? If this interpretation is accepted, what is the significance of the 'sameness' of the constrained problems?

6.12 BASIC PROGRAM

The BASIC program presented here gives the first- and second-order partial derivatives of a quadratic function of two variables. The program also states the type of stationary value for functions where the sufficient second-order condition is satisfied. For problems in which the discriminant is zero, the user is referred to the Appendix. The program will run in QuickBasic or in Q-Basic. A specimen program output is as follows.

The function specified is:

$f(x, y) = 4x^2 + 5xy - 6y^2 + 7x + 8y + 9$

The first order partial derivatives are:

$fx = 8x + 5y + 7$

$fy = 5x - 12y + 8$

The second order partial derivatives are:

$fxx = 8$

$fyy = -12$

$fxy = 5 = f_{yx}$

Nature of stationary value of function...
Stationary value is a saddle point.

The full program listing is as follows.

Program 6.1

```
'                        QUADALL.BAS

DECLARE SUB getfunction ()
DECLARE SUB display ()
DECLARE SUB derivatives ()

COMMON SHARED a, b, c, d, e, f

CLS

PRINT 'The program gives the first- and second-order partial'
PRINT 'derivatives of a quadratic function of two variables.'
PRINT 'Information is given about the nature of any stationary'
PRINT 'value for the function.'
PRINT '--------------------------------------------------------------------'
PRINT
PRINT 'The form of the function is...'
PRINT
PRINT 'f(x, y) = ax² + bxy + cy² + dx + ey + f'
```

CALL getfunction

PRINT 'Press any key to continue...'

DO WHILE INKEY$ = ''

LOOP

CALL display

VIEW PRINT 6 TO 24

CALL derivatives

END

SUB derivatives

PRINT
PRINT 'The first-order partial derivatives are:'
PRINT

PRINT 'fx = ';

IF a < > 0 THEN

    PRINT 2*a; 'x';

END IF

IF b > 0 THEN

    b$ = ' + '

    PRINT b$; b; 'y';

ELSEIF b < 0 THEN

    b$ = ' − '

    PRINT b$; ABS(b); 'y';

END IF

IF d > 0 THEN

    d$ = ' + '

    PRINT d$; d

ELSEIF d < 0 THEN

```
        d$ = ' - '
    PRINT d$; ABS(d)
END IF
PRINT
PRINT 'fy =';
IF b <> 0 THEN
    PRINT b; 'x';
END IF
IF c > 0 THEN
    c$ = ' + '
    PRINT c$; 2*c; 'y';
ELSEIF c < 0 THEN
    c$ = ' - '
    PRINT c$; 2*ABS(c); 'y';
END IF
IF e > 0 THEN
    e$ = ' + '
    PRINT e$; e
ELSEIF e < 0 THEN
    e$ = ' - '
    PRINT e$; ABS(e)
END IF
PRINT
PRINT 'The second-order partial derivatives are:'
PRINT

PRINT 'fxx = '; 2*a
PRINT
PRINT 'fyy = '; 2*c
PRINT
PRINT 'fxy = '; b; ' = fyx'
```

```
PRINT
PRINT 'Nature of stationary value of function...'
PRINT

discriminant = 4*a*c - b^2

SELECT CASE discriminant

    CASE IS < 0

        PRINT 'Stationary value is a saddle point.'

    CASE IS > 0

        IF a > 0 THEN

            PRINT 'Stationary value is a minimum.'

        ELSEIF a < 0 THEN

            PRINT 'Stationary value is a maximum.'

        END IF

    CASE 0

        PRINT 'For nature of stationary value, if any,'
        PRINT 'refer to Appendix.'

END SELECT

PRINT '-------------------------------------------------------------------------------'

END SUB


SUB display

CLS

PRINT 'The function specified is:'
PRINT

PRINT 'f(x, y) = ';

IF a <> 0 THEN

    PRINT a; 'x^2';

END IF

IF b > 0 THEN
```

```
        b$ = ' + '
        PRINT b$; b; 'xy';
ELSEIF b < 0 THEN
        b$ = ' - '
        PRINT b$; ABS(b); 'xy';
END IF
IF c > 0 THEN
        c$ = ' + '
        PRINT c$; c; 'y²';
ELSEIF c < 0 THEN
        c$ = ' - '
        PRINT c$; ABS(c); 'y²';
    END IF
    IF d > 0 THEN
    d$ = ' + '
    PRINT d$; d; 'x';
ELSEIF d < 0 THEN
    d$ = ' - '
    PRINT d$; ABS(d); 'x';
END IF
IF e > 0 THEN
    e$ = ' + '
    PRINT e$; e; 'y';
ELSEIF e < 0 THEN
    e$ = ' - '
    PRINT e$; ABS(e); 'y';
```

```
END IF

IF f > 0 THEN

    f$ = ' + '

    PRINT f$; f

ELSEIF f < 0 THEN

    f$ = ' - '

    PRINT f$; ABS(f)

END IF

PRINT
PRINT '--------------------------------------------------------------------------------'
PRINT

PRINT 'Press any key to continue...'

DO WHILE INKEY$ = ''

LOOP

END SUB

SUB getfunction

PRINT
PRINT 'Please supply the coefficients in the function...'
PRINT

INPUT 'Coefficient of x squared'; a
PRINT

INPUT 'Coefficient of xy'; b
PRINT

INPUT 'Coefficient of y squared'; c
PRINT

INPUT 'Coefficient of x'; d
PRINT

INPUT 'Coefficient of y'; e
PRINT

INPUT 'Constant term'; f
PRINT

END SUB
```

ADDITIONAL PROBLEMS

1 A division of a company is allocated a budget of £610k for expenditure on its inputs. Given this restriction, the division is required to maximize its profits. The levels of input used are represented by x and y, and the budget constraint is

$$8x + 5y = 610$$

The division's profit, z, as a function of input levels is

$$z = 360x + 100y - 2x^2 - 2.5y^2$$

(i) What is the maximum profit that can be earned given the expenditure constraint?
(ii) How much would the firm wish to spend on inputs if there was no budgetary constraint?

2 A firm manufacturing two products, q_1 and q_2, wishes to identify the profits maximizing output levels for the two goods. The market demand curves for the products are

$$p_1 = 320 - q_1 \quad \text{and} \quad p_2 = 190 - 0.5q_2$$

Costs are, respectively,

$$c_1 = 20q_1 + 0.5q_1^2 \quad \text{and} \quad c_2 = 10q_2 + q_2^2$$

Find the profits maximizing output levels.

3 Suppose now that the firm in exercise 2 faces changed demand and cost conditions, such that on the demand side

$$p_1 = 430 - q_1 \quad \text{and} \quad p_2 = 400 - 0.5q_2$$

while costs are now related and are defined in the *joint cost function*

$$C = 20q_1 + 10q_2 + 0.5q_1^2 + q_2^2 + 2q_1q_2$$

What will be the new profits maximizing output levels for the firm?

4 In classical stock control models two important decision variables are the size of stock replenishment (here q) and the re-order level (here r). In a particular case, a firm's stock related costs are given by

$$C = 4q + \frac{10{,}000}{q} + 2r^2 - 80r$$

Find the optimal size of replenishment and the best re-order level.

REFERENCES AND FURTHER READING

1 Chiang, A. C. (1984) *Fundamental Methods of Mathematical Economics* (Third Edition), McGraw-Hill.
2 Finney, R. L. and Thomas, G. B., Jr (1990) *Calculus*, Addison-Wesley.
3 Rich, A., Rich, J. and Stoutemyer, D. (1989) *Derive: User Manual* (Third Edition), Soft Warehouse.
4 Jeffrey, A. (1989) *Mathematics for Engineers and Scientists* (Fourth Edition), Van Nostrand Reinhold.
5 Weber, J. E. (1982) *Mathematical Analysis: Business and Economic Applications* (Fourth Edition), Harper and Row.

SOLUTIONS TO EXERCISES

Exercises 6.2

1   (i) *On* the plane. The values of x, y and z satisfy the equation.
   (ii) *Above* the plane. The z value in the co-ordinate triple is above the value (200) that is given by the function using the x and y co-ordinates of the point.
   (iii) Above the plane.
   (iv) Above the plane.
   (v) Below the plane.
   (vi) On the plane.

2   (i) The distance is

$$[(10-2)^2 + (20-16)^2 + (30-38)^2]^{1/2} = 12$$

   (ii) The distance is

$$[(2-(-2))^2 + (18-6)^2 + (-5-1)^2]^{1/2} = 14$$

3 The equations of the contours are, respectively,

   (i) $y = 240 - 6x$
   (ii) $y = -6x$
   (iii) $y = 140 - 6x$

4 By use of substitution:

   (i) $x = 1$ $y = 2$ $z = 3$
   (ii) $x = 20$ $y = -5$ $z = 30$

Exercises 6.3

1   (i) $z = 100x - 4x^2 + 450$
   (ii) $z = 500 - 0.5y^2$

2 (i) With y stated as an explicit function of x, the equation of the contour is

$$y = \frac{1000 - 6x + 9x^2}{10 - 3x}$$

(ii) With y as an explicit function of x, the numerator of the ratio factors as $(10 - 3x)^2$ with the result that the contour is the straight line

$$y = 10 - 3x$$

3 (i) All of the terms are of degree two, so that the function is homogeneous of degree two.
(ii) Not homogeneous — due to the degree three term xyw.
(iii) Homogeneous of degree two.
(iv) Not homogeneous due to the $-2x + 1$ component of the expansion of $(x - 1)^2$.

Exercises 6.4

1 The slope is -2 in the fundamental x direction and 0.5 in the y direction.

2 (i) $\dfrac{\partial z}{\partial x} = 2x - 4y$

$\dfrac{\partial z}{\partial y} = -4x + 1$

(ii) (a) x direction: $2(10) - 4(2) = 12$
y direction: $-4(10) + 1 = -39$
(b) x direction: $2(-2) - 4(7) = -32$
y direction: $-4(-2) + 1 = 9$

3 The partial derivatives are

$f_x = 2xy^3$
$f_y = 3x^2y^2$

and at (5, 4, 1600) these functions evaluate as

$f_x = 640$
$f_y = 1200$

4 (i) $f_x = 7x^6 + y^7 + 7x^6y^7$
$f_y = 7xy^6 + 7x^7y^6$
(ii) $f_x = 9900x^{98}y^{101}$
$f_y = 10{,}100x^{99}y^{100}$

(iii) $f_x = 16x^3 - 9x^2y^4 + 16x^3y^3$
$f_y = 10y^4 - 12x^3y^3 + 12x^4y^2$

(iv) $\dfrac{\partial q}{\partial l} = 0.7l^{-0.3}k^{0.4}$

$\dfrac{\partial q}{\partial k} = 0.4l^{0.7}k^{-0.6}$

(v) $\dfrac{\partial q}{\partial l} = al^{a-1}k^b$

$\dfrac{\partial q}{\partial k} = bl^a k^{b-1}$

(vi) $\dfrac{\partial z}{\partial x} = mnx^{m-1}y^{n-1}$

$\dfrac{\partial z}{\partial y} = n(n-1)x^m y^{n-2}$

(vii) $\dfrac{\partial z}{\partial x} = \dfrac{1}{y}$

$\dfrac{\partial z}{\partial y} = -\dfrac{x}{y^2}$

(viii) $\dfrac{\partial z}{\partial x} = 5x^4 y - 2yx^{-3}$

$\dfrac{\partial z}{\partial y} = x^5 + x^{-2}$

5 (i) $f_x = 30x^2 y^4 w^2 - 8x^3 y^3 - 10xw^4$
$f_y = 40x^3 y^3 w^2 - 6x^4 y^2 - 12y^2 w^2$
$f_w = 20x^3 y^4 w - 20x^2 w^3 + 8y^3 w$

(ii) $\dfrac{\partial Q}{\partial L} = rAL^{r-1}K^s M^t$

$\dfrac{\partial Q}{\partial K} = sAL^r K^{s-1} M^t$

$\dfrac{\partial Q}{\partial M} = tAL^r K^s M^{t-1}$

Functions of more than one variable 431

6 The marginal profits on x and y are the first-order partial derivatives of the function with respect to these variables. So:

marginal profit on $x = 500 + 4y - 6x$
marginal profit on $y = 400 + 4x - 4y$

7 (i) $f_x = (3x^2y - 8x^3y^3)(12xy^3 + 2y^2)$
$\qquad + (6x^2y^3 + 2xy^2)(6xy - 24x^2y^3)$
$f_y = (3x^2y - 8x^3y^3)(18x^2y^2 + 4xy)$
$\qquad + (6x^2y^3 + 2xy^2)(3x^2 - 24x^3y^2)$

(ii) $f_x = Ay(1 + Bxy) + By(1 + Axy)$
$f_y = Ax(1 + Bxy) + Bx(1 + Axy)$

8 (i) $f_x = \dfrac{y(x-y) - xy}{(x-y)^2}$
$\qquad = \dfrac{-y^2}{(x-y)^2}$
$f_y = \dfrac{x(x-y) + xy}{(x-y)^2}$
$\qquad = \dfrac{x^2}{(x-y)^2}$

(ii) $f_x = \dfrac{(y^2 - 5xy)(4xy^2 - 3x^2) - (2x^2y^2 - x^3)(-5y)}{(y^2 - 5xy)^2}$
$f_y = \dfrac{(y^2 - 5xy)(4x^2y) - (2x^2y^2 - x^3)(2y - 5x)}{(y^2 - 5xy)^2}$

9 (i) $f_x = 6(6x^4y^5 - 3x^4y^3)^5(24x^3y^5 - 12x^3y^3)$
$f_y = 6(6x^4y^5 - 3x^4y^3)^5(30x^4y^4 - 9x^4y^2)$

(ii) $f_x = d(ax^2 + bxy + cy^2)^{d-1}(2ax + by)$
$f_y = d(ax^2 + bxy + cy^2)^{d-1}(bx + 2cy)$

10 $f_x = (y^3 - w^4x^2)(3x^2 - 2xy^2w^2) + (x^3 - x^2y^2w^2)(-2w^4x)$
$f_y = (y^3 - w^4x^2)(-2x^2yw^2) + (x^3 - x^2y^2w^2)(3y^2)$
$f_w = (y^3 - w^4x^2)(-2x^2y^2w) + (x^3 - x^2y^2w^2)(-4w^3x^2)$

Exercises 6.5

1. (i) The first-order direct derivative with respect to x is

$$\frac{\partial z}{\partial x} = 5x^4 y^2$$

so that the required second derivative is

$$\frac{\partial^2 z}{\partial x^2} = 20x^3 y^2$$

(ii) $\dfrac{\partial^2 z}{\partial x^2} = -160x^3 y^3 - 84x^5 y$

(iii) $\dfrac{\partial^2 z}{\partial x^2} = 30xy^3 - 12y^2 w^2$

2. (i) $\dfrac{\partial^2 z}{\partial y^2} = 2x^5$

(ii) $\dfrac{\partial^2 z}{\partial y^2} = 504 y^7 - 48 x^5 y$

(iii) $\dfrac{\partial^2 z}{\partial y^2} = 30 x^3 y - 12 x^2 w^2$

3. (i) $\dfrac{\partial^2 z}{\partial x\, \partial y} = 10 x^4 y = \dfrac{\partial^2 z}{\partial y\, \partial x}$

(ii) $\dfrac{\partial^2 z}{\partial x\, \partial y} = -120 x^4 y^2 - 14 x^6 = \dfrac{\partial^2 z}{\partial y\, \partial x}$

(iii) $\dfrac{\partial^2 z}{\partial x\, \partial y} = 45 x^2 y^2 - 24 xyw^2 - w = \dfrac{\partial^2 z}{\partial y\, \partial x}$

4. (i) $\dfrac{\partial z}{\partial x} = 6x^2 - 4y^2 + 10xy + 2x + 7y - 3$

so that at $x = 1$, $y = 2$ the slope is 25.

(ii) $\dfrac{\partial z}{\partial y} = 3y^2 - 8xy + 5x^2 - 6y + 7x + 8$

which evaluates to $+4$ at $x = 1$ and $y = 2$.

(iii) $\dfrac{\partial^2 z}{\partial y\, \partial x} = -8y + 10x + 7 = \dfrac{\partial^2 x}{\partial x\, \partial y}$

Functions of more than one variable 433

(iv) $\dfrac{\partial^2 z}{\partial x^2} = 12x + 10y + 2 = 34$ at $x = 1, y = 2$

(v) $\dfrac{\partial^2 z}{\partial y^2} = 6y - 8x - 6 = -2$ at $x = 1, y = 2$

(vi) $\dfrac{\partial^3 z}{\partial x^3} = 12$

(vii) $\dfrac{\partial^3 z}{\partial y^3} = 6$

5 (i) $f_{xx} = 60x^4 y^7 - 60x^3 y^2 + 4y^4$
   $f_{xy} = 84x^5 y^6 - 30x^4 y + 16xy^3$
   $f_{yy} = 84x^6 y^5 - 6x^5 + 24x^2 y^2$

(ii) (a) $f_{xy} = 20x^4 y^3 w^6$
     (b) $f_{xw} = 30x^4 y^4 w^5$
     (c) $f_{yw} = 24x^5 y^3 w^5$

6 (i) $\dfrac{\partial^3 z}{\partial x^3} = 60x^2 y^6 - 24y^2$

(ii) $\dfrac{\partial^3 z}{\partial x \, \partial y^2} = 150x^4 y^4 - 24x^2 + 144xy^2$

(iii) $\dfrac{\partial^4 z}{\partial x^2 \, \partial y^2} = 600x^3 y^4 - 48x + 144y^2$

(iv) $\dfrac{\partial^5 z}{\partial x \, \partial y^4} = 1800x^4 y^2 + 288x$

7 (i) $f_w = 3x^4 y^4 - 20x^5 w^4 - 30y^6 w^5$
  (ii) $f_{yw} = 12x^4 y^3 - 180y^5 w^5$
  (iii) $f_{xyw} = 48x^3 y^3$
  (iv) $f_{www} = -240x^5 w^2 - 600y^6 w^3$

Exercises 6.6

1 (i) At the first order, the conditions are

$$f_x = 20 + 8y - 2x = 0$$
$$f_y = 64 + 8x - 64y = 0$$

from which, at the second order, it works out that

$$f_{xx} = -2 \quad f_{yy} = -64 \quad f_{xy} = 8$$

and, since the discriminant

$$(-2)(-64) - (8)^2 > 0$$

is positive, the stationary value is a maximum of the function.

(ii) At the first order, the conditions are

$$f_x = 4x - 16 - 2y = 0$$
$$f_y = 2y - 20 - 2x = 0$$

and, at the second order,

$$f_{xx} = 4 \quad f_{yy} = 2$$

No further work is needed since it is clear from the direct second-order derivatives that the conditions for a maximum are not satisfied.

(iii) The first-order conditions are

$$f_x = 5 + y - 0.5x = 0$$
$$f_y = -4 + x - y = 0$$

and, at the second order,

$$f_{xx} = -0.5 \quad f_{yy} = -1$$

So far so good. But since $f_{xy} = 1$, the discriminant evaluates as

$$(-0.5)(-1) - (1)^2 < 0$$

Thus the conditions for a maximum are *not* fulfilled in this case.

2 (i) The first-order conditions

$$f_x = -4 + y - 2x = 0$$
$$f_y = 5 + x - y = 0$$

solve for $x = 1$ and $y = 6$. At the second order,

$$f_{xx} = -2 \quad f_{yy} = -1 \quad f_{xy} = 1$$

and the condition on the discriminant

$$(-2)(-1) - (1)^2 > 0$$

is satisfied. Therefore, the stationary value is a maximum and at $x = 1, y = 6$:

$$f(x, y) = 100 - 4 + 30 + 6 - 1 - 18$$
$$= 113$$

(ii) The first-order conditions
$$f_x = 5 + 2y - 0.5x = 0$$
$$f_y = -16 + 2x - 16y = 0$$
solve for $x = 12$ and $y = 0.5$. At the second order,
$$f_{xx} = -0.5 \quad f_{yy} = -16 \quad f_{xy} = 2$$
and the condition on the discriminant
$$(-0.5)(-16) - (2)^2 > 0$$
is satisfied. Therefore, the stationary value is a maximum and at $x = 12$, $y = 0.5$:
$$f(x, y) = 100 + 60 - 8 + 12 - 36 - 2 = 126$$

3 The first-order conditions for a local maximum are
$$\frac{\partial z}{\partial x} = -21 + 2xy + y^2 - 6x^2 = 0$$

$$\frac{\partial z}{\partial y} = 576 + x^2 + 2xy - 24y^2 = 0$$

Substitution of the values $x = 2$ and $y = 5$ in these conditions confirms that they are satisfied. At the second order, the direct derivatives
$$\frac{\partial^2 z}{\partial y^2} = 2y - 12x = -14 < 0$$

$$\frac{\partial^2 z}{\partial y^2} = 2x + 2x - 48y = -226 < 0$$

are the appropriate sign at $x = 2$ and $y = 5$. The mixed derivative
$$\frac{\partial^2 z}{\partial x \partial y} = 2x + 2y = 14$$

at $x = 2$, $y = 5$. So the discriminant condition
$$(-14)(-226) - (14)^2 > 0$$

is also fulfilled, and a maximum at $x = 2$, $y = 5$ is confirmed.

4 (i) The first-order conditions are
$$f_x = -14 - 2.5y + 3x = 0$$
$$f_y = 4 - 2.5x + 4y = 0$$
which solve for $x = 8$ and $y = 4$. At the second order,
$$f_{xx} = 3 \quad f_{yy} = 4 \quad f_{xy} = -2.5$$

These values fulfil the discriminant condition since

$$(3)(4) - (2.5)^2 > 0$$

Thus the combination $x = 8$ and $y = 4$ produces a minimum of $f(x, y)$. Substitution of these values into $f(x, y)$ shows the minimum value to be 48.

(ii) The first-order conditions are

$$f_x = 6 - 3y + x = 0$$
$$f_y = -54 - 3x + 18y = 0$$

which solve for $x = 6$ and $y = 4$. At the second order,

$$f_{xx} = 1 \quad f_{yy} = 18 \quad f_{xy} = -3$$

These values fulfil the discriminant condition since

$$(1)(18) - (-3)^2 > 0$$

Thus the combination $x = 6$ and $y = 4$ produces a minimum of $f(x, y)$. Substitution of these values into $f(x, y)$ shows the minimum value to be

$$f(x, y) = 126 + 36 - 216 - 72 - 18 + 144$$
$$= 0$$

5 The first-order conditions are

$$f_x = 3x^2 - 288 - xy - 2y^2 = 0$$
$$f_y = -0.5x^2 + 42 - 4xy + 48y^2 = 0$$

which solve for $x = 10$ and $y = 1$. For these values, the second-order conditions work out as

$$f_{xx} = 6x - y$$
$$= 59 > 0$$
$$f_{yy} = -x - 4x + 96y$$
$$= 46 > 0$$

so that the conditions on the direct second-order derivatives are fulfilled. The value of the mixed derivative

$$f_{xy} = -x - 4y$$
$$= -14$$

leads to satisfaction of the condition on the discriminant since

$$(59)(46) - (-14)^2 > 0$$

Exercises 6.7

1. (i) At the first order, for *any* stationary value of the function $f(x, y)$:

    $f_x = 8 + 2y - 4x = 0$
    $f_y = -5 + 2x - 0.5y = 0$

    These conditions solve for $x = 3$, $y = 2$. At the second order,

    $f_{xx} = -4 \qquad f_{yy} = -0.5$

    Since both of the direct second-order derivatives are negative, the stationary value could be either a maximum or a saddle point, a minimum being ruled out. The cross partial derivative is

    $f_{xy} = 2$

    and the discriminant evaluates as

    $(-4)(-0.5) - (2)^2 = -2 < 0$

    This being negative, the location is therefore distinguished as a saddle point.

   (ii) The first-order conditions for a stationary value are

    $f_x = 15 - 3y - 1.5x = 0$
    $f_y = 12 - 3x + 3y = 0$

    which solve for the values $x = 6$ and $y = 2$. At the second order,

    $f_{xx} = -1.5 \qquad f_{yy} = 3$

    which means that the point must be a saddle point since regardless of the value of f_{xy} (which is, in fact, -3) the discriminant is negative, as required for a saddle point.

   (iii) The first-order conditions for a stationary value are that

    $f_x = 4x + 8y = 0$
    $f_y = y + 8x = 0$

    which equations solve for $x = 0$ and $y = 0$. And at the second order,

    $f_{xx} = 4 \qquad f_{yy} = 1 \qquad f_{xy} = 8$

    which means that the point must be a saddle point since the discriminant is negative:

    $(4)(1) - (8)^2 = -60 < 0$

2 (i) At the first order,

$$f_x = -220 - 14y + 10x = 0$$
$$f_y = 310 - 14x + 20y = 0$$

which conditions solve for $x = 15$ and $y = -5$. At the second order,

$$f_{xx} = 10 \qquad f_{yy} = 20$$

thus eliminating the possibility of a maximum. The cross partial derivative

$$f_{xy} = -14$$

so that the discriminant evaluates as

$$(10)(20) - (196) = 4 > 0$$

which rules out the saddle point possibility and identifies the point as a minimum.

(ii) At the first order,

$$f_x = 15 - 7y - 6x = 0$$
$$f_y = 17 - 7x - 8y = 0$$

which conditions solve for $x = -1$ and $y = 3$. At the second order,

$$f_{xx} = -6 \qquad f_{yy} = -8$$

thus eliminating the possibility of a minimum. The cross partial derivative

$$f_{xy} = -7$$

so that the discriminant evaluates as

$$(-6)(-8) - (-7)^2 = -1 < 0$$

which identifies the point as a saddle point.

(iii) At the first order,

$$f_x = 34 - y - 5x = 0$$
$$f_y = -10 - x + 4y = 0$$

which conditions solve for $x = 6$ and $y = 4$. At the second order,

$$f_{xx} = -5 \qquad f_{yy} = 4$$

Since the direct derivatives are of opposite sign, the stationary value in this case *must* take the form of a saddle point.

Functions of more than one variable 439

3   (i) Saddle point at $x = -4$, $y = 6$
   (ii) Saddle point at $x = 3$, $y = 1$
   (iii) Minimum at $x = 2$, $y = 8$
   (iv) Saddle point at $x = 2$, $y = 3$
   (v) Maximum at the origin
   (vi) Saddle point at $x = 2$, $y = -12$

Exercises 6.8

1   (i) For a stationary value of the function $f(x,y)$, at the first order it is required that

$$f_x = 24x + 12y + 8 = 0$$
$$f_y = 12x + 6y - 10 = 0$$

An attempt to solve these equations simultaneously requires the equation of two unequal constants. They are therefore inconsistent, so that no point fulfils the conditions for a stationary value.

  (ii) At the first order,

$$f_x = 60 - 5x + 0.5y = 0$$
$$f_y = 64.5 + 7y + 0.5x = 0$$

which solve uniquely for $x = 11$, $y = -10$. At the second order,

$$f_{xx} = -5 \qquad f_{yy} = 7$$

which derivatives, being of opposite sign, imply that the discriminant is negative and thus a saddle point is produced by the values $x = 11$, $y = -10$.

  (iii) At the first order,

$$f_x = -40x + 20y - 50 = 0$$
$$f_y = -10y + 20x + 30 = 0$$

Multiplication of f_y by 2 and adding the two equations produces the result

$$10 = 0$$

so that the equations are inconsistent, and no stationary value of $f(x,y)$ exists in this case.

2   (i) At the first order,

$$f_x = 4.5x - 9y - 9 = 0$$
$$f_y = 18y - 9x + 18 = 0$$

440 Mathematics for business, finance and economics

which conditions can both be rearranged as the equation

$$y = 0.5x - 1$$

so that all points satisfying this condition produce a stationary value of the function (which turn out to be minima).

(ii) At the first order,

$$f_x = x + 2y - 5 = 0$$
$$f_y = 4y + 2x - 8 = 0$$

which equations can be rearranged as

$$x + 2y = 5$$
$$x + 2y = 4$$

which are clearly inconsistent, and the function $f(,y)$ therefore has no stationary value.

Exercises 6.9

1 The slope of the constraint line is $-1/3$ as may be seen from rearrangement of the constraint as

$$y = \frac{55}{3} - \frac{1}{3}x$$

The derivative of the objective function with respect to x is

$$f_x = 22 + 3y - 3x$$

and with respect to y is

$$f_y = 33 + 3x - 4.5y$$

Tangency between the contours of the function and the constraint requires that

$$-\frac{f_x}{f_y} = -\frac{1}{3}$$

so therefore

$$-\frac{22 + 3y - 3x}{33 + 3x - 4.5y} = -\frac{1}{3}$$

Hence,

$$66 + 9y - 9x = 33 + 3x - 4.5y$$

which rearranges as

$$12x - 13.5y = 33$$

into which the constraint relationship

$$x = 55 - 3y$$

may be substituted with the outcome that

$$y = \frac{38}{3} \quad \text{and} \quad x = 17$$

which values produce a maximum value for $f(x,y)$ of 643.5.

2 First find the value of the function $f(x,y)$ at the unconstrained optimum. At the first order, the conditions are

$$f_x = 20 - 0.2x = 0$$
$$f_y = 16 - 0.1y = 0$$

which equations solve for $x = 100$ and $y = 160$. At the second order,

$$f_{xx} = -0.2 \quad f_{yy} = -0.1 \quad f_{xy} = 0$$

and so

$$f_{xx}f_{yy} - (f_{xy})^2 > 0$$

and the stationary value is a maximum. At this point,

$$f(100, 160) = 2280$$

With the constraint included, the equilibrium condition is then

$$\frac{20 - 0.2x}{16 - 0.1y} = \frac{5}{2}$$

which can be restated as

$$x = 1.25y - 100$$

while from the constraint

$$x = \frac{325}{5} - \frac{2y}{5}$$
$$= 65 - 0.4y$$

and so

$$1.25y - 100 = 65 - 0.4y$$

So with $y = 100$ as a result, and $x = 25$,

$$f(25, 100) = 1537.5$$

so the reduction in $f(x,y)$ is

$$2280 - 1537.5 = 742.5$$

Note that if the function $f(x, y)$ represents profit, then £742.5 is the maximum price worth paying in order to remove the constraint.

Exercises 6.10

1 From the constraint,

$$y = 40 - 1.6x$$

which when substituted into z produces

$$z^* = 100x + 50(40 - 1.6x) + 2.5x(40 - 1.6x)$$
$$= 120x - 4x^2 + 2000$$

So the conditions for a constrained maximum are

$$\frac{dz^*}{dx} = 120 - 8x = 0$$

$$\frac{d^2z^*}{dx^2} = -8 < 0$$

From the first-order condition $x = 15$ and therefore from the constraint $y = 16$, which values produce $z = 2900$ as a maximum.

2 From the constraint,

$$y = 100 + 2x$$

Substitution for y in z yields

$$z^* = 20(100 + 2x) - 0.1x^2$$
$$= 2{,}000 + 40x - 0.1x^2$$

so the first and second derivatives are

$$\frac{dz^*}{dx} = 40 - 0.2x = 0$$

$$\frac{d^2z^*}{dx^2} = -0.2 < 0$$

From the first-order condition $x = 200$, so from the constraint $y = 500$. The maximum value of z is 6000.

3 From the constraint,

$$y = 20 - 0.5x$$

Substitution into z gives

$$z^* = 50x^2 - x^4 + 1000$$

the derivatives of which are

$$\frac{dz^*}{dx} = 100x - 4x^3 = 0$$

$$\frac{d^2z^*}{dx^2} = 100 - 12x^2 < 0$$

The first-order condition can be reorganized as

$$x(100 - 4x^2) = 0$$

which solves for

$$x = 0 \qquad x = 5 \qquad x = -5$$

At the second order, the condition for a maximum is satisfied for the values $x = 5$ and $x = -5$, both of which give a maximum value of z, subject to the constraint, of 1625.

4 From the first constraint,

$$x = 80 - 10y - 5w \qquad (1)$$

Inserting this relationship into the second constraint gives

$$-2(80 - 10y - 5w) + 30y + 2.5w = 115$$

which simplifies to the relationship

$$w = 22 - 4y$$

Inserting this into (1) produces

$$x = 10y - 30$$

Substituting into z for w and x in terms of y and simplification gives

$$z^* = 320y - 40y^2$$

and so

$$\frac{dz^*}{dy} = 320 - 80y = 0$$

$$\frac{d^2z^*}{dy^2} = -80 < 0$$

The first-order condition solves for $y = 4$, which corresponds to a maximum as the second-order condition is automatically satisfied. From the earlier equations, if $y = 4$ then $x = 10$ and $w = 6$, giving a constrained maximum of z of 640.

5 From the constraint,

$$w = \frac{x}{3} + \frac{2y}{5} - 2$$

which, substituted into the objective function, gives

$$f^* = 1000 - \frac{5x^2}{3} + 4xy + 10x - 3y^2$$

which is to be maximized with respect to both x and y. At the first order,

$$f_x^* = -\frac{10x}{3} + 4y + 10 = 0$$

$$f_y^* = 4x - 6y = 0$$

which equations solve simultaneously for $x = 15$ and $y = 10$ (so that, by implication, $w = 7$). At the second order,

$$f_{xx}^* = -\frac{10}{3} < 0 \qquad f_{yy}^* = -6 < 0 \qquad f_{xy}^* = 4$$

so that the discriminant condition

$$(-10/3)(-6) - (4)^2 = 4 > 0$$

is satisfied and the function achieves a maximum value of

$$f(w, x, y) = 1000 - 5(7)(15) + 6(15)(10) - 3(10)^2$$
$$= 1075$$

6 (i) With the right-hand side equal to 120, the problem solves for

$$x = 240 \qquad y = 240 \qquad f(x, y) = 57{,}600$$

With the right-hand side equal to 121, the problem solves for

$$x = 242 \qquad y = 242 \qquad f(x, y) = 58{,}564$$

so that the marginal valuation of the resource represented by the right-hand side is 964, since

$$58{,}564 - 57{,}600 = 964$$

(ii) With the right-hand side equal to 100, the problem solves for

$$x = 12 \qquad y = 17.5 \qquad f(x, y) = 522$$

With the right-hand side equal to 101, the problem solves for

$$x = 12.2 \qquad y = 17.625 \qquad f(x, y) = 528.25625$$

so that the marginal valuation of the right-hand side resource is

$$528.25625 - 522.5 = 5.75625$$

Functions of more than one variable 445

(iii) With the right-hand side equal to 100, the problem solves for

$$x = 75 \qquad y = 50 \qquad f(x, y) = 12{,}600$$

With the right-hand side equal to 101, the problem solves for

$$x = 75.5 \qquad y = 50 \qquad f(x, y) = 12{,}700$$

so that the marginal valuation of the resource is 100.

(iv) With the right-hand side equal to 600;

$$x = 255 \qquad y = 90 \qquad z = 6{,}975{,}000$$

With the right-hand side equal to 601,

$$x = 255.425 \qquad y = 90.15 \qquad z = 6{,}998{,}269.375$$

so the marginal resource valuation is 23,269.375.

Additional problems

1 (i) From the constraint,

$$y = 122 - 1.6x$$

which, when substituted into the profits expression, gives the result

$$z^* = 1176x - 8.4x^2 - 25{,}010$$

so

$$\frac{dz^*}{dx} = 1176 - 16.8x = 0$$

$$\frac{d^2 z^*}{dx^2} = -16.8 < 0$$

The first-order condition solves for $x = 70$, which from the constraint means that $y = 10$. With these values, the profit made is $z = 16{,}150$.

(ii) The expenditure on inputs here would be that corresponding to the input usage at the unconstrained maximum. In the absence of a constraint, the firm would seek to maximize

$$z = 360x + 100y - 2x^2 - 2.5y^2$$

So,

$$\frac{\partial z}{\partial x} = 360 - 4x = 0$$

$$\frac{\partial z}{\partial y} = 100 - 5y = 0$$

which conditions solve for

$x = 90$ and $y = 20$

so that expenditure on the inputs x and y would be

$8(90) + 5(20) = 820$

2 The profits function is

$$\pi = (320 - q_1)q_1 + (190 - 0.5q_2)q_2 - (20q_1 + 0.5q_1^2) - (10q_2 + q_2^2)$$

which simplifies to

$$\pi = 300q_1 - 1.5q_1^2 + 180q_2 - 1.5q_2^2$$

The first-order conditions are

$$\frac{\partial \pi}{\partial q_1} = 300 - 3q_1 = 0$$

$$\frac{\partial \pi}{\partial q_2} = 180 - 3q_2 = 0$$

At the second order,

$$\frac{\partial^2 \pi}{\partial q_1^2} = -3 \qquad \frac{\partial^2 \pi}{\partial q_2^2} = -3 \qquad \frac{\partial^2 \pi}{\partial q_1\, \partial q_2} = 0$$

So that *any* stationary value is a maximum. The first-order conditions solve for

$q_1 = 100$ and $q_2 = 60$

which are the profits maximizing output levels.

3 The profits function is

$$\pi = 410q_1 - 1.5q_1^2 - 2q_1q_2 - 390q_2 - 1.5q_2^2$$

for a maximum of which the necessary first-order conditions are

$$\frac{\partial \pi}{\partial q_1} = 410 - 3q_1 - 2q_2 = 0$$

$$\frac{\partial \pi}{\partial q_2} = 390 - 3q_2 - 2q_1 = 0$$

and, at the second order,

$$\frac{\partial^2 \pi}{\partial q_1^2} = -3 \qquad \frac{\partial^2 \pi}{\partial q_2^2} = -3 \qquad \frac{\partial^2 \pi}{\partial q_1 \partial q_2} = -2$$

so that the discriminant will be positive. The first-order conditions solve for

$$q_1 = 90 \text{ and } q_2 = 70$$

which are the profit maximizing output levels.

4 $$C_q = 4 - \frac{10{,}000}{q^2} = 0$$

$$C_r = 4r - 80 = 0$$

At the second order,

$$C_{qq} = \frac{20{,}000}{q^3} \qquad C_{rr} = 4 \qquad C_{qr} = 0$$

From the first-order conditions

$$q = (10{,}000/4)^{1/2}$$
$$= 50$$

at which value the direct second derivative C_{qq} is positive as required at the minimum. The discriminant is clearly positive. From the condition $C_r = 0$ we obtain $r = 20$. Thus the firm should replenish stock in batches of 50 and should place orders for replenishment when stock on hand has fallen to the level of 20.

NOTES

1 Highly sophisticated packages such as *Mathematica*, *Math Cad* or *Maple* are available in addition to intermediate range products such as *Derive* or the Student edition of *Math Cad*.
2 Points which do lie on the same plane are *coplanar* and points which lie on the same straight line are *collinear*.

3 As stated in Chapter 5, the term *critical point* also applies to points where the derivative does not exist.
4 For discussion of these cases see the Appendix.
5 Again, we shall not at this point consider cases where the discriminant is zero. These will be reviewed in the Appendix. The conditions expressed as strict inequalities are sufficient but not necessary.
6 Conditions (6.16a) and (6.16b) are sufficient but not strictly necessary (because of the exclusion of the equality possibility for purposes of simplification).
7 Although differentiable functions are not always involved here, the principle carries over.
8 For a proof see Jeffrey (1989).
9 Though not quite as alike as the two figures might suggest, due to vertical scaling.
10 There being the same number of equations as there are variables.
11 Where the constant 13.5 is added merely to give a minimum value for z of zero.
12 It is a *quadratic programming* problem if in addition to (6.20) the variables x and y are sign restricted. We consider sign restrictions in the next chapter.
13 Differentials are useful in numerical methods and aspects of integral calculus.
14 If there was a non-unique optimum, the real choices facing the decision-maker would be unchanged if x and y were interchanged, though x and y need not take the same values in any one of the equally good solutions. For example, in the problem 'maximize $z = 1000 - x^2 y^2$ subject to $x + y = 10$', both $(0, 10)$ and $(10, 0)$ produce the maximum of 1000 for z.
15 Assuming, that is, that the constraint equations are independent and consistent.

Chapter 7

Constrained Optimization with Lagrange Multipliers

7.1	Constrained optimization – Lagrange multipliers (i)	450
7.2	Interpretation of the Lagrange multiplier	461
7.3	Sign restricted and bounded variables	464
7.4	Lagrange multipliers (ii) Inequality constraints	484
7.5	Lagrange multipliers (iii) Inequality constraints with sign requirements	492
7.6	The Kuhn–Tucker conditions	498
7.7	Economic application – multi-product monopoly	510
7.8	Concluding remarks	515
	Additional problems	515
	References and further reading	517
	Solutions to exercises	517

This chapter extends the discussion of constrained optimization problems introduced through the tangency approach and the method of substitution in Chapter 6. The more widely applicable Lagrange multiplier method is described and is applied in a variety of models. Following a second look at the problems solved by substitution in Chapter 6, emphasis is given in later sections to the need to meet sign requirements and other restrictions on the decision variables.

The discussion builds towards description of the Kuhn–Tucker conditions. Cases considered along the way also have their own relevance. An application to the multi-product firm is described in Section 7.7.

By the end of the chapter, you will have seen what is involved in the solution of a range of constrained optimization problems, and be able to tackle many one- and two-constraint optimization problems yourself. You will be familiar with the interpretation of Lagrange multipliers and understand equilibrium requirements in a number of models.

7.1 CONSTRAINED OPTIMIZATION – LAGRANGE MULTIPLIERS (I)

In the substitution method discussed in Chapter 6, the constraint was included within a revised objective function and hence a problem in constrained optimization was converted into a simpler unconstrained optimization problem. With the method of substitution, the constraint was included in such a way that it became implicit in the modified form of objective function. This method is *asymmetric* in that it singles out one variable for special treatment – and to be able to do this it must be possible to solve for one variable in terms of the others. There will be cases in which this cannot be done, but it would be advantageous to retain the principle of converting the problem to one in which an unconstrained stationary value is sought via inclusion of the constraint into a revised objective function. So, is there another way of adapting the original objective function to guarantee the satisfaction of the constraint or constraints? The answer is provided by the *Lagrange multiplier method*. Consider again the first of the two examples worked through in Chapter 6. This was

maximize $F = 4x + 4y + 2xy - x^2 - 1.5y^2$
subject to $2x + y = 20$

but this time we shall *add* a variable to the objective function. First, the constraint is expressed in implicit form as

$20 - 2x - y = 0$

and the modified objective function incorporates the left-hand side of the constraint in implicit form and is written as

$$L = F + j(20 - 2x - y) \qquad (7.1)$$

in which the new unknown, j, is the *Lagrange multiplier*, and the function L itself is called the *Lagrangian function*. Note that L is a function of x, y and j and that, crucially, provided that the term

$j(20 - 2x - y)$

is zero, L will have the same value as F. But how does the Lagrangian help in the solution of the constrained maximization problem? To see how the method works, differentiate L partially with respect to x, y and j and set the derivatives to zero to produce a stationary value of L. The Lagrangian L in full is

$L(x, y, j) = 4x + 4y + 2xy - x^2 - 1.5y^2 + j(20 - 2x - y)$

and, at the first order, the conditions for a stationary value of L are

$$L_x = \frac{\partial L}{\partial x} = 4 + 2y - 2x - 2j = 0 \qquad (7.2)$$

$$L_y = \frac{\partial L}{\partial y} = 4 + 2x - 3y - j = 0 \tag{7.3}$$

$$L_j = \frac{\partial L}{\partial j} = 20 - 2x - y = 0 \tag{7.4}$$

Note that (7.4) means that *the constraint must be satisfied*, which in turn means that the stationary value of L and the corresponding value of the original function are *the same*. More than this, it can be shown that subject to certain conditions (principally the concavity of the objective function F) equations (7.2)–(7.4) will produce a maximum of F subject to the constraint $2x + y = 20$.[1] While this result is by no means obvious, some confirmation can be given in the present case by solving the first-order conditions to corroborate the outcome achieved by the method of substitution. Using equation (7.2) to solve for the multiplier j in terms of x and y gives

$$j = 2 + y - x$$

and substitution of this result into (7.3) produces

$$4 + 2x - 3y - (2 + y - x) = 0$$

which simplifies to

$$2 + 3x - 4y = 0$$

which must now be solved simultaneously with equation (7.4). From (7.4) the value of y in terms of x is

$$y = 20 - 2x$$

which, when substituted into the equation

$$2 + 3x - 4y = 0$$

gives

$$2 + 3x - 4(20 - 2x) = 0$$

i.e.

$$11x = 78$$

so that the optimal value of x is

$$x = 78/11$$

and therefore

$$y = 64/11 \quad \text{and} \quad z = 364/11$$

which confirms the result obtained by the method of substitution. Note that the value of the Lagrange multiplier j is 8/11, the significance of which will

452 Mathematics for business, finance and economics

be considered in due course. What guarantee is there that the result produced by this method is a constrained maximum of F? There are two questions in one here: Is the point a *local* maximum of F subject to the constraint? Is the point a constrained *global* maximum? In fact, further conditions can be set out at the second order,[2] the significance of which is to require the concavity of the objective function in the vicinity of the stationary value. Thus second-order conditions address the 'local' question but not the global one. A *sufficient* second-order condition[3] in the case of a function of two variables maximized subject to a single equality constraint proves useful. Given the problem

maximize $f(x,y)$
subject to $g(x,y) = 0$

the Lagrangian for which will be

$$L(x,y,j) = f(x,y) + j[g(x,y)]$$

then, at the second order, it is sufficient that the value E defined by

$$E = L_{xx}g_y^2 - 2L_{xy}g_xg_y + L_{yy}g_x^2 \qquad (7.5)$$

should be negative. In (7.5)

$$g_x = \frac{\partial g}{\partial x} \quad \text{and} \quad g_y = \frac{\partial g}{\partial y}$$

where all components of (7.5) are evaluated at the x, y and j values satisfying the first-order conditions.[4] Because of the way the two constraint derivatives g_x and g_y enter into the condition (either squared or multiplied together) it does not matter whether we write $g(x,y)$ as

$$20 - 2x - y$$

or as

$$2x + y - 20$$

If, in (7.5), the value of E turns out to be positive, then a constrained minimum is produced; if $E = 0$ then no statement can be made on the basis of this test about the nature of the stationary value.[5] Now use (7.5) in the current example, in which we can write[6]

$$g(x,y) = 20 - 2x - y = 0$$

for which

$$g_x = -2 \quad \text{and} \quad g_y = -1$$

and from the Lagrangian

$$L_{xx} = -2 \qquad L_{yy} = -3 \qquad L_{xy} = 2$$

Substitution of these values into (7.5) gives

$$E = (-2)(-1)^2 - 2(2)(-2)(-1) + (-3)(-2)^2$$
$$= -2 - 8 - 12 = -22 < 0$$

which is the required sign for a constrained local maximum.

The interpretation of second-order conditions can give useful insight, for example in the neoclassical theories of the firm and of consumer behaviour. But for calculation of optimal values, the second-order conditions can be more trouble than they are worth. In general it can be a major task to find solutions to the *first*-order conditions. We shall assume where necessary that the objective function has at least local concavity (for maximization) or convexity (for minimization) properties. That is, if there is a single solution to the first-order conditions in a maximization problem, the objective function will be such that the point produces a maximum.[7] We now apply the Lagrange multiplier approach to another of the problems solved by the method of substitution in Chapter 6. The problem was

maximize $z = xy$
subject to $x + y = 12$

for which the Lagrangian is:

$$L = xy + j(12 - x - y)$$

and for which the first-order conditions are

$$L_x = y - j = 0$$
$$L_y = x - j = 0$$
$$L_j = 12 - x - y = 0$$

which equations solve for

$x = 6 \quad y = 6 \quad j = 6$

with the resulting value of $z = 36$, confirming the result obtained by the method of substitution. Note that the value of the Lagrange multiplier is approximately equal to the change in the value of z (6.25) that we had previously calculated if an additional unit of the right-hand side resource had been available. Now consider a further numerical example: the objective function to be maximized is

$$F = 60x + 80y - 2xy - x^2 - 2y^2$$

with the constraint

$2x + y = 40$

The Lagrangian function will therefore be

$$L = 60x + 80y - 2xy - x^2 - 2y^2 + j(40 - 2x - y)$$

and the first-order conditions are

$$L_x = \frac{\partial L}{\partial x} = 60 - 2y - 2x - 2j = 0$$

$$L_y = \frac{\partial L}{\partial y} = 80 - 2x - 4y - j = 0$$

$$L_j = \frac{\partial L}{\partial j} = 40 - 2x - y = 0$$

From the partial derivative of L with respect to x, we obtain the relationship

$$j = 30 - y - x$$

which, when substituted into the derivative of L with respect to y, produces

$$50 - x - 3y = 0$$

which can be solved simultaneously with the implicit form of the constraint (i.e. the derivative L_j equated to zero) to obtain the optimal values of x and y. After appropriate rearrangement of the two equations for the use of elimination we have

$$\begin{aligned} 2x + 6y &= 100 \\ 2x + y &= 40 \\ \hline 5y &= 60 \end{aligned}$$

so that

$$y = 12 \qquad x = 14 \qquad j = 4$$

and with these values the value of the objective function gives $F = 980$. Checking the second-order condition (7.5) we have

$$L_{xx} = -2 \qquad L_{xy} = -2 \qquad L_{yy} = -4 \qquad g_x = -2 \qquad g_y = -1$$

so that (7.5) evaluates as

$$\begin{aligned} E &= (-2)(-1)^2 - 2(-2)(-2)(-1) + (-4)(-2)^2 \\ &= -2 + 8 - 16 \\ &= -10 < 0 \end{aligned}$$

and the point represents a constrained maximum. There is a further important advantage in using the Lagrange multiplier method to solve constrained optimization problems. This lies in the information conveyed by the value of the Lagrange multiplier at the optimum. The value of j

shows the rate at which the optimal value of the function F changes as the right-hand side of the original constraint varies. This is the marginal valuation, at a point, of the resource represented by the constraint and is a useful guide to the price worth paying at the margin for further units of the scarce resource. How all this works out in the present case can be seen by re-solving the problem with the right-hand side of the constraint standing at 41. The first-order conditions are changed only by the substitution of 41 for 40 where this occurs, and at the point where elimination is reached the equations to solve become

$$\begin{aligned}2x + 6y &= 100\\ 2x + y &= 41\\ \hline 5y &= 59\end{aligned}$$

It follows that

$$y = 11.8, \quad x = 14.6 \quad \text{and} \quad F = 983.8$$

So the change in F resulting from the extra unit of resource is 3.8. It turns out in this example that j represents an *upper bound* on the change in F following a unit increase in the right-hand side of the constraint.[8] On the basis of the value of j in this case we can therefore rule out any proposal that would cost *more than j* to relax the constraint by one unit. However, without actually carrying out the calculations, it is not possible to evaluate proposals that would cost less than j to relax the constraint. It is instructive to see the pattern of results for the problem as the right-hand side of the constraint is changed. Details for each variable and for the objective are shown in Table 7.1.

Various patterns are evident from Table 7.1. For example note how the values of x, y, j and ΔF move by constant increments as the right-hand side value (b) goes up. The x and y values at $b = 50$ are those at the *unconstrained* maximum of the function. When the right-hand side is increased beyond 50 and since with an equality constraint resources cannot be left unused, the objective function is forced down. This means that j becomes negative, as a penalty is then incurred by an increase in the value of b.

Minimization problems

Minimization of a function of several variables can also be carried out by the Lagrange multiplier method. Where, as here, the problems are those involving equality constraints and variables which are not restricted in sign, the first-order conditions for a constrained minimum are the same as those for a constrained maximum. At the second order, as already stated, a constrained minimum requires that

$$E = L_{xx}g_y^2 - 2L_{xy}g_xg_y + L_{yy}g_x^2 > 0$$

Table 7.1

b	x	y	j	F
35	11	13	6	955
36	11.6	12.8	5.6	960.8
37	12.2	12.6	5.2	966.2
38	12.8	12.4	4.8	971.2
39	13.4	12.2	4.4	975.8
40	14	12	4	980
41	14.6	11.8	3.6	983.8
42	15.2	11.6	3.2	987.2
43	15.8	11.4	2.8	990.2
44	16.4	11.2	2.4	992.8
45	17	11	2	995
46	17.6	10.8	1.6	996.8
47	18.2	10.6	1.2	998.2
48	18.8	10.4	0.8	999.2
49	19.4	10.2	0.4	999.8
50	20	10	0	1000
51	20.6	9.8	−0.4	999.8
52	21.2	9.6	−0.8	999.2
53	21.8	9.4	−1.2	998.2
54	22.4	9.2	−1.6	996.8
55	23	9	−2	995

As an example of minimization using the Lagrange multiplier approach, consider the problem

$$\text{minimize } F = 0.5x^2 + 8y^2 + 4xy - 80x - 240y + 5000$$

subject to the constraint:

$$2.5x + 5y = 75$$

The Lagrangian function is

$$L = 0.5x^2 + 8y^2 + 4xy - 80x - 240y + 5000 + j(75 - 2.5x - 5y)$$

and the first-order conditions are

(i) $L_x = x + 4y - 80 - 2.5j = 0$

(ii) $L_y = 16y + 4x - 240 - 5j = 0$

(iii) $L_j = 75 - 2.5x - 5y = 0$

From equation (i)
$$2.5j = x + 4y - 80$$
so that
$$5j = 2x + 8y - 160$$
which, when inserted into (ii), produces
$$16y + 4x - 240 - 2x - 8y + 160 = 0$$
which simplifies to
$$8y + 2x - 80 = 0$$
with the result
$$x = 40 - 4y$$
which when substituted into (iii) produces
$$75 - 100 + 10y - 5y = 0$$
so that
$$5y = 25$$
and therefore
$$y = 5$$
It then follows from substitutions into the earlier equations that
$$x = 20 \quad \text{and} \quad j = -16$$
Now check the satisfaction (or otherwise) of the condition on E at the second order. In this case we have
$$L_{xx} = 1 \qquad L_{xy} = 4 \qquad L_{yy} = 16 \qquad g_x = -2.5 \qquad g_y = -5$$
so that E evaluates as
$$E = (1)(-5)^2 - 2(4)(-2.5)(-5) + 16(-2.5)^2$$
$$= 25 - 100 + 100$$
$$= 25 > 0$$
which is of the required sign for a constrained minimum of F. With $x = 20$ and $y = 5$, the corresponding minimum value of $F = 3000$. The negative value of the Lagrange multiplier in this case is interesting. It means that, as the right-hand side of the constraint is increased, a *lower* value of F is achievable. Again, the results from varying the right-hand side of the constraint are revealing. For right-hand side values between 70 and 80 the results are shown in Table 7.2.

Table 7.2

RHS	x	y	j	F
70	16	6	−16	3080
71	16.8	5.8	−16	3064
72	17.6	5.6	−16	3048
73	18.4	5.4	−16	3032
74	19.2	5.2	−16	3016
75	20	5	−16	3000
76	20.8	4.8	−16	2984
77	21.6	4.6	−16	2968
78	22.4	4.4	−16	2952
79	23.2	4.2	−16	2936
80	24	4	−16	2920

Since in this example the value of the Lagrange multiplier does not change, j is the actual change in F resulting from a unit increase in the right-hand side of the constraint. Prompted by this output, the question arises as to whether there is a 'best' value of the right-hand side in this case. You are invited to explore this question. We conclude this section with a numerical example illustrating the caution necessary in problems where the objective function is not globally concave. In the example, we shall see that the Lagrange multiplier method produces a constrained local maximum (and also a constrained local minimum point). But the objective function is, in fact, unbounded despite the constraint. The problem is this:

maximize $z = 1000x + 200y - x^2y - 55x^2$
subject to $2x + y = 20$

The Lagrangian function and first-order conditions are

$L = 1000x + 200y - x^2y - 55x^2 + j(20 - 2x - y)$

(i) $L_x = 1000 - 2xy - 110x - 2j = 0$

(ii) $L_y = 200 - x^2 - j = 0$

(iii) $L_j = 20 - 2x - y = 0$

From condition (ii)

$j = 200 - x^2$

Substitution into (i) produces

(i)' $1000 - 2xy - 110x - 400 + 2x^2$

From (iii)

$$y = 20 - 2x$$

and substitution into (i)' results in

$$1000 - 2x(20 - 2x) - 110x - 400 + 2x^2 = 0$$

which reduces to

$$x^2 - 25x + 100 = 0$$

which factors as

$$(x - 5)(x - 20) = 0$$

When $x = 5$, $y = 10$, $j = 175$ and the value of $z = 6365$. To find the nature of this point, consider the second-order condition. We have here

$$L_{xx} = -2y - 110 \quad L_{xy} = -2x \quad L_{yy} = 0$$

and

$$g_x = -2 \quad g_y = -1$$

so that, given L_{xx} and L_{xy}, E will be a function of x and y:

$$E = (-2y - 110)(-1)^2 - 2(-2x)(-2)(-1)$$

When $x = 5$ and $y = 10$, E evaluates as

$$E = -130 + 40$$
$$= -90 < 0$$

so the point is identified as a local constrained maximum. Now, when

$$x = 20 \quad y = -20 \quad j = -200 \quad z = 2000$$

the value of E at this point is

$$E = -70 + 160$$
$$= 90 > 0$$

so that this point is a local minimum of z subject to the constraint. But the function's highest point is not at $x = 5$, nor is its lowest point at $x = 20$. For higher x values beyond the constrained local minimum at $x = 20$, the function increases without limit. For x values below $x = 5$, the function is unbounded from below. An unbounded outcome is easier to detect in the substitution method particularly when, as here, the objective is a function of two variables. Substituting

$$y = 20 - 2x$$

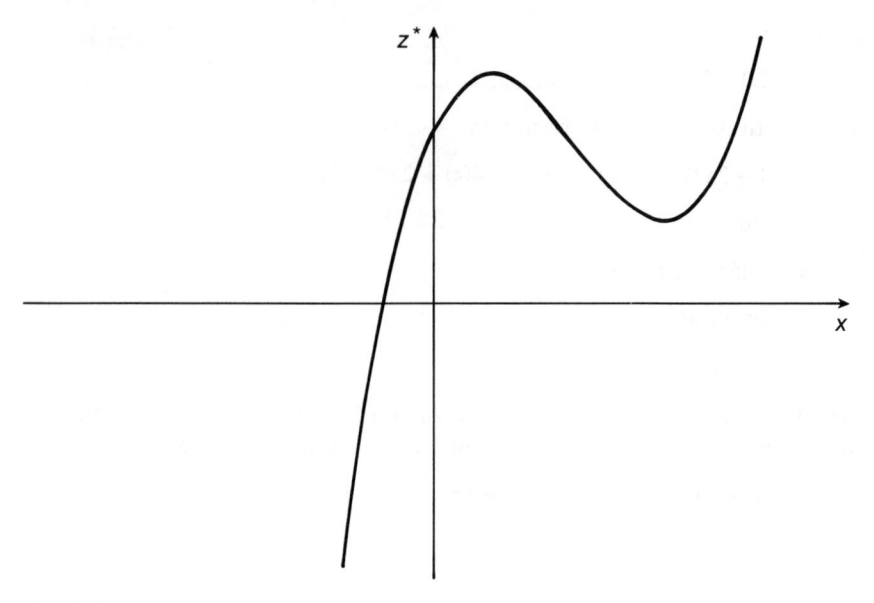

Figure 7.1

into the expression for z gives

$$z^* = 2x^3 - 75x^2 + 600x + 4000$$

in which the positive coefficient of the leading term reveals the unboundedness, x being made arbitrarily large since there is no requirement for y to be non-negative. The graph of z^* is shown in Figure 7.1.

Practical problems, particularly those in which profit or revenue forms the objective function, do not usually have unbounded outcomes! This characteristic may mean that the function used to model the real situation has all the properties that it is convenient to assume. But it may mean that, although the objective function itself has no finite maximum,[9] there are other restrictions on the values of x and y which keep the function finite. Sign requirements are a particular instance and these are the subject of a later section.

EXERCISES 7.1

1 Using the Lagrange multiplier method, find the maximum value of the function

$$f(x, y) = 20x + 15y - xy - x^2 - 0.5y^2$$

subject to the constraint

$$x + 2y = 20$$

Use the sufficient second-order condition (7.5) to confirm that a maximum value has been obtained.

2 For the function
$$f(x,y) = 1200x + 400y - 20xy - 40x^2 - 2.5y^2$$
and given the resource constraint
$$10x + 5y = b$$
(i) find the maximum value of $f(x,y)$ when the value of b is 150;
(ii) find the value at the margin of an extra unit of the scarce resource.

3 Minimize the value of
$$f(x,y) = 2x^2 + 6xy + 4.5y^2 - 100x - 200y$$
subject to $2.5x + 1.25y = 40$

Find the effect on the objective of a unit increase in the right-hand side value.

4 Find the values of x and y which maximize the function
$$f(x,y) = 44x + 101y + 9xy + 5x^2 - 8y^2$$
subject to the constraint
$$3x + 2y = 29$$

What is the effect on the objective of a unit increase in the right-hand side value?

5 Find the maximum value of the function
$$f(x,y) = x^3 y$$
subject to the constraint
$$x + y = 20$$

7.2 INTERPRETATION OF THE LAGRANGE MULTIPLIER

In the case of the maximization of a function of two variables subject to a linear equality constraint, we can write the problem as

maximize $z = f(x,y)$
subject to $ax + cy = b$

462 Mathematics for business, finance and economics

The Lagrangian function and first-order conditions are therefore

$$L = f(x, y) + j(b - ax - cy)$$

(i) $\dfrac{\partial L}{\partial x} = \dfrac{\partial f}{\partial x} - ja = 0$

(ii) $\dfrac{\partial L}{\partial x} = \dfrac{\partial f}{\partial y} - jc = 0$

(iii) $\dfrac{\partial L}{\partial j} = b - ax - cy = 0$

Solving condition (i) for j gives

$$j = \dfrac{1}{a} \dfrac{\partial f}{\partial x}$$

while from condition (ii) we obtain

$$j = \dfrac{1}{c} \dfrac{\partial f}{\partial y}$$

Now consider these terms in detail. The partial derivative $\partial f/\partial x$ is the rate at which the objective function rises as x is increased. The ratio $1/a$ is the rate of change of x as the resource level is increased.[10] For example if $a = 2$ then x can be increased at the rate of half a unit per unit of resource transferred. The product

$$\dfrac{1}{a} \dfrac{\partial f}{\partial x}$$

can now be interpreted in the following way. It is:

> The rate at which the achievement of the objective changes, as further resource is applied to x.

Condition (i) therefore sets the value of the Lagrange multiplier to this rate. A similar interpretation results from (ii), the Lagrange multiplier being set at the rate at which the achievement of objective changes as further resource is applied to y. The equalization of these rates through j at the *optimum* means that the same rate of change of the objective is secured if a marginal variation in resourcing is applied through either variable. The necessity of this condition for a position to be optimal can be seen by contrast with a situation where there is disparity between the rates. For example, suppose that at some particular point the following values obtained:

$$\dfrac{1}{a} \dfrac{\partial f}{\partial x} = 5 \quad \text{while} \quad \dfrac{1}{c} \dfrac{\partial f}{\partial y} = 7$$

These rates apply *in either direction* at the margin, and it may help to think in terms of a small finite change. If a unit of resource is *withdrawn* from use in x, then on this account the value of the objective function will fall by approximately 5. But if this same unit of resource is now applied to increase the level of y, then on this account the value of the objective function will increase by about 7. So the net effect would be an approximate improvement of 2, so that the original point could not be optimal. There will always be scope for such beneficial transfers whenever there is a divergence between the marginal rates. For optimality there must be a *uniform rate* – and the value of j is that rate. Thus the Lagrange multiplier can be seen as a mediator between the competing claims on the scarce resource. In any one equation, j can be seen as carrying information about the value of the resource in an alternative use. Thus the effect of the multiplier is to ensure that, at the margin, the resource is allocated optimally between the possible uses. In other words, the use of resource is maximally efficient: it is *economized*.[11]

In summary, by comparison with the method of substitution, the Lagrange multiplier approach to constrained optimization problems has a number of advantages. Amongst the most important of these are the facts that the method can be used with a wider variety of constraints, and that useful management information is conveyed by the value of the Lagrange multiplier at the optimum. In a later section, we shall return to Lagrange multipliers and consider their use in problems with inequality constraints. First we shall address the question of how to take account of sign restricted decision variables. The procedures developed will also be needed later in problems which include resource constraints.

EXERCISES 7.2

1 A decision-maker has the objective of finding the maximum value of the function

$$f(x, y) = 4x^2 + 16xy + 5y^2$$

and, in so doing, is restricted by the following constraint on overall expenditure:

$$2x + 3y = 30$$

(i) Find the maximum achievable value of the objective given the financial constraint.

(ii) Show that, at the optimum, the rate of improvement of the objective function for extra resource expended on x is the same as the rate of improvement of the objective for extra resource expended on y.

(iii) With the original level of resource, show that at a sub-optimal position such as $x = 3$, $y = 8$, the rates of improvement of objective for extra expenditure on x and y are unequal.

7.3 SIGN RESTRICTED AND BOUNDED VARIABLES

Many problems involve variables with no specific endpoints to their range of possible values. The occurrence of negative values, or very high positive values, of the decision variables need not be meaningless. Such problems arise in a number of ways. For example, there are variables such as profit (loss), net present value (plus or minus) or the balance of payments (surplus or deficit) which can naturally be of either sign. Then there are models that are expressed in terms of relative or absolute *changes* in the values of variables which would normally be sign restricted. Examples are provided by changes in stock levels, an index of prices or a measure of money supply. Changes expressed in terms of proportions can involve very large values when the base is close to zero.

But there are also many models that use physical quantities such as manufacturing output, gross domestic product or prices – measures which cannot be negative. And there are other models where both upper and lower limits to the range of variation of variables need to be recognized. For example, an upper bound on short-term production levels is set by plant capacity. If one of the decision variables is probability, p, the value of p must be in the range $0 \leq p \leq 1$. In this section we consider optimization models involving variables restricted in range from below and/or from above. The most common practical limitation on decision variables is the requirement of non-negativity, and it is with this subject that we begin.

Since the explicit introduction of sign requirements inevitably makes any model less convenient, it is legitimate to enquire if, in a particular situation, the sign requirements can be ignored or circumvented. One way to bypass the sign requirements is to select the objective function appropriately, choosing a function for which negative values of the decision variables will not arise.[12] An alternative approach is to discard any negative values that are produced.[13] This is frequently a reasonable procedure, as for example in the square root model of stock control, when it is known that a positive root will also be generated. But these avoidance devices are not always enough, and non-negativity requirements may need to be built in directly. One of the advantages of a mathematical programming approach is the explicit incorporation of sign requirements – as we saw in linear programming. To begin with, consider the problem of maximizing a function of a single, sign restricted, variable, there being no other constraints. The problem is to

maximize $y = f(x)$
subject to $x \geqslant 0$ (7.6)

There are two points to note concerning (7.6). First, the sign requirement can be seen as a form of *inequality constraint*. Second, the sign requirement is also a particular instance of a *lower bound*. Problems expressed in terms of variables limited from below at levels other than zero can be reformulated in terms of sign restricted variables. In the following discussion, the default assumption is that a finite maximum of $f(x)$ exists for a finite value of x within the domain, and that the function is continuous and differentiable[14] for all values of $x \geqslant 0$. On this basis, cases such as the rational function

$$f(x) = \frac{1}{1-x}$$

which is unbounded as x approaches unity, are ruled out. However, we shall make reference to such possibilities from time to time. As is often the case, diagrams prove useful in examining problems of this nature. Given the assumptions made, three distinct cases can be distinguished.

7.3.1 An interior optimum

Figure 7.2 shows an example in which the function $f(x)$ is such that a maximum is produced for strictly positive x. The optimum here is a turning point of the function in the positive quadrant, so in this case the sign requirement on x has no bearing on the optimum. The maximum occurs

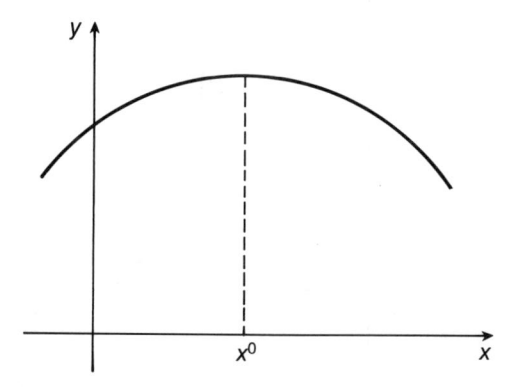

Figure 7.2

466 Mathematics for business, finance and economics

when $x = x^0$. Two qualitative properties of this interior optimum should be noted:

$$x > 0 \quad \text{and} \quad \frac{dy}{dx} = 0$$

Note that our original assumptions ruled out the interior situations shown in Figures 7.3(a) and 7.3(b). Figure 7.3(a) shows a *cusp* while in Figure 7.3(b) there is an asymptote in the positive quadrant. We shall not, in the main, be concerned with these possibilities.[15]

7.3.2 A boundary optimum

In Figure 7.4, the function has negative slope for all permitted values of x. It follows that the maximum value of the function in the domain must occur at the (lower) boundary, i.e. at $x = 0$.

Figure 7.5 shows a case where the overall optimum is at the boundary, but where there is also a local maximum at a turning point of the function within the domain. However, the endpoint produces the *global* maximum for the specified domain.

Two qualitative properties of the boundary maximum case should be noted:

$$x = 0 \quad \text{and} \quad \frac{dy}{dx} < 0$$

When we have considered the third case we shall draw these results together.

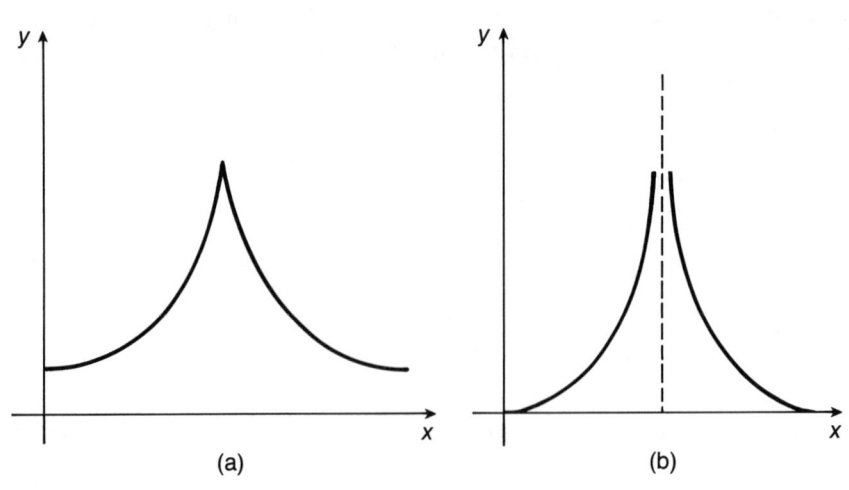

Figure 7.3

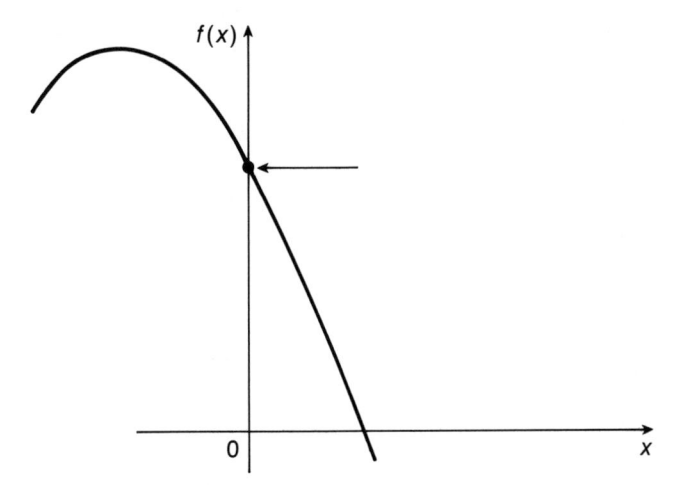

Constrained optimization with Lagrange multipliers 467

Figure 7.4

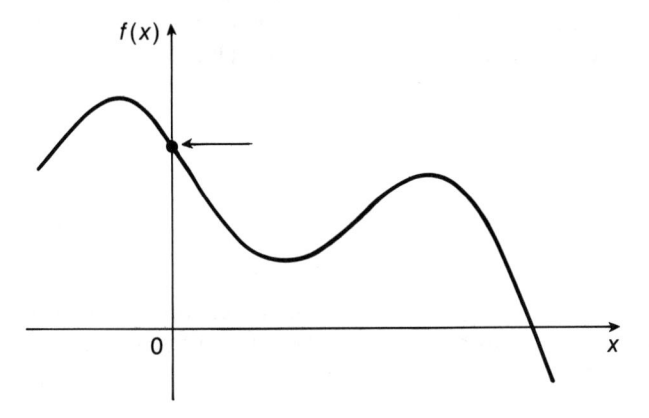

Figure 7.5

7.3.3 The 'degenerate' case

Figure 7.6 shows a possible outcome where the turning point of the function occurs precisely at the endpoint of the domain.

The word *degenerate* describes limiting, coincidental or singular states where special conditions apply. This may represent the compounding of two normally distinct situations. As is evident, in this case the turning point maximum and the boundary maximum coincide. The qualitative conditions here are

$$x = 0 \quad \text{and} \quad \frac{dy}{dx} = 0$$

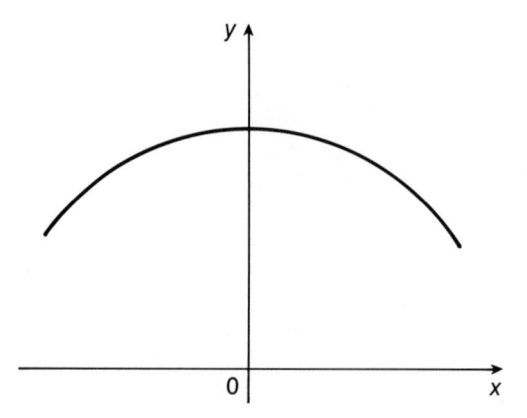

Figure 7.6

The results can now be summarized as follows. In all cases dy/dx is either zero or negative. Furthermore, the derivative is negative *only* if $x = 0$. At the sign restricted optimum, the value of x itself is either zero or positive, the positive outcome occurring *only* if dy/dx is zero. Thus in all cases

(a) $\dfrac{dy}{dx} \leq 0$

(b) $x \geq 0$ (7.7)

(c) $x \dfrac{dy}{dx} = 0$

The properties (7.7) are *necessary conditions* for a sign restricted local maximum. To solve particular problems, we make use of these conditions by using the following four-step procedure.

1 Find the derivative dy/dx.
2 Test the satisfaction of the conditions for $x = 0$. With $x = 0$ assumed, (b) and (c) *must* be satisfied. What remains, therefore, is to check the sign of dy/dx at $x = 0$. If all of the conditions are satisfied, record the values of x and y.
3 Test the satisfaction of the conditions for $dy/dx = 0$. Since under this assumption (a) and (c) must be satisfied, this means that any non-negative roots of $dy/dx = 0$ should now be found. Record any such values of x and the corresponding y values.
4 Select the position corresponding to the best value of y recorded under 2 and 3. Given the properties assumed for the objective function, this position is the sign restricted optimum.

Note that in general both 2 and 3 can give results satisfying all of the conditions. There may therefore be several possibilities for the overall optimum. In step 4 we distinguish between these points by enumeration.[16] In a problem of this size it is simple enough to check the behaviour of the function for extreme positive values for x. Examination of the sign of the leading coefficient should provide the assurance required. Consider an example:

maximize $y = 20 + 8x - x^2$
subject to $x \geqslant 0$

First find the derivative as required by step 1. It will also be helpful to set out the conditions in full for ease of reference. It is required that:

$$\frac{dy}{dx} = 8 - 2x \leqslant 0$$

$$x \geqslant 0$$

$$x \frac{dy}{dx} = 0$$

Now go to step 2 and try $x = 0$. To 'try' this value of x means to see if the condition required for the derivative is satisfied at $x = 0$. Clearly it is not, since $dy/dx = 8 > 0$. Step 3 is next. Here we try $dy/dx = 0$. Since with $dy/dx = 0$ conditions (b) and (c) *must* be satisfied, to try this possibility means to look for non-negative x for which $dy/dx = 0$. In fact $x = 4$, so that this value satisfies the full conditions, and the point $x = 4$, $y = 36$ is recorded. Step 4 here is trivial as only one point has been identified under 2 and 3. So (4, 36) is the sign restricted maximum — as a sketch of the parabola will confirm. For a less straightforward exercise in sign restricted maximization consider the problem

maximize $y = 21x^2 - 2x^3 - 60x + 100$

subject to $x \geqslant 0$

For step 1 the first derivative is

$$\frac{dy}{dx} = 42x - 6x^2 - 60 \leqslant 0$$

Moving to step 2 and testing the sign of the derivative dy/dx at $x = 0$ gives

$$\frac{dy}{dx} = -60 < 0$$

so the condition on the derivative is satisfied and $x = 0$, $y = 100$ is recorded as a possible optimum. Now move to step 3 and try $dy/dx = 0$. Satisfaction of conditions (b) and (c) is ensured, so the only question is that of the sign of x. The quadratic solves for $x = 2$ and $x = 5$ which values both satisfy $x \geq 0$. The familiar second-order condition *could* be used to show that $x = 2$ gives a relative minimum and $x = 5$ represents a local maximum. But in step 4 inserting all three candidate values into the expression for y gives the results

x	y
0	100
2	48
5	75

from which it is clear that the sign restricted maximum in this case occurs when $x = 0$. Examination of the sign of the leading coefficient or a sketch of the function will confirm this result. The picture is as shown in Figure 7.7.

Now consider the problem

maximize $z = 100 - 25x + 1.25x^2$
subject to $x \geq 0$

The derivative is

$$\frac{dz}{dx} = -25 + 2.5x$$

At $x = 0$, $dz/dx = -25$ so $x = 0$ is a candidate value for a sign restricted maximum. When $dz/dx = 0$, $x = 10$, so this is also a candidate position. We choose between $x = 0$ and $x = 10$ on the basis of the value of z:

x	z
0	100
10	-25

It might appear that the value $x = 0$ produces the sign restricted maximum. But this is not so, as in this case the function, z, is unbounded for large, positive values of x. This is revealed by the positive leading coefficient. Use of the second-order condition that $d^2z/dx^2 < 0$ would have identified the turning point at $x = 10$ as a local minimum (and thus removed it from further consideration) but would not, of course, have applied at $x = 0$. If there is reason to suppose that the objective function may be unbounded in the direction of optimization, it is advisable to apply the method with circumspection. The context of the application may give useful guidance in this respect.

Now consider the conditions required for sign restricted *minima*. Here again, we must allow for the possibility of boundary as well as interior optima. As sketch diagrams will confirm, it emerges that the only difference from (7.7) in the necessary first-order conditions is that the slope of the function at a boundary minimum *should not be negative*. If the endpoint

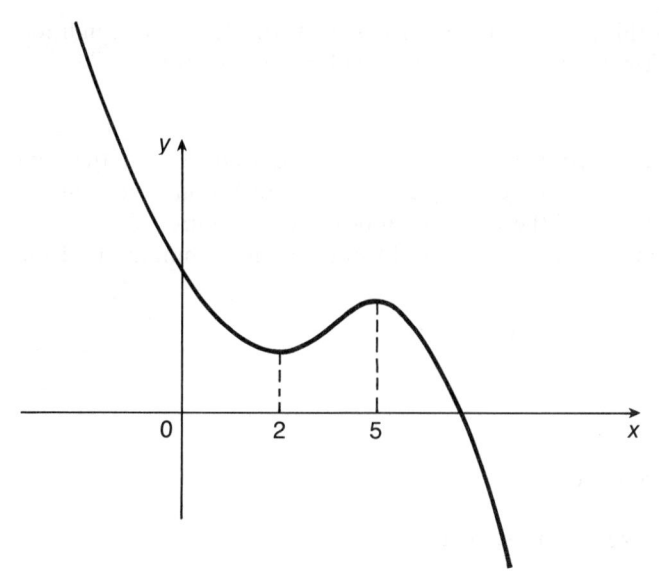

Constrained optimization with Lagrange multipliers

Figure 7.7

is to give a minimum the function must not be going down as we move into the feasible region. Conditions at the first order are

$$\frac{dy}{dx} \geq 0$$

$$x \geq 0 \qquad (7.8)$$

$$x \frac{dy}{dx} = 0$$

Consider an example:

minimize $y = 2x^3 - 294x + 100$
subject to $x \geq 0$

The conditions for a sign restricted minimum are

$$\frac{dy}{dx} = 6x^2 - 294 \geq 0$$

$$x \geq 0$$

$$x \frac{dy}{dx} = 0$$

First try $x = 0$. At this point $dy/dx = -294$ so that the derivative condition is not satisfied. Now try $dy/dx = 0$. The quadratic solves for

$x = 7$ and $x = -7$

which latter value of course breaks the sign requirement. So the sign restricted minimum of the function is at $x = 7$ at which point the value of y is -1272. The graph of the function appears as in Figure 7.8.

As a further example consider the problem of finding a minimum of the quartic

$$y = 3x^4 - 28x^3 + 24x^2 + 144x + 500$$

subject to

$$x \geqslant 0$$

Necessary conditions are

$$\frac{dy}{dx} = 12x^3 - 84x^2 + 48x + 144 \geqslant 0$$

$$x \geqslant 0$$

$$x \frac{dy}{dx} = 0$$

At the endpoint $x = 0$ the derivative is positive, so this is a possible sign restricted minimum. Equating the derivative to zero to find possible interior

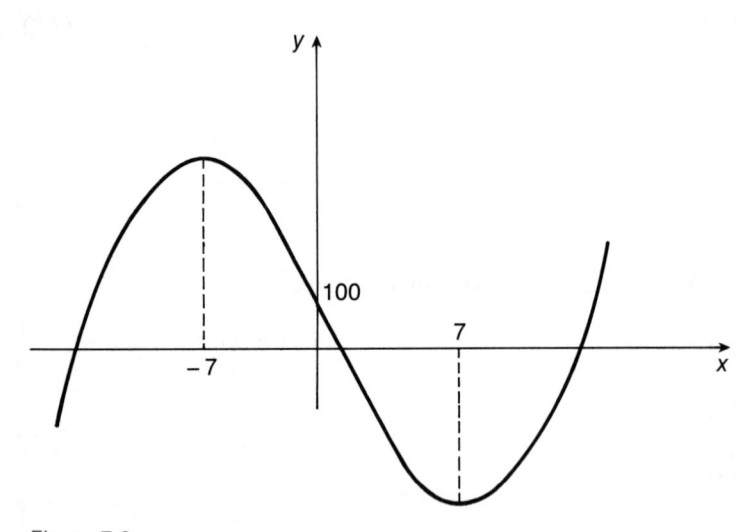

Figure 7.8

optima under step 3 means that a cubic equation has to be solved. The roots emerge as

$$x = -1 \qquad x = 2 \qquad x = 6$$

as may be confirmed by substitution into the derivative. After eliminating the negative root, the function is now evaluated at the permissible values of x. The results are

x	y
0	500
2	708
6	68

so the sign restricted minimum for the function occurs at $x = 6$. Now consider the maximization of a function of two independent variables, with sign requirements on each variable and no other restrictions. In general, the problem is to

maximize $f(x, y)$
subject to $x \geq 0 \qquad y \geq 0$

Necessary conditions here are a generalization of the conditions for a function of a single variable:[17]

$$\begin{aligned} f_x &\leq 0 \\ xf_x &= 0 \\ x &\geq 0 \\ f_y &\leq 0 \\ yf_y &= 0 \\ y &\geq 0 \end{aligned} \qquad (7.9)$$

Since each variable, the derivative of the function with respect to the variable, or both must be zero, a possible procedure to find solutions to the conditions (7.9) would be to try the allowable combinations in a systematic way. We could, for example, proceed as follows:

	Try:		Check for:	
(1)	$x = 0$	$y = 0$	$f_x \leq 0$	$f_y \leq 0$
(2)	$x = 0$	$f_y = 0$	$f_x \leq 0$	$y \geq 0$
(3)	$f_x = 0$	$y = 0$	$x \geq 0$	$f_y \leq 0$
(4)	$f_x = 0$	$f_y = 0$	$x \geq 0$	$y \geq 0$

Consider an example:

maximize $f(x, y) = 15x - 10y - 5xy - 2x^2 - 4y^2$
subject to $x \geq 0 \qquad y \geq 0$

The derivatives are

$$f_x = 15 - 5y - 4x$$
$$f_y = -10 - 5x - 8y$$

First, try the combination $x = 0$, $y = 0$. If it transpires that $f_x \leqslant 0$ and $f_y \leqslant 0$, then this is a candidate position for the sign restricted optimum. When $x = 0$ and $y = 0$, $f_x = 15$ and $f_y = -10$. The outcome for f_x means that this point cannot be optimal. The results for this and the other combinations are shown in Table 7.3. It is evident that the conditions are only fulfilled in case (3), and the sign restricted maximum of the function is $f(3.75, 0) = 28.125$. Note that case (4) of Table 7.3 corresponds to the unrestricted maximum (for which $f(x, y) = 150$). Note also in this case that the function does not increase without limit for x and y in the non-negative orthant. Since neither x nor y can be negative, the term $-5xy$ cannot be positive. Neither can the terms $-2x^2$ and $-4y^2$. Since these three are all the terms of degree two, the sign restricted optimum obtained under (3) must be the global optimum. Note that the number of situations that would need to be enumerated in this approach (four here) will in general be 2^n where n is the number of sign restricted variables. In some problems involving several independent variables, some variables may be restricted in sign whilst others are not. If it is required to maximize

$$f(x_1, x_2, \ldots, x_n)$$

then if some variable x_i is sign restricted it will be required that

$$f_{xi} \leqslant 0$$
$$x_i f_{xi} = 0$$
$$x_i \geqslant 0$$

If some variable x_j is *not* sign restricted, then the normal condition

$$f_{xj} = 0$$

is all that is required at the first order. Following the procedure of Table 7.3, if m of the n variables in the problem are sign restricted, then

Table 7.3

(1) $x = 0$	$y = 0$	$f_x = 15$	$f_y = -10$
(2) $x = 0$	$f_y = 0$	$f_x = 21.25$	$y = -1.25$
(3) $f_x = 0$	$y = 0$	$x = 3.75$	$f_y = -28.75$
(4) $f_x = 0$	$f_y = 0$	$x = 10$	$y = -5$

there will be 2^m distinct cases to enumerate. Consider an example. Find the maximum value of the function

$$f(x, y) = 40x - 50y + 5xy - 3x^2 - 2.5y^2$$

when

(i) x alone is sign restricted;
(ii) y alone is sign restricted;
(iii) both x and y are sign restricted;
(iv) neither x nor y are sign restricted.

Take case (i) first. The derivatives are

$$f_x = 40 + 5y - 6x$$
$$f_y = -50 + 5x - 5y$$

and the necessary conditions here are

$$f_x \leqslant 0$$
$$xf_x = 0$$
$$x \geqslant 0$$
$$f_y = 0$$

Trying $x = 0$ first of all, we then have

$$f_x = 40 + 5y \leqslant 0$$
$$f_y = -50 - 5y = 0$$

so that, from $f_y = 0$, $y = -10$ and therefore $f_x = -10$, and the point

$$x = 0 \qquad y = -10 \qquad f(x, y) = 250$$

is a possible optimum. Now trying $f_x = 0$, we then require

$$f_x = 40 + 5y - 6x = 0$$
$$f_y = -50 + 5x - 5y = 0$$

which solve for

$$x = -10 \qquad \text{and} \qquad y = -20$$

which is not feasible in problem (i). So the optimum here is

$$x = 0 \qquad y = -10 \qquad f(x, y) = 250$$

Now consider problem (ii). The requirements here are

$$f_x = 0$$
$$f_y \leqslant 0$$
$$yf_y = 0$$
$$y \geqslant 0$$

Trying $y = 0$ first of all, it will then be required that

$$f_x = 40 - 6x = 0$$
$$f_y = -50 + 5x \leqslant 0$$

From $f_x = 0$, the value of x is $x = 20/3$ and therefore $f_y = -50/3$ so that this is a possible optimum. Now we have already seen that when $f_y = 0$ and $f_x = 0$, negative values of both variables result. So the optimum in case (ii) is

$$x = 20/3 \qquad y = 0 \qquad f(x, y) = 400/3$$

Now consider case (iii). The results are shown in Table 7.4. The optimal solution is as for case (ii). Under case (iv), we have already seen that the unrestricted optimum occurs at

$$x = -10 \qquad y = -20 \qquad f(x, y) = 300$$

Sign restrictions are an example of variables with *lower bounds*. In some cases a variable may have a lower bound other than zero as, for example, when a manufacturer wants to ensure that the order of a favoured customer is fulfilled. But in such cases the variable may be transformed, without distorting the problem, to one for which the lower bound is zero. The problem can then be solved in the new sign restricted variable and the results re-expressed in terms of the original variable. For a numerical example, consider the following problem:

maximize $z(x) = 160x - 2x^2$
subject to $x \geqslant 50$

To solve this problem, define a new variable w such that

$w = x - 50 \qquad$ so that $\qquad x = w + 50$

Now re-express the problem in w:

maximize $z(w) = 160(w + 50) - 2(w + 50)^2$
subject to $w + 50 \geqslant 50$

which simplifies to

maximize $z(w) = 3000 - 40w - 2w^2$
subject to $w \geqslant 0$

Table 7.4

$x = 0$	$y = 0$	$f_x = 40$	$f_y = -50$
$x = 0$	$f_y = 0$	$f_x = -10$	$y = -10$
$f_x = 0$	$y = 0$	$x = 20/3$	$f_y = -50/3$
$f_x = 0$	$f_y = 0$	$x = -10$	$y = -20$

Constrained optimization with Lagrange multipliers 477

for which the conditions for a sign restricted optimum are

$$\frac{dz}{dw} = -40 - 4w \leqslant 0$$

$$w \geqslant 0$$

$$w \frac{dz}{dw} = 0$$

When $w = 0$, $dz/dw = -40$, so the conditions are fulfilled. When $dz/dw = 0$, $w = -10$, which violates the conditions. So the optimum is at $w = 0$, corresponding to $x = 50$ and $z = 3000$. The problem is graphed in Figure 7.9.

Notice that the unrestricted maximum at $x = 40$ corresponds to the point where $w = -10$. Variables can be restricted from above as well as from below. We previewed this situation in Chapter 5 when considering the absolute maximum of a function over a restricted domain. If a variable is restricted only from above, the maximum of a function of this variable will occur either at the endpoint or at a turning point below the endpoint. This parallels the sign restricted case. We can get the necessary conditions by transforming the problem into one in a sign restricted variable. Consider an example:

maximize $z(x) = 160x - 2x^2$
subject to $x \leqslant 30$

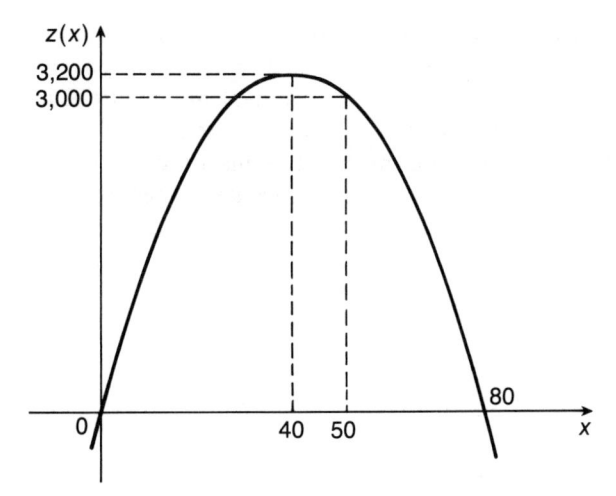

Figure 7.9

Define a new variable, v, such that

$$v = 30 - x$$

Clearly:

$$v \geqslant 0$$

Now since

$$x = 30 - v$$

substitution of this relationship into the original problem gives

maximize $z(v) = 160(30 - v) - 2(30 - v)^2$
subject to $30 - v \leqslant 30$

which simplifies to

maximize $z(v) = 3000 - 40v - 2v^2$
subject to $v \geqslant 0$

for which the necessary conditions are

$$\frac{dz}{dv} = -40 - 4v \leqslant 0$$

$$v \frac{dz}{dv} = 0$$

$$v \geqslant 0$$

which conditions are fulfilled for $v = 0$, giving a constrained optimum at the endpoint meaning that the optimum is at $x = 30$. The problem may be transformed as above, but our purpose is to give equivalent conditions in terms of the original variable. To find these conditions, first consider the requirement $dz/dv \leqslant 0$. What does this mean in terms of dz/dx? The *composite function rule* can answer this question. Since v is a function of x

$$\frac{dz}{dx} = \frac{dz}{dv} \frac{dv}{dx}$$

but, since

$$v = 30 - x$$

it follows that

$$\frac{dv}{dx} = -1$$

so therefore

$$\frac{dz}{dx} = -\frac{dz}{dv}$$

Constrained optimization with Lagrange multipliers 479

So, since the conditions in terms of v require that $dz/dv \leq 0$, this means that in terms of x

$$\frac{dz}{dx} \geq 0$$

Now consider the requirement that

$$v \frac{dz}{dv} = 0$$

We can now write this as

$$-v \frac{dz}{dx} = 0$$

i.e.

$$v \frac{dz}{dx} = 0$$

which means that

$$(30 - x) \frac{dz}{dx} = 0$$

so that the full conditions in terms of x are

$$\frac{dz}{dx} \geq 0$$

$$(30 - x) \frac{dz}{dx} = 0$$

$$(30 - x) \geq 0$$

In general, therefore, given the problem

maximize $z(x)$
subject to $x \leq b$

it is necessary that

$$\frac{dz}{dx} \geq 0$$

$$(b - x) \frac{dz}{dx} = 0 \qquad (7.10)$$

$$(b - x) \geq 0$$

which is the form of conditions that we require. Now suppose that a variable is restricted both from above and from below. Without loss of generality, we can take the lower bound to be zero.[18] Therefore, in terms of a single variable, the problem is to

maximize $z(x)$
subject to $x \geq 0$
$x \leq b$

Necessary conditions at the first order include elements of both (7.7) and (7.10), which are economically combined using the *Kuhn–Tucker conditions* to be described in Section 7.6. For the present we describe an implementation of the conditions that will enable solution of problems on a modest scale. The following procedure is essentially that described in Chapter 5 for global maximization of a function over a restricted domain. There are three possible outcomes:

1 an optimum at the lower bound;
2 an optimum at the upper bound;
3 an interior optimum.[19]

Local constrained optima may occur at all three positions, but the requirement on the derivative (non-positive, non-negative or zero) will vary between the positions. For a local optimum at the lower bound

$$\frac{dz}{dx} \leq 0$$

whereas, at the upper bound, it is required that

$$\frac{dz}{dx} \geq 0$$

and, of course, at a turning point

$$\frac{dz}{dx} = 0$$

The following five-step procedure is useful.

1 Find dz/dx.
2 Try $x = 0$. If $dz/dx \leq 0$, record $x = 0$ as a possible optimum.
3 Try $x = b$. If $dz/dx \geq 0$, record $x = b$ as a possible optimum.
4 Solve $dz/dx = 0$. Roots in the interval $0 \leq x \leq b$ should be recorded as possible optima.
5 Evaluate z at all of the points recorded under 2–4. The greatest value of z identifies the optimum position.[20]

Consider a numerical example. Find the maximum value of

$$z = 500 - 240x + 39x^2 - 2x^3$$

subject to

$$x \geq 0 \quad \text{and} \quad x \leq 10$$

Applying the procedure, first find the derivative

$$\frac{dz}{dx} = -240 + 78x - 6x^2$$

At $x = 0$, $dz/dx = -240$. So $x = 0$ is one possibility to record. At $x = 10$, $dz/dx = -60$, ruling out the upper bound as a possible optimum. Solving $dz/dx = 0$ gives the values $x = 5$ and $x = 8$. The function is now evaluated at the three positions:

x	z
0	500
5	25
8	52

So the maximum occurs at $x = 0$. There are several special cases of interest. One of these is two-variable constrained optimization problems involving a linear equality constraint and sign requirements on both variables. Interest in this case arises because many examples of constrained optimization in microeconomic theory are two-variable problems with an equality constraint. But the usual contexts given (consumption, production or resource levels) imply that negative values of the decision variables would be meaningless. There is not much extra work involved explicitly to take account of realistic signs for the variables, since it turns out that such problems can be re-expressed as problems in a single variable restricted from above and from below. Consider the function

$$z = 100x - x^2 - 112y - 4y^2 + 2xy$$

for which a maximum is sought with respect to the linear constraint

$$5x + y = 90$$

Without sign requirements, the maximum value of $z = 1920$ is achieved when $x = 20$ and $y = -10$. Now include sign requirements on both x and y. So, in addition to the linear constraint above, it is also required that

$$x \geq 0 \quad \text{and} \quad y \geq 0$$

Now employ substitution. From the constraint,

$$y = 90 - 5x$$

but now since $y \geqslant 0$ it must be the case that

$$90 - 5x \geqslant 0$$

i.e.

$$x \leqslant 18$$

So, after substitution for y, the problem is now to maximize

$$z = 100x - x^2 - 112(90 - 5x) - 4(90 - 5x)^2 + 2x(90 - 5x)$$
$$= 4440x - 111x^2 - 42480$$

subject to

$$x \geqslant 0 \quad \text{and} \quad x \leqslant 18$$

to which problem the method above can now be applied. The derivative is

$$\frac{dz}{dx} = 4440 - 222x$$

At $x = 0$, the derivative $dz/dx = 4440 > 0$, so this is not a possible maximum. At the upper bound $x = 18$, $dz/dx = 444$, so this position *is* a possible constrained maximum. Setting $dz/dx = 0$ gives $x = 20$, so this point is ruled out. The constrained maximum is therefore

$$x = 18 \quad y = 0 \quad z = 1476$$

The total effect of the sign requirements can be costed. It is the difference between the values of the objective function achieved without and with the sign requirements. That is,

$$1920 - 1476 = 444$$

If there is discretion as to whether variables are or are not sign restricted (as for example with dividend changes) then the price of not exercising the discretion can be calculated as above. This is valuable management information.

EXERCISES 7.3

1. (i) Maximize $z = 24x - 1.5x^2 + 120$
   subject to $x \geqslant 0$
   (ii) Maximize $z = 100 - 21x - 3.5x^2$
   subject to $x \geqslant 0$

2. (i) Maximize $z = 1000 + 120x + 27x^2 - x^3$
   subject to $x \geqslant 0$
   (ii) Maximize $z = 8000 - 1440x^2 + 128x^3 - 3x^4$
   subject to $x \geqslant 0$

3 (i) Minimize $z = 1.2x^2 - 12x + 30$
   subject to $x \geq 0$
  (ii) Minimize $z = 7x^2 + 84x + 50$
   subject to $x \geq 0$

4 (i) Minimize $z = x^3 - 13.5x^2 - 30x + 650$
   subject to $x \geq 0$
  (ii) Minimize $z = 0.75x^4 - 110x^3 + 1500x^2$
   subject to $x \geq 0$

5 For the function
$$f(x, y) = 100 - 300x + 40y + 10xy - 30x^2 - y^2$$
find the maximum value of $f(x, y)$ subject to the conditions that
   (i) $x \geq 0$
  (ii) $y \geq 0$
 (iii) $x \geq 0$ and $y \geq 0$
 (iv) both x and y unrestricted in sign

6 (i) Maximize $z = 500 + 150x - x^2$
   subject to $x \leq 50$
  (ii) Maximize $z = 1000 + 2000x - 8x^2$
   subject to $x \leq 100$
 (iii) Maximize $z = 1600x - 2x^2$
   subject to $x \leq 450$

7 (i) Find the maximum value of
$$z = 2500 - 528x + 45x^2 - x^3$$
   subject to $x \geq 0$ and $x \leq 20$
  (ii) Maximize $z = 97.5x^2 - 900x - x^3$
   subject to $x \geq 0$ and $x \leq 50$
 (iii) Maximize $z = 2x^3 - 105x^2 + 1500x - 1000$
   subject to $x \geq 0$ and $x \leq 30$

8 For the function
$$f(x, y) = 100 - x^2 - 2.5xy - 2y^2 + 10x + 2y$$
find
   (i) the unrestricted maximum
  (ii) the maximum subject to the constraint
$$2x + y = 20$$
 (iii) the maximum subject to the condition
$$2x + y = 20$$
and the sign requirements $x \geq 0$ and $y \geq 0$.

7.4 LAGRANGE MULTIPLIERS (II) – INEQUALITY CONSTRAINTS

In this section we show how the method of Lagrange multipliers can be used with weak inequality rather than strict equality constraints. In order to isolate the effect of inequality resource constraints, sign requirements will not be included until the next section. Weak inequalities are the most common form of limitation arising in business, financial and economic applications. Resource constraints are almost always 'less than or equal to' rather than 'strictly equal to' or, less commonly still, in the 'strictly less than' guise. In Chapter 3 we saw linear examples of weak inequality restrictions, and how linear programming problems in these constraints could be solved. Sign requirements and upper bounds on variables can themselves be seen as a form of weak inequality constraint. For example, in a problem involving the non-negative variables x_1 and x_2, the sign requirements could be written in the form

$$-1x_1 + 0x_2 \leq 0$$
$$0x_1 - 1x_2 \leq 0 \tag{7.11}$$

The conditions for optima with variables restricted both from above and from below will provide a route into the conditions required for more general inequality constraints. Consider a problem in a single variable (not sign restricted) with a single inequality constraint. This simplified structure will be used to identify optimality requirements in terms of first-order Lagrangian conditions. Suppose that the maximum of

$$F = 600 + 72x - 3x^2$$

is required subject to the constraint

$$4x \leq 40$$

F may be profit expressed in terms of a variable input, x, on which expenditure is limited to a maximum of 40. Set up the Lagrangian as for an equality constraint:

$$L = 600 + 72x - 3x^2 + j(40 - 4x)$$

for which

$$\frac{\partial L}{\partial x} = 72 - 6x - 4j$$

and

$$\frac{\partial L}{\partial j} = 40 - 4x$$

What are the requirements on these derivatives at an optimum? First it is

Constrained optimization with Lagrange multipliers 485

clear that, since $\partial L/\partial j$ is a rearrangement of the constraint, this derivative must be non-negative. But what of $\partial L/\partial x$? To begin an informal approach to the answer to this question, let us look at the sketch of the problem as shown in Figure 7.10.

From Figure 7.10, we can see that the problem is one of maximizing a function subject to an upper bound on the value of the decision variable. What happens to the function at the endpoint for x is crucial in determining whether this value produces a maximum, or whether an interior point will be optimal. Suppose that x was set at its endpoint value. The value of the function F at the endpoint could then be expressed in terms of the endpoint. The behaviour of this function as the endpoint varies can be found from its derivative. Let b represent the value of the right-hand side of the constraint (so with the original data $b = 40$); then at the endpoint $x = b/4$. Now substitute this value into F. The result is

$$F = 600 + 72(b/4) - 3(b/4)^2$$
$$= 600 + 18b - 3b^2/16$$

Differentiating,

$$\frac{\partial F}{\partial b} = 18 - \frac{6b}{16}$$

Equating $\partial F/\partial b$ to zero gives $b = 48$. This value of b maximizes F given x equal to $b/4$. If b is less than 48, $\partial F/\partial b$ will be positive, indicating an increase in F if the constraint is relaxed. The value of $\partial F/\partial b$ is the *rate* of increase of F given b. As seen in Chapter 6, the rate of increase of the maximum value of F as the constraint is relaxed is also given by the

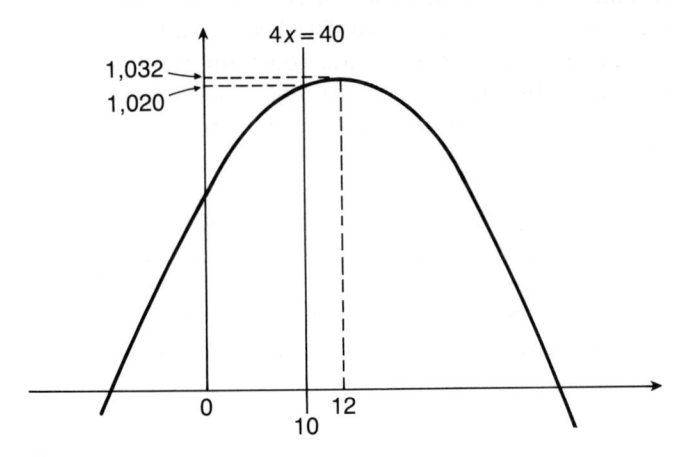

Figure 7.10

Lagrange multiplier j. Now consider various values of b. With $b = 48$ (i.e. $x = 12$)

$$\frac{\partial F}{\partial b} = 18 - \frac{6(48)}{16} = 0 = j$$

and, with $x = 12$ and $j = 0$,

$$\frac{\partial L}{\partial x} = 72 - 6(12) - 4(0) = 0$$

Now suppose that b is set at a value less than 48 – say $b = 40$ corresponding to $x = 10$. Then:

$$\frac{\partial F}{\partial b} = 18 - \frac{6(40)}{16} = 3 = j$$

and with $j = 3$ and $x = 10$

$$\frac{\partial L}{\partial x} = 72 - 6(10) - 4(3) = 0$$

In either case the derivative $\partial L/\partial x$ is equal to zero. *If* a value of b greater than 48 was selected and *if* it was required that $x = b/4$ with no option to satisfy the resource constraint as an inequality, then the resulting value of j would be negative, but still this value of j along with the corresponding value of $x = b/4$ would produce $\partial L/\partial x = 0$. With the proviso that there is the option to satisfy the resource constraint as an inequality, the minimum value of j will be zero, consistent with the fact that marginal variations in the right-hand side value would then have no effect on the optimum. In this case we are able to set $\partial F/\partial x = 0$ (corresponding to the interior and unconstrained optimum position) and, with j at zero, $\partial L/\partial x = 0$ in these circumstances too. So in all cases we may equate this derivative to zero.

How can it be determined whether the resource constraint should hold as an equality or an inequality? Introduce a *slack variable*[21] into the constraint to produce an equality for all values of x. We can now write

$$4x + s = b$$

in which the slack variable s represents any excess of resource availability (the constraint right-hand side, b) over the resource requirement ($4x$). For the original weak inequality constraint to be satisfied, the slack variable s must be non-negative. Recall that at the optimum the Lagrange multiplier j will be either positive or zero. With the inclusion of a slack variable, the conditions become

$$72 - 6x - 4j = 0$$
$$40 - 4x - s = 0 \qquad (7.12)$$

given the requirements that

$$s \geq 0 \quad \text{and} \quad j \geq 0$$

Conditions (7.12) have *two* equations in *three* unknowns, but the consequence of an infinity of possible solutions does not follow. This is because of an important relationship that exists between j and s at any optimum: *at least one of s and j will be zero.*[22] We have seen that j represents the value (the effect on the objective) of extra resource (increased b) at the margin. But such effect *must* be zero if at the optimum not all of the resource was used (i.e. if $s > 0$). On the other hand, if at the optimum $j > 0$ then it *must* be the case that $s = 0$. This is because if s was not zero and some resource therefore remained unused, with $j > 0$ it must have been possible to increase F by using more of the resource, in which case the original position could not have represented a maximum. *Both s and j could be zero* if the optimum was such that the turning point of F occurred at the boundary imposed by the constraint. In this event it is only a relaxation of the constraint (an increase of resource) that is correctly valued by j. Any tightening of the constraint (decrease of resource) would reduce F. Care is needed in interpreting the multipliers in such degenerate situations.

It is now required to find a solution to

$$72 - 6x - 4j = 0$$
$$b - 4x - s = 0$$
$$sj = 0$$
$$s \geq 0 \quad j \geq 0$$

in which b is the given value of the constraint right-hand side. Note that the product term sj makes this a *non-linear* system and that the sign requirements on s and j mean that linear inequalities as well as simultaneous equations are involved. In a problem of this size, a possible solution procedure is as follows.

1 Set up the Lagrangian, L.
2 Obtain the derivatives of L.
3 Set out the conditions including slack variable and sign requirements.
4 Try setting $s = 0$. If feasible values of x and j emerge, this will be a *possible* constrained optimum; if not, this position is ruled out of consideration.
5 Now try $j = 0$. If feasible values of x and s result, then record the positions as possible optima.
6 If more than one possibility is recorded under 4 and 5, the objective function is evaluated at the values of x in all the potential optima.

For an exhaustive process to be practicable, the list of possible optima must not be too long. Let us apply this enumerative approach to the present problem for three different values of the right-hand side:

(i) $b = 40$ (ii) $b = 56$ (iii) $b = 48$

(i) b = 40

First try $s = 0$. This means that

$$40 - 4x - 0 = 0$$

so that $x = 10$. Inserting $x = 10$ in the derivative condition produces

$$72 - 6(10) - 4j = 0$$

so that $j = 3$. All the conditions are therefore satisfied and the position is a possible optimum. The value of the objective function here is $F = 1020$. Now try $j = 0$. The x derivative condition is then

$$72 - 6x - 4(0) = 0$$

which solves for $x = 12$ and upon substitution into $\partial L/\partial j$ produces

$$40 - 4(12) - s = 0$$

which solves for $s = -8$. As this is not a feasible outcome, at $b = 40$ there is only one possibility for the optimal position:

$x = 10$ $j = 3$ $s = 0$ $F = 1020$

which must therefore be the optimum.

(ii) b = 56

Setting $s = 0$ in the resource constraint would require that

$$56 - 4x - 0 = 0$$

so $x = 14$. Insertion of this value of x into the $\partial L/\partial x$ equation results in

$$72 - 6(14) - 4j = 0$$

which solves for $j = -3\frac{2}{3}$ and rules out this position as a possible optimum. Next, trying $j = 0$ gives

$$72 - 6x - 4(0) = 0$$

which solves for $x = 12$. Insertion of this value for x into the second constraint produces

$$56 - 4(12) - s = 0$$

so $s = 8$. This is the only combination of values satisfying all requirements and must therefore be the optimum which, in full, is where

$x = 12$ $j = 0$ $s = 8$ $F = 1032$

(iii) $b = 48$

Trying $s = 0$,

$$48 - 4x - 0 = 0$$

so that $s = 12$ which, when inserted into the x derivative condition, produces

$$72 - 6(12) - 4j = 0$$

so that $j = 0$. This is a permissible outcome and produces the one candidate for optimality:

$$s = 0 \quad j = 0 \quad x = 12 \quad F = 1032$$

The reason that both s and j are zero here is that with the right-hand side set at 48 there is *just* sufficient resource to reach the unconstrained optimum. Note that although resource increases are of no value, any reduction in resource availability would cut into F.

The above procedure is summarized as follows. Given the single variable problem

$$\text{maximize } z(x)$$

subject to the requirement that

$$ax + c \leqslant b$$

in which a, b and c are real valued constants, the Lagrangian is

$$L(x) = z(x) + j(b - ax - c)$$

and conditions at the first order can be expressed as

$$z_x - aj = 0$$
$$b - ax - c - s = 0$$
$$sj = 0$$
$$s \geqslant 0 \quad j \geqslant 0$$

Solutions to these necessary conditions for a local constrained maximum are then sought, employing steps 1–6 above or by other means. We now extend this approach to two decision variables, and in so doing we shall broaden the form of constraint into one which does not translate directly into upper bounds. Consider the problem

$$\text{maximize } F = 80x - 2.5x^2 + 40y - 4y^2 + 2xy$$
$$\text{subject to } 4x + 2y \leqslant 80$$

The Lagrangian is
$$L = 80x - 2.5x^2 + 40y - 4y^2 + 2xy + j(80 - 4x - 2y)$$
and the first-order conditions are

(i) $\dfrac{\partial L}{\partial x} = 80 - 5x + 2y - 4j = 0$

(ii) $\dfrac{\partial L}{\partial y} = 40 - 8y + 2x - 2j = 0$

(iii) $\dfrac{\partial L}{\partial j} = 80 - 4x - 2y \geqslant 0$

In (iii) a slack variable can be added to give the requirement as an equality:
$$80 - 4x - 2y - s = 0$$
To these three conditions should be added
$$sj = 0$$
and the requirements that
$$s \geqslant 0 \qquad j \geqslant 0$$
Trying $s = 0$, it emerges from (iii) that
$$y = 40 - 2x$$
Substitution of this expression for y into condition (i) produces
$$80 - 5x + 2(40 - 2x) - 4j = 0$$
i.e.
$$160 - 9x - 4j = 0$$
while substitution for y into (ii) results in
$$40 - 8(40 - 2x) + 2x - 2j = 0$$
which in turn simplifies to
$$-280 + 18x - 2j = 0$$
and therefore the resulting two simultaneous equations in x and j are
$$18x - 2j = 280$$
$$9x + 4j = 160$$
which solve for
$$x = 16 \quad \text{and} \quad j = 4$$

with the consequence that $y = 8$ and, from substitution of the x and y values into the objective function, $F = 960$. This position is a possible optimum. Now try setting the Lagrange multiplier to zero. With $j = 0$, conditions (i) and (ii) require

$$5x - 2y = 80$$
$$-2x + 8y = 40$$

which solve for $x = 20$ and $y = 10$. Substituting these values into the constraint in equality form produces

$$80 - 4(20) - 2(10) - s = 0$$

so $s = -20$, indicating violation of the original constraint and the fact that the position is not feasible.[24] Thus the optimum is

$$x = 16 \quad y = 8 \quad j = 4 \quad F = 960$$

We have confined discussion to the case of a single linear constraint in one or two variables with a 'friendly' quadratic objective function. But these conditions can be broadened in more general cases involving several constraints, an increased number of variables and a wider range of objective functions and constraints. But the *solution* of the first-order conditions in larger and more complex problems is another matter. It is one thing to *describe* and interpret the solution in terms of the conditions that apply; it is quite another to *obtain* the values of variables in the solution.

EXERCISES 7.4

1 Find the maximum value taken by the function

$$z = x^3 + 7.5x^2 - 150x + 1000$$

subject to the constraint

$$5x + 40 \leqslant 100$$

2 Find the maximum value of

$$f(x, y) = 40x - 8x^2 + 4xy + 80y - 5y^2$$

subject to

$$2x + 4y \leqslant 40$$

3 Find the maximum value of

$$f(x, y) = 120x - 2.5x^2 + 3xy - 60y - 3.5y^2$$

subject to

$$3x + 2y \leqslant 35$$

4 Find the maximum value of

$$f(x, y) = 100x - 3x^2 + 4xy - 55y - 2.5y^2$$

subject to

$$2x + 5y \leqslant 90$$

7.5 LAGRANGE MULTIPLIERS (III) – INEQUALITY CONSTRAINTS WITH SIGN REQUIREMENTS

The considerations of Sections 7.3 and 7.4 can now be brought together. We shall look at the maximization of a quadratic function of two variables subject to a linear constraint and sign restrictions. Later, a second linear constraint will be added and a solution procedure will be discussed. First consider the problem

maximize $F = 100 + 8x + 16y - x^2 - y^2$
subject to $x + 2y \leqslant 15$
$\quad\quad x \geqslant 0 \quad\quad y \geqslant 0$

The Lagrangian is formed in the usual way,

$$L = 100 + 8x + 16y - x^2 - y^2 + j(15 - x - 2y)$$

and the first-order conditions can be written as

$L_x = 8 - 2x - j \leqslant 0$
$L_y = 16 - 2y - 2j \leqslant 0$
$L_j = 15 - x - 2y \geqslant 0$
$xL_x = 0 \quad\quad yL_y = 0 \quad\quad jJ_j = 0$
$x \geqslant 0 \quad\quad y \geqslant 0 \quad\quad j \geqslant 0$

The inequality form of the conditions for the derivatives with respect to the decision variables follows from the sign restrictions on the variables – as we saw in Section 7.3. The derivative with respect to the Lagrange multiplier, as ever, restates the constraint and so is also in inequality form. The required zero value for the products of the variables and their corresponding derivatives follows from the sign restrictions. Finally, the sign requirements themselves are restated. For solution purposes, the addition of slack variables makes a systematic search for solutions to the conditions somewhat easier. With sign restricted slack variables t and u introduced, the conditions can be written as

$$8 - 2x - j + t = 0$$
$$16 - 2y - 2j + u = 0$$
$$15 - x - 2y - s = 0$$
$$tx = 0 \quad uy = 0 \quad sj = 0$$
$$x \geqslant 0 \quad y \geqslant 0 \quad j \geqslant 0 \quad t \geqslant 0 \quad u \geqslant 0 \quad s \geqslant 0$$

Note that the requirement that the products of variables and their corresponding slacks should be zero has precisely the same implications as the derivative times variable form – in each case at least one member of the pairing must be zero. These are *complementary slackness* conditions. We shall now consider means by which feasible solutions may be found. In small problems it is possible to enumerate the combinations of variables to be set at zero. In the present case, with three pairs of linked variables, there will be $2^3 = 8$ combinations of variables with zero values. But in a problem with n variables and m constraints, the number of combinations that would have to be worked out under complete enumeration is 2^{m+n}, a value which soon assumes impractical proportions. For the current example, Table 7.5 shows the values of non-zero variables and the objective function value for each set of zero variables.

Note that in this problem there is only *one* feasible solution – the constrained optimum. Cases (b), (c), (d), (e), (f) and (h) all violate the sign requirements and are ruled out on these grounds. In addition, cases (b) and (f) violate the sign requirements in such a manner as to be infeasible in terms of the original resource constraint. Case (g) gives the requirement to equate 15 with zero. Although (g) is feasible in terms of x and y, the unrealizable equality means that the point does not produce a solution to

Table 7.5

Zero group	Solution values (non-zero)						Objective F	Comment
	x	t	y	u	s	j		
(a) t u s	3		6			2	175	
(b) t u j	4		8		−5		180	Infeasible
(c) t y s	15			−60		−22	−5	Signs
(d) t y j	4			−16	11		116	Signs
(e) x u s		−7.5	7.5			0.5	163.75	Signs
(f) x u j		−8	8		−1		164	Infeasible
(g) x y s		*		*		*	100	No solution
(h) x y j		−8		−16	15		100	Signs

494 Mathematics for business, finance and economics

the conditions and therefore cannot represent a constrained optimum. The remaining equations in case (g) are

$$8 - j + t = 0$$
$$16 - 2j + u = 0$$

for which, with three variables and two equations, there is no determinate solution. However, since both x and y are zero, any set of values for t, u and j here will correspond to a value of F equal to 100. Case (b) produces the unconstrained optimum. This is so since $j = 0$ and therefore the right-hand side of the constraint is not restricting the achievement of objective. Furthermore, since t and u are also zero, the derivatives with respect to x and y are simply the derivatives of F, and their satisfaction as equalities is a necessary condition for the unconstrained optimum. At the constrained optimum, case (a), the Lagrange multiplier j is positive ($s = 0$ and the resource constraint is satisfied as an equality). The value of j indicates the instantaneous rate of change of the maximum of F as the resource constraint is relaxed. It is a revealing exercise to match up the cases of Table 7.5 with a graph showing the kind of point to which the case corresponds. Figure 7.11 illustrates the location of the points as they each would be in a problem in which they were feasible. Also shown are the contours that would result in the point being optimal.[25]

Case (a) at the tangency point marked as T1 shows the optimum in the current problem with contours sloped appropriately. Case (b) shows how

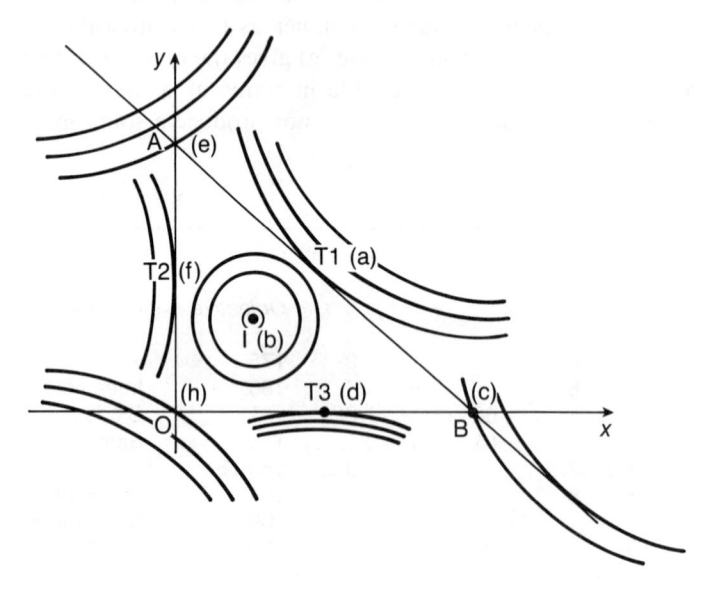

Figure 7.11

Constrained optimization with Lagrange multipliers   495

an interior optimum (corresponding to an unconstrained maximum) would appear. As evidenced by the negative value of s, in the current problem the (b) position lies outside the feasible region above and to the right of T1. Case (c) represents a corner point solution since two constraints (the resource constraint and the sign requirement on y) are satisfied as strict equalities. The corresponding straight lines must therefore intersect and result in a corner of the feasible region. Corner point optima can be seen as 'frustrated tangency' representing the nearest feasible position to a point of tangency which cannot be reached due to sign restrictions on the decision or slack variables.

Case (d) gives a point such as T3 on the diagram where the contours are tangential to an *axis*. In the case of T3, the axis is the x axis (the equation of which is $y = 0$). Case (e) is corner A of the feasible region. In the current problem the negativity of t means that this point cannot be the constrained optimum. Case (f) corresponds to an axis tangency point such as T2. In the current problem (not shown as such in the diagram) this tangency position lies above the corner A and thus is infeasible in terms of the constraint. This fact is evidenced by the value of $s = -1$. Case (h) is the origin. This is also a corner point which could be optimal but is not so in the current problem. Case (g) as we have seen, while feasible in terms of x and y, does not produce a solution to the conditions.

The information in Table 7.5 can be mobilized in another way. All the 'solutions'[26] to the conditions in which $t = 0$ are those for which x is not sign restricted (if $t = 0$, $L_x = 0$ as required for non sign restricted variables). As we can see, in all of the solutions in which $t = 0$ (i.e. (a)–(d) inclusive) x is at a positive level. We can therefore conclude that the imposition of a sign restriction on x has no effect in this particular problem. Similarly, the solutions in which $u = 0$ show the effects of the removal of the sign restriction on y – these are the zero group combinations (a), (b), (e) and (f). Again, in this problem, the sign requirement on y makes no difference, as in all of (a), (b), (e) and (f) y takes a positive value.

Solutions in which $j = 0$ correspond to the non-appliance of the constraint. So if we also want the best sign restricted solution without the resource constraint, Table 7.5 should be inspected for the best, feasible solution that emerges from the four combinations involving $j = 0$, i.e. (b), (d), (f) and (h). The unconstrained optimum (b) is achievable in this case. The difference in the value of the objective function between (b) and (a) is $180 - 175 = 5$. This difference is attributable to the constraint, and represents the maximum price worth paying for its removal as an effective restriction. Now consider a further example. Find the maximum of

$$f(x, y) = 180x + 5xy - 4x^2 - 100y - 2y^2$$
subject to $5x + 4y \leqslant 100$
$$x \geqslant 0 \quad y \geqslant 0$$

The Lagrangian is

$$L = 180x + 5xy - 4x^2 - 100y - 2y^2 + j(100 - 5x - 4y)$$

and the necessary conditions are

$$L_x = 180 + 5y - 8x - 5j \leqslant 0$$
$$L_y = 5x - 100 - 4y - 4j \leqslant 0$$
$$L_j = 100 - 5x - 4y \geqslant 0$$
$$xL_x = 0 \qquad yL_y = 0 \qquad jL_j = 0$$
$$x \geqslant 0 \qquad y \geqslant 0 \qquad j \geqslant 0$$

With slack variables, these conditions can be expressed as

$$180 + 5y - 8x - 5j + t = 0$$
$$5x - 100 - 4y + u = 0$$
$$100 - 5x - 4y - s = 0$$
$$tx = 0 \qquad uy = 0 \qquad sj = 0$$
$$x \geqslant 0 \quad y \geqslant 0 \quad j \geqslant 0 \quad t \geqslant 0 \quad u \geqslant 0 \quad s \geqslant 0$$

Table 7.6 shows the results (to two decimal places) of the eight combinations of zero valued variables. Of these cases, (c) alone satisfies the conditions and produces the constrained maximum of $f(x,y)$. So the solution is

$$x = 20 \qquad y = 0 \qquad j = 4 \qquad f(x,y) = 2000$$

In this problem, the constrained maximum is a corner point.[27] Case (a) would be the best position if the sign requirement on y was dropped. The difference in the value of the objective function between (a) and (c) is interpretable as the cost of the sign restriction on y, given the constraint and

Table 7.6

Zero group	x	t	y	u	s	j
(a) t u s	20.75		−0.93			1.87
(b) t u j	31.43		14.29		−114.29	
(c) t y s	20			16		4
(d) t y j	22.5			−12.5	−12.5	
(e) x u s		−55	25			−50
(f) x u j		−55	−25		200	
(g) x y s			No solution			
(h) x y j		−180		100	100	

its current right-hand side value. Position (b) gives the unconstrained maximum of the function where

$$f(x,y) = \frac{14{,}800}{7} \approx 2114.29$$

Since both x and y happen to be non-negative at this point, the difference between the objective function values at (b) and (c), 114.29, can be seen as the cost of the constraint. Examination of those combinations of zero variables in Table 7.6 in which t is required to be zero shows that the removal of the sign restriction on x would make no difference to the optimum. However, the removal of the sign restriction on y would allow point (a) to be achieved. At this point $f(x, y) = 2007.48$, so that the cost of this sign requirement (given the level of the resource constraint) is 7.48.

In the next section, the discussion is taken a stage further. The conditions for constrained optimization involving differentiable functions of several variables are stated.

EXERCISES 7.5

1 Given the function

$$f(x, y) = 32x - x^2 + 16y - y^2 + 150$$

(i) Find the maximum value of the function subject to the constraint

$$2x + y \leqslant 30$$

and the sign requirements

$$x \geqslant 0 \qquad y \geqslant 0$$

(ii) What improvement in $f(x, y)$ would be possible if the constraint was removed?

2 For the function

$$f(x, y) = 63x - 0.8x^2 - 35y - 0.4y^2 + xy$$

(i) Find the maximum value of the function subject to the constraint

$$x + 0.8y \leqslant 35$$

and the sign requirements

$$x \geqslant 0 \qquad y \geqslant 0$$

(ii) What improvement in $f(x,y)$ would be possible if the constraint was removed? What, therefore, may be said to be the 'cost' of the constraint to the decision-maker?
(iii) With the constraint still in place, what improvement in the function $f(x,y)$ would be possible if the sign requirement on x was removed?
(iv) With the constraint and the sign requirement on x still in place, what improvement in $f(x,y)$ could be gained if it was possible to remove the sign requirement on y?

7.6 THE KUHN–TUCKER CONDITIONS

The problems studied so far can be seen as special cases of the general conditions towards which we have been working. Objective functions are not always quadratic and constraints, while taking the form of weak inequalities, will not always be linear. But if the objective function and the left-hand side of the constraints are differentiable, and certain regularity conditions are satisfied, then conditions developed below will apply. The constraints should not produce singularities such as an outward-pointing cusp at the optimal point. *Constraint qualification conditions* could be introduced to provide against such situations.[28] We shall present optimality conditions in the context of a two-variable, two-constraint model, but the conditions generalize to problems with more variables and constraints. But numerical examples will be confined to problems with quadratic objective functions and linear constraints. The two-variable, two-constraint problem can be expressed in the following way:

$$\begin{aligned}
&\text{maximize } F = F(x_1, x_2) \\
&\text{subject to } g_1(x_1, x_2) \leqslant b_1 \\
&\phantom{\text{subject to }} g_2(x_1, x_2) \leqslant b_2 \\
&\phantom{\text{subject to }} x_1 \geqslant 0 \quad x_2 \geqslant 0
\end{aligned} \qquad (7.13)$$

System (7.13) is the *non-linear programming* model.[29] Necessary conditions are obtained by forming the Lagrangian:

$$L = F(x_1, x_2) + j_1 [b_1 - g_1(x_1, x_2)] + j_2 [b_2 - g_2(x_1, x_2)]$$

At the first order:

$$\frac{\partial L}{\partial x_1} = \frac{\partial F}{\partial x_1} - j_1 \frac{\partial g_1}{\partial x_1} - j_2 \frac{\partial g_2}{\partial x_1} \leqslant 0$$

$$\frac{\partial L}{\partial x_2} = \frac{\partial F}{\partial x_2} - j_1 \frac{\partial g_1}{\partial x_2} - j_2 \frac{\partial g_2}{\partial x_2} \leqslant 0$$

$$x_1 \frac{\partial L}{\partial x_1} + x_2 \frac{\partial L}{\partial x_2} = 0$$

$$x_1 \geqslant 0 \qquad x_2 \geqslant 0$$

$$\frac{\partial L}{\partial j_1} = b_1 - g_1(x_1, x_2) \geqslant 0 \qquad (7.14)$$

$$\frac{\partial L}{\partial j_2} = b_2 - g_2(x_1, x_2) \geqslant 0$$

$$j_1 \frac{\partial L}{\partial j_1} + j_2 \frac{\partial L}{\partial j_2} = 0$$

$$j_1 \geqslant 0 \qquad j_2 \geqslant 0$$

These conditions characterize local maxima of F subject to the constraints and sign requirements, and are known as the *Kuhn–Tucker* or *Karush–Kuhn–Tucker* conditions.[30] Excepting constraints producing extraordinary circumstances at endpoints, the Kuhn–Tucker conditions are *necessary* for a local maximum, but are not in general *sufficient* conditions.[31] However, if a solution procedure identifies all points satisfying the Kuhn–Tucker conditions then, at least in principle, the global maximum can be found by running the points through the objective function. Finding *any* solutions to such conditions can be a major task. There may well be a large, possibly even infinite, number of solutions to the conditions, in which case enumeration cannot be used. But at least the conditions can check the credentials of a claimed optimum – the point *must* satisfy the Kuhn–Tucker conditions. Possibly limited usefulness at a high generality does not condemn an approach. It is likely that other ways of seeking exact, global optima will also be difficult to use. Simpler circumstances can often be assumed. For example, if

1 the objective function is concave,
2 the constraint functions g are convex,

then the Kuhn–Tucker conditions are both necessary and sufficient for a global optimum. This means that 'all' that is required is a feasible solution to the system (7.14). Fortunately in economic and business models it is often acceptable to assume 1 and 2. Such problems are called *convex programming* problems.

The Kuhn–Tucker conditions are necessary for a maximum of the original function F, subject to the constraints. This does not correspond to a maximum of the Lagrangian function in all directions. The constrained optimum of F represents a *saddle point* of the Lagrangian function, being a maximum in the x_i directions and a minimum with respect to the multipliers. This is shown up by the required signs for the partial derivatives. Those with respect to the x variables must be non-positive while the derivatives with respect to the multipliers must be non-negative. Let us now see how the Kuhn–Tucker conditions turn out in some simple cases. In an earlier section, we considered single variable problems with an upper and lower (zero) bound. We can now set out the required conditions as a special case of the Kuhn–Tucker conditions. For the problem

maximize $z(x)$
subject to $x \geq 0$ and $x \leq b$

The Lagrangian is

$$L = z(x) + j(b - x)$$

and the conditions are

$$L_x = \frac{dz}{dx} - j \leq 0$$

$$xL_x = 0$$

$$x \geq 0$$

$$L_j = b - x \geq 0$$

$$jL_j = 0$$

$$j \geq 0$$

Note that the product of x and L_x is required to be zero and not $x\,dz/dx$. Now consider a numerical example. Suppose that the problem is

maximize $F = x^2 - 10x + 50$
subject to $x \leq 8$ $x \geq 0$

The Lagrangian and Kuhn–Tucker conditions are

$$L = x^2 - 10x + 50 + j(8 - x)$$

$$L_x = 2x - 10 - j \leq 0$$

$$xL_x = 0$$

$$x \geq 0$$

$$L_j = 8 - x \geq 0$$

$$jL_j = 0$$

$$j \geq 0$$

Constrained optimization with Lagrange multipliers

and, with slack variables introduced in the complementary slackness form of these conditions, it is required that

$$2x - 10 - j + t = 0$$
$$xt = 0$$
$$x \geqslant 0 \quad t \geqslant 0$$
$$8 - x - s = 0$$
$$js = 0$$
$$j \geqslant 0 \quad s \geqslant 0$$

Enumeration produces the results shown in Table 7.7.

The constrained maximum here occurs at the $x = 0$ endpoint. The interior turning point (ii) is a *minimum*, the objective function in this case being convex. The problem is graphed in Figure 7.12.

Table 7.7

Zero group	Solution values (non-zero)		Objective
(i) t s	$x = 8$	$j = 6$	$F = 34$
(ii) t j	$x = 5$	$s = 3$	$F = 25$
(iii) x s	No solution		
(iv) x j	$t = 10$	$s = 8$	$F = 50$

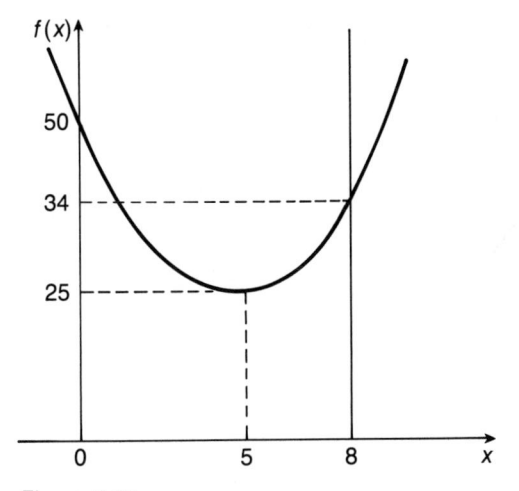

Figure 7.12

502 Mathematics for business, finance and economics

Now consider a different objective function maximized over the same feasible region. The problem is to

maximize $F = 50 - x^2 + 10x$
subject to $x \leqslant 8$
$x \geqslant 0$

The necessary first-order conditions in complementary slackness form are

$-2x + 10 - j + t = 0$
$xt = 0$
$x \geqslant 0 \qquad t \geqslant 0$
$8 - x - s = 0$
$js = 0$
$j \geqslant 0 \qquad s \geqslant 0$

The outcome of enumeration is set out in Table 7.8. In this case the constrained maximum occurs at the interior point (ii) as shown in Figure 7.13.

Table 7.8

Zero group	Solution values (non-zero)		Objective
(i) t s	x = 8	j = −6	F = 66
(ii) t j	x = 5	s = 3	F = 75
(iii) x s	No solution		
(iv) x j	t = −10	s = 8	F = 50

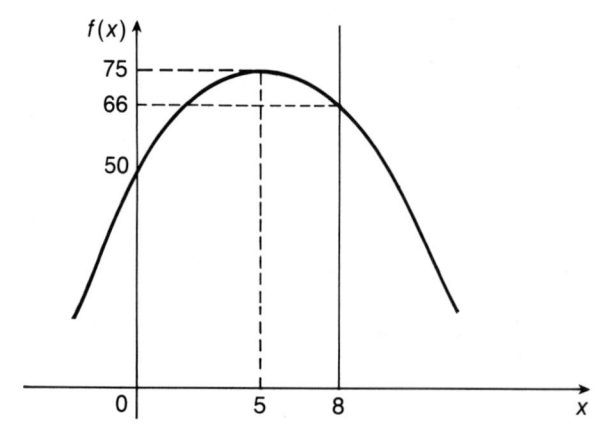

Figure 7.13

One of the advantages of using enumeration where practicable is that significant business decisions are often taken by an executive or committee on the basis of professional advice. Quite frequently in this process the board is presented with a list of possibilities from which one is to be selected and acted upon. This will not *inevitably* be the course of action showing the greatest value of the objective function (although this option *should* be on the list) as wider considerations may need to be taken into account. The enumeration approach has the advantage of providing a comprehensive listing from which the options may be drawn.

As we have seen, some of the positions highlighted by enumeration which do not satisfy the Kuhn–Tucker conditions provide better values of the objective function than some points for which the conditions are satisfied. This allows a value to be placed on the relaxation or removal of binding constraints. Such points may not show up as well in a procedure concentrating exclusively on points fulfilling the Kuhn–Tucker conditions. Now consider a more substantial problem in two decision variables and in which there are two linear constraints and a quadratic objective function. The problem is

maximize $F = 10x - 0.1x^2 + 20y - 0.2y^2$
subject to $5x + y \leqslant 200$
$x + 2y \leqslant 90$
$x \geqslant 0 \qquad y \geqslant 0$

Before solving this problem, we will solve the *unconstrained* problem to gain a point of reference. We will then look at the feasible region, discuss the possible locations of the optimal solution to the constrained problem, and set out the Kuhn–Tucker conditions. In the unconstrained case, the first-order conditions are

$F_x = 10 - 0.2x = 0$
$F_y = 20 - 0.4y = 0$

which solve for

$x = 50, \qquad y = 50 \qquad$ and therefore $\qquad F = 750$

As may be confirmed, the second order conditions for a maximum are satisfied at this position. Now let us look at the constraints. Since the constraints are linear, the feasible region will be as for a linear programming problem and is shown in Figure 7.14.

For an arbitrary objective function, the possible positions of the optimum within the feasible region are the corner points O, A, B and C, a point of tangency between a contour of the objective function and constraint 1 (shown as T1 in the figure), a point of tangency between a contour and constraint 2 (shown as T2), tangencies with the axes corresponding to T3

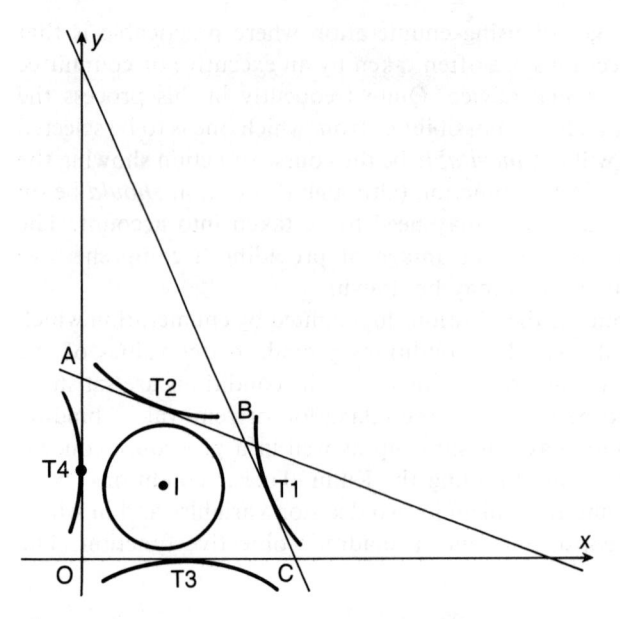

504 Mathematics for business, finance and economics

Figure 7.14

and T4 and an interior optimum marked as I. Not all of these possibilities relate to the objective function in the current example. The tangency position for one constraint may be infeasible in terms of the other and the tangencies within the axes may lie outside of the feasible region. Now set up the Lagrangian and obtain the Kuhn–Tucker conditions. With two constraints, two multipliers will be required, but the flexibility of the formulation easily accommodates this requirement. The Lagrangian and Kuhn–Tucker conditions are

$$L = 10x - 0.1x^2 + 20y - 0.2y^2 + j(200 - 5x - y) + k(90 - x - 2y)$$
$$L_x = 10 - 0.2x - 5j - k \leqslant 0$$
$$L_y = 20 - 0.4y - j - 2k \leqslant 0$$
$$xL_x = 0 \qquad yL_y = 0$$
$$x \geqslant 0 \qquad y \geqslant 0$$
$$L_j = 200 - 5x - y \geqslant 0$$
$$L_k = 90 - x - 2y \geqslant 0$$
$$jL_j = 0 \qquad kL_k = 0$$
$$j \geqslant 0 \qquad k \geqslant 0$$

Constrained optimization with Lagrange multipliers

In complementary slackness form, the conditions produce the following equations, of which the optimum will be a feasible solution:

$$10 - 0.2x - 5j - k + t = 0$$
$$20 - 0.4y - j - 2k + u = 0$$
$$200 - 5x - y - s = 0$$
$$90 - x - 2y - r = 0$$
$$tx = 0 \quad uy = 0 \quad sj = 0 \quad rk = 0$$
$$x, t, u, y, s, j, r, k \geq 0$$

One approach to solving these equations is enumeration. This would be a fairly time consuming process, but could be organized around the $2^4 = 16$ sets of values of the zero variables using a tabular layout similar to those above. The optimal solution that emerges is

$$x = 30 \quad y = 30 \quad k = 4 \quad s = 20$$
$$t = 0 \quad u = 0 \quad r = 0 \quad j = 0$$

These values produce a constrained optimal value of the objective function of $F = 630$. The location of the optimal solution, a tangency position in this problem, is shown in Figure 7.15.

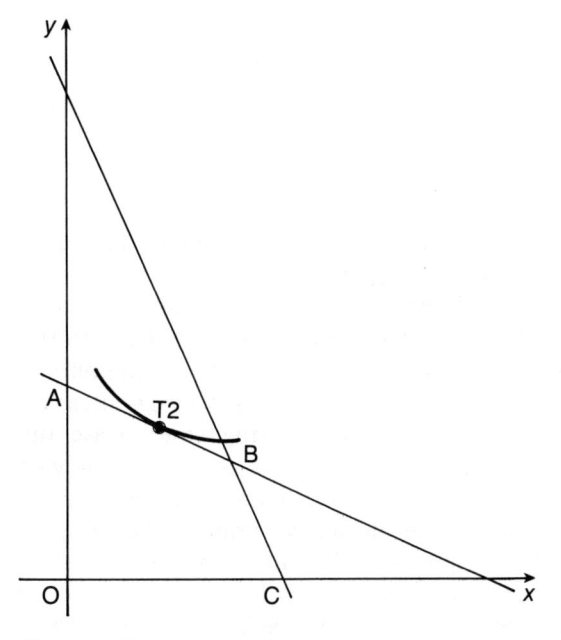

Figure 7.15

The tangency position is with constraint 2 and occurs strictly below constraint 1 (corresponding to $j = 0$). The implication of this is that even though the amount of resource 1 is less than the amount needed to reach the unconstrained optimum, further units of resource 1 are *not* worth having *without* a significant increase in resource 2. This is important information and is a situation which could not have occurred in linear programming where the objective function as well as the constraints are linear. Incidentally, the (corner) solution at the intersection of the resource constraints corresponding to full utilization of both resources (i.e. $s = 0$ and $r = 0$) would produce a value of F of 627.04. Now consider a further example:

$$\text{maximize } f(x, y) = 120x - x^2 + 200y - 2y^2$$
$$\text{subject to } x + 2y \leqslant 130$$
$$2x + y \leqslant 152$$
$$x \geqslant 0 \quad y \geqslant 0$$

The Lagrangian is

$$L = 120x - x^2 + 200y - 2y^2 + j(130 - x - 2y) + k(152 - 2x - y)$$

for which conditions at the first order, in complementary slackness form, are

$$120 - 2x - j - 2k + t = 0$$
$$200 - 4y - 2j - k + u = 0$$
$$130 - x - 2y - s = 0$$
$$152 - 2x - y - r = 0$$
$$tx + uy + js + kr = 0$$
$$x, y, t, u, j, k, r, s \geqslant 0$$

The sixteen combinations of zero-valued variables and the resulting values of the other variables are shown in Table 7.9.

The only complete solution to the conditions, and the constrained optimum, is case (ii), which results in a value for $f(x, y)$ of 8300. In this problem, the optimal position is a point of tangency between contours of the objective function and the first constraint. Amongst the sign-infeasible 'solutions' of Table 7.9, there are three noteworthy cases. Case (iii) represents the tangency position of contours of the objective function with the second constraint. The fact that this gives a resource consumption requirement that is infeasible in terms of the first constraint is shown by the negative value of s. Case (i) corresponds to the full use of both resources – the constraints are satisfied as strict equalities. Case (iv) is the unconstrained, global, optimum. The problem is graphed in Figure 7.16, in which the optimum is indicated as point T.

Constrained optimization with Lagrange multipliers 507

Table 7.9

Zero group	Solution values (non-zero)							
	x	t	y	u	j	s	k	r
(i) tusr	58		36		36		−16	
(ii) tusk	50		40		20			12
(iii) tujr	52		48			−18	8	
(iv) tujk	60		50			−30		−18
(v) tysr				No solution				
(vi) tysk	130			−480	−140			−108
(vii) tyjr	76			−216		54	−16	
(viii) tyjk	60			−200		70		32
(ix) xusr				No solution				
(x) xusk		−150	65	−30				87
(xi) xujr		−1036	152			−174	−408	
(xii) xujk		−120	50				30	102
(xiii) xysr				No solution				
(xiv) xysk				No solution				
(xv) xyjr				No solution				
(xvi) xyjk		−120		−200		130		152

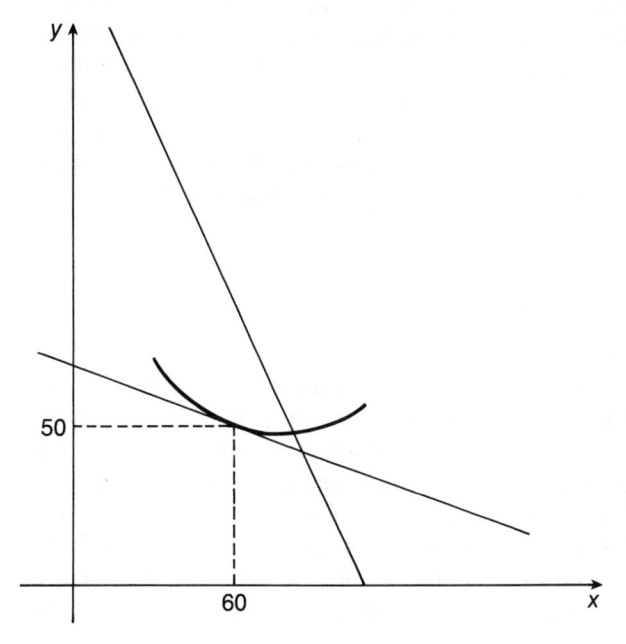

Figure 7.16

An enumeration approach nears its practical limit in the case of two sign restricted variables and two linear constraints. For larger problems involving linear constraints, if a quadratic objective function is retained, a solution can be found using *quadratic programming*,[32] for which there are solution procedures based on the simplex method. In the quadratic programming approach (e.g. in the method due to Wolfe) the first-order conditions in complementary slackness form and the sign requirements together make up the *constraints*. A means can be found within this framework to provide for the complementary slackness conditions *proper* (i.e. $xt = 0$, $yu = 0$ etc.). In the quadratic programming approach, a modified form of objective is used solely to enable a solution to the Kuhn–Tucker conditions to be found. This can be done on the basis of the simplex method. For problems involving more general forms of objective function or constraints, methods such as the gradient search procedure[33] would be necessary. But as with linear programming, non-linear programming problems do not have to be solved manually. Personal computer software enables problems of quite significant size to be solved. For example, the accompanying software package to Hillier and Lieberman (1990) can solve problems with up to six linear constraints, three decision variables and a polynomial objective function in which the exponent for each variable in each term is a non-negative integer up to 9. At the start of this section, we mentioned that the Kuhn–Tucker conditions apply so long as the binding constraints at the optimum meet certain requirements. Consideration of constraint qualification conditions is beyond the scope of this text, but we can illustrate the kind of circumstances where the Kuhn–Tucker conditions do *not* apply by reference to an example produced by Kuhn and Tucker themselves (see Kuhn and Tucker, 1975). The problem was this:

maximize x
subject to $y - (1-x)^3 \leq 0$
$\quad\quad\quad x \geq 0 \quad\quad y \geq 0$

The problem is graphed in Figure 7.17.

The feasible region for this problem is the (non-convex) area OAB. As the variable y does not enter into the objective function, the contours are vertical lines. The optimum is located at point B where $x = 1$, $y = 0$ and $F = 1$. Note the sharp cusp at the optimum where the constraint boundary has a stationary point of inflection at the point of contact (and tangency) with the x axis. This is precisely the kind of situation in which the Kuhn–Tucker conditions do not apply, and which would be identified by failure to meet a constraint qualification condition. The Lagrangian function is

$L = x + j[(1-x)^3 - y]$

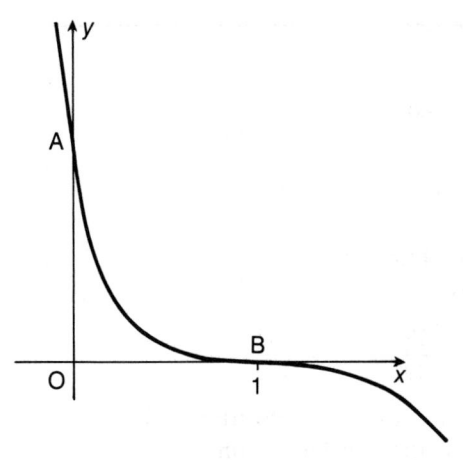

Figure 7.17

and the Kuhn–Tucker conditions (in complementary slackness form) are

$$1 - 3j(1-x)^2 + t = 0$$
$$-j + u = 0$$
$$xt = 0 \quad yu = 0$$
$$x \geqslant 0 \quad y \geqslant 0 \quad t \geqslant 0 \quad u \geqslant 0$$
$$(1-x)^3 - y - s = 0$$
$$sj = 0$$
$$j \geqslant 0 \quad s \geqslant 0$$

The point to note about these conditions is that they are not satisfied at the optimum. If $x = 1$, then by complementary slackness $t = 0$, and insertion of these values into the derivative condition would result in the unfulfillable requirement to equate zero and one.

The general non-linear programming problem is an n-variable, m-constraint version of the model (7.13). While the Kuhn–Tucker conditions characterize the exact solution to these problems, there is no general algorithm that will guarantee to find the global optimum in all such problems. Algorithms *do* exist for a number of particular cases (amongst which are included linear programming and quadratic programming) and there are procedures which give good approximate solutions.[34]

EXERCISES 7.6

1 For the following problem:

$$\text{maximize } f(x,y) = 40x - x^2 + 60y - 2y^2$$
$$\text{subject to } 5x + 4y \leq 120$$
$$10x + y \leq 140$$
$$x \geq 0 \quad y \geq 0$$

(i) Set out the Kuhn–Tucker conditions in full.
(ii) Confirm that the set of values

$$x = \frac{88}{7} \quad y = \frac{100}{7} \quad j = \frac{192}{490} \quad k = \frac{632}{490}$$

satisfies the conditions. Verify that this position, which is the optimum, is a corner point of the feasible region.
(iii) What increase in the value of $f(x,y)$ would be possible if the constraints were completely removed?

2 Set out the Kuhn–Tucker conditions for the following problem:

$$\text{maximize } f(x,y) = 12{,}000x - 0.2x^6 + 2{,}000{,}000y - y^8$$
$$\text{subject to } 3x + 7y \leq 21$$
$$2x^2 + y \leq 20$$
$$x \geq 0 \quad y \geq 0$$

3 For the problem:

$$\text{maximize } f(x,y) = 100x - 2x^2 + 60y - y^2$$
$$\text{subject to } 2x + y \leq 65$$
$$0.5x + y \leq 38$$
$$x \geq 0 \quad y \geq 0$$

(i) Give the Kuhn–Tucker conditions in complementary slackness form.
(ii) Find the constrained optimum.
(iii) What is the unconstrained optimum? What, therefore, could be said to be the cost of the constraints as a whole?

7.7 ECONOMIC APPLICATION – MULTI-PRODUCT MONOPOLY

A model in which an extreme value of a quadratic function is sought subject to linear restrictions on decision variables is that of the *multi-product monopolist.* Suppose that a firm makes two products, for which the demand conditions are given by

$$p(x) = 75 - x$$
$$p(y) = 74 - 2y$$
(7.15)

where x and y are the quantities produced, and where $p(x)$ and $p(y)$ are the respective prices. The products have *independent demand functions* as shown in the demand relationships[35] of (7.15); i.e. there is no x term in the equation for $p(y)$ nor is there a y term in the equation for $p(x)$. We shall assume a linear technology although a quadratic programming approach requires only the weaker condition that the products are *technically independent*[36] and give rise to quadratic or linear cost functions. The linkage between the decisions on production levels occurs through the common consumption of scarce resources. Specifically, suppose that two resources are in restricted supply. Table 7.10 gives details of maximum resource availability, given resource prices and constant per unit resource requirements for the two products.

From the data of Table 7.10, the costs of production for each product can be found. Each unit of x requires 5 units of resource 1 at £3 per unit and 10 units of resource 2 at £2 per unit, so

unit cost of $x = 5(3) + 10(2) = 35$

Similarly, for y,

unit cost of $y = 4(3) + 1(2) = 14$

Thus for any values of x and y the production costs for the two products will be

$$C(x) = 35x \quad \text{and} \quad C(y) = 14y$$

The demand conditions imply that the revenues from the products, $R(x)$ and $R(y)$, will be given by

$$R(x) = px = 75x - x^2$$

and

$$R(y) = py = 74y - 2y^2$$

Table 7.10

	Availability	Requirements x	y	Price per unit
Resource 1	120	5	4	£3
Resource 2	140	10	1	£2

The unit profits from the two products, $\pi(x)$ and $\pi(y)$, will be

$$\pi(x) = R(x) - C(x) = 40x - x^2$$

and

$$\pi(y) = R(y) - C(y) = 60y - 2y^2$$

The firm wishes to select values of x and y so as to maximize overall profit from the two products, F, where

$$F = \pi(x) + \pi(y)$$

Given that negative production levels are meaningless, and that there are limits to the usages of resource, the firm's problem is to set the decision variables x and y so as to

maximize $F = 40x - x^2 + 60y - 2y^2$
subject to $5x + 4y \leqslant 120$
$10x + y \leqslant 140$
$x \geqslant 0 \quad y \geqslant 0$

which is a problem of the form that we have been examining. The Lagrangian is

$$L = 40x - x^2 + 60y - 2y^2 + j(120 - 5x - 4y) + k(140 - 10x - y)$$

and the Kuhn–Tucker conditions in complementary slackness form are

$$40 - 2x - 5j - 10k + t = 0$$
$$60 - 4y - 4j - k + u = 0$$
$$120 - 5x - 4y - s = 0$$
$$140 - 10x - y - r = 0$$
$$tx = 0 \quad uy = 0 \quad sj = 0 \quad rk = 0$$
$$x, t, u, y, s, j, r, k \geqslant 0$$

The following values for the variables give a solution to the Kuhn–Tucker conditions, and therefore represent the constrained optimum:

$$x = \frac{88}{7} \approx 12.57 \qquad y = \frac{100}{7} \approx 14.29$$

and

$$j = \frac{192}{490} \approx 0.39 \qquad k = \frac{632}{490} \approx 1.29$$

$$t = 0 \quad u = 0 \quad s = 0 \quad r = 0$$

Both resources are used to the full, and the optimal solution in this case is the corner point given by the two resource constraints as equalities. The value of the objective function at the optimum is

$$F = \frac{38,896}{49} \approx 793.80$$

For comparative purposes, the unconstrained maximum occurs when

$$x = 20 \qquad y = 15 \qquad F = 850$$

The full 'cost' imposed by the constraints as a whole is therefore

$$850 - 793.8 = 56.2$$

The problem is graphed in Figure 7.18, in which the constrained optimum is point P and the unconstrained optimum is U.

As the objective function in this case is strictly concave, the values of the Lagrange multipliers, j and k, are *upper bounds* on the increases in the objective function following unit increases in the resource availabilities. So the Lagrange multipliers can be used to *rule out*, on grounds of cost, specific proposals to increase resource levels. For example, suppose that there was the opportunity to take five further units of the second resource for £17, i.e. at a cost of £7 over and above the normal cost of £10. This opening would not be taken up, since the increase in profit if the resources were obtained at the original prices is less than the cost premium for the additional units. That is,

$$5\frac{632}{490} = 6.45 < 7$$

But a proposal to secure the extra resources for £15.5, i.e. at a total premium of £5.5, could neither be rejected nor accepted on the basis of the value of k. Further investigation is required in this case.[37] The actual changes in F following unit increases in one resource alone are as follows:

resource 1 → 121 $\Delta F = \dfrac{279}{1225} \approx 0.2278$

resource 2 → 141 $\Delta F = \dfrac{1514}{1225} \approx 1.2359$

It can be shown that, at the optimum in this problem, each Lagrange multiplier is an increasing function of the level of the *other* resource.[38] This effect, which cannot occur under linear programming, becomes evident when a quadratic programming approach is used. It can also be shown that, as the result of the equality of second-order cross partial derivatives, each multiplier is affected in the same (linear) way by the level of the other

y↑

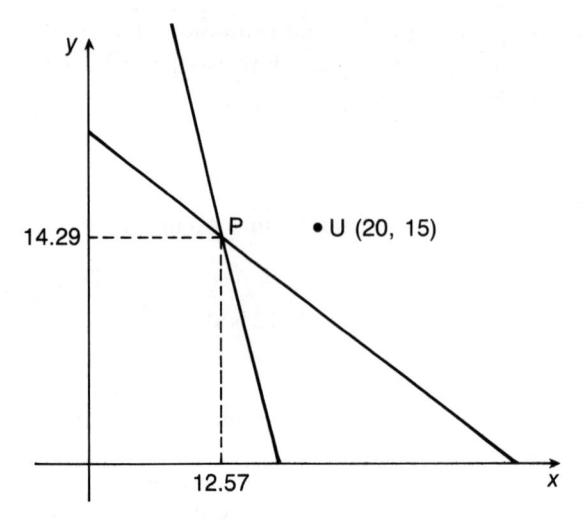

14.29 ┆-------- P   • U (20, 15)

           12.57   x

Figure 7.18

resource. If either resource is increased by one unit, the multiplier for the other resource increases by the amount:

$$\frac{83.2}{490} \approx 0.17$$

Outside of linear optimization problems the marginal valuation of a resource will usually be linked to the levels of other scarce resources.

EXERCISES 7.7

1 A two-product firm faces demand conditions given by the following equations:

$$p(x) = 95 - 0.5x$$
$$p(y) = 85 - 0.5y$$

in which x and y are the quantities produced and sold at prices $p(x)$ and $p(y)$. The following resource constraints must also be observed:

$$x + y \leqslant 100 \text{ (resource 1)}$$
$$x + 2y \leqslant 140 \text{ (resource 2)}$$

Each unit of resource 1 costs £5 and each unit of resource 2 costs £10. Negative production levels are meaningless. The firm's objective is the maximization of profit.

(i) Set out a constrained optimization model of the problem.

(ii) Find the optimal level of production for each good and the maximum profit made.

7.8 CONCLUDING REMARKS

There is a great deal more to the subject of constrained optimization than considered here. For example, solution procedures for non-linear programming problems are beyond the scope of this book. We have focused on the characteristics of optimal solutions, and stressed the value of interpretation of equilibrium conditions in business applications. We progressively developed discussion towards the Kuhn–Tucker conditions, but in so doing noted the interest of particular cases. Insight can also be gained from the solutions to numerical examples and we have used complete enumeration, essentially a 'non-method' of solution, to obtain results, to enable consideration of optimal solutions and other solutions and to assess the impact of the constraints.

The structure of constrained optimization problems is characteristic of a wide range of business, economic and financial problems. Important though it is, it is not simply the *solution*, or the method of solution, of constrained optimization problems that is significant – there is considerable value in knowing the equilibrium conditions to a problem. The conditions can frequently be given contextual interpretation and current values of decision variables can be checked against the conditions to establish whether or not the position is optimal.[39] An understanding of the meaning of equilibrium conditions in particular applications can assist this process and can give valuable insight in its own right.

ADDITIONAL PROBLEMS

1 Given the objective function

$$z = 240x + 1180y + 2xy - 4x^2 - 8y^2$$

in which x and y are unrestricted in sign,

(i) find the optimal values of x, y, the Lagrange multiplier, and the maximum value of z given the constraint

$$4x + y = 300$$

(ii) Find the optimal values of x, y, the Lagrange multiplier, and the maximum value of z given the constraint

$$4x + y = 400$$

Compare the value of the objective function with that achieved under (i). What does the value of the Lagrange multiplier tell you in this case?

(iii) Find the unconstrained maximum of the function and the corresponding best value to which the right-hand side of the constraint could have been set.

2 Given the objective function

$$z = 600x - x^2 + 1440y - y^2$$

in which x and y are unrestricted in sign,

(i) find the maximum value of the function subject to the constraint

$$x + y \leqslant 600$$

(ii) Find the maximum value of the function subject to the constraint

$$x + 2y \leqslant 936$$

(iii) Find the maximum value of the function subject to the pair of constraints:

$$x + y \leqslant 600$$
$$x + 2y \leqslant 936$$

(iv) Find the unbounded maximum of the function.

3 For non sign restricted variables x and y:

(i) Find the maximum value of the function

$$z = 64x - 4x^2 + 24y - y^2$$

subject to the constraints

$$2x + y \leqslant 20$$
$$2x + 2y \leqslant 30$$

(ii) What is the increase in z if the first constraint is removed?
(iii) With the second constraint in place, what is the maximum increase in the value of the right-hand side of the constraint that would be worth having?
(iv) What is the unconstrained maximum of z?

REFERENCES AND FURTHER READING

1. Borowski, E. J. and Borwein, J. M. (1989) *Dictionary of Mathematics*, Collins.
2. Chiang, A. C. (1984) *Fundamental Methods of Mathematical Economics* (Third Edition), McGraw-Hill.
3. Finney, R. L. and Thomas, G. B., Jr (1990) *Calculus*, Addison-Wesley.
4. Hillier, F. S. and Lieberman, G. J. (1990) *Introduction to Operations Research* (Fifth Edition), McGraw-Hill.
5. Intriligator, M. D. (1971) *Mathematical Optimisation and Economic Theory*, Prentice Hall.
6. Jeffrey, A. (1989) *Mathematics for Engineers and Scientists* (Fourth Edition), Van Nostrand Reinhold.
7. Kuhn, H. W. and Tucker, A. W. (1951) *Nonlinear Programming*, in *Proceedings of the Second Berkeley Symposium on Mathematical Statistics and Probability* (Editor: J Neyman), University of California Press.
8. Rich, A., Rich, J. and Stoutemyer, D. (1989) *Derive: User Manual* (Third Edition), Soft Warehouse.
9. Silberberg, E. (1978) *The Structure of Economics: Mathematical Analysis*, McGraw-Hill.
10. Walsh, G. R. (1975) *Methods of Optimization*, Wiley.
11. Weber, J. E. (1982) *Mathematical Analysis: Business and Economic Applications* (Fourth Edition), Harper and Row.
12. Wilkes, F. M. (1983) *Capital Budgeting Techniques* (Second Edition), Wiley.
13. Wisniewski, M. and Dacre, T. (1990) *Mathematical Programming: Optimization Models for Business and Management Decision Making*, McGraw-Hill.

SOLUTIONS TO EXERCISES

Exercises 7.1

1 The Lagrangian is

$$L(x, y, j) = 20x + 15y - xy - x^2 - 0.5y^2 + j(20 - x - 2y)$$

so that the conditions at the first order are

$$L_x = 20 - y - 2x - j = 0$$
$$L_y = 15 - x - y - 2j = 0$$
$$L_j = 20 - x - 2y = 0$$

From L_x,

$$j = 20 - y - 2x$$

which, when substituted into L_y produces

$$15 - x - y - 2(20 - y - 2x) = 0$$

which can be rearranged as the equation

$$6x + 2y = 50 \qquad (1)$$

Now writing the L_j constraint as

$$x + 2y = 20$$

and subtracting from (1) produces

$$5x = 30$$

so that $x = 6$ and, by substitution back, $y = 7$ and $j = 1$. The maximum value of $f(x, y)$ is therefore

$$20(6) + 15(7) - (6)(7) - (6)^2 - 0.5(7)^2 = 122.5$$

The sufficient second-order condition was that

$$E = L_{xx}(g_y)^2 - 2L_{xy}g_xg_y + L_{yy}(g_x)^2 < 0$$

In the present case,

$$L_{xx} = -2 \quad L_{yy} = -1 \quad L_{xy} = -1$$
$$g_x = -1 \quad g_y = -2$$

so that E evaluates as

$$E = -2(-2)^2 - 2(-1)(-1)(-2) + (-1)(-1)^2$$
$$= -8 + 4 - 1 = -5 \leq 0$$

as required at a constrained maximum.

2 (i) The Lagrangian is:

$$L(x, y, j) = 1200x + 400y - 20xy - 40x^2 - 2.5y^2$$
$$+ j(150 - 10x - 5y)$$

and conditions at the first order are

$$L_x = 1200 - 20y - 80x - 10j = 0$$
$$L_y = 400 - 20x - 5y - 5j = 0$$
$$L_j = 150 - 10x - 5y = 0$$

which conditions solve for

$$x = 5 \quad y = 20 \quad j = 40 \quad f(x, y) = 10{,}000$$

(ii) First-order conditions would be unchanged except for 151 replacing 150 in L_j. The solution values then are

$$x = 4.9 \quad y = 20.4 \quad j = 40 \quad f(x, y) = 10{,}040$$

so that the value of an extra unit of resource is 40.

3 The Lagrangian is
$$L = 2x^2 + 6xy + 4.5y^2 - 100x - 200y + j(40 - 2.5x - 1.25y)$$
and conditions at the first order are
$$L_x = 4x + 6y - 100 - 2.5j = 0$$
$$L_y = 6x + 9y - 200 - 1.25j = 0$$
$$L_j = 40 - 2.5x - 1.25y = 0$$
For the second-order condition,
$$L_{xx} = 4 \qquad L_{xy} = 6 \qquad L_{yy} = 9 \qquad g_x = -2.5 \qquad g_y = 1.25$$
so that
$$E = 4(-1.25)^2 - 2(6)(-2.5)(-1.25) + 9(2.5)^2$$
$$= 6.25 - 37.5 + 56.25$$
$$= 25$$
and therefore the solution of the first-order conditions will correspond to a minimum. The conditions solve for
$$x = 5.25 \qquad y = 21.5 \qquad j = 20 \qquad f(x, y) = -2012.5$$

4 The Lagrangian in this case is
$$L = 44x + 101y + 9xy + 5x^2 - 8y^2 + j(29 - 3x - 2y)$$
and at the first order
$$L_x = 44 + 9y + 10x - 3j = 0$$
$$L_y = 101 + 9x - 16y - 2j = 0$$
$$L_j = 29 - 3x - 2y = 0$$
which conditions solve for
$$x = 7 \qquad y = 4 \qquad j = 50 \qquad f(x, y) = 1081$$
At the second order,
$$L_{xx} = 10 \qquad L_{xy} = 9 \qquad L_{yy} = -16 \qquad g_x = -3 \qquad g_y = -2$$
and so
$$E = 10(-2)^2 - 2(9)(-3)(-2) - 16(-3)^2$$
$$= 40 - 108 - 144 = -212 \leqslant 0$$
so that a constrained maximum occurs. With 30 as the right-hand side value, the results are (approximate values):
$$x = 7.311 \qquad y = 4.033 \qquad j = 51.137 \qquad f(x, y) = 1131.57$$

5 The Lagrangian is
$$L = x^3 y + j(20 - x - y)$$
so that the first-order conditions are
$$L_x = 3x^2 y - j = 0$$
$$L_y = x^3 - j = 0$$
$$L_j = 20 - x - y = 0$$
From L_x and L_y, eliminating j, we obtain
$$x^3 - 3x^2 y = 0 \tag{1}$$
from L_j
$$y = 20 - x$$
Substituting this relation into (1) gives
$$x^3 - 3x^2(20 - x) = 0$$
i.e.
$$4x^3 - 60x^2 = 0$$
which factors as
$$x^2(4x - 60) = 0$$
which is satisfied only for $x = 0$ and $x = 15$. At the second order,
$$L_{xx} = 6xy \qquad L_{xy} = 3x^2 \qquad L_{yy} = 0$$
Also
$$g_x = -1 \quad \text{and} \quad g_y = -1$$
so that, for a maximum,
$$E = (6xy)(-1)^2 - 2(3x^2)(-1)(-1) < 0$$
When $x = 0$, $E = 0$, and the sufficient form of this second-order condition is not fulfilled (in fact, the graph of $x^3(20 - x)$ has a point of inflection at the origin). When $x = 15$, $y = 5$, and E evaluates as
$$E = 6(15)(5)(-1)^2 - 2(3)(15)^2(-1)(-1)$$
$$= 450 - 1350$$
$$= -900 < 0$$
so a constrained maximum is produced, and the function takes the value 16,875.

Exercises 7.2

1 (i) The Lagrangian is

$$L = 4x^2 + 16xy + 5y^2 + j(30 - 2x - 3y)$$

and at the first order

$$L_x = 8x + 16y - 2j = 0$$
$$L_y = 16x - 10y - 3j = 0$$
$$L_j = 30 - 2x - 3y = 0$$

which conditions solve for

$$x = 10.5 \qquad y = 3 \qquad j = 66 \qquad f(x,y) = 990$$

At the second order,

$$L_{xx} = 8 \qquad L_{xy} = 16 \qquad L_{yy} = 10 \qquad g_x = -2 \qquad g_y = -3$$

So,

$$E = 8(-3)^2 - 2(16)(-2)(-3) - 10(-2)^2$$
$$= 72 - 192 + 40$$
$$= -80 < 0$$

and so a constrained maximum is confirmed.

(ii) The rates of improvement of objective for an additional unit expended on each of x and y are respectively

$$f_x \frac{1}{r_x} \qquad \text{and} \qquad f_y \frac{1}{r_y}$$

where r_x and r_y are the unit costs of x and y, so that in this problem $r_x = 2$ and $r_y = 3$. In the present case we have

$$f_x = 8x + 16y$$

which evaluates at the optimum as

$$f_x = 8(10.5) + 16(3)$$
$$= 132$$

and

$$f_y = 16x + 10y$$
$$= 198$$

at the optimum. So therefore at the optimum

$$f_x \frac{1}{r_x} = 132\left(\frac{1}{2}\right) = 66$$

$$f_y \frac{1}{r_y} = 198\left(\frac{1}{3}\right) = 66$$

which values are equal – and, of course, equal to the Lagrange multiplier, j, as required.

(iii) At the suboptimal position where $x = 3$ and $y = 8$, the partial derivatives evaluate as

$$f_x = 8(3) + 16(8) = 152$$
$$f_y = 16(3) + 10(8) = 128$$

so that

$$f_x \frac{1}{r_x} = 152\left(\frac{1}{2}\right) = 76$$

$$f_y \frac{1}{r_y} = 128\left(\frac{1}{3}\right) = 42.67$$

which disparity can be used to improve the objective by reducing expenditure on y and increasing expenditure on x until the optimal balance is achieved.

Exercises 7.3

1 (i) The derivative is

$$\frac{dz}{dx} = 24 - 3x$$

so, trying $x = 0$, $dz/dx = 24 > 0$ so that $x = 0$ cannot produce the sign restricted optimum. Now trying $dz/dx = 0$, here $x = 8$, so that the conditions are fulfilled and $z = 216$.

(ii) The derivative is

$$\frac{dz}{dx} = -21 - 7x$$

so, trying $x = 0$, $dz/dx = -21 < 0$ as required, so that $x = 0$ could produce the sign restricted optimum. Now trying $dz/dx = 0$, the resulting value is $x = 8$. Thus the conditions are fulfilled only at $x = 0$ where $z = 100$.

Constrained optimization with Lagrange multipliers 523

2 (i) The derivative is

$$\frac{dz}{dx} = 120 + 54x - 3x^2$$

When $x = 0$, $dz/dx = 120 > 0$ so that $x = 0$ cannot produce the sign restricted optimum. Now try $dz/dx = 0$. The equation solves for $x = -2$ or $x = 20$. Thus the conditions are fulfilled only at this point and the sign restricted maximum of z is 6200.

(ii) The derivative here is

$$\frac{dz}{dx} = -2880x + 384x^2 - 12x^3$$

At $x = 0$, the derivative $dz/dx = 0$ so that in this case a turning point occurs at the boundary and $x = 0$ could produce the sign restricted optimum. The other values of x which satisfy $dz/dx = 0$ are found by solving the quadratic

$$-2880 + 384x - 12x^2 = 0$$

which solves for $x = 12$ and $x = 20$. The possibilities to be enumerated are then

x	y
0	8,000
12	$-40,384$
20	$-24,000$

so the sign restricted maximum is at $x = 0$.

3 (i) The derivative is:

$$\frac{dz}{dx} = 2.4x - 12$$

Trying $x = 0$, $dz/dx = -12$ so that $x = 0$ cannot produce the sign restricted minimum. Now trying $dz/dx = 0$, this solves for $x = 5$ which is acceptable and at which point

$$z = 1.2(25) - 12(5) + 30$$
$$= 0$$

so that the constrained minimum is zero at $x = 5$.

(ii) The derivative is

$$\frac{dz}{dx} = 14 + 84$$

At $x = 0$, $dz/dx = 84 > 0$ as required, so that $x = 0$ could produce the sign restricted minimum. Now trying $dz/dx = 0$, the resulting value of $x = -6$. Thus the conditions are fulfilled only at $x = 0$ where $z = 50$.

4 (i) The derivative here is

$$\frac{dz}{dx} = 3x^2 - 27x - 30$$

so, first trying $x = 0$, $dz/dx = -30 < 0$, thus ruling out the origin as the sign restricted minimum. Now trying $dz/dx = 0$, the equation solves for $x = -1$ and $x = 10$. Thus the conditions are fulfilled only at $x = 10$ giving $z = 0$ as the constrained minimum of the function.

(ii) The derivative here is

$$\frac{dz}{dx} = 3x^3 - 330x^2 - 3000x$$

At $x = 0$, $dz/dx = 0$, so that this is a possible sign restricted optimum. The other values of x which satisfy $dz/dx = 0$ are the roots of the quadratic

$$3x^2 - 330x - 3000 = 0$$

i.e.

$$x^2 - 110x - 1000 = 0$$

which solves for $x = 10$ and $x = 100$. Enumerating the possibilities:

x	z
0	0
10	47,500
100	$-20,000,000$

so that the function takes its sign restricted minimum at $x = 100$.

5 The derivatives are

$$f_x = -300 + 10y - 60x$$
$$f_y = 40 + 10x - 2y$$

(i) We require that

$$f_x \leq 0$$
$$xf_x = 0$$
$$x \geq 0$$
$$f_y = 0$$

Trying $x = 0$, $f_y = 0$ solves for $y = 20$, at which value $f_x = -100$, so this is a possible optimum. With $f_x = 0$, $f_x = 0$ and $f_y = 0$ solve for $x = -10$ and $y = -30$ which is infeasible. So the optimum here is

$$x = 0 \qquad y = 20 \qquad f(x, y) = 500$$

(ii) It is necessary that

$$f_x = 0$$
$$f_y \leq 0$$
$$yf_y = 0$$
$$y \geq 0$$

Trying $x = 0$, $f_x = 0$ solves for $x = -5$ (which is acceptable) at which value f_y emerges as -10. So a possible optimum is $y = 0$, $x = -5$, $f(x, y) = 850$. Trying $f_y = 0$ along with $f_x = 0$ produces the infeasible position already identified in (i). So the optimum is

$$x = -5 \qquad y = 0 \qquad f(x, y) = 850$$

(iii) The outcomes are

(1) $x = 0$ $\qquad y = 0$ $\qquad f_x = -300$ $\qquad f_y = 40$
(2) $x = 0$ $\qquad f_y = 0$ $\qquad f_x = -100$ $\qquad y = 20$
(3) $f_x = 0$ $\qquad y = 0$ $\qquad x = -5$ $\qquad f_y = -10$
(4) $f_x = 0$ $\qquad f_y = 0$ $\qquad x = -10$ $\qquad y = -30$

so the solution with both variables sign restricted is (2), i.e.

$$x = 0 \qquad y = 20 \qquad f(x, y) = 500$$

(iv) The conditions required are

$$f_x = 0 \qquad f_y = 0$$

which, as we have already seen, solve for

$$x = -10 \qquad y = -30 \qquad f(x, y) = 1000$$

which is the overall maximum of the function.

6 (i) The conditions are

$$\frac{dz}{dx} \geq 0$$

$$(50 - x)\frac{dz}{dx} = 0$$

$$50 - x \geq 0$$

The derivative is

$$\frac{dz}{dx} = 150 - 2x$$

Trying $50 - x = 0$, $dz/dx = 50 > 0$, so that this point fulfils the necessary conditions. Trying $dz/dx = 0$, $x = 75$, so $50 - 75 < 0$ and therefore this is not a possible constrained optimum (it is, of course, the unconstrained maximum). The solution is therefore

$$x = 50 \quad \text{when} \quad z = 5500$$

(ii) The derivative is

$$\frac{dz}{dx} = 2000 - 16x$$

When $100 - x = 0$, $dz/dx \geqslant 0$, so that this is a possible optimum. When $dz/dx = 0$, $100 - x = -25$, thus ruling out this possibility. The constrained optimum is therefore at $x = 100$ when $z = 121,000$.

(iii) The derivative is

$$\frac{dz}{dx} = 1600 - 4x$$

At $x = 450$, $dz/dx = -200$, so ruling out this position as a possible optimum. When $dz/dx = 0$, $x = 400$, so that $450 - x = 50 > 0$ and this is the constrained optimum (and is also the unconstrained optimum).

7 (i) The derivative is

$$\frac{dz}{dx} = -528 + 90x - 3x^2$$

Following the five-step procedure: at $x = 0$, $dz/dx = -528$, so this is a possible optimum. At $x = 20$, $dz/dx = 72$, so this too is a possible constrained optimum. Now trying $dz/dx = 0$, the roots are $x = 8$ and $x = 22$. Clearly $x = 22$ is infeasible. Evaluating z at the points recorded as possibilities:

x	z
0	2500
8	644
20	1940

so that $x = 0$ is the optimum.

(ii) The derivative is

$$\frac{dz}{dx} = 195x - 900 - 3x^2$$

At $x = 0$, $dz/dx = -900$, so this is a possible optimum. At $x = 50$, $dz/dx = 1350$, so this is also a possible constrained optimum. Now $dz/dx = 0$ solves for $x = 5$ and $x = 60$. Ruling out $x = 60$ and evaluating z at the three recorded points:

x	z
0	0
5	$-2,187.5$
50	73,750

so that $x = 50$ is the optimum.

(iii) The derivative is

$$\frac{dz}{dx} = 6x^2 - 210x + 1500$$

At $x = 0$, $dz/dx = 1500$, so this is not a possible optimum. At $x = 30$, $dz/dx = 600$, so this position is a candidate. When $dz/dx = 0$, $x = 10$ or $x = 25$. Evaluating z at the three possibilities:

x	z
10	5500
25	2125
30	3500

Therefore $x = 10$ is the constrained optimum.

8 (i) The derivatives are

$$f_x = -2x - 2.5y + 10$$
$$f_y = -2.5x - 4y + 2$$

which when equated to zero solve for

$$x = 20 \qquad y = -12$$

giving

$$f(x, y) = 188$$

which is the optimal position.

(ii) The Lagrangian is

$$L = 100 - x^2 - 2.5xy - 2y^2 + 10x + 2y + j(20 - 2x - y)$$

and the first-order conditions are

$$L_x = -2x - 2.5y + 10 - 2j = 0$$
$$L_y = -2.5x - 4y + 2 - j = 0$$

which solve for

$$x = 14.5 \qquad y = -9 \qquad j = 1.75$$

giving $f(x, y) = 181$. At the second order,

$$L_{xx} = -2 \qquad L_{yy} = -4 \qquad L_{xy} = -2.5$$

and from the constraint

$$g_x = -2 \qquad g_y = -1$$

so that the second-order condition

$$E = L_{xx}g_y^2 - 2L_{xy}g_xg_y + L_{yy}g_x^2$$
$$= -2(-1)^2 - 2(-2.5)(-2)(-1) - 4(-2)^2$$
$$= -2 + 10 - 16$$
$$= -8 < 0$$

as required for a constrained maximum.

(iii) Using the information from the constraint that

$$y = 20 - 2x$$

since $y \geq 0$, it follows that this can be expressed in terms of x as $x \leq 10$. Now substituting for y in $f(x, y)$ produces

$$z = 116x - 4x^2 - 660$$

for which the derivative is

$$\frac{dz}{dx} = 116 - 8x$$

Trying $x = 0$, $dz/dx = 116 > 0$, so that this cannot be the optimal position. Trying $dz/dx = 0$, the equation solves for $x = 14.5$ which is (now) infeasible. Trying $x = 10$, $dz/dx = 36 > 0$ as required for the upper bound to be a possible optimum. Since this is the only possibility, the solution is

$$x = 10 \qquad y = 0 \qquad f(x, y) = 100$$

Constrained optimization with Lagrange multipliers 529

Exercises 7.4

1 The Lagrangian is
$$L = x^3 + 7.5x^2 - 150x + 1000 + j(100 - 5x - 40)$$
and at the first order the conditions are that
$$3x^2 + 15x - 150 - 5j = 0$$
$$100 - 5x - 40 - s = 0$$
$$sj = 0$$

Trying $s = 0$, $x = 12$ and $j = 92.4$, so that this point is a possible optimum. Now trying $j = 0$, the L_x equation solves for $x = -10$ and $x = 5$. At $x = -10$, $s = 110$ so that this point too is a possible optimum (recall that there are no sign requirements). At $x = 5$, $s = 35$ so this point also cannot be ruled out at the first order. Evaluating z at the points obtained:

x	z
12	2008
5	562.5
−10	2250

so the constrained maximum is at $x = -10$.

2 The Lagrangian is
$$L = 40x - 8x^2 + 4xy + 80y - 5y^2 + j(40 - 2x - 4y)$$
and the first-order conditions can be expressed as
$$40 - 16x + 4y - 2j = 0$$
$$4x + 80 - 10y - 4j = 0$$
$$40 - 2x - 4y - s = 0$$
$$sj = 0$$
$$s \geq 0 \quad j \geq 0$$

Trying $s = 0$, from the third equation $x = 20 - 2y$ and substitution into and solution of the first two equations for y and j leads to the outcome
$$x = 4 \quad y = 8 \quad j = 4 \quad z = 480$$
Trying $j = 0$, the first two equations can then be written as
$$16x - 4y = 40$$
$$-4x + 10y = 80$$
which solve for $y = 10$ and $x = 5$, which imply that $s = -10$. The

constraint would therefore be violated at this position. The constrained optimum is then

$$x = 4 \quad y = 8 \quad j = 4 \quad z = 480$$

3 The Lagrangian is

$$L = 120x - 2.5x^2 + 3xy - 3.5y^2 - 60y + j(35 - 3x - 2y)$$

and the first-order conditions can be expressed as

$$120 - 5x + 3y - 3j = 0$$
$$3x - 7y - 60 - 4j = 0$$
$$35 - 3x - 7y - s = 0$$
$$sj = 0$$
$$s \geqslant 0 \quad j \geqslant 0$$

Trying $s = 0$, the conditions solve for

$$x = 15 \quad y = -5 \quad j = 10 \quad f(x, y) = 1225$$

Trying $j = 0$, the first two equations solve for

$$x = 330/13 \quad y = 30/13$$

but the constraint related equation then produces

$$s = -595/13$$

and the constraint would therefore be violated at this position. The constrained optimum is then

$$x = 15 \quad y = -5 \quad j = 10 \quad f(x, y) = 1225$$

4 The Lagrangian is

$$L = 100x - 3x^2 + 4xy - 2.5y^2 - 55y + j(90 - 2x - 5y)$$

and the first-order conditions can be expressed as

$$100 - 6x + 4y - 2j = 0$$
$$4x - 5y - 55 - 5j = 0$$
$$90 - 2x - 5y - s = 0$$
$$sj = 0$$
$$s \geqslant 0 \quad j \geqslant 0$$

Trying $s = 0$, the conditions solve for

$$x = 23 \quad y = 8.8$$

but

$$j = -1.4$$

meaning that this position cannot be optimal. Trying $j = 0$, the conditions solve for

$$x = 20 \quad y = 5 \quad s = 25$$

This is the only point fulfilling the conditions, and therefore produces the constrained maximum of $f(x, y) = 862.5$.

Exercises 7.5

1 (i) The Lagrangian is

$$L = 32x + 16y - x^2 - y^2 + 150 + j(30 - 2x - y)$$

and the first-order conditions can be expressed as

$$32 - 2x - 2j + t = 0$$
$$16 - 2y - j + u = 0$$
$$30 - 2x - y - s = 0$$
$$tx = 0 \quad uy = 0 \quad sj = 0$$
$$x \geqslant 0 \quad y \geqslant 0 \quad j \geqslant 0 \quad t \geqslant 0 \quad u \geqslant 0 \quad s \geqslant 0$$

Enumerating the various combinations of variables set to zero, the results are as follows.

	Solution values (non-zero)					
Zero group	x	t	y	u	s	j
(a) t u s	12		6			4
(b) t u j	16		8		−10	
(c) t y s	15			−15		1
(d) t y j	16			−16	−2	
(e) x u s		−120	30			−44
(f) x u j		−32	8		22	
(g) x y s			No solution			
(h) x y j		−32		−16	30	

From the table, the optimum position is (a) at which point

$$x = 12 \quad y = 6 \quad j = 4$$

giving

$$f(x, y) = 450$$

since this is the only complete solution to the conditions.

(ii) Here the unconstrained optimum (b) can be achieved (since the sign requirements happen to be met at (b)) and $f(x, y) = 470$. So the 'cost' of the constraint is

$$470 - 450 = 20$$

2   (i) The Lagrangian is

$$L = 63x - 0.8x^2 - 35y - 0.4y^2 + xy + j(35 - x - 0.8y)$$

and the first-order conditions can be expressed in complementary slackness form as

$$63 - 1.6x + y - j + t = 0$$
$$-35 - 0.8y + x - 0.8j + u = 0$$
$$35 - x - 0.8y - s = 0$$
$$tx = 0 \quad uy = 0 \quad sj = 0$$
$$x \geqslant 0 \quad y \geqslant 0 \quad j \geqslant 0 \quad t \geqslant 0 \quad u \geqslant 0 \quad s \geqslant 0$$

Enumerating the various combinations of variables set to zero, the results are as follows (approximate values in case (a)).

		Solution values (non-zero)				
Zero group	x	t	y	u	s	j
(a) t u s	36.31		−1.64			3.27
(b) t u j	55		25		−40	
(c) t y s	35			5.6		7
(d) t y j	39.375			−4.375	−4.375	
(e) x u s		−19.25	43.75			−87.5
(f) x u j		−19.25	−43.75		70	
(g) x y s			No solution			
(h) x y j		−63		35	35	

From the table, the constrained optimum position is identified as (c) since this is the only complete solution to the conditions. At this point:

$$x = 35 \quad y = 0 \quad j = 7 \quad f(x,y) = 1225$$

(ii) Without the requirement to satisfy the constraint, the best position is (b), where the sign requirements happen still to be met and where

$$x = 55 \quad y = 25 \quad f(x,y) = 1295$$

and the constraint therefore 'costs'

$$1295 - 1225 = 70$$

(iii) No change in $f(x,y)$.

(iv) Combination (a) now becomes feasible. At this point

$$f(x,y) \approx 1229.58$$

So the benefit of the removal of the sign restriction on y is 4.58.

Exercises 7.6

1 (i) The Lagrangian is

$$L = 40x - x^2 + 60y - 2y^2 + j(120 - 5x - 4y) + k(140 - 10x - y)$$

and the Kuhn–Tucker conditions are

$$L_x = 40 - 2x - 5j - 10k \leqslant 0$$
$$L_y = 60 - 4y - 4j - k \leqslant 0$$
$$xL_x + yL_y = 0$$
$$x \geqslant 0 \quad y \geqslant 0$$
$$L_j = 120 - 5x - 4y \geqslant 0$$
$$L_k = 140 - 10x - y \geqslant 0$$
$$jL_j + kL_k = 0$$
$$j \geqslant 0 \quad k \geqslant 0$$

(ii) When $x = 88/7$, $y = 100/7$, $j = 192/490$, $k = 632/490$, so that the first four conditions are satisfied. We also obtain

$$L_j = 0 \quad L_k = 0$$

so that the last four conditions are also clearly satisfied.

(iii) The unconstrained optimum is when

$$f_x = 40 - 2x = 0$$
$$f_y = 60 - 4y = 0$$

i.e. where $x = 20$ and $y = 15$, giving $f(x, y) = 850$.
At the constrained optimum

$$f(x, y) = \frac{38{,}896}{49} \approx 793.80$$

so that the possible improvement in $f(x, y)$ is 56.2.

2 The conditions are

$$L_x = 12{,}000 - 1.2x^5 - 3j - 4xk \leqslant 0$$
$$L_y = 2{,}000{,}000 - 8y^7 - 7j - k \leqslant 0$$
$$xL_x + yL_y = 0$$
$$x \geqslant 0 \quad y \geqslant 0$$
$$L_j = 21 - 3x - 7y \geqslant 0$$
$$L_k = 20 - 2x^2 - y \geqslant 0$$
$$jL_j + kL_k = 0$$
$$j \geqslant 0 \quad k \geqslant 0$$

3 The Lagrangian is

$$L = 100x - 2x^2 + 60y - y^2 + j(65 - 2x - y)$$
$$+ k(38 - 0.5x - y)$$

for which conditions at the first order, in complementary slackness form, are

$$100 - 4x - 2j - 0.5k + t = 0$$
$$60 - 2y - j - k + u = 0$$
$$65 - 2x - y - s = 0$$
$$38 - 0.5x - y - r = 0$$
$$tx + uy + js + kr = 0$$
$$x, y, t, u, j, k, r, s \geqslant 0$$

The 16 combinations of zero-valued variables, and the resulting values of the other variables, are as follows.

Zero group	Solution values (non-zero)							
	x	t	y	u	j	s	k	r
(i) tusr	18		29		18		−16	
(ii) tusk	20		25		10			3
(iii) tujr	24		26			−9	8	
(iv) tujk	25		30			−15		−4.5
(v) tysr					No solution			
(vi) tysk	32.5				−75 −15			21.75
(vii) tyjr	76				−468	−87	−408	
(viii) tyjk	25				−60	150		25.5
(ix) xusr					No solution			
(x) xusk		−280	65		−70			−27
(xi) xujr		−108	38			27	−16	
(xii) xujk		−100	30			35		8
(xiii) xysr					No solution			
(xiv) xysk					No solution			
(xv) xyjr					No solution			
(xvi) xyjk		−100			−60	65		38

The optimal solution is (ii), at which point $f(x, y) = 2075$. The unconstrained optimum is (iv), where $f(x, y) = 2150$. The cost of the constraints as a whole is therefore 75.

Exercises 7.7

1 (i) The costs for the products are

unit cost of $x = 1(5) + 1(10) = 15$
unit cost of $y = 1(5) + 2(10) = 25$

so

$$C(x) = 15x \qquad C(y) = 25y$$

Revenues from the sale of the goods are

$$R(x) = xp(x) = 95x - 0.5x^2$$
$$R(y) = yp(y) = 85y - 0.5y^2$$

so that overall profit is

$$F = R(x) - C(x) + R(y) - C(y)$$
$$= 95x - 0.5x^2 - 15x + 85y - 0.5y^2 - 25y$$
$$= 80x - 0.5x^2 + 60y - 0.5y^2$$

536 Mathematics for business, finance and economics

Thus the problem in full is

maximize $F = 80x - 0.5x^2 + 60y - 0.5y^2$
subject to $x + y \leq 100$
$x + 2y \leq 140$
$x \geq 0 \quad y \geq 0$

for which the Lagrangian is

$$L = 80x - 0.5x^2 + 60y - 0.5y^2 + j(100 - x - y) + k(140 - x - 2y)$$

and the conditions in complementary slackness form are

$80 - x - j - k + t = 0$
$60 - y - j - 2k + u = 0$
$100 - x - y - s = 0$
$140 - x - 2y - r = 0$
$tx + uy + js + kr = 0$
$x, y, t, u, j, k, r, s \geq 0$

The results of applying the combinations of zero variables are as follows.

	Solution values (non-zero)							
Zero group	x	t	y	u	j	s	k	r
(i) tusr	60		40		20		0	
(ii) tusk	60		40		20			0
(iii) tujr	68		36			−4	12	
(iv) tujk	80		60			−40		−60
(v) tysr				No solution				
(vi) tysk	100			−80	−20			40
(vii) tyjr	140			−180		−40	−60	
(viii) tyjk	80			−60			20	60
(ix) xusr				No solution				
(x) xusk		−120	100	−40				−60
(xi) xujr		−85	70			30	−5	
(xii) xujk		−80	60				40	20
(xiii) xysr				No solution				
(xiv) xysk				No solution				
(xv) xyjr				No solution				
(xvi) xyjk		−80		−60			100	140

So cases (i) and (ii) produce the same complete solution to the conditions. This phenomenon simply means that the corner point at the intersection of constraints 1 and 2 is also a point of tangency (with the first constraint). The optimal production levels for the two products are therefore

$$x = 60 \quad \text{and} \quad y = 40$$

giving a maximum overall profit of $F = 4600$. Note that the unconstrained optimum is case (iii) (the values of j and k both being zero), and profit here would be $F = 5000$.

Additional Problems

1 (i) The Lagrangian is

$$L = 240x + 1180y + 2xy - 4x^2 - 8y^2 + j(300 - 4x - 2y)$$

and, at the first order, the conditions are

$$L_x = 240 + 2y - 8x - 4j = 0$$
$$L_y = 1180 + 2x - 16y - 2j = 0$$
$$L_j = 300 - 4x - 2y = 0$$

These conditions solve for

$$x = 37.25 \quad y = 75.5 \quad j = 23.25$$

which result in

$$z = 52{,}502.5$$

Checking this solution at the second order, we have

$$L_{xx} = -8 \quad L_{yy} = -16 \quad L_{xy} = 2$$

and from the constraint

$$g_x = -4 \quad \text{and} \quad g_y = -2$$

so that

$$E = (-8)(-2)^2 - 2(2)(-4)(-2) + (-16)(-2)^2$$
$$= -32 - 32 - 64$$
$$= -128 < 0$$

as is required for a maximum.

(ii) The only change to the first-order conditions is that

$$L_j = 400 - 4x - 2y = 0$$

and the conditions now solve for

$$x = 58.5 \quad y = 83 \quad j = -15.5$$

which result in

$$z = 52{,}890$$

While the value of z is greater than with the right-hand side at 300 under (i), the negative value of the Lagrange multiplier tells us that the right-hand side is now *too* great (given the requirement to satisfy the constraint as a strict equality) and z would in fact increase if the right-hand side was reduced.

(iii) With the constraint removed, the first-order conditions are

$$Z_x = 240 + 2y - 8x = 0$$
$$Z_y = 1180 + 2x - 16y = 0$$

while, at the second order,

$$Z_{xx} = -8 \quad Z_{yy} = -16 \quad Z_{xy} = 2$$

and therefore since

$$(-8)(-16) - (2)^2 = 124 > 0$$

the solution to the first-order conditions

$$x = 50 \quad y = 80$$

results in

$$z = 53{,}200$$

which represents the optimal solution. In terms of a strict equality constraint, the 'ideal value' of the right-hand side is that which allows the achievement of this position. That is,

$$\text{RHS} = 4(50) + 2(80) = 360$$

2 The answers to all parts of this question can be obtained from the enumerated solutions to the Kuhn–Tucker conditions for the problem with both constraints included, certain of the requirements being ignored in respect of cases (i), (ii), (iii) and (iv). So, setting up the Lagrangian,

$$L = 600x - x^2 + 1440y - 2y^2$$
$$+ j(600 - x - y) + k(936 - x - 2y)$$

and the Kuhn–Tucker conditions can be expressed as

$$600 - 2x - j - k = 0$$
$$1440 - 4y - j - 2k = 0$$
$$600 - x - y - s = 0$$
$$936 - x - 2y - r = 0$$
$$js + kr = 0$$
$$j, k, r, s \geqslant 0$$

The four combinations of zero variables result in the following.

Zero group	Solution values (non-zero)						
	x	y	j	s	k	r	z
(i) j k	300	360		−40		−84	349,200
(ii) j r	272	332		−4	56		346,848
(iii) s k	260	340	80			−4	346,800
(iv) s r	264	336	48		24		346,752

For part (i) it is only required that $s \geqslant 0$. Since the second constraint does not apply here, r can take any value. Thus case (iii) with $z = 346,800$ is the best solution in which $s \geqslant 0$.

For part (ii), it is only required that $r \geqslant 0$. Since the first constraint does not apply here, s can take any value. Thus case (ii) with $z = 346,848$ is the best solution in which $r \geqslant 0$.

For part (iii), a complete solution to the conditions is required, in which all of the sign requirements are observed. Case (iv) is the only possibility and in this case $z = 346,752$.

For part (iv), the sign restrictions on s and r can both be ignored. Case (i) gives the best outcome, with $z = 349,200$.

3 (i) The Lagrangian is

$$L = 64x - 4x^2 + 24y - y^2 + j(20 - 2x - y) + k(30 - 2x - 2y)$$

and the Kuhn–Tucker conditions can be expressed as

$$64 - 8x - 2j - 2k = 0$$
$$24 - 2y - j - 2k = 0$$
$$20 - 2x - y - s = 0$$
$$30 - x - 2y - r = 0$$
$$js + kr = 0$$
$$j, k, r, s \geqslant 0$$

The four combinations of zero variables result in the following.

Zero group	Solution values (non-zero)						
	x	y	j	s	k	r	z
(i) j k	8	12		−8		−10	400
(ii) j r	7	8		−2	4		380
(iii) s k	6	8	8			2	368
(iv) s r	5	10	20		−8		360

The maximum of z is 368 when both constraints must be satisfied.
(ii) If the first constraint is removed, the value of s can be ignored, but r must still be non-negative. The best attainable position is therefore (ii).
(iii) Case (ii) is the best available position with $r \geqslant 0$. Here, $s = -2$, indicating that a right-hand side of 22 for the first constraint would allow this position to be achieved.
(iv) The unconstrained optimum arises in case (i) where $z = 400$.

NOTES

1 The stationary point of L is *not* a maximum of L in all directions. With inequality constraints it is more easily seen that a *saddle point* of L occurs at the constrained optimum of F. With equality constraints the saddle point is degenerate. The saddle point property will be considered further in Section 7.5.
2 Discussion of second-order conditions for an optimum of a function of several variables subject to an equality constraint is contained in the Appendix.
3 Sufficient given satisfaction of first-order conditions.
4 For further discussion see Silberberg (1978).
5 See the Appendix, Section three.
6 This condition will be revisited following discussion of determinants in Chapter 10.
7 With global concavity if there were several solutions to first-order conditions the solutions would all be global maxima (and would constitute a convex set). In the absence of global concavity some of the points may be minima, so that for those which are relative maxima, only local concavity would apply.
8 The value of j will not always represent an upper bound. In the case of the problem 'max xy subject to $x + y = 12$' we saw that the Lagrange multiplier *undervalued* the effect of a unit increment in the right-hand side resource.
9 As, for example, in linear programming.
10 Since the constraint is linear, $1/a$ is also the *number* of units of x that can result from an extra unit of resource. But it is the *rate* of change interpretation that is important.

11 A common misuse of the phrase 'to economize' is to take it to mean 'to minimize the use of'. The true meaning as in the present context is 'to make the *most efficient* use of'.
12 Awareness of the properties of the objective function is needed even when explicit provision is made for sign requirements. Specifically, we need to be sure that the objective function does not improve without bound for high, positive x.
13 We eliminated such values as meaningless in the linear demand, quadratic supply model of Chapter 4.
14 For a formal definition of differentiability see Borowski and Borwein (1989). For our purposes it suffices to interpret this term as meaning simply 'can be differentiated'.
15 One example of a problem involving a cusp will be given later in the chapter.
16 Note that the second-order condition $d^2y/dx^2 < 0$ could be used to eliminate turning points which are local minima.
17 These conditions are a special case of the Kuhn–Tucker conditions described in Section 7.6.
18 After any necessary transformation of the variable.
19 We are excluding functions with such properties as a vertical asymptote within $0 \leqslant x \leqslant b$. Recall also that a turning point may occur at the boundary.
20 Subject to the usual proviso concerning finite maxima.
21 Reference the discussion of slack variables and their role in linear programming in Chapter 3.
22 And the addition of sign restrictions gives a unique solution in this case.
23 Informing us, incidentally, that if the constraint had been an equality originally, then the objective function would *improve* as the right-hand side is reduced.
24 Note that the x and y values associated with $j = 0$ correspond to the unconstrained optimum, and that the value of s represents the shortfall from the minimum value of the right-hand side of the constraint that will achieve this position.
25 Of course, only one such set of contours applies in any particular case.
26 By 'solutions' we mean the values taken by the other variables, not all of which will be feasible in terms of the original problem.
27 Equivalent to point B in Figure 7.11.
28 Constraint qualification conditions are not covered here. The reader is referred to Walsh (1975). However, we shall provide an example of a feasible region with a cusp.
29 See Intriligator (1971) for a description in this context.
30 The conditions were derived independently by W. Karush in 1939 and Kuhn and Tucker in 1951.
31 A necessary condition is required – a point cannot be optimal if the condition does not hold, but it still may not be optimal if it does. A sufficient condition guarantees optimality but may not be needed. For example, $x > 10$ is a *necessary* condition for $x > 100$ while $x > 1000$ is *sufficient* for $x > 100$. For a formal definition of necessary conditions and sufficient conditions see Borowski and Borwein (1989).
32 The solution to the current problem was found in this way. For illustration of the use of quadratic programming interpreted in the context of an investment problem, see Wilkes (1983).
33 For a description of the gradient search procedure, the Frank–Wolfe algorithm and other methods, see Hillier and Lieberman (1990).
34 See Hillier and Lieberman (1990). For a classification and description of programming models, see Intriligator (1971).

35 The demand *function* states quantity demanded as a function of price.
36 That is, y does not appear in the production function for x and vice versa.
37 It turns out not to be worthwhile.
38 For further details see Wilkes (1983).
39 However, rounding errors can cause difficulties in recognizing solutions to the Kuhn–Tucker conditions.

Chapter 8

Integration

8.1	Introduction	544
8.2	Rules for integration	547
8.3	Application to the marginal analysis of the firm	552
8.4	Differential equations	555
8.5	Integration by substitution	567
8.6	The definite integral	571
8.7	Numerical integration	584
8.8	Concluding remarks	591
	Additional problems	592
	References and further reading	593
	Solutions to exercises	594

This chapter considers integration. The process of integration, as its name implies, is essentially the reverse of differentiation. We examine how a function with a specified derivative may be found. Thus functions may be 'recovered' given knowledge of their derivatives.

The chapter presents the concept of the antiderivative, gives fundamental rules for finding both indefinite and definite integrals, and introduces the method of integration by substitution. The methods are then applied to cost and revenue functions in the theory of the firm. Differential equations and numerical methods are introduced.

By the end of this chapter, you will be able to find the indefinite integral for polynomial functions and selected rational and other functions. You will know how to evaluate the definite integral in these cases, find areas under curves, and be able to solve some of the simpler types of differential equation.

8.1 INTRODUCTION

When we considered differentiation, we usually began with a specified function and applied the most appropriate procedure for differentiation. But the original function could itself be a derivative. There are many occasions when it would be useful to be able to recover the original function that gave rise to the derivative. To do this, we shall need the derivative and the value of the function itself for a given value of x. For example, knowledge of the marginal revenue function (and the fact that the total revenue function goes through the origin) will allow us not only to reconstruct the total revenue function, but also to uncover the product demand curve. The process by which an original function is recovered from its derivative is called *integration*. In addition to 'function recovery', in functions of a single variable integration can be seen as the process of finding the area under the curve defined by the derivative expression over a given part of the domain. This represents the evaluation of the *definite integral*.

We begin with the process of finding a function that produces a known derivative. Suppose that we commence with the term

$$3x^2$$

What expression has this term as its derivative? This is not too difficult to answer, as we have frequently come across terms such as $3x^2$ when considering differentiation. The solution might appear to be

$$x^3$$

but further thought will reveal that there are other possible functions which have $3x^2$ as their derivative. For instance the expression

$$x^3 + 1$$

would do just as well, as it too has $3x^2$ as its derivative. But so also do the following expressions:

$$x^3 + 74, \qquad x^3 - 20, \qquad x^3 + \pi, \qquad x^3 - 2\sqrt{2} \qquad (8.1)$$

In fact, x^3, $x^3 + 1$ and all of the functions in (8.1) are amongst the *antiderivatives* of the term $3x^2$. But note that an expression such as

$$x^3 + x$$

can *not* be an antiderivative of $3x^2$. All antiderivatives of $3x^2$ must have the form

$$x^3 + k$$

where k is a constant, and in graphical terms they differ only by the location of the intercept. The antiderivatives are graphed in Figure 8.1, from which

Integration 545

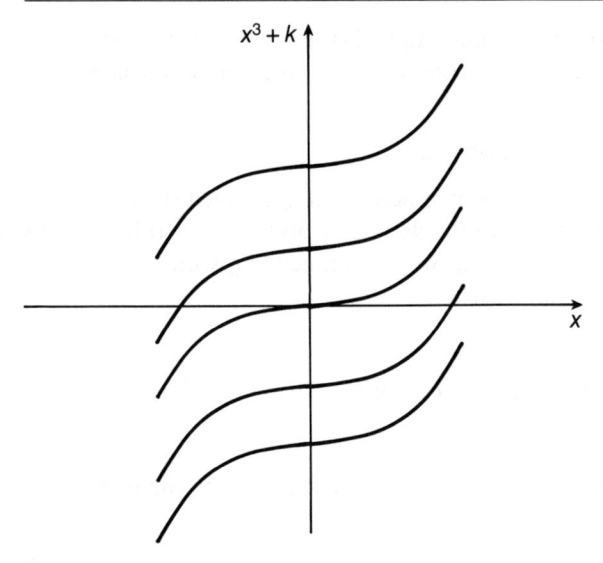

Figure 8.1

it is evident that at any given value of x, say x^0, all antiderivatives have the same slope.

In an application, it may be known that the antiderivative takes a specific value for given x. This information can be used to identify the value of k relevant to the given context. For example, if the antiderivative of $3x^2$ was known to take the value 950 when x took the value 10, then it must be the case that $k = -50$, and the original relationship must therefore have been

$$y = x^3 - 50$$

Fixing a value of the constant of integration to fit an application is an important procedure to which we shall return in Section 8.4. The expression

$$x^3 + k$$

which covers all antiderivatives[1] is the *indefinite integral* of $f(x) = 3x^2$. The indefinite integral is represented in the following notation:

$$\int 3x^2 \, dx = x^3 + k \tag{8.2}$$

The *integral sign*

$$\int$$

was chosen by Leibniz as the symbol to represent integration. It takes the form of an archaic 'S' – the initial letter of 'Summa', and reflecting the area

546   Mathematics for business, finance and economics

under the curve aspect of integration.[2] In (8.2) the symbol 'dx' means 'with respect to x' (just as it did in differentiation) so the whole left-hand side of (8.2) reads

'The integral of $3x^2$ with respect to x'.

The expression to be integrated, which here is $3x^2$, is called the *integrand*. Using Leibniz's notation (for both the derivative and the integral) the steps can be summarized in the following way. Commencing with

$$\frac{dy}{dx} = 3x^2$$

we then interpret dy and dx as *distinct entities*, and write[3]

$$dy = 3x^2 \, dx$$

from which the original function, y, is recovered by taking the integral of both sides:

$$y = \int dy = \int 3x^2 \, dx = x^3 + k$$

Where no specific function is stated, the integrand is usually written as $f(x)$ and the indefinite integral as $F(x) + k$, as in[4]

$$\int f(x) \, dx = F(x) + k$$

As in any process, construction is inherently more difficult than separation, so it is no surprise to discover that integration is harder than differentiation. Fortunately, many of the rules for differentiation have their counterparts in integration and in the following section we introduce the first of these methods.

EXERCISES 8.1

1 Which of the following are antiderivatives of $4x^3$?

(a) $x^4 + 100$
(b) $x^4 - \pi$
(c) $x^4 - 0.01x$
(d) $(x^2 - 1)^2$
(e) $(x^2 - 1)(x^2 + 1)$

2 Which of the following are antiderivatives of $x^4 - x$?

(a) $0.2x^5 - 0.5x^2$

(b) $\dfrac{x^2(0.4x^3 - 1)}{2}$

(c) $0.4x^5 - x^2$

(d) $0.2x^5 - 0.5x^2 - 100\pi$

8.2 RULES FOR INTEGRATION

Four useful and relatively simple rules for integration are introduced in this section. The rules find ready application to the integration of power functions.[5] The first of the rules is very simple.

The constant function rule

This rule applies when the integrand is a constant. The constant function rule is:

$$\int c \, dx = cx + k \tag{8.3}$$

where, in (8.3), the constant c is any real number. As examples of the rule, consider the integral of the number 7 with respect to x,

$$\int 7 \, dx = 7x + k$$

or the integral of $-\sqrt{2}$ with respect to x,

$$\int -\sqrt{2} \, dx = -\sqrt{2}x + k$$

We now turn to the more general rule of which the constant function rule is a special case.

The power rule

The power rule applies when the integrand is a variable raised to a given power. The rule is

$$\int x^n \, dx = \frac{x^{n+1}}{n+1} + k \tag{8.4}$$

where in (8.4) n is any real number *except -1*. The validity of the rule can be confirmed by differentiation. The exponent, $n + 1$, when brought down will cancel with the denominator of the ratio. We can get a long way with this straightforward but important rule, so we now give several examples.

First, consider the integral of the cubing function. Here, $n = 3$, and the power rule gives

(i) $$\int x^3 \, dx = \frac{x^4}{4} + k$$

Example (i) illustrates the application of the power rule when n is a positive integer. The power rule also applies if n is any negative integer *other than* -1. For example:

(ii) $$\int x^{-3} \, dx = -\frac{x^{-2}}{2} + k$$

In this example the power rule is applied directly, but note that what we have shown is that

$$\int \frac{1}{x^3} \, dx = -\frac{1}{2x^2} + k$$

It is much easier to handle a negative exponent than a reciprocal. The questions arise as to why $n = -1$ is excluded and what, if anything, is the integral of x^{-1}. An attempt to apply (8.4) to this case would result in a division by zero error, the resulting ratio being $x^0/0$. At this stage, we merely point out that x^{-1} (i.e. $1/x$) *does* have an integral, but the integral is not a power function, and consideration is deferred to the next chapter. Now consider the use of the power rule with a rational exponent. Letting $n = \frac{1}{2}$:

(iii) $$\int x^{1/2} \, dx = \tfrac{2}{3} x^{3/2} + k$$

Once again, the rule is simply applied directly, inserting $\frac{1}{2}$ for n. Note that this example shows that

$$\int \sqrt{x} \, dx = \frac{2\sqrt{x^3}}{3} + k$$

It is far easier to work out the integral with a rational exponent than to try to operate with radical signs. The power rule applies to irrational exponents. Suppose that $n = \sqrt{2}$:

(iv) $$\int x^{\sqrt{2}} \, dx = \frac{x^{\sqrt{2}+1}}{\sqrt{2}+1} + k$$

The rule also applies to transcendental exponents. So, for example,

$$\int x^\pi \, dx = \frac{x^{\pi+1}}{\pi + 1} + k$$

The power rule is a simple, important and frequently used rule for finding indefinite integrals. Now consider the integration of functions consisting of other functions or simply of several terms.

The sum–difference rule

This is an equivalent rule for integration to the sum–difference rule for differentiation, and the rule is just as easily applied. The sum–difference rule for integration can be stated as

$$\int [f(x) \pm g(x)] \, dx = \int f(x) \, dx \pm \int g(x) \, dx \tag{8.5}$$

The rule reads as follows: the integral of the sum of (or difference between) functions is the sum of (or difference between) the integrals. As a first example consider

$$\int (4x^3 + 5x^4) \, dx = \int 4x^3 \, dx + \int 5x^4 \, dx$$
$$= x^4 + k_1 + x^5 + k_2$$
$$= x^4 + x^5 + k$$

Note that only *one* constant of integration is necessary in the application of the sum–difference rule, as the sum of, or difference between, two arbitrary constants (k_1 and k_2) is itself an arbitrary constant (k). As a second example of the application of the sum–difference rule consider

$$\int [(6x^2 - 7) + 12x^3] \, dx = \int (6x^2 - 7) \, dx + \int 12x^3 \, dx$$
$$= 2x^3 - 7x + 3x^4 + k$$

In working out this indefinite integral, note that both the power rule and the constant function rule have been applied. The sum–difference rule greatly expands the range of application of the power and constant function rules. The sum–difference rule applies regardless of the number of terms or functions that are involved. An important consequence of the sum–difference rule is that polynomials can be integrated term by term. For example:

$$\int (5x^4 - 2x^3 + 9x^2 - 14x + 13) \, dx$$
$$= x^5 - 0.5x^4 + 3x^3 - 7x^2 + 13x + k$$

A multi-term integrand need not be a polynomial; powers of x which are not positive integers may be involved. For example:

$$\int (3x^{0.5} - \pi^2 x^{\pi - 1} - 2x^{-3}) \, dx = 2x^{1.5} - \pi x^\pi + x^{-2} + k$$

The constant multiple rule

This rule states that the integral of a constant multiple of a function is that multiple of the integral of the function. That is,

$$\int cf(x) \, dx = c \int f(x) \, dx \tag{8.6}$$

The usefulness of the constant multiple rule is that a constant may be taken outside of the integral sign. As a numerical example consider

$$\int (45x^2 + 60x^3) \, dx$$

The number 15, which is the highest common factor of 45 and 60, can be taken outside of the integral sign, and it is therefore required to find

$$15 \int (3x^2 + 4x^3) \, dx = 15(x^3 + x^4) + k$$

$$= 15x^3 + 15x^4 + k$$

in which it will be noted that there is no need to multiply the constant of integration by 15. Before considering further methods of integration, we examine the application of integration to marginal analysis in microeconomics and apply the techniques so far developed in that context.

EXERCISES 8.2

1 Find the following indefinite integrals:

(i) $\int 3 \, dx$

(ii) $\int -7.5 \, dx$

(iii) $\int \sqrt{3} \, dx$

(iv) $\int \pi \, dx$

2 Find the following indefinite integrals:

(i) $\int x^9 \, dx$

(ii) $\int x^{-9} \, dx$

(iii) $\int \dfrac{1}{x^5} \, dx$

(iv) $\int x^{1/4} \, dx$

(v) $\int x^{0.2} \, dx$

(vi) $\int x^{-0.4} \, dx$

(vii) $\int \dfrac{1}{\sqrt{x}} \, dx$

(viii) $\int x^{\sqrt{3}} \, dx$

(ix) $\int x^{2\pi} \, dx$

3 Integrate the following expressions with respect to x:

(i) $4x^3 + 5x^4$
(ii) $6x^5 - 2.5x^4 + x^3$
(iii) $20x^4 - 3x^{-4} + 4x^{-5}$
(iv) $5x^{1.5} + x^{-11}$
(v) $\pi^3 x^{\pi-1} + (2\pi + 1)x^{2\pi} + k$

4 Find:

(i) $\int 12x^2 \, dx$

(ii) $\int (9250x^4 - 125x^9) \, dx$

5 Find:

(i) $\int (ax^2 + bx + c) \, dx$

(ii) $\int mx^m \, dx$

(iii) $\int [(n+1)x^n - (m-1)x^{m-2}] \, dx$

(iv) $\int (3t^2 + 2t) \, dt$

(v) $-3 \int x^{-4} \, dx$

8.3 APPLICATION TO THE MARGINAL ANALYSIS OF THE FIRM

Marginal cost, MC, is defined as the rate of change of a firm's total costs as output is varied. Thus $MC(x)$ is the first derivative of cost with respect to output level, x. This derivative is the same if the total figure for costs is total *variable* costs, $VC(x)$, or the overall total including fixed costs, $TC(x)$. The relationship is

$$\frac{dTC(x)}{dx} = \frac{dVC(x)}{dx} = MC(x)$$

Working in terms of total variable costs, the indefinite integral of the expression for marginal cost will be total variable costs plus the constant of integration:

$$\int MC(x) \, dx = VC(x) + k$$

Here is a case where an appropriate value can be chosen for the constant of integration to correspond to the level of fixed costs. For example, suppose that, as a result of observing the cost consequences of output level variations, a firm finds that its marginal costs are

$$MC(x) = 4x + 20$$

from which it follows that

$$\int MC(x) \, dx = 2x^2 + 20x + k = \text{total costs} \qquad (8.7)$$

The firm would use one observation for its total costs to find the specific value of k relevant to its own situation.[6] Thus if at an output level of $x = 100$ units total costs are known to be £23,000, then k must take a value such that

$$2(100)^2 + 20(100) + k = 23{,}000$$

so that

$$k = 1000$$

and costs in full can be written as

$$\text{total costs} = 2x^2 + 20x + 1000$$

Similar analysis can be carried out on the revenue side. Marginal revenue, MR, is defined as the rate of change of the firm's total sales revenue as output level is varied. Thus $\text{MR}(x)$ is the first derivative of total revenue, $R(x)$, with respect to output, x. That is,

$$\text{MR}(x) = \frac{dR(x)}{dx}$$

So total revenue will be the integral of marginal revenue with respect to output:

$$\int \text{MR}(x)\, dx = R(x) + k$$

However, as there are no 'fixed revenues', the appropriate value of k in this context is zero. With this presumption, if in a particular case

$$\text{MR}(x) = 100 - 6x$$

then total revenue is given by

$$R(x) = \int (100 - 6x)\, dx$$
$$= 100x - 3x^2$$

Now, since total revenue is the product of price, p, and volume of sales, x, the equation of the demand curve for the product can be deduced:

$$R = px = 100x - 3x^2$$

so that

$$p = \frac{R}{x} = 100 - 3x \tag{8.8}$$

The procedure of integration has allowed the demand curve to be obtained on the basis of observations on the revenue consequences of output variations at the margin. For a further example, suppose that cost and revenue information was available for a firm as functions of the output level, x, as follows:

$$\text{marginal cost} = 140 + 2x$$
$$\text{marginal revenue} = 320 - 4x$$
$$\text{total cost} = £2000 \quad \text{when} \quad x = 10$$

and that it is required to determine the maximum profit level and the price to charge in order to achieve this. The profits maximizing *output* level could be found by equating marginal revenue and marginal cost. But in order to establish the level of profit and price at the optimum, it is necessary to obtain the cost, revenue and profit functions in addition to the demand curve relationship. On the income side, total revenue $TR(x)$ is given by

$$\int MR(x) \, dx = \int (320 - 4x) \, dx$$

$$= 320x - 2x^2$$

Dividing by x gives

$$\text{price} = 320 - 2x$$

In terms of costs, total costs $TC(x)$ are given by

$$TC(x) = \int MC(x) \, dx$$

$$= \int (140 + 2x) \, dx$$

$$= 140x + x^2 + k$$

and since $TC(x) = 2000$ when $x = 10$, the level of fixed costs, k, must satisfy

$$140(10) + (10)^2 + k = 2000$$

so that

$$k = 500$$

So,

$$TC(x) = 140x + x^2 + 500$$

Profits, $\pi(x)$, are then given by

$$\pi(x) = TR(x) - TC(x)$$
$$= 320x - 2x^2 - 140x - x^2 - 500$$
$$= 180x - 3x^2 - 500$$

Now, maximizing $\pi(x)$ with respect to x,

$$\frac{d\pi}{dx} = 180 - 6x = 0$$

which condition solves for $x = 30$. At the second order,

$$\frac{d^2\pi}{dx^2} = -6$$

Integration 555

which satisfies the condition for a maximum. The level of profit at $x = 30$ is

$\pi = 180(30) - 3(900) - 500$
$= 2200$

and from the demand curve the profits maximizing level of price is

$p = 320 - 2(30)$
$= 260$

EXERCISES 8.3

1. A firm's marginal revenue function is

   $MR(x) = 2000 - 10x$

   What is the relationship between price and quantity in this case?

2. With x representing the level of output, and given the following information on a firm's costs,

   marginal cost $= 100 + x + 0.3x^2$
   total cost $= 4000$ when $x = 20$

   find the firm's total cost function.

3. Given the following information on a firm's costs and revenues,

   $MC(q) = 210 + 5q$
   $MR(q) = 480 - 10q$
   $TC(q) = 2780$ when $q = 10$

   in which $MC(q)$ is marginal cost, $TC(q)$ is total costs, $M(q)$ is marginal revenue and q is output level, find:

   (i) the product price that would produce maximum revenue;
   (ii) the maximum profit at the corresponding price.

8.4 DIFFERENTIAL EQUATIONS

In dynamic models the natural way to express the relationship between variables is often through *rates of change*. Where change is continuous, rates of change will be represented by derivatives, which will be included in the equations defining the model. For example, situations involving steady rates of growth (or contraction) perhaps of the gross domestic product of an economy, its total population or pollution level, may be suitable for modelling in this way. In this section we consider only the simplest type of differential equations. Remarks are grouped under the following heads:

1 general considerations;
2 first-order linear differential equations (with constant coefficients);
3 higher order differential equations (special cases);
4 a first-order, second-degree example.

General considerations

An equation which contains one or more derivatives of a function is called a *differential equation*. Even an expression such as

$$\frac{dy}{dx} = f(x)$$

which could be any one of the derivatives obtained in Chapter 5, is a simple example of a differential equation. The equation may appear as above, but could also arise in *implicit form* as

$$f(x) - \frac{dy}{dx} = 0$$

so the equation

$$6x^2 - \frac{dy}{dx} + 10 = 0 \qquad (8.9)$$

is a simple example of a differential equation expressed in implicit form. There are many different types of differential equation.[7] Where the function is that of a single variable, the equation is an *ordinary differential equation* – in contrast to the *partial differential equations* which arise from functions of several variables. In general, ordinary differential equations can be written in the following way:

$$f\left(x, y, \frac{dy}{dx}, \frac{d^2y}{dx^2}, \ldots, \frac{d^n y}{dx^n}\right) = 0 \qquad (8.10)$$

The equation

$$\frac{dy}{dx} + 3x^2 + \frac{d^3 y}{dx^3} - \left(\frac{d^2 y}{dx^2}\right)^2 = 0 \qquad (8.11)$$

is an example of an ordinary differential equation. Equation (8.11) is of *order* three, since it contains the third derivative and no higher derivatives. The *degree* of the differential equation is the power to which the *highest order* derivative is raised.[8] Thus equation (8.11) is of the *first* degree despite the fact that the second derivative is squared. The degree is given by the term involving the third derivative, and this is raised only to the first power. By contrast, the equation

$$7x^2 + 2x + \left(\frac{d^2y}{dx^2}\right)^3 = 0$$

is of the third degree, since the highest order derivative is cubed. A differential equation of the first degree in which the coefficients are either constants or functions of the independent variable is called a *linear differential equation*. The derivatives in the equation may be of any order:

$$\frac{dy}{dx}, \frac{d^2y}{dx^2}, \frac{d^3y}{dx^3}, \text{ etc.}$$

but note that these are not themselves raised to a power other than one. The equation

$$5x^3 - \frac{dy}{dx} + 7x^2 + 3\frac{d^2y}{dx^2} = 0$$

is an example of a linear differential equation, in this case having constant coefficients. In contrast, equation (8.11) is a non-linear differential equation because of the squaring of the second derivative term. A *first-order differential equation*, as its name implies, contains only the first derivative, and can be written in general as

$$\frac{dy}{dx} = f(x, y)$$

in which the nature of $f(x, y)$ determines solvability.[9] The equation

$$\frac{dy}{dx} = 3x^2 + 7\left(\frac{dy}{dx}\right)^2$$

is an example of a first-order differential equation of the second degree. The equation

$$7x^4 + 3\frac{d^2y}{dx^2} + \left(\frac{dy}{dx}\right)^2$$

is an example of a second-order differential equation of the first degree. Differential equations are important in the study of dynamic processes. There are numerous examples of the use of differential equations in the natural sciences and engineering, including the study of radioactive decay,[10] fluid dynamics and meteorology. In economics, many models of economic growth are expressed in terms of differential equations, while in finance, optimization models using continuous compounding of interest give rise to differential equations.

First-order linear differential equations (constant coefficients)

Equation (8.9) is an example of this type of differential equation. Such equations constitute a special category for which there are simple solution procedures.[11] The *solution* of an ordinary differential equation takes the form

$$y = F(x) + k$$

in which there are *no derivatives*, and which must be consistent with the original equation. Consider equation (8.9). This can now be seen as a linear first-order differential equation in constant coefficients, which can be written more conveniently as

$$\frac{dy}{dx} = 6x^2 + 10$$

Once this rearrangement has been carried out, it becomes clear that the solution can be obtained by taking the integral. Therefore, we can write

$$y = F(x) + k = \int (6x^2 + 10) \, dx$$

$$= 2x^3 + 10x + k$$

and the equation

$$y = 2x^3 + 10x + k$$

is the *general solution* of the differential equation and includes all antiderivatives. In a practical application of differential equations, there will usually be at least one observation of the value of y for given x. For instance, in the current example, if we know that

when $x = 4$, $y = 183$

then for the equation to be satisfied it must be the case that

$$183 = 2(4)^3 + 10(4) + k$$

with the result that

$$k = 15$$

We can therefore write

$$y = 2x^3 + 10x + 15$$

which is called the *particular solution* to the differential equation (8.9). In the particular solution, a specific value is assigned to the constant of integration in the general solution. This is done as a result of particular information linking a figure for y to a specific value of x.[12] Such information is referred to as a *boundary condition* or, in problems where the independent variable is time t and where a value of $F(t) + k$ is specified at $t = 0$, the boundary condition is called an *initial condition*.

Now consider some further examples. First, find the particular solution to

$$1.5x^2 - 4x^3 - 8x + \frac{dy}{dx} + 10 = 0$$

given the boundary condition that [13]

$$y = 300 \quad \text{when} \quad x = 4$$

Reorganizing the original equation,

$$\frac{dy}{dx} = 4x^3 - 1.5x^2 + 8x - 10$$

so, with integration,

$$y = \int (4x^3 - 1.5x^2 + 8x - 10)\, dx$$

and the general solution therefore is

$$y = x^4 - 0.5x^3 + 4x^2 - 10x + k$$

Using the boundary condition in the general solution, when $x = 4$, $y = 300$, so that

$$300 = (4)^4 - 0.5(4)^3 + 4(4)^2 - 10(4) + k$$
$$= 256 - 32 + 64 - 40 + k$$

so that

$$k = 52$$

and the particular solution is therefore

$$y = x^4 - 0.5x^3 + 4x^2 - 10x + 52$$

As a third example, find the particular solution to

$$16x^3 + 12x^2 - 40x - 2\frac{dy}{dx} - 100 = 0$$

with the boundary condition that

$$y = 8 \quad \text{when} \quad x = 2$$

Minor rearrangement produces

$$\frac{dy}{dx} = 8x^3 + 6x^2 - 20x - 50$$

so

$$y = \int (8x^3 + 6x^2 - 20x - 50) \, dx$$
$$= 2x^4 + 2x^3 - 10x^2 - 50x + k$$

which is the general solution of the equation. Now, using the given boundary condition,

$$8 = 2(2)^4 + 2(2)^3 - 10(2)^2 - 50(2) + k$$
$$= 32 + 16 - 40 - 100 + k$$

so

$$k = 100$$

and the particular solution is

$$y = 2x^4 + 2x^3 - 10x^2 - 50x + 100$$

Further examples of first-order linear differential equations will be considered following coverage of exponential and logarithmic functions in Chapter 9.

Higher order differential equations (special cases)

Under this heading, we shall consider briefly certain initial value problems[14] that involve *second* derivatives. For example, the equation

$$\frac{d^2y}{dx^2} - 12x + 8 = 0 \tag{8.12}$$

is a *second-order* differential equation of the first degree. But because the equation contains *only* the second derivative, it is much easier to deal with than second-order differential equations including both first- and second-order derivatives. Nevertheless, two arbitrary constants will be included in the general solution to a second-order equation, and to find a particular solution observations for *two* values of x will be needed.[15] As we shall see, these observations can take various forms. Suppose that the boundary conditions are

when $x = 2$, $y = 20$

and

when $x = 3$, $y = 44$

Integration 561

Now consider the solution to (8.12). Rearrangement produces the more convenient form

$$\frac{d^2y}{dx^2} = 12x - 8$$

Now to begin finding the solution, we use the fact that

$$\int \frac{d^2y}{dx^2} \, dx = \frac{dy}{dx} + \text{constant}$$

and in the present case we have

$$\int (12x - 8) \, dx = 6x^2 - 8x + \text{constant}$$

which result we write as

$$\frac{dy}{dx} = 6x^2 - 8x + k_1 \qquad (8.13)$$

In (8.13), the constant of integration is specified as k_1, since a second constant of integration will also be needed. k_1 is an unknown constant that the boundary conditions will be used to determine in the particular solution. But (8.13) itself is not a solution of the differential equation, since it still contains a derivative. Integration is again needed to complete the elimination of all derivatives and obtain the solution. The result is

$$y = \int (6x^2 - 8x + k_1) \, dx$$

so that

$$y = 2x^3 - 4x^2 + k_1 x + k_2 \qquad (8.14)$$

which is the general solution of the equation. The boundary conditions can now be used to find the particular solution. These will give rise to two simultaneous linear equations in k_1 and k_2 which will then be solved. Substitution into equation (8.14) when $x = 3$ produces

$$44 = 54 - 36 + 3k_1 + k_2$$

and, when $x = 2$,

$$20 = 16 - 16 + 2k_1 + k_2$$

Simplification and rearrangement results in the simultaneous equations

$$3k_1 + k_2 = 26$$
$$2k_1 + k_2 = 20$$

so $\quad k_1 \quad = 6$

so that

$$k_2 = 8$$

and so the particular solution is

$$y = 2x^3 - 4x^2 + 6x + 8 \qquad (8.15)$$

The particular solution (8.15) could have been obtained with data for just *one* value of x, provided that in addition to y the value of dy/dx at this point was also given. For example, in the present case, the information

$$y = 2 \quad \text{when} \quad x = 2, \quad \text{and} \quad \frac{dy}{dx} = 14$$

would have produced the result (8.15). Consider a further example. Find the particular solution of the differential equation:

$$6x^2 - 0.5 \frac{d^2 y}{dx^2} - 6x + 5 = 0$$

given the boundary conditions

$$\frac{dy}{dx} = 22 \quad \text{when} \quad x = 2$$

and

$$y = 8 \quad \text{when} \quad x = 1$$

Rearrangement of the equation produces

$$\frac{d^2 y}{dx^2} = 12x^2 - 12x + 10$$

and integration then results in

$$\frac{dy}{dx} = 4x^3 - 6x^2 + 10x + k_1$$

Now using the information concerning the value of dy/dx,

$$22 = 32 - 24 + 20 + k_1$$

which solves for $k_1 = 6$, so that

$$\frac{dy}{dx} = 4x^3 - 6x^2 + 10x - 6$$

upon which, integration produces

$$y = x^4 - 2x^3 + 5x^2 - 6x + k_2$$

and using the boundary condition

$$8 = 1 - 2 + 5 - 6 + k_2$$

from which emerges the value $k_2 = 10$, and the required particular solution is therefore

$$y = x^4 - 2x^3 + 5x^2 - 6x + 10$$

The problem of recovering the original function given only the form of its nth derivative could be solved in a similar fashion. Note that this would require n boundary conditions, e.g. for given x, the values of the original function and its derivatives down to the $(n - 1)$th derivative.[16] Consider an illustration of the process when $n = 3$. Suppose that the differential equation is

$$12x - \frac{d^3 y}{dx^3} - 1.5 = 0$$

and boundary conditions are specified as, when $x = 1$,

$$\frac{d^2 y}{dx^2} = -1.5$$

$$\frac{dy}{dx} = 0.25$$

$$y = 22.25$$

Rearrangement of the equation gives

$$\frac{d^3 y}{dx^3} = 12x - 1.5$$

and integration produces

$$\frac{d^2 y}{dx^2} = 6x^2 - 1.5x + k_1$$

which, when $x = 1$, is equal to -1.5, from which it follows that $k_1 = -6$. Thus:

$$\frac{d^2 y}{dx^2} = 6x^2 - 1.5x - 6$$

Further integration then produces

$$\frac{dy}{dx} = 2x^3 - 0.75x^2 - 6x + k_2$$

which, when $x = 1$, takes the value 0.25, and therefore $k_2 = 5$, so that

$$\frac{dy}{dx} = 2x^3 - 0.75x^2 - 6x + 5$$

Integration then produces

$$y = 0.5x^4 - 0.25x^3 - 3x^2 + 5x + k_3$$

which, when $x = 1$, is equal to 22.5, so that $k_3 = 20$, and the required particular solution to the differential equation is therefore

$$y = 0.5x^4 - 0.25x^3 - 3x^2 + 5x + 20$$

A first-order, second-degree example

We conclude the section with a very simple example of a non-linear differential equation. Consider the equation

$$2\left(\frac{dy}{dx}\right)^2 - 8x^2 + 112x - 392 = 0$$

which is an ordinary differential equation of the *first order* and *second degree*. First consider the general solution. Rearrangement produces

$$\left(\frac{dy}{dx}\right)^2 = 4x^2 - 56x + 196$$

and factoring the quadratic gives

$$\left(\frac{dy}{dx}\right)^2 = 4(x - 7)^2$$

Taking the square root of both sides produces

$$\frac{dy}{dx} = \pm 2(x - 7)$$

and therefore the integrals of $2x - 14$ and $14 - 2x$ will represent general solutions. That is:

(i) $y = x^2 - 14x + k$

and

(ii) $y = 14x - x^2 + k$

are both general solutions. The context of an application may suggest which of (i) and (ii) best fits the situation. For example, if y represents the total costs of production and if x is output level with typical values around 50 (say), then (i) is the appropriate form. Variable costs, $14x - x^2$, would, unrealistically, be given as negative by (ii) for values of x in excess of 14.

Working on the assumption that (i) is more appropriate, then given the boundary condition that

$$y = 2000 \quad \text{when} \quad x = 45$$

the constant of integration (representing fixed costs in this context) emerges from (i) as $k = 325$. So the appropriate particular solution is

$$y = x^2 - 14x + 325$$

In presenting this illustration of a higher degree differential equation, it cannot be overemphasized that non-linear differential equations are usually much more difficult to solve.

EXERCISES 8.4

1 Give the degree and order of the following differential equations:

(i) $10x^3 - 3x + \left(\dfrac{d^3 y}{dx^3}\right)^2 = 0$

(ii) $\dfrac{d^2 y}{dx^2} + 4 \dfrac{dy}{dx} - 7 = 0$

(iii) $20x - \left(\dfrac{d^2 y}{dx^2}\right)^4 + \left(\dfrac{d^3 y}{dx^3}\right)^3$

(iv) $6x^7 - 12 \dfrac{dy}{dx} + 100 = 0$

(v) $100x^2 + \left(\dfrac{dy}{dx}\right)^4 - 17x^5 = 0$

(vi) $27x^3 - \dfrac{d^4 y}{dx^4} + \dfrac{d^3 y}{dx^3} + 10 \dfrac{dy}{dx} = 0$

2 Find the general solutions to the following differential equations:

(i) $\dfrac{dy}{dx} = 8x^3 + 9x^2 + 5x + 10$

(ii) $-2x^3 - 4.5x^2 + \dfrac{dy}{dx} - 7 = 0$

(iii) $100 - \dfrac{dy}{dx} + 10x^9 = 0$

(iv) $27x^2 - x^3 - 0.5 \dfrac{dy}{dx} + 17 = 0$

566 Mathematics for business, finance and economics

3 Find particular solutions to the following differential equations with the given boundary conditions:

(i) $\dfrac{dy}{dx} = 5x^4 - 4x^3 + 3x^2 + 1$

Boundary condition: $y = 300$ when $x = 3$

(ii) $x^3 - 1.5x^2 - \dfrac{dy}{dx} + 4x + 10 = 0$

Boundary condition: $y = 90$ when $x = 4$

(iii) $24x^3 - 90x^2 + 3\dfrac{dy}{dx} - 6x - 15 = 0$

Boundary condition: $y = 100$ when $x = 5$

(iv) $2.5x^4 + 3x^3 - 0.25\dfrac{dy}{dx} - 5x + 5 = 0$

Boundary condition: $y = 112$ when $x = 2$

4 Find particular solutions in the following cases:

(i) $10 - \dfrac{dy}{dx} + 2x^{-3} - 3x^{-2} = 0$

Boundary condition: $y = 35$ when $x = 0.2$

(ii) $10.5x^6 - 0.5\dfrac{dy}{dx} - 10x^{-6} = x + 16$

Boundary condition: $y = 0$ when $x = 1$

5 Find particular solutions to the following differential equations:

(i) $10 - \dfrac{d^2y}{dx^2} - 30x = 0$

Boundary conditions: $y = 0$ when $x = 3$
$y = -210$ when $x = 5$

(ii) $6x - 0.25\dfrac{d^2y}{dx^2} = 3$

Boundary conditions:
when $x = 6$ $y = 672$
$\dfrac{dy}{dx} = 362$

6 Find the particular solution to the third-order differential equation

$$6 - 0.1 \frac{d^3y}{dx^3} = 0$$

given that, when $x = 1$,

$$\frac{d^2y}{dx^2} = 70$$

$$\frac{dy}{dx} = 60$$

$$y = 50$$

7 Given the second-degree equation

$$0.9x^2 - 0.1\left(\frac{dy}{dx}\right)^2 - 18x + 90 = 0$$

and the information that

(a) the variable y represents the costs of production with output, x, typically in the range 50–150 units;
(b) when the output level is 100, costs in total are 20,000;

find the appropriate particular solution.

8.5 INTEGRATION BY SUBSTITUTION

Integration can sometimes be simplified by a change of variable in the integrand. An appropriate substitution is determined, the simpler expression in the new variable is integrated and the substitution is then reversed. This is the essence of the *method of substitution* – one of the most useful methods of integration which has its roots in the power and chain rules for differentiation.

Consider an example. It is possible to work out the indefinite integral

$$\int (x^2 - 7)^4 2x \, dx$$

by expanding the bracket and using the sum–difference rule. But the exercise is easier if an appropriate substitution – a change of variable – is made. Suppose that we let

$$u = x^2 - 7$$

568 Mathematics for business, finance and economics

With this change only, the integrand would then be

$$\int u^4 2x \, dx$$

which looks worse than the original expression! But what is being sought is an expression *entirely* in terms of u. To achieve this, the substitution must be carried through to any remaining parts of terms in x, and du and dx need to be related. Now since $u = x^2 - 7$,

$$\frac{du}{dx} = 2x \qquad (8.16)$$

Interpreting du/dx in (8.16) as a *ratio*, we can write

$$du = 2x \, dx \qquad (8.17)$$

In (8.17), $2x \, dx$ is the *differential*[17] of u. The differential has many uses,[18] and its utility here is that (8.17) can be used to express the integrand solely in terms of u. When $2x \, dx$ is replaced by du, the simplicity of the resulting integral is evident:

$$\int u^4 \, du = \frac{u^5}{5} + k \qquad (8.18)$$

The final step is to convert the right-hand side of (8.18) back to an expression in x, the role of u as an intermediate device being complete. Making this substitution, the result is

$$\int (x^2 - 7)^4 2x \, dx = \frac{(x^2 - 7)^5}{5} + k$$

Now consider a further example of integration by substitution. Suppose that it is required to find the following indefinite integral:

$$\int (x^2 + 2x - 4)(x + 1) \, dx$$

We attempt as the substitution to set u equal to the content of the major bracketed component of the integrand.[19] In doing this, the hope is that the remainder of the expression will relate conveniently to du. This turns out to be the case here. Making the substitution

$$u = x^2 + 2x - 4$$

and following the steps of the first example above, the integrand initially becomes

$$u(x + 1) \, dx$$

Taking the derivative:

$$\frac{du}{dx} = 2x + 2$$

We can write

$$du = (2x + 2) \, dx$$

or more to the point

$$(x + 1) \, dx = \frac{du}{2}$$

The integrand and indefinite integral are

$$\int \frac{u}{2} \, du = \frac{u^2}{4} + k$$

which on re-expression in x becomes

$$\frac{(x^2 + 2x - 4)^2}{4} + k$$

The integrand is simple enough for the result to be checked by expansion and integration of the original expression directly in terms of x. This is left as an exercise for the interested reader. We now consider the first of two examples in which the method of substitution comes clearly into its own. Let us find the indefinite integral

$$\int 0.5x(x^2 + 10)^{1/2} \, dx$$

For the substitution, let

$$u = x^2 + 10$$

so that

$$du = 2x \, dx \quad \text{or} \quad 0.5x \, dx = \frac{du}{4}$$

Substitution in the integrand produces

$$\int u^{1/2} 0.5x \, dx = \int \frac{u^{1/2}}{4} \, du = \frac{1}{4} \int u^{1/2} \, du$$

$$= \frac{1}{4} \left(\frac{2}{3} u^{3/2} \right) + k$$

$$= \frac{u^{3/2}}{6} + k$$

$$= \frac{(x^2 + 10)^{3/2}}{6} + k$$

For the second example consider the integral of a rational expression:

$$\int \frac{(x^{0.2} - 1)^4}{x^{0.8}} \, dx$$

The integrand can be re-expressed as

$$x^{-0.8}(x^{0.2} - 1)^4$$

As the substitution try

$$u = x^{0.2} - 1$$

from which will result

$$du = 0.2 x^{-0.8} \, dx$$

so that

$$x^{-0.8} \, dx = 5 \, du$$

In terms of the substitute variable, the required integral has now become

$$\int 5u^4 \, du = u^5 + k$$

$$= (x^{0.2} - 1)^5 + k$$

The substitution method is only of value if the substitution simplifies the integrand. Not all possible substitutions have this effect. The approach is likely to be productive when the integrand can be expressed as a constant multiple of the product of a function of the substitute variable and its derivative, i.e.

$$kf(u) \frac{du}{dx} \tag{8.19}$$

In the simpler cases $k = 1$. It will be seen that the examples we have considered above are of the form (8.19). The substitution approach is less likely to be of help if the substitution results in a variable multiple (e.g. x or x^2) in place of k. Finally, note that:

1 more than one possible substitution may be effective;
2 a *sequence* of substitutions may be needed for complicated expressions;
3 there may be *no* substitution that is effective.

In respect of 3 there are many deceptively simple looking expressions for which the integral is not an algebraic function: for example $(x^3 + 1)^{1/2}$.[20]

EXERCISES 8.5

1 Use integration by substitution to find the following integrals:

(i) $\int (10 + x^2)^3 2x \, dx$

(ii) $\int (2x^2 + 6x + 15)^3 (2x + 3) \, dx$

(iii) $\int 4x(x^2 - 15)^{1/2} \, dx$

(iv) $\int (6x^2 - 8x + 20)(3x - 2) \, dx$

2 Find the following indefinite integrals:

(i) $\int 3x(5x^2 - 19)^{1/2} \, dx$

(ii) $\int -4(5 + x)^{-5} \, dx$

(iii) $\int \dfrac{dx}{(x - 5)^2}$

(iv) $\int \dfrac{-6}{(3x + 4)^2} \, dx$

3 For $d \neq 0$, $b \neq 0$, $n \neq 1$ confirm that

$$\int \frac{d}{(bx + c)^n} \, dx = \frac{d}{b(1 - n)(bx + c)^{n-1}} + k$$

4 Find:

(i) $\int \dfrac{2x + 7}{(x^2 + 7x - 8)^5} \, dx$

(ii) $\int \dfrac{6x^2 + 4x + 2}{(x^3 + x^2 + x + 1)^3} \, dx$

8.6 THE DEFINITE INTEGRAL

While the indefinite integral of a function of x is a function, the *definite integral* is a number. This number is the difference between the values of the indefinite integral worked out for two values of x. These values of x are the endpoints of an interval in the domain, and are called the *limits of integration*. A value for the definite integral may be all that is required in a particular application. This can be advantageous, as many expressions have no antiderivative that can be expressed in terms of elementary functions.[21] In such cases the *definite* integral may be sought by numerical methods. As we shall see, the value taken by the definite integral of a function of a single variable corresponds to the *area under the curve* of the integrand between the limits of integration. Indeed, it was the problem of finding the area enclosed by a curved boundary that gave rise to the integral calculus. The definite integral of a function $f(x)$ between the limits of integration a and b is written as

$$\int_{x=a}^{x=b} f(x)\, dx$$

for which

$$\int_{x=a}^{x=b} f(x)\, dx = [F(x)+k]^{\text{for }x=b} - [F(x)+k]^{\text{for }x=a} \qquad (8.20)$$

where, in (8.20), $F(x)+k$ is the indefinite integral; a is the *lower* and b the *upper* limit of integration. Note that the constant of integration, k, will cancel as the integral is evaluated. For the first example, consider the definite integral of $f(x)=2x$ from $x=2$ to $x=5$:

$$\int_{x=2}^{x=5} 2x\, dx = [x^2+k]^{\text{for }x=5} - [x^2+k]^{\text{for }x=2}$$
$$= 25+k - [4+k]$$
$$= 21$$

In graphical terms, the definite integral is the difference between the heights (distances above the x axis) of the curve $F(x)=x^2+k$ at $x=5$ and $x=2$. This is illustrated in Figure 8.2, which is drawn for $k=0$.

In the theory of the firm, if $f(x)=2x$ represented marginal cost, then $F(x)$ would be total variable cost. The definite integral then represents the change in total variable costs when output increases from $x=2$ to $x=5$. The output increase could only be justified in profit terms if there was at least an equivalent increase in total revenue between $x=2$ and $x=5$. Consider some further examples.

(i) $\displaystyle\int_{x=2}^{x=3} (x^3-1)\, dx = [0.25x^4-x]^{x=3} - [0.25x^4-x]^{x=2}$

$$= [0.25(81)-3] - [0.25(16)-2]$$
$$= 17.25 - 2$$
$$= 15.25$$

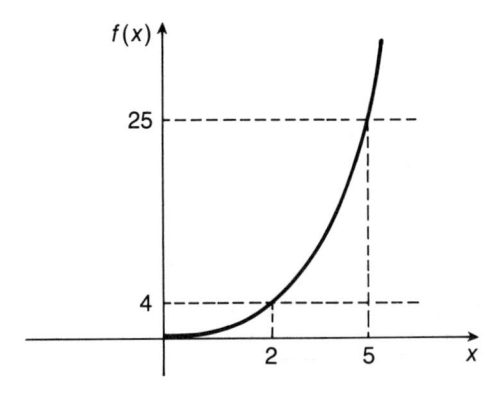

Figure 8.2

(ii) $\int_{x=1}^{x=4} (6x^2 - 4x + 4)\, dx = [2x^3 - 2x^2 + 4x]^{x=4} - [2x^3 - 2x^2 + 4x]^{x=1}$

$= 128 - 32 + 16 - [2 - 2 + 4]$

$= 108$

The lower limit of integration is frequently zero, as in the next example.

(iii) $\int_{x=0}^{x=8} 0.625 x^4\, dx = [0.125 x^5]^{x=8} - 0$

$= 0.125(32{,}768)$

$= 4096$

Neither the limits of integration nor the definite integral have to be positive numbers. This is illustrated in the following case.

(iv) $\int_{x=-3}^{x=5} (4x - 20)\, dx = [2x^2 - 20x]^{x=5} - [2x^2 - 20x]^{x=-3}$

$= -50 - 78$

$= -128$

Figure 8.3 shows the graph of $F(x) = 2x^2 - 20x$.

In Figure 8.3 note that the height of $F(x)$ is positive at $x = -3$ and negative at $x = 5$; hence the difference will be negative as the position at the upper limit of integration is taken first. Now consider an integrand with a rational exponent.

(v) $\int_{x=1}^{x=4} 3(\sqrt{x})\, dx = \int_{x=1}^{x=4} 3x^{1/2}\, dx$

$= [2x^{3/2}]^{x=4} - [2x^{3/2}]^{x=1}$

$= 16 - 2$

$= 14$

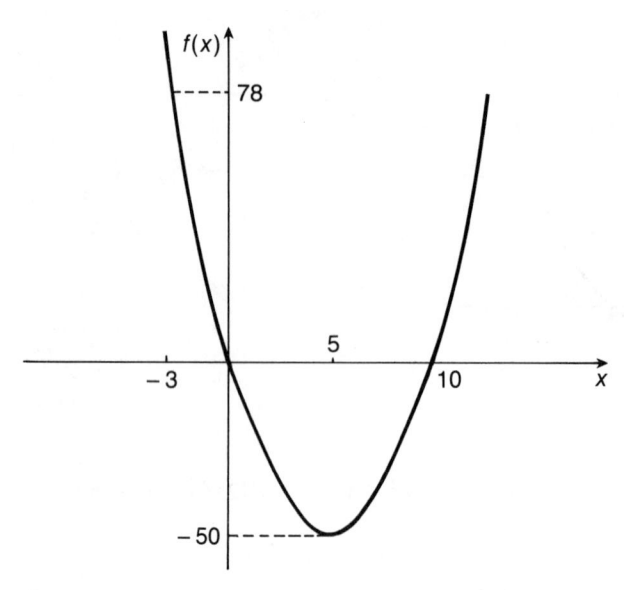

574 Mathematics for business, finance and economics

Figure 8.3

Where the upper or lower limit of integration is infinite, we speak of an *improper integral*. For example, if the definite integral

$$\int_{x=a}^{x=U} f(x)\,dx$$

approaches the value D as the upper limit of integration, U, increases without limit, then the improper integral

$$\int_{x=a}^{x=\infty} f(x)\,dx$$

exists and its value is equal to D. For example:

(vi) $\int_{x=1}^{x=\infty} \dfrac{50}{x^2}\,dx = \left[-\dfrac{50}{x}\right]_{x=1}^{x=\infty}$

$= 50$

Improper integrals in which one or both limits of integration are infinite often arise in applications involving probability. An integral is also said to be improper if the value of the integrand becomes infinite at or between the limits of integration. For example:

(vii) $\int_{x=0}^{x=4} \dfrac{10}{\sqrt{x}}\,dx = [20\sqrt{x+k}]^{x=4} - [20\sqrt{x+k}]^{x=0}$

$= 40$

In example (vii), although $20\sqrt{x}$ has a definite value at $x = 0$, the integrand, $10/\sqrt{x}$, is not defined at the lower limit. In cases where the integrand becomes infinite at a point *between* the limits of integration, the improper integral, where it exists, can be obtained by dividing the range of integration into two parts in which the problematic value of x is the upper limit of integration in one part and the lower limit in the other. The required overall value is the sum of the two (still improper) integrals.[22] It was stated at the outset that the definite integral could be seen as the area 'under' the integrand function between the upper and lower limits of integration. We can now illustrate this property in the case of the function $f(x) = 2x$ between the limits $x = 2$ and $x = 5$. We saw from Figure 8.2 that the definite integral was the difference, 21, in the height of x^2 at $x = 5$ and $x = 2$. In terms of the integrand, $2x$, 21 is the area between the interval $x = 2$ to $x = 5$ and $f(x) = 2x$. This is illustrated in Figure 8.4.

In Figure 8.4, the area of interest is that of the trapezium made up of the rectangle, with area R, and the triangle, with area T. The area of the rectangle is base times height, so therefore

$$R = (5 - 2)f(2) = 12$$

The triangular region has an area of one-half of the base times height, so that

$$T = \tfrac{1}{2}(5 - 2)[f(5) - f(2)]$$
$$= 1.5(10 - 4) = 9$$

So the area between the line and the x axis above the interval is given by

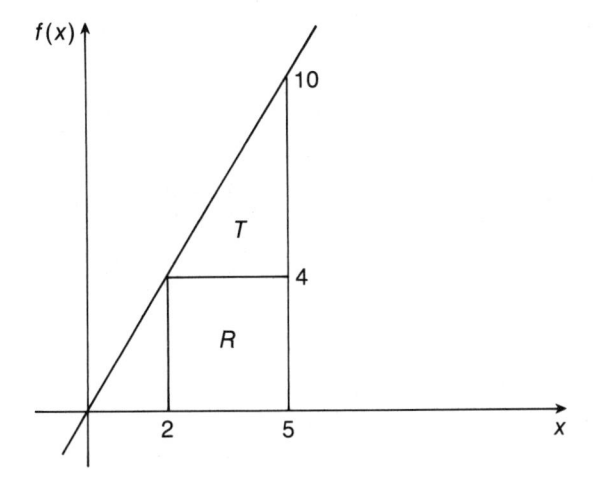

$$\text{area} = R + T$$
$$= 12 + 9 = 21$$

Figure 8.4

The fact that, whatever the limits of integration, the definite integral is the area under the curve can easily be confirmed in the case of $f(x) = 2x$. With limits of integration a and b the definite integral is

$$\int_{x=a}^{x=b} 2x \, dx = [x^2]^{x=b} - [x^2]^{x=a}$$
$$= b^2 - a^2$$

The rectangular area is

$$R = (b-a)f(a)$$
$$= (b-a)2a$$
$$= 2ba - 2a^2$$

The triangular area will be

$$T = \tfrac{1}{2}(b-a)[f(b) - f(a)]$$
$$= \tfrac{1}{2}(b-a)(2b - 2a)$$
$$= (b-a)^2$$
$$= b^2 - 2ba + a^2$$

So the trapezoidal area as a whole is

$$\text{area} = R + T$$
$$= 2ba - 2a^2 + b^2 - 2ba + a^2$$
$$= b^2 - a^2$$

In a context of marginal and total variable costs, total variable cost at any output is the *area under the marginal cost curve* from the origin to the said output level. The total area can be seen as the sum of successive trapezoidal columns with unit base, these areas being the incremental costs of successive units of output. We now set out rules for, and properties of, definite integrals.

(i) $$\int_{x=a}^{x=b} cf(x) \, dx = c \int_{x=a}^{x=b} f(x) \, dx \qquad (8.21)$$

The constant multiple rule applies to definite as well as indefinite integrals. This fairly obvious property can be very useful on occasion. For example, consider the definite integral

$$\int_{x=10}^{x=20} \frac{\pi}{2} x \, dx = \pi \int_{x=10}^{x=20} \frac{x}{2} \, dx$$
$$= \pi([0.25x^2 + k]^{\text{for } x=20} - [0.25x^2 + k]^{\text{for } x=10})$$
$$= \pi([100 + k] - [25 + k]) = 75\pi$$

The next rule is also familiar from rules for indefinite integrals:

(ii) $$\int_{x=a}^{x=b} [f(x) \pm g(x)] \, dx = \int_{x=a}^{x=b} f(x) \, dx \pm \int_{x=a}^{x=b} g(x) \, dx \quad (8.22)$$

The sum–difference rule is applied in the following example.

$$\int_{x=0}^{x=2} (2x + 3x^2) \, dx = \int_{x=0}^{x=2} 2x \, dx + \int_{x=0}^{x=2} 3x^2 \, dx$$

After integration of both sides, the definite integral is evaluated as

$$(x^2 + x^3)^{x=2} - (x^2 + x^3)^{x=0}$$
$$= [(x^2)^{x=2} - (x^2)^{x=0}] + [(x^3)^{x=2} - (x^3)^{x=0}]$$

i.e.

$$12 = 4 + 8$$

The third rule is

(iii) $$\int_{x=a}^{x=b} f(x) \, dx = -\int_{x=b}^{x=a} f(x) \, dx \quad (8.23)$$

It follows from this fairly obvious property that

$$\int_{x=a}^{x=a} f(x) = 0$$

We now turn to a very useful result to which we referred in the discussion of improper integrals. For a point c within the range of integration,

$$a \leqslant c \leqslant b$$

the definite integral can be broken down into two (or more) parts:

(iv) $$\int_{x=a}^{x=b} f(x) \, dx = \int_{x=a}^{x=c} f(x) \, dx + \int_{x=c}^{x=b} f(x) \, dx \quad (8.24)$$

In addition to possible use with improper integrals, we shall see two further situations in which this rule is useful. But first consider two examples of integrands that could be evaluated without the rule. Given the requirement to find

$$\int_{x=1}^{x=3} 2x \, dx$$

the result follows directly as

$$(x^2)^{x=3} - (x^2)^{x=1} = 9 - 1 = 8$$

The same outcome could have been obtained by selecting (say) $c = 2$ and evaluating the definite integral in two parts as follows:

$$\int_{x=1}^{x=2} 2x \, dx + \int_{x=2}^{x=3} 2x \, dx$$
$$= (x^2)^{x=2} - (x^2)^{x=1} + [(x^2)^{x=3} - (x^2)^{x=2}]$$
$$= 3 + 5 = 8$$

Now find

$$\int_{x=2}^{x=8} (0.75x^2 + 6x + 10) \, dx$$
$$= (0.25x^3 + 3x^2 + 10x)^{x=8} - (0.25x^3 + 3x^2 + 10x)^{x=2}$$
$$= 400 - 34$$
$$= 366$$

With, for example, the value of c set at 6, the result is

$$(0.25x^3 + 3x^2 + 10x)^{x=6} - (0.25x^3 + 3x^2 + 10x)^{x=2}$$
$$+ (0.25x^3 + 3x^2 + 10x)^{x=8} - (0.25x^3 + 3x^2 + 10x)^{x=6}$$
$$= 230 - 34 + 400 - 230$$
$$= 196 + 170$$
$$= 366$$

The rule really comes into its own in two types of problem that follow. The first case is *piecewise-defined* functions, first encountered in Chapter 4. Clearly, the place at which the range of integration is divided, c, would coincide with the point where the definition of the function changes, as in the problem

$$\int_{x=3}^{x=8} f(x) \, dx$$

where

$$f(x) = \begin{cases} 2x + 5 & \text{for } x \leq 6 \\ 5x - 10 & \text{for } x > 6 \end{cases}$$

Figure 8.5 shows the graph of the function and the area under the curve corresponding to the definite integral.

The best place to set the break point is at $c = 6$. Excluding the constant of integration, the indefinite integral will be

$$F(x) = \begin{cases} x^2 + 5x & \text{for } x \leq 6 \\ 2.5x^2 - 10x & \text{for } x > 6 \end{cases}$$

f(x)

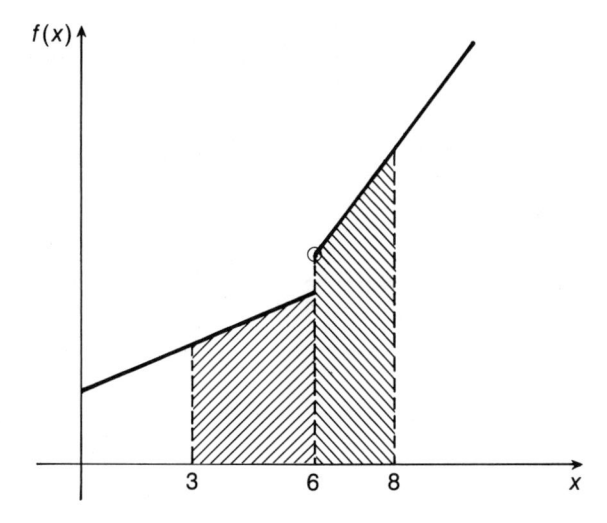

Figure 8.5

3   6   8   *x*

so that the definite integral will then be

$$(x^2 + 5x)^{x=6} - (x^2 + 5x)^{x=3} + (2.5x^2 - 10x)^{x=8} - (2.5x^2 - 10x)^{x=6}$$
$$= 66 - 24 + 80 - 30$$
$$= 92$$

The second situation in which (8.24) is of real value is when the graph of the integrand crosses the x axis between the limits of integration. Consider the problem of finding the definite integral

$$\int_{x=4}^{x=6} (x-5) \, dx = (0.5x^2 - 5x)^{x=6} - (0.5x^2 - 5x)^{x=4}$$
$$= (18 - 30) - (8 - 20)$$
$$= 0$$

This result is correct and may indeed be the relevant value in some contexts.[23] But as is clear from Figure 8.6, there is *some* space between the integrand and the x axis.

The problem is that a negative sign is attached to areas under the x axis. The positive and negative components offset each other and in this case cancel altogether. In a particular application we may need to know the total area regardless of sign. This value can be found by working out the integral

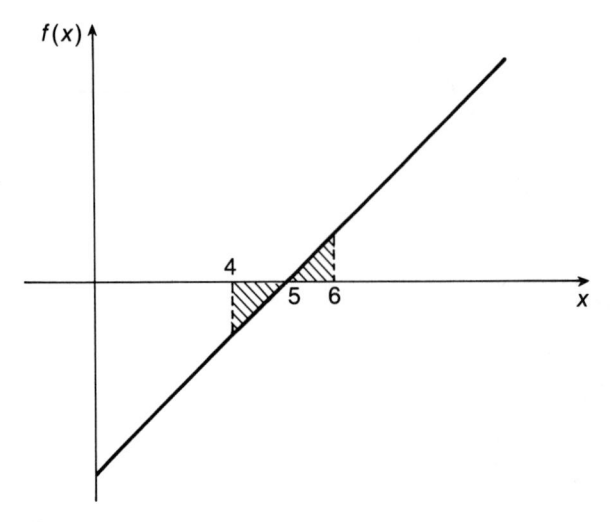

580 Mathematics for business, finance and economics

Figure 8.6

either side of the zero of the function and *ignoring sign*. In the present example this is found as follows:

$$\int_{x=4}^{x=5} (x-5) \, dx = (0.5x^2 - 5x)^{x=5} - (0.5x^2 - 5x)^{x=4}$$

$$= -12.5 - (-12)$$
$$= (-)0.5$$

Where the function is above the axis,

$$\int_{x=5}^{x=6} (x-5) \, dx = (0.5x^2 - 5x)^{x=6} - (0.5x^2 - 5x)^{x=5}$$

$$= -12 - (-12.5)$$
$$= (+)0.5$$

The absolute values of the two areas are now added to give the total area enclosed between the function and the x axis. This is one unit. Consider a further example that will also illustrate the fact that the rule (8.24) can be extended to involve as many subdivisions of the range of integration as may be necessary. We shall find

$$\int_{x=0}^{x=16} (3x^2 - 36x + 60) \, dx$$

If one or both roots of the quadratic fall between the limits of integration, then subdivision of the interval will be required. The roots are $x = 2$ and

$x = 10$ and the absolute value of the area is found by dividing the range of integration into the sections

0–2   2–10   10–16

The indefinite integral (constant omitted) is

$$\int (3x^2 - 36x + 60) \, dx = x^3 - 18x^2 + 60x$$

Therefore

$$\int_{x=0}^{x=2} = (x^3 - 18x^2 + 60x)^{x=2} - (x^3 - 18x^2 + 60x)^{x=0}$$

$$= 56 - 0 = 56$$

$$\int_{x=2}^{x=10} = (x^3 - 18x^2 + 60x)^{x=10} - (x^3 - 18x^2 + 60x)^{x=2}$$

$$= -200 - 56 = -256$$

$$\int_{x=10}^{x=16} = (x^3 - 18x^2 + 60x)^{x=16} - (x^3 - 18x^2 + 60x)^{x=10}$$

$$= 448 - (-200) = 648$$

So the total area between the curve and the x axis, regardless of sign, is

$$56 + 256 + 648 = 960$$

If sign had been taken into account, the result would have been

$$\int_{x=0}^{x=16} = (x^3 - 18x^2 + 60x)^{x=16} - (x^3 - 18x^2 + 60x)^{x=0}$$

$$= 448 - 0$$

$$= 448$$

Where it is required to find the area enclosed between two curves, rather than the area between one curve and the x axis, the area can be obtained with the use of definite integrals if the functions concerned are non-negative over the relevant interval. If the two curves are those of $y = f(x)$ and $y = g(x)$, and the limits between which the area is required are a and b, then the definite integral needed is

$$\int_{x=a}^{x=b} [f(x) - g(x)] \, dx$$

provided that $f(x) \geq g(x)$ between a and b. If it is more convenient, the integral could be expressed as

$$\int_{x=a}^{x=b} f(x)\,dx - \int_{x=a}^{x=b} g(x)\,dx \tag{8.25}$$

Figure 8.7 graphs a possible situation for parabolic $f(x)$ and linear $g(x)$. The enclosed area is shaded.

In Figure 8.7, it is presumed that the required area is the entire region between the curves, although the definite integral (8.25) is not restricted to this case. For a numerical exercise, suppose that:

$$f(x) = 125 + 20x - x^2$$

and

$$g(x) = 80 + 8x$$

Apart from the steepness of the curves, the problem graphs as shown in Figure 8.7. It is assumed that the entire region is required, so the first task is to find the points of intersection of the curves. These points give the two limits of integration, a and b. Begin by equating the values of the functions:

$$f(x) = g(x)$$
$$125 + 20x - x^2 = 80 + 8x$$

so that

$$x^2 - 12x - 45 = 0$$

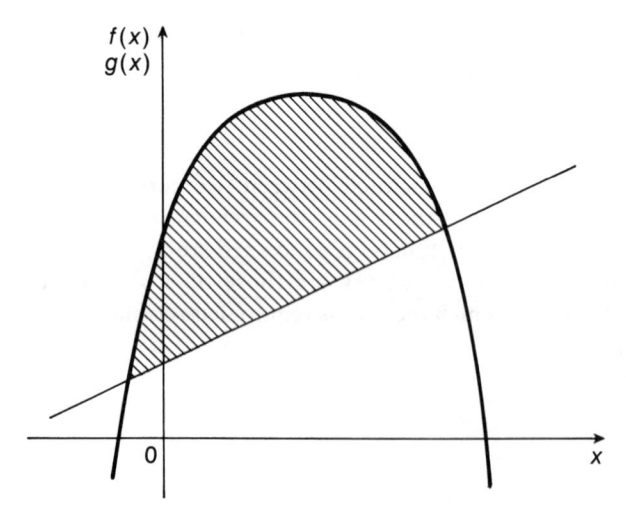

Figure 8.7

which conveniently factors as

$$(x+3)(x-15) = 0$$

so that the limits of integration are

$$a = -3 \quad \text{and} \quad b = +15$$

Carrying out the integration in the form of (8.25),

$$\int_{x=-3}^{x=15} (-x^2 + 12x + 45)\, dx$$

$$= [-x^3/3 + 6x^2 + 45x]^{x=15} - [-x^3/3 + 6x^2 + 45x]^{x=-3}$$

$$= 900 - [-72] = 972$$

So the area enclosed between the curves is 972. In problems involving areas enclosed between curves, care must be taken if the curves cross within the range of integration. This is the reason for the stipulation that $f(x) \geq g(x)$ between a and b. If this is not the case, the range of integration can be broken up into intervals within which one function is never less than the other.

EXERCISES 8.6

1 Evaluate the following definite integrals:

(i) $\displaystyle\int_{3}^{7} 4x\, dx$

(ii) $\displaystyle\int_{-5}^{-1} (6x - 3)\, dx$

(iii) $\displaystyle\int_{0}^{20} (0.9x^2 + x + 10)\, dx$

(iv) $\displaystyle\int_{4}^{16} 12x^{0.5}\, dx$

2 Find the following definite integrals:

(i) $\displaystyle\int_{0.5}^{\infty} \frac{100}{x^2}\, dx$

(ii) $\displaystyle\int_{4}^{25} \frac{25}{\sqrt{x}}\, dx$

(iii) $\displaystyle\int_{2}^{5} 3\pi x^2\, dx$

(iv) $\int_{-10}^{0} (x^4 + 3x^2 + 20) \, dx$

3 Find the area enclosed between the x axis and the following curves between the given limits:

  (i) $f(x) = 3x^2 + 2x + 5$

  between the limits $x = 0$ and $x = 10$.

  (ii) $f(x) = x^3 + 10x$

  between the limits $x = -10$ and $x = -2$.

  (iii) $f(x) = 1.5x^2 + 12x + 30$

  between the limits $x = -5$ and $x = 5$.

4 Find the total area under the curves given by (i) and (ii) below, between the given limits, and regardless of sign:

  (i) $f(x) = x - 10$ between $x = 5$ and $x = 15$
  (ii) $f(x) = 3x^2 - 18x + 15$ between $x = 0$ and $x = 8$

5 Evaluate:

  (i) $\int_{0}^{40} f(x) \, dx$ where $f(x) = \begin{cases} 20 & \text{for } x \leq 20 \\ x & \text{for } x > 20 \end{cases}$

  (ii) $\int_{5}^{15} f(x) \, dx$ where $f(x) = \begin{cases} 4x - 20 & \text{for } x \leq 10 \\ x + 10 & \text{for } x > 10 \end{cases}$

8.7 NUMERICAL INTEGRATION

We have evaluated a definite integral by finding a formula for an antiderivative of the function, working out the value of the antiderivative at the endpoints of the range, and taking the difference. But it is not always convenient or even possible to proceed in this way. Some indefinite integrals are very difficult to obtain, while for other expressions there is *no* elementary formula for the antiderivative.[24] Numerical methods compute an approximation to the definite integral and in this section we examine two methods for the evaluation of definite integrals: the *trapezium rule* and *Simpson's rule*. First consider the trapezium rule. Suppose that we require the definite integral for a function for which a section of the curve is shown in Figure 8.8.

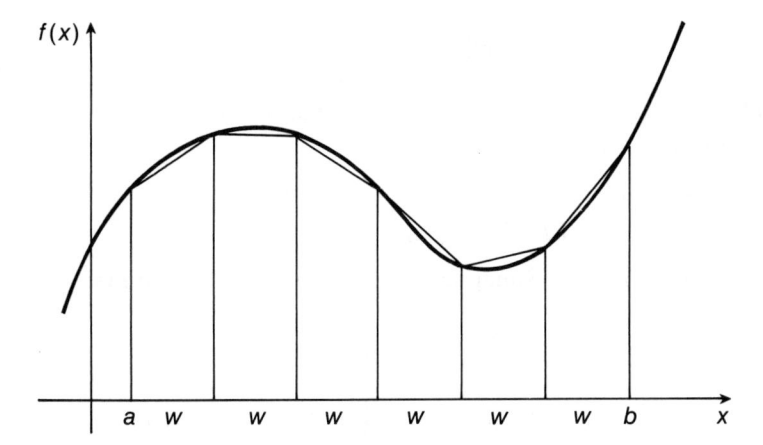

Figure 8.8

In Figure 8.8, the limits of integration are a (lower limit) and b upper limit. The interval a, b has been divided into a number of sections or *steps* – here, arbitrarily, six. The section width or *step size* is w given by

$$w = \frac{b-a}{6}$$

In each step a trapezium is formed, the base of which is the step width, the parallel sides are the ordinates at the beginning and endpoint of the section, while the fourth side is the straight line segment connecting the tops of the ordinates.[25] The area of the trapezium approximates the area under the curve in each section, the accuracy of the approximation within each step being the better, the smaller the step size is. With $f(x)$ being the value of the function (the height of the ordinate) at any point, the areas of the trapezia are

$$T_1 = \frac{w}{2} [f(a) + f(a+w)]$$

$$T_2 = \frac{w}{2} [f(a+w) + f(a+2w)]$$

$$T_3 = \frac{w}{2} [f(a+2w) + f(a+3w)]$$

$$T_4 = \frac{w}{2} [f(a+3w) + f(a+4w)]$$

$$T_5 = \frac{w}{2} [f(a+4w) + f(a+5w)]$$

$$T_6 = \frac{w}{2}[f(a+5w)+f(a+6w)]$$

So the total area of the trapezia and the approximation to the value of the definite integral is T where

$$T = 0.5w[f(a) + 2f(a+w) + 2f(a+2w) + 2f(a+3w)$$
$$+ 2f(a+4w) + 2f(a+5w) + f(a+6w)]$$

The trapezium rule for approximating definite integrals is as follows.

Trapezium rule

$$\int_a^b f(x)\,dx \approx \frac{w}{2}[f(a) + 2f(a+w) + 2f(a+2w) + \cdots + f(a+mw)]$$

where m is the number of steps into which the interval a, b is apportioned. For an example of the use of the trapezium rule, the rule will be used to approximate the area under the curve $y = x^2$ between $x = 1$ and $x = 4$. We shall arbitrarily divide the interval into six steps, so that the step size w will be

$$w = \frac{4-1}{6} = 0.5$$

Using the trapezium rule:

$$T = \frac{0.5}{2}[f(1) + 2f(1.5) + 2f(2) + 2f(2.5) + 2f(3) + 2f(3.5) + f(4)]$$
$$= 0.25[(1)^2 + 2(1.5)^2 + 2(2)^2 + 2(2.5)^2 + 2(3)^2 + 2(3.5)^2 + (4)^2]$$
$$= 0.25[1 + 4.5 + 8 + 12.5 + 18 + 24.5 + 16]$$
$$= 0.25(84.5) = 21.125$$

Now, in the case of a function such as $f(x) = x^2$, obtain the exact value of the definite integral by analytical means:

$$\int_1^4 x^2\,dx = \tfrac{1}{3}[(4)^3 - (1)^3] = 21$$

So in this case the trapezium rule is out by just under 0.6 per cent. Given the shape of the function $y = x^2$, it can be seen that since the area of each trapezium is slightly greater than the true area under the curve in its section, the result must be an overall overestimate. With other functions, the result may be either an overestimate or an underestimate of the true value. It can be shown that accuracy increases as the step size is reduced.[26] Table 8.1 gives the results of the trapezium rule for the current example and for a range of step sizes.

Table 8.1

Steps	T	Error	Percentage error
1	25.50000	4.5000	21.4286
2	22.12500	1.1250	5.3571
3	21.50000	0.5000	2.3810
4	21.28125	0.2813	1.3393
5	21.18000	0.1800	0.8572
6	21.12500	0.1250	0.5952
10	21.04500	0.0450	0.2143
15	21.02000	0.0200	0.0952
20	21.01125	0.0112	0.0536
25	21.00720	0.0072	0.0343
50	21.00180	0.0018	0.0086
100	21.00045	0.0004	0.0021
200	21.00011	0.0001	0.0005
300	21.00005	0.0000	0.0002

The increase in accuracy as the number of subdivisions of the interval increases is evident. The error is less than 1 per cent for 5 steps, and is less than one-tenth of 1 per cent if 15 steps had been used. But in most cases where the trapezium rule would be used, it will not be possible to calculate the precise percentage error. The inherent error can be shown to be of the form

$$|f''(c)|\frac{(b-a)w^2}{12}$$

where c is some point in the interval (a, b). If the maximum value that $f''(c)$ takes in (a, b) is M, we shall then have an upper bound for the error, and can write

$$|E_T| \leq \frac{M(b-a)w^2}{12}$$

In the present case, where $f(x) = x^2$, the second derivative is constant at $f''(x) = 2$, and the error formula gives

$$|E_T| \leq \frac{2(4-1)(0.5)^2}{12} = 0.125$$

which is the exact discrepancy. Needless to say, this will not always be the case. Now consider another example. Use the trapezium rule with five subdivisions of the interval to estimate the definite integral

$$\int_{x=1}^{x=6} (x^3 - 9x^2 + 24x)\,dx$$

With

$$w = (b-a)/5 = 1$$

the trapezium rule gives

$$T = 0.5[f(1) + 2f(2) + 2f(3) + 2f(4) + 2f(5) + f(6)]$$
$$= 0.5[16 + 40 + 36 + 32 + 40 + 36]$$
$$= 100$$

Before calculating the exact figure for comparative purposes, consider the use of the error formula in this case. The second derivative is

$$f''(x) = 6x - 18$$

We are concerned with the absolute value of $f''(x)$ over the interval $(1, 6)$. Since $f''(x)$ is linear here, the maximum absolute size must be at one of the endpoints of the interval. The values are

x	$f''(x)$
1	−12
6	18

So the modulus of the derivative takes its greatest value at the upper end of the interval, and using $M = 18$ in the error formula then gives

$$|E_T| \leq \frac{18(6-1)(1)^2}{12} = 7.5$$

Now the exact value of the integral can again be obtained in this case:

$$\int_{x=1}^{x=6} (x^3 - 9x^2 + 24x)\, dx$$
$$= [0.25x^4 - 3x^3 + 12x^2]^{x=6} - [0.25x^4 - 3x^3 + 12x^2]^{x=1}$$
$$= [324 - 648 + 432] - [0.25 - 3 + 12]$$
$$= 108 - 9.25 = 98.75$$

So the true error, 1.25, is under 17 per cent of the upper limit given by $|E_T|$. As an exercise, the reader may confirm that if the upper limit of integration had been 5 instead of 6, then with the interval divided into four sections the trapezium rule gives precisely the correct value of 72 for the definite integral.[27]

Simpson's rule differs from the trapezium rule in that a parabolic rather than a linear function is used to approximate the curve in each section of the interval. The formula[28] is as follows.

Simpson's rule[29]

$$\int_a^b f(x) \, dx \approx \frac{w}{3} \{f(a) + 4f(a+w) + 2f(a+2w) + 4f(a+3w)$$
$$+ 2f(a+4w) + \cdots + 2f[a+(m-2)w]$$
$$+ 4f[a+(m-1)w] + f(b)\}$$

in which w is defined as for the trapezium rule and the number of steps, m, must be an even number. There is an error formula for Simpson's rule. If some upper bound, M, can be placed on the value of the *fourth* derivative of $f(x)$ in the interval (a, b), then the modulus of the error, E_S, from the use of Simpson's rule is bounded as

$$|E_S| \leq \frac{M(b-a)w^4}{180}$$

Now consider the use of Simpson's rule to approximate the definite integral

$$\int_{x=0}^{x=3} x^4 \, dx$$

Using $m = 6$ subsections, we have

$$w = \frac{2-0}{6} = 0.5$$

The area given by the rule is then

$$S = \frac{0.5}{3} [0 + 4(0.5)^4 + 2(1)^4 + 4(1.5)^4 + 2(2)^4 + 4(2.5)^4 + (3)^4]$$

$$= \frac{0.5}{3} [0.25 + 2 + 20.25 + 32 + 156.25 + 81]$$

$$= \frac{0.5}{3} [291.75] = 48.625$$

For comparative purposes, the exact area can again be calculated in this case:

$$\int_{x=0}^{x=3} x^4 \, dx = [0.2x^5]^{x=3} - [0.2x^5]^{x=0}$$

$$= 48.6$$

Thus the error by Simpson's rule in this case is 0.025, or just over 0.05 per cent of the true figure – and this for a division into only six sections. To find the upper bound on $|E_S|$ that would have been given by the error formula, we must first obtain the upper bound, M, on the value taken by the fourth derivative in the interval. In the present case this is easy, since the fourth derivative is a constant:[30]

$$f^{iv}(x) = 24$$

so $M = 24$ and therefore

$$|E_S| \leqslant \frac{24(3-0)(0.5)^4}{180} = 0.025$$

which is equal to the actual error in this case. For a further comparison, we shall estimate the area under the same curve using the trapezium rule. This is

$$T = \frac{0.5}{2} [f(0) + 2f(0.5) + 2f(1) + 2f(1.5) + 2f(2)$$

$$+ 2f(2.5) + f(3)]$$
$$= 0.25 [0 + 0.125 + 2 + 10.125 + 32 + 78.125 + 81]$$
$$= 0.25 [203.375]$$
$$= 50.84375$$

representing an error of just over 4.6 per cent, given the chosen number of divisions of the interval. It is natural to ask if Simpson's rule gives better results than the trapezium rule. While this is broadly the case, a categorical answer is not possible. A look at the factors in the error formulae helps to explain this equivocation. We are comparing

$$|E_T| \leqslant \frac{M(b-a)w^2}{12} \quad \text{and} \quad |E_S| \leqslant \frac{M(b-a)w^4}{180}$$

Although the M values in the formulae are different – relating to the second and fourth derivatives respectively – in a case where these values are comparable, the coefficients of $M(b-a)$ are respectively

$$\frac{w^2}{12} \quad \text{and} \quad \frac{w^4}{180}$$

so that, *if* $w < 1$, the error limit by Simpson's rule must be lower than that for the trapezium rule – with nothing to choose between the computational labour required.[31]

EXERCISES 8.7

1 Using $m = 6$ steps, estimate the value of the definite integral

$$\int_{x=2}^{x=5} (6x^2 + 2) \, dx$$

   (i) by use of the trapezium rule;
   (ii) by use of Simpson's rule.
   (iii) Calculate the exact value of the integral by analytical means.

2 For the definite integral

$$\int_{x=1}^{x=3} (x^4 - 0.5x^2 + 1) \, dx$$

and with $m = 4$ steps:

   (i) Estimate the value of the integral using the trapezium rule.
   (ii) What bound may be placed on the error measure $|E_T|$?
   (iii) Estimate the value of the integral using Simpson's rule.
   (iv) What bound may be placed on the error measure $|E_S|$?
   (v) What is the exact value of the integral?

8.8 CONCLUDING REMARKS

The main purpose of this chapter has been to introduce the concept of an integral and enable polynomial and rational expressions to be integrated using some simple rules. More advanced devices for integration include:

1 further analytical techniques (such as integration by parts and integration by partial fractions);
2 the use of standard formulae (where the form of the integrand is such that the indefinite integral is already known);

3 forming approximations to more complex expressions using Taylor series;
4 the use of more sophisticated numerical methods, such as variants of Simpson's rule, to obtain the value of definite integrals;
5 the use of computer software packages (such as *Mathematica*, *MathCad*, *Maple* or *Derive*) which not only evaluate definite integrals but also find expressions for indefinite integrals where these exist.

The indefinite integral is likely to be of value when information is available on the derivative of a function of interest. As we have seen, the total revenue function and the demand curve can be recovered from the marginal revenue function, as may expressions for total cost and unit cost given knowledge of the marginal cost function. The definite integral also has many uses, e.g. in finance in the calculation of present values under continuous compounding and in microeconomics in estimating producer or consumer surplus. This concludes our discussion of integration for the time being, but the subject is revisited for exponential and logarithmic functions in the following chapter.

ADDITIONAL PROBLEMS

1 Find the following indefinite integrals:

(i) $\int 3x^2(x^3 + 10)^4 \, dx$

(ii) $\int x(x^2 + 1)^{0.5} \, dx$

(iii) $\int (x^4 + 10)4x^3 \, dx$

2 Find particular solutions to the following differential equations:

(i) $24x^2 - 4\dfrac{dy}{dx} + 8x + 12 = 0$

given that $y = 40$ when $x = 2$

(ii) $5x - 0.5\dfrac{d^2y}{dx^2} = 0$

given that, when $x = 3$, $\dfrac{dy}{dx} = 50$

$y = 70$

3 With x representing the level of output, marginal costs of production for a firm are given by

$$MC = 9x^2 - 4x + 20$$

while the firm's marginal revenue is given by

$$MR = 1000 - 4x$$

At an output level of 5 units, the firm makes a profit of £4400.

(i) How many units could the firm sell if price was set at £600?
(ii) What is the level of fixed costs?

4 Regardless of sign, find the total area between the curve

$$f(x) = 180x - 3x^2 - 1500$$

and the x axis between $x = 0$ and $x = 60$.

5 Using six subsections of the interval, estimate the value of the definite integral

$$\int_{x=1}^{x=4} 2(x^4 + 1) \, dx$$

(i) using the trapezium rule,
(ii) using Simpson's rule.

Having also found the exact value of the integral, give the actual percentage error in each case.

REFERENCES AND FURTHER READING

1 Chiang, A. C. (1984) *Fundamental Methods of Mathematical Economics* (Third Edition), McGraw-Hill.
2 Churchhouse, R. F. (1978) *Numerical Analysis: A First Year Course*, University College Cardiff Press.
3 Courant, R. (1937) *Differential and Integral Calculus* (Second Edition), Volume 1, Blackie.
4 Finney, R. L. and Thomas, G. B., Jr (1990) *Calculus*, Addison-Wesley.
5 Jeffrey, A. (1989) *Mathematics for Engineers and Scientists* (Fourth Edition), Van Nostrand Reinhold.
6 Mizrahi, A. and Sullivan, M. (1988) *Mathematics for Business and Social Sciences* (Fourth Edition), Wiley.
7 Rich, A., Rich, J. and Stoutemyer, D. (1989) *Derive: User Manual* (Third Edition), Soft Warehouse.
8 Weber, J. E. (1982) *Mathematical Analysis: Business and Economic Applications* (Fourth Edition), Harper and Row.

SOLUTIONS TO EXERCISES

Exercises 8.1

1 Cases (a), (b) and (c) are antiderivatives of $4x^3$ since each represents $x^4 \pm$ constant, and which would therefore produce $4x^3$ on differentiation. Cases (c) and (d) contain a term in x with a non-zero coefficient which would not vanish on differentiation.

2 Cases (a), (b) and (d) are antiderivatives of $x^4 - x$. Case (c) is not an antiderivative of $x^4 - x$ since its derivative is $2x^4 - 2x$.

Exercises 8.2

1 (i) $3x + k$
   (ii) $-7.5x + k$
   (iii) $\sqrt{3}x + k$
   (iv) $\pi x + k$

2 (i) $\dfrac{x^{10}}{10} + k$
   (ii) $-\dfrac{x^{-8}}{8} + k$
   (iii) $-\dfrac{x^{-4}}{4} + k$, i.e. $-\dfrac{1}{4k^4} + k$
   (iv) $\dfrac{4}{5} x^{5/4} + k$
   (v) $\dfrac{x^{1.2}}{1.2} + k$
   (vi) $\dfrac{x^{0.6}}{0.6} + k$
   (vii) $2x^{1/2} + k$, i.e. $2\sqrt{x} + k$
   (viii) $\dfrac{x^{\sqrt{3}+1}}{\sqrt{3}+1} + k$
   (ix) $\dfrac{x^{2\pi+1}}{2\pi + 1} + k$

3 (i) $x^4 + x^5 + k$
   (ii) $x^6 - 0.5x^5 + 0.25x^4 + k$
   (iii) $4x^5 + x^{-3} - x^{-4} + k$
   (iv) $2x^{2.5} - 0.1x^{-10} + k$
   (v) $\pi^2 x^\pi + x^{2\pi+1} + k$

4 (i) $12 \int x^2 \, dx = 12 \frac{x^3}{3} = 4x^2$

(ii) $125 \int (2x^4 - x^9) \, dx = 185 \left(\frac{2x^5}{5} - \frac{x^{10}}{10} \right)$
$= 50x^5 - 12.5x^{10}$

5 (i) $\frac{a}{3} x^3 + \frac{b}{2} x^2 + cx + k$

(ii) $\frac{m}{m+1} x^{m+1} + k$

(iii) $x^{n+1} - x^{m-1} + k$
(iv) $t^3 + t^2 + k$
(v) $x^{-3} + k$

Exercises 8.3

1 $\int (2000 - 10x) \, dx = 2000x - 5x^2 = $ total revenue

since in this context the appropriate value of k is zero. Therefore:

$$\text{price} = \frac{2000x - 5x^2}{x} = 2000 - 5x$$

2 Total cost $TC(x)$ is given by

$$\int (100 + x + 0.3x^2) \, dx$$

so that

$$TC(x) = 100x + 0.5x^2 + 0.1x^3 + k$$

Now when $x = 20$, $TC(x) = 4000$, so that

$TC = 100(20) + 0.5(400) + 0.1(8000) + k$
$= 4000$

and so

$3000 + k = 4000$

Therefore

$k = 1000$

and the total cost function is

$TC(x) = 100x + 0.5x^2 + 0.1x^3 + 1000$

3 (i) Total revenue, TR, is maximized when MR = 0, i.e. when
$$480 - 10q = 0$$
i.e.
$$q = 48$$
To find the corresponding value of price, TR will be given by
$$\int MR(q) \, dq = \int (480 - 10q) \, dq$$
$$= 480q - 5q^2$$
the constant of integration being zero in this context. So, since
$$TR = pq$$
$$p = 480 - 5q$$
and when $q = 48$, $p = 240$.

(ii) The firm's profit, π, is given by
$$\pi = TR - TC$$
Now total costs will be
$$TC = \int MC(q) \, dq$$
$$= \int (210 + 5q) \, dq$$
$$= 210q + 2.5q^2 + k$$
When $q = 10$:
$$TC = 210(10) + 2.5(100) + k = 2780$$
and so
$$k = 430$$
Profit will then be
$$\pi = 480q - 5q^2 - 210q - 2.5q^2 - 430$$
$$= 270q - 7.5q^2 - 430$$
and
$$\frac{d\pi}{dq} = 270 - 15q = 0$$

which solves for $q = 18$. Checking at the second order,

$$\frac{d^2\pi}{dq^2} = -15 < 0$$

which confirms the maximum. The level of profit when $q = 18$ is

$$\pi = 270(18) - 7.5(18)^2 - 430$$
$$= 2000$$

and the profit maximizing price is

$$p = 480 - 5(18) = 390$$

Exercises 8.4

1. (i) Second degree, third order.
   (ii) First degree (linear), second order.
   (iii) Third degree, third order.
   (iv) First degree, first order.
   (v) Fourth degree, first order.
   (vi) First degree, fourth order.

2. (i) $y = 2x^4 + 3x^3 + 2.5x^2 + 10x + k$
   (ii) $y = 0.5x^4 + 1.5x^3 + 7x + k$
   (iii) $y = 100x + x^{10} + k$
   (iv) $y = 18x^3 - 0.5x^4 + 34x + k$

3. (i) The general solution is

$$y = \int (5x^4 - 4x^3 + 3x^2 - 2x + 1) \, dx$$
$$= x^5 - x^4 + x^3 - x^2 + x + k$$

   and the boundary condition requires that

$$300 = (3)^5 - (3)^4 + (3)^3 - (3)^2 + 3 + k$$

   so that $k = 117$, and the particular solution is therefore

$$y = x^5 - x^4 + x^3 - x^2 + x + 117$$

   (ii) The general solution is

$$y = 0.25x^4 - 0.5x^3 + 2x^2 + 10x + k$$

   into which insertion of the boundary condition requirements produces

$$90 = 84 + k$$

so that $k = 6$, and the particular solution is therefore
$$y = 0.25x^4 - 0.5x^3 + 2x^2 + 10x + 6$$
(iii) $y = -2x^4 + 10x^3 + x^2 + 5x + 50$
(iv) $y = 2x^5 + 3x^4 - 10x^2 + 20x$
and in this case, note that $k = 0$.

4 (i) The general solution here is
$$y = 10x - x^{-2} + 3x^{-1} + k$$
i.e.
$$y = 10x - \frac{1}{x^2} + \frac{3}{x} + k$$
When $x = 0.2$, $y = 35$, i.e.
$$35 = 20 - \frac{1}{0.4} + \frac{3}{0.2} + k$$
i.e.
$$35 = 20 - 2.5 + 15 + k$$
so that $k = 2.5$, and the particular solution is
$$y = 10x - x^{-2} + 3x^{-1} + 2.5$$
(ii) The general solution is
$$y = 4x^{-5} + 3x^7 - x^2 - 32x + k$$
Inserting the given boundary conditions that when $x = 1$, $y = 0$,
$$0 = 4 + 3 - 1 - 32 + k$$
so that $k = 26$. The particular solution is therefore
$$y = 4x^{-5} + 3x^7 - x^2 - 32x + 26$$

5 (i) Since
$$\frac{d^2y}{dx^2} = 10 - 30x$$
$$\frac{dy}{dx} = 10x - 15x^2 + k_1$$
and
$$y = 5x^2 - 5x^3 + k_1 x + k_2$$

Using the boundary conditions gives two simultaneous equations in k_1 and k_2:

$$3k_1 + k_2 = 90$$
$$5k_1 + k_2 = 290$$
$$-2k_1 = -290$$

So $k_1 = 100$ and hence $k_2 = -210$, and the particular solution is

$$y = 5x^2 - 5x^3 + 100x - 210$$

(ii) The differential equation rearranges as

$$\frac{d^2y}{dx^2} = 24x - 12$$

and so

$$\frac{dy}{dx} = 12x^2 - 12x + k_1$$

Using the boundary condition to find k_1,

$$362 = 12(36) - 12(6) + k_1$$

so that $k_1 = 2$ and

$$\frac{dy}{dx} = 12x^2 - 12x + 2$$

which, upon integration, produces

$$y = 4x^3 - 6x^2 + 2x + k_2$$

and using the boundary condition on y,

$$672 = 4(216) - 6(36) + 2(60) + k_2$$

means that $k_2 = 12$ and the particular solution is therefore

$$y = 4x^3 - 6x^2 + 2x + 12$$

6 The equation rearranges as

$$\frac{d^3y}{dx^3} = 60$$

and integrating

$$\frac{d^2y}{dx^2} = 60x + k_1$$

so, with the given information, $k_1 = 10$, and integration gives

$$\frac{dy}{dx} = 30x^2 + 10x + k_2$$

The fact that this derivative is 60 when $x = 1$ means that $k_2 = 20$. Integration then gives

$$y = 10x^3 + 5x^2 + 20x + k_3$$

which is 50 when $x = 1$, so that $k_3 = 15$ and the particular solution is

$$y = 10x^3 + 5x^2 + 20x + 15$$

7 Reorganization gives

$$\left(\frac{dy}{dx}\right)^2 = 9x^2 - 180 + 900$$

so that

$$\frac{dy}{dx} = \pm 3(x - 10)$$

So that with integration, either

$$y = 1.5x^2 - 30x + k$$

or

$$y = -1.5x^2 + 30x + k$$

But, with the output range given, the variable component (representing variable costs) is only plausible if positive, so that, in this context,

$$y = 1.5x^2 - 30x + k$$

Now when $x = 100$, $y = 20{,}000$, so

$$20{,}000 = 15{,}000 - 3000 + k$$

and k, representing fixed costs, is 8000; the relevant particular solution is

$$y = 1.5x^2 - 30x + 8000$$

Exercises 8.5

1   (i) Try the substitution

$$u = 10 + x^2$$

so that $du = 2x$, and the integral in terms of u is

$$\int u^3 \, du = 0.25u^4 + k$$

and therefore, in terms of the original variable x,

$$\int (10+x^2)^3 2x \, dx = 0.25(10+x^2)^4 + k$$

(ii) Use the substitution

$$u = 2x^2 + 6x + 15$$

so that

$$du = (4x+6) \, dx$$

i.e.

$$(2x+3) \, dx = \frac{du}{2}$$

The integral in terms of u is

$$\int \frac{u^3}{2} \, du = \frac{u^4}{8} + k$$

$$= 0.125(2x^2 + 6x + 15)^4 + k$$

(iii) Using the substitution

$$u = (x^2 - 15)$$

it follows that

$$du = 2x \, dx$$

so that

$$4x \, dx = 2 \, du$$

and the integral in terms of u is

$$\int 2u^{1/2} \, du = \frac{4u^{3/2}}{3} + k$$

which, in terms of x, is

$$\frac{4(x^2-15)^{3/2}}{3} + k$$

(iv) Using the substitution

$$u = (6x^2 - 8x + 20)$$

it follows that

$$du = (12x - 8) \, dx$$

so that
$$0.25\,du = (3x - 2)\,dx$$
and the integral required is
$$\int 0.25u\,du = 0.125u^2 + k$$
which, in terms of x, is
$$0.125(6x^2 - 8x + 20)^2 + k$$

2 (i) Setting $u = 5x^2 - 19$
$$du = 10x\,dx$$
so that
$$3x\,dx = 0.3\,du$$
and the integral required is
$$\int 0.3u^{1/2}\,du = 0.2u^{3/2} + k$$
$$= 0.2(5x^2 - 19)^{3/2} + k$$

(ii) Let $u = 5 + x$. Then $du = dx$ and the required integral is
$$\int -4u^{-5}\,du = u^{-4} + k$$
$$= (5 + x)^{-4} + k$$

(iii) The required integral is
$$\int (x - 5)^{-2}\,dx$$
and letting $u = x - 5$, $du = dx$ and the integral is
$$\int u^{-2}\,du = -u^{-1} + k$$
$$= -(x - 5)^{-1} + k$$
$$= -\frac{1}{x - 5} + k$$

(iv) Let $u = 3x + 4$. Then $du = 3\,dx$ so that $dx = du/3$ and the required integral is

$$\int -2u^{-2}\,du = 2u^{-1} + k$$

$$= 2(3x + 4)^{-1} + k$$

$$= -\frac{2}{3x + 4} + k$$

3 Let $u = bx + c$ so that

$$dx = \frac{du}{b}$$

and the required integral is

$$\int \frac{d}{bu^n}\,du = \int \frac{d}{b} u^{-n}\,du$$

$$= \frac{(d/b)u^{-n+1}}{1 - n} + k$$

$$= \frac{d}{b(1 - n)(bx + c)^{n-1}} + k$$

4 (i) Let $u = x^2 + 7x - 8$. Then

$$du = (2x + 7)\,dx$$

and the required integral is

$$\int \frac{du}{u^5} = -0.25u^{-4} + k$$

$$= -\frac{1}{(x^2 + 7x - 8)^4} + k$$

(ii) Let $u = x^3 + x^2 + x + 1$ so that

$$2\,du = (6x^2 + 4x + 2)\,dx$$

and the required integral is

$$\int 2u^{-3}\,du = -u^{-2} + k$$

$$= -\frac{1}{(x^3 + x^2 + x + 1)^2} + k$$

Exercises 8.6

1 (i) $[2x^2 + k]^{\text{for } x=7} - [2x^2 + k]^{\text{for } x=3}$
   $= 98 - 18 = 80$

Note: From this point on the constant of integration will be omitted.

   (ii) $[3x^2 - 3x]^{\text{for } x=-1} - [3x^2 - 3x]^{\text{for } x=-5}$
   $= 6 - 90 = -84$

   (iii) $[0.3x^3 + 0.5x^2 + 10x]^{\text{for } x=20} - 0$
   $= 2800$

   (iv) $[8x^{1.5}]^{\text{for } x=16} - [8x^{1.5}]^{\text{for } x=4}$
   $= 512 - 64 = 448$

2 (i) $\left[-\dfrac{100}{x}\right]^{\text{for } x=\infty} - \left[-\dfrac{100}{x}\right]^{\text{for } x=0.5}$
   $0 = 0 - (-200) = 200$

   (ii) $[50\sqrt{x}]^{\text{for } x=25} - [50\sqrt{x}]^{\text{for } x=4}$
   $= 250 - 100 = 150$

   (iii) $\pi([x^3]^{\text{for } x=5} - x[x^3]^{\text{for } x=2}) = 117\pi$

   (iv) $[0.2x^5 + x^3 + 20x]^{x=0} - [0.2x^5 + x^3 + 20x]^{x=-10}$
   $= 0 - [-20{,}000 - 1000 - 200]$
   $= 21{,}200$

3 (i) The required area is given by the definite integral
   $$\int_{x=0}^{x=10} (3x^2 + 2x + 5)\, dx$$
   $= [x^3 + x^2 + 5x]^{x=10} - [x^3 + x^2 + 5x]^{x=0}$
   $= 1000 + 100 + 50 - 0$
   $= 1150$

   (ii) The area is given by the definite integral
   $$\int_{x=-10}^{x=-2} (x^3 + 10x)\, dx$$
   $= [0.25x^4 + 5x^2]^{x=-2} - [0.25x^4 + 5x^2]^{x=-10}$
   $= 24 - 3000$
   $= -2976$

(iii) The area is given by the definite integral

$$\int_{x=-5}^{x=5} (1.5x^2 + 12x + 30)\, dx$$
$$= [0.5x^3 + 6x^2 + 30x]^{x=5} - [0.5x^3 + 6x^2 + 30x]^{x=-5}$$
$$= 362.5 - [-62.5]$$
$$= 425$$

4 (i) In this case, the definite integral is split into two parts

$$\int_5^{10} (x-10)\, dx + \int_{10}^{15} (x-10)\, dx$$

and in each case the absolute value of the outcome is taken. Therefore, between 5 and 10 the integral evaluates as

$$[0.5x^2 - 10x]^{x=10} - [0.5x^2 - 10x]^{x=5}$$
$$= 50 - 100 - [12.5 - 50]$$
$$= -12.5$$

The area regardless of sign in this interval is 12.5. In the second part of the interval, the curve is above the axis and the area is given by

$$[0.5x^2 - 10x]^{x=15} - [0.5x^2 - 10x]^{x=10}$$
$$= 12.5$$

so that the total area is $12.5 + 12.5 = 25$.

(ii) First find the roots of

$$f(x) = 3x^2 - 18x + 15$$

in order to find the points where the function crosses the x axis. $f(x) = 0$ for $x = 1$ and $x = 5$. So the interval should be subdivided at 1 and 5; the total area is the sum of the absolute values of the definite integrals

$$\int_0^1 f(x)\, dx + \int_1^5 f(x)\, dx + \int_5^8 f(x)\, dx$$

The individual areas are

$$[x^3 - 9x^2 + 15x]^{x=1} - [x^3 - 9x^2 + 15x]^{x=0} = 7$$
$$[x^3 - 9x^2 + 15x]^{x=5} - [x^3 - 9x^2 + 15x]^{x=1} = -32$$
$$[x^3 - 9x^2 + 15x]^{x=8} - [x^3 - 9x^2 + 15x]^{x=5} = 81$$

So the required total area is

$$7 + 32 + 81 = 120$$

5 (i) The definite integral required is

$$\int_{x=0}^{x=20} 20 \, dx + \int_{x=20}^{x=40} x \, dx$$

which evaluates as

$[20x]^{x=20} - [20x]^{x=0} + [0.5x^2]^{x=40} - [0.5x^2]^{x=20}$
$= 400 - 0 + 800 - 200$
$= 1000$

(ii) The required area is

$$\int_{5}^{10} (4x - 20) \, dx + \int_{10}^{15} (x + 10) \, dx$$

$= [2x^2 - 20x]^{x=10} - [2x^2 - 20x]^{x=5}$
$\quad + [0.5x^2 + 10x]^{x=15} - [0.5x^2 + 10x]^{x=10}$
$= 50 + 112.5$
$= 162.5$

Exercises 8.7

1 (i) Using the trapezium rule with $m = 5$ (and therefore with $w = 0.5$) the definite integral is evaluated as

$T = 0.25 \, [f(2) + 2f(2.5) + 2f(3) + 2f(3.5) + 2f(4)$
$\quad + 2f(4.5) + f(5)]$
$= 0.25 \, [26 + 79 + 112 + 151 + 196 + 247 + 152]$
$= 0.25 \, [963] = 240.75$

(ii) By Simpson's rule:

$$S = \frac{0.5}{3} \, [f(2) + 4f(2.5) + 2f(3) + 4f(3.5) + 2f(4)$$
$\quad + 4f(4.5) + f(5)]$

$= \frac{0.5}{3} \, [26 + 158 + 112 + 302 + 196 + 494 + 152]$

$= \frac{0.5}{3} \, [1440]$

$= 240$

(iii) $\int_{x=2}^{x=5} (6x^2 + 2)\,dx = [2x^3 + 2x]^{x=5} - [2x^3 + 2x]^{x=2}$

$= [250 + 10] - [16 + 4]$
$= 240$

Note the precise accuracy of Simpson's rule in the case of a quadratic function.

2 (i) Using the trapezium rule as specified:

$m = 4$ and so $w = (3-1)4 = 0.5$

and

$T = 0.25[f(1) + 2f(1.5) + 2f(2) + 2f(2.5) + f(3)]$
$= 0.25[1.5 + 9.875 + 30 + 73.875 + 77.5]$
$= 0.25[192.75]$
$= 48.1875$

(ii) The error in the use of the trapezium rule has an upper bound of

$$|E_T| \leqslant \frac{M(b-a)w^2}{12}$$

in which M is the maximum value taken by the second derivative within the interval. The second derivative is $12x^2 - 1$, which is clearly at its greatest value in the interval at $x = 3$. So $m = 109$ and

$$|E_T| \leqslant \frac{109(3-1)0.25}{12}$$

$= 4.54$

(iii) Using Simpson's rule:

$S = \dfrac{0.5}{3}[f(1) + 4f(1.5) + 2f(2) + 4f(2.5) + 2f(3)]$

$= \dfrac{0.5}{3}[1.5 + 19.75 + 30 + 147.75 + 77.5]$

$= \dfrac{0.5}{3}[276.5]$

$= 46.08\overline{33}$

(iv) In this case the fourth derivative is constant at 24, which therefore becomes the value for M. The error formula gives

$$|Es| \leqslant \frac{24(3-1)(0.5)^4}{180}$$

$$= \frac{3}{180}$$

$$= 0.01\overline{66}$$

(v) The exact value of the integral is $691/15 = 46.0\overline{66}$.

Additional problems

1 (i) $\dfrac{(x^3+10)^5}{5} + k$

(ii) $\dfrac{(x^2+1)^{1.5}}{3} + k$

(iii) $\dfrac{(x^4+10)^2}{2} + k$

2 (i) $y = 2x^3 + x^2 + 3x + 14$

(ii) The first derivative is

$$\frac{dy}{dx} = 5x^2 + k_1$$

which is 50 when $x = 3$ so that $k_1 = 5$ and therefore

$$y = \tfrac{5}{3}x^3 + 5x + k_2$$

which takes the value 70 when $x = 3$ so that $k_2 = 10$ and the particular solution is

$$y = \tfrac{5}{3}x^3 + 5x + 10$$

3 (i) $TR(x) = \int MR(x) \, dx = 1000x - 2x^2$

so that

$p = 1000 - 2x$

and when $p = 600$ the level of sales will be $x = 200$.

(ii) Total costs $TC(x)$ are given by

$$TC(x) = \int MC(x) \, dx = 3x^3 - 2x^2 + 20x + k$$

Profit $\pi(x)$ is

$\pi(x) = TR(x) - TC(x)$
$\quad\quad = 980x - 3x^3 - k$
$\quad\quad = 4400$ when $x = 5$

so that fixed costs are $k = 125$.

4 First, establish whether the curve crosses the x axis between the limits of integration of 0 and 60. The quadratic

$180x - 3x^2 - 1500 = 0$

solves for

$x = 10$ and $x = 50$

so that the integral

$$\int f(x) \, dx = 90x^2 - x^3 - 1500x$$

must be evaluated in three segments: 0–10, 10–50 and 50–60. Now,

$[90x^2 - x^3 - 1500x]^{10} - [90x^2 - x^3 - 1500x]^0 = -7000$
$[90x^2 - x^3 - 1500x]^{50} - [90x^2 - x^3 - 1500x]^{10} = 32{,}000$
$[90x^2 - x^3 - 1500x]^{60} - [90x^2 - x^3 - 1500x]^{50} = -7000$

and so the total area between the curve and the x axis, regardless of sign, is

$7000 + 32{,}000 + 7000 = 46{,}000$

5 The exact value of the integral is

$$\int_{x=1}^{x=4} (2x^4 + 2)\,dx = [0.4x^5 + 2x]^4 - [0.4x^5 + 2x]^1$$

$$= [409.6 + 8] - [0.4 + 2]$$
$$= 415.2$$

With six subsections of the interval, by the trapezium rule the integral is approximated as

$$T = 0.25\,[f(1) + 2f(1.5) + 2f(2) + 2f(2.5) + 2f(3) + 2f(3.5) + f(4)]$$
$$= 0.25\,[4 + 24.25 + 68 + 160.25 + 328 + 604.25 + 514]$$
$$= 0.25\,[1702.75] = 425.6875$$

so the percentage error here is just under 2.53 per cent.
(ii) By Simpson's rule,

$$S = 0.513\,[f(1) + 4f(1.5) + 2f(2) + 4f(2.5) + 2f(3) + 4f(3.5) + f(4)]$$
$$= 0.513\,[4 + 48.5 + 68 + 320.5 + 328 + 1208.5 + 514]$$
$$= 0.513\,[2491.5] = 415.25$$

so that the percentage error here is only 0.012 per cent.

NOTES

1. And therefore is sometimes called the *general antiderivative* (of $3x^2$).
2. The area can be found as the limit of a process of *summation*.
3. dy and dx are called *differentials*. See Section 8.5.
4. In $F(x) + k$, $F(x)$ is sometimes called a *particular integral*.
5. And as we shall see in Chapter 9, to exponential functions.
6. Problems of this nature are sometimes called *initial value problems*, particularly where the given values correspond to the commencement of a process. We consider several such problems in Section 8.4.
7. A comprehensive type classification of differential equations would include many further categories than those mentioned here – for example, *separable*, *exact* and *homogeneous* differential equations.
8. After any necessary rationalization to remove fractional powers.
9. Linear first-order differential equations are solvable, as are separable and homogeneous equations.
10. Which we illustrate in Chapter 9, following coverage of exponential functions.
11. For more advanced material on differential equations in an economic context see Chiang (1984).
12. It is possible that an ordinary differential equation has a *singular solution* that cannot be obtained from the general solution by any choice of values for the arbitrary constants.
13. A boundary condition is sometimes written in the form $F(4) = 300$, in which $F(x)$ is taken to include the constant of integration. However, for consistency with our earlier usage of $F(x)$, we shall express the boundary conditions in x and y.
14. We are using the term in a wider sense not restricted to time processes.
15. In general, the number of arbitrary constants involved is the same as the order of the differential equation.
16. Or some combination representing n independent pieces of information.
17. The differential of a function $u = f(x)$, designated du, is defined as $f'(x)\,dx$.
18. For example in obtaining linear approximations of functions and error estimation.
19. The word 'attempt' is used with justification. The first, or subsequent, attempts may not work and considerable judgement can be needed in choosing effective substitutions.
20. The integral of $(x^3 + 1)^{1/2}$ is an elliptic function.
21. Elementary functions are those that can be built up from polynomial, rational, exponential and trigonometric functions and their inverses.
22. In the manner shown later in this section.
23. For example when the surpluses and deficits of business units are to be aggregated.
24. Recall that 'elementary' means made up from polynomial, rational, exponential and trigonometric functions and their inverses.
25. UK usage of the term *trapezium* means a four-sided figure (quadrilateral) with two parallel sides of unequal length. The US term for this figure is a *trapezoid*. In the UK a trapezoid refers to a quadrilateral in which neither pair of sides is parallel; in US usage, this defines a trapezium! Consequently the US terminology for the trapezium rule is the *trapezoidal rule*.
26. So long as the function has a continuous second derivative, and up to the point where rounding errors of computation overtake any theoretical increase in accuracy with reduced step size.

27 The function in this interval produces exact cancellation of the overestimates and underestimates of the individual section areas.
28 For a derivation of the formula for Simpson's rule see Finney (1990).
29 The *Derive* mathematical software package uses an adaptation of Simpson's rule to estimate definite integrals to a high degree of accuracy.
30 Things are not always so convenient. Where Simpson's rule is used of necessity, the least upper bound can rarely be found. The best obtainable value has to be used.
31 This is negligible in computer implementation, even though computing time increases sharply with the precision required.

Chapter 9

Exponential and Logarithmic Functions

9.1	Exponential functions: introduction	614
9.2	The natural exponential function and its derivative	622
9.3	Integration of natural exponential functions	629
9.4	Natural logarithmic functions	635
9.5	The derivative of natural logarithmic functions	643
9.6	Elasticity	648
9.7	Natural logarithmic functions and integration	654
9.8	Integration by parts	658
9.9	Logarithmic and exponential functions to bases other than e	663
9.10	Aggregate sales curves	671
9.11	BASIC programs	675
	Additional problems	678
	References and further reading	680
	Solutions to exercises	680

This chapter broadens the range of functions to which techniques so far developed can be applied and new areas of application are opened up. The relevance of exponential and logarithmic functions to processes involving growth and appreciation, and decay and depreciation is demonstrated. The relationship of logarithmic functions to elasticity of demand is covered in Section 9.6, while Section 9.10 gives a marketing application.

An objective of the chapter is to show how exponential and logarithmic functions can be differentiated and integrated. Integration by parts is valuable where functions contain terms with both power and exponential components, and is introduced in Section 9.8. Further examples of differential equations are given and the chapter concludes with relevant BASIC programs.

By the end of the chapter, you should be able to carry out calculus operations on simple exponential and logarithmic functions. You will gain useful insight into business, financial and economic concepts based on functions of this nature and be aware of important properties of models based on such functions.

9.1 EXPONENTIAL FUNCTIONS: INTRODUCTION

While polynomial and rational functions have many applications and are widely used, there is another important type of function which is used to model many situations, particularly those in which change and growth are involved. These are *exponential functions*. In polynomial expressions, the exponents of the variables are constants. Exponential functions have variable *exponents*. For example, with x as the variable

$$y = a^x \qquad (9.1)$$

is the simplest example of an exponential function. In equation (9.1), a is the *base* and is a positive constant not equal to one. The variable x is the *exponent*. The behaviour and appearance of the function (9.1) depends on whether the base, a, is greater or less than unity. Consider the case where $a > 1$, specifically the function

$$y = 2^x$$

Let us look at some values of $y = 2^x$ for various x. These are shown in Table 9.1.

As is clear from the table, successive unit increases in x produce successive doubling of the value of the function. This factor of multiplication is the value of the base. Conversely, unit *decreases* in x result in a halving in the value of the function, the multiple here being the reciprocal of the base. The function $y = 2^x$ is graphed in Figure 9.1.

Figure 9.1 is typical of all exponential functions with a base greater than one. Features of the function and its graph are that it is

1 positive everywhere
2 always increasing
3 increasing at an increasing rate

Such curves are sometimes called *growth curves*. They are used to model expansion and increase – for example of populations, atmospheric CO_2 or outstanding debt – at a *compound rate*. As one example, for a developed economy, a typical sustainable annual rate of growth in real gross domestic product (GDP) is around 2.5 per cent. Thus

$$\text{GDP}(1993) = \text{GDP}(1992)(1.025)$$

Table 9.1

x	0	1	2	3	4	5	6	−1	−2	−3	−4
2^x	1	2	4	8	16	32	64	0.5	0.25	0.125	0.0625

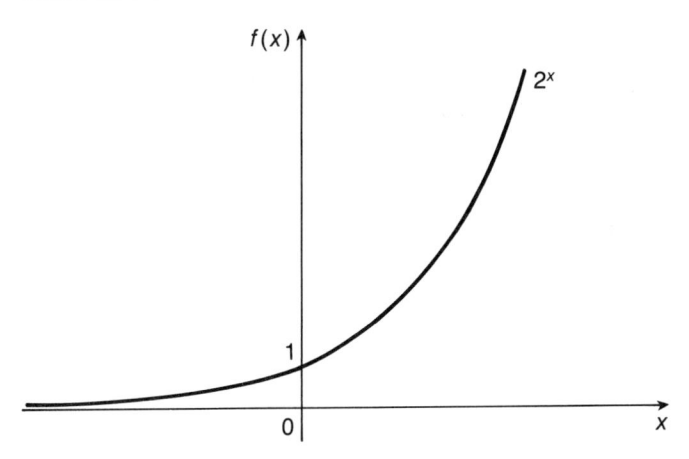

Figure 9.1

or, with reference to an initial year zero,[1] GDP in some year t would be given by

$$GDP_t = GDP_0(1.025)^t \qquad (9.2)$$

In equation (9.2) the base $a = 1.025$. The base of the exponential function is one plus the rate of growth as a decimal. If in general the rate of growth had been $100r$ per cent then the relationship would have been

$$GDP_t = GDP_0(1 + r)^t$$

Now consider an exponential function in which the base is less than one: specifically

$$y = (\tfrac{1}{2})^x$$

Table 9.2 shows values of the function for various x.

Note that as x increases by unit increments the value of the function is progressively halved. Conversely, as x is decreased, the value of the function doubles. The graph of $y = (\tfrac{1}{2})^x$ is shown in Figure 9.2.

The graph is seen to be the reflection in the y axis of the graph of 2^x and it typifies exponential functions with a base less than one, i.e. $0 < a < 1$. Properties of the graph are that it is

1 positive everywhere
2 always decreasing
3 decreasing at a decreasing rate

Exponential functions with base less than one produce *decay curves* and can be used to model processes of decay or contraction. It is a characteristic of decay curves that, if x represents time, the time required to reduce y to half

Table 9.2

x	0	1	2	3	4	−1	−2	−3	−4	−5
$(\frac{1}{2})^x$	1	0.5	0.25	0.125	0.0625	2	4	8	16	32

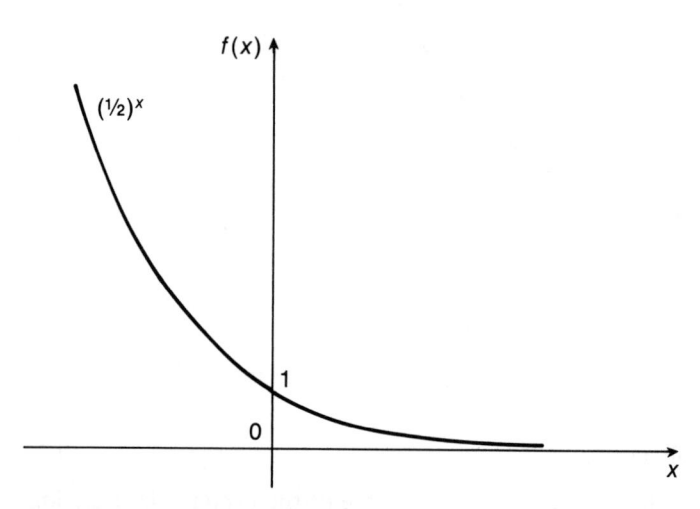

Figure 9.2

its value is the same regardless of the initial value of y. Note also that the process of exponential decay could just as well be modelled using the form of function

$$y = a^{-x}$$

where $a > 1$. This follows from the fact that

$$\left(\frac{1}{a}\right)^x = \frac{1}{a^x} = a^{-x}$$

In terms of the examples we have been using

$$(\tfrac{1}{2})^x = y = 2^{-x}$$

as may be confirmed as necessary by drawing up a table for 2^{-x} and matching it entry for entry with $(\frac{1}{2})^x$. There is an obvious relevance of decay curves to physical processes such as radioactive decay. Within economics and finance, a decay curve can be used to model the *depreciation* of assets, economic output during periods of recession or, for the domain $x \geqslant 0$, demand. For example suppose that during a protracted recession GDP is contracting at $1\frac{1}{2}$ per cent per annum in real terms. After t years of recession GDP will be given by

$$\text{GDP}_t = \text{GDP}_0(0.985)^t$$

In the context of depreciation, suppose that each year the retail value of a car decreases by 20 per cent. The value of the car after t years will then be

$$V_t = V_0(0.8)^t$$

where V_0 is the original value of the car. Thus, if the vehicle cost £7500 initially, then after 4 years its value would be V_4 where

$$V_4 = 7500(0.8)^4$$
$$= 7500(0.4096)$$
$$= 3072$$

Consider growth curves again. One of the most important financial applications of exponential functions is in compound interest calculations. Suppose that a bank deposit account pays 10 per cent interest annually compounded. Suppose that the sum deposited initially, the *principal*, is £1. By the end of one year, the investor will have £1.10 in the account. This can be thought of as the amount on deposit at the start of the year, £1, plus the interest on this sum, 0.1(£1). Now suppose that the whole £1.1 is left on deposit. There is no reason to treat this investor less well than a new customer who makes a deposit of £1.10 at this time. Therefore, for the second year, interest must be earned on the entire sum − principal and first year interest − of £1.1. This is the key point in compounding: that interest itself earns interest. At the end of the second year, the investor will have what the year was started with, £1.1, plus the interest on this sum, 0.1(£1.1). The total is £1.21. Table 9.3 records the growth of the original £1 invested.

It is clear from Table 9.3 that after n years the original £1 has become £$(1 + 0.1)^n$ at 10 per cent interest. It is evident that £1 invested for n years at a rate of $100x$ per cent would have become £$(1 + x)^n$. The *future value*, FV_n, of a principal sum £P after n years at $100x$ per cent is then

$$\text{FV}_n = P(1 + x)^n \tag{9.3}$$

In writing (9.3) as an exponential function of n, we have used x to represent an unspecified component of the base. The reason for this notational role for x will become evident. Consider an example of the use of (9.3). The future value of a sum of £75 invested for 12 years at an 8 per cent compound interest rate is

$$\text{FV}_{12} = £75(1 + 0.08)^{12}$$
$$= £75(2.51817)$$
$$= £188.86$$

Table 9.3

Principal	One year	Two years	Three years
1	$1 + 0.1(1)$ $= 1.1$	$1.1 + 0.1(1.1)$ $1.1(1 + 0.1)$ $= (1 + 0.1)^2$ $= 1.21$	$1.21 + 0.1(1.21)$ $= 1.21(1 + 0.1)$ $= (1 + 0.1)^3$ $= 1.331$

If £75 had been invested for 8 years at 12 per cent the result would have been

$$FV_8 = £75(1 + 0.12)^8$$
$$= £75(2.47596)$$
$$= £185.70$$

This use of (9.3) is all very well, but it may be argued that in practice most financial institutions pay interest more frequently than once a year on most accounts. Let us see how the formula can be adjusted to allow for more frequent payments. If the quoted rate of interest was $100x$ per cent but in fact interest was paid half yearly at $100(x/2)$ per cent, the result is two compounding periods a year and, for investors, a relatively rare example of benign understatement by the bank! In these circumstances, £1 on deposit for 1 year at $100x$ per cent nominal becomes

$$\left(1 + \frac{x}{2}\right)^2 = 1 + x + \frac{x^2}{4} > 1 + x$$

So the future value is increased if this is worked out using half the nominal interest rate compounded twice as frequently. For example, if a declared 12 per cent annual interest is in practice implemented as twice yearly compounding at 6 per cent, the end of year value of £1 will be

$$(1 + 0.12/2)^2 = 1.1236$$

which is equivalent to an annually compounded rate of 12.36 per cent. But what if interest was compounded quarterly? At 12 per cent the result would be

$$(1 + 0.12/4)^4 = 1.1255 \text{ or } 12.55\%$$

If interest was compounded monthly, as with most credit cards, the result would be

$$(1 + 0.12/12)^{12} = 1.1268 \text{ or } 12.68\% \,^2$$

In some major financial transactions interest is charged on a daily basis, so the equivalent annual rate in this case would be given by

$$(1 + 0.12/365)^{365} = 1.1275 \text{ or } 12.75\%$$

So, for example, a sum of £850 borrowed initially at a nominal 12 per cent but with daily compounding would have grown, by the end of the year, to

$$850(1 + 0.12/365)^{365} \approx 958$$

What would happen if interest was compounded more frequently still? Is there a limit to the value that the equivalent annual rate would take? The answer is provided by the limiting value of

$$\left(1 + \frac{x}{n}\right)^n \qquad (9.4)$$

as n increases without limit. In so doing, *continuous compounding* is approached. In working towards a value for (9.4), we start with the simpler case where $x = 1$. Table 9.4 shows values of $(1 + 1/n)^n$ for selected values of n.

As may be surmised from the values of Table 9.4, the value of $(1 + 1/n)^n$ approaches the figure 2.71828 (to five decimal places) as n continues to increase. When stated as a *limit* (a value that an expression approaches ever more closely as the independent variable increases) this is written as

$$\lim_{n \to \infty} \left(1 + \frac{1}{n}\right)^n = 2.71828\ 18284\ 59045\ldots = e \qquad (9.5)$$

The resulting number e, sometimes known as the *natural number*, is a transcendental number and is an important special constant. Now what of $(1 + x/n)^n$? It can be shown that the limit of this term as n increases is e raised to the power of x. That is,

$$\lim_{n \to \infty} \left(1 + \frac{x}{n}\right)^n = e^x \qquad (9.6)$$

$y = e^x$ is an important and useful exponential function. But before considering some of the properties of e^x, an answer can now be given to the question of to where the process of ever more frequent compounding would lead. Continuous compounding at 12 per cent gives an end of year result for a principal of £1 of

$$e^{0.12} \approx 1.1275 \text{ or } 12.75\%$$

Two years of continuous compounding at $100x$ per cent per annum produces a future value of e^{2x}. So for example with $x = 0.12$, after 2 years £1 would become

$$£e^{2(0.12)} \approx £1.2712$$

Table 9.4

n	1	2	5	10	25	100
$(1+1/n)^n$	2	2.25	2.48832	2.59374	2.66584	2.70481

n	1000	10,000	100,000	500,000	1,000,000
$(1+1/n)^n$	2.71692	2.71815	2.71827	2.71828	2.71828

In general, if a principal of £P is invested at 100r per cent per annum continuously compounded, with no withdrawals, the future value after t years is

$$PV(t) = Pe^{rt}$$

So, for example, a principal of £500 deposited for 2 years at 12 per cent continuously compounded would produce a future value of

$$FV(2) = 500e^{0.24}$$
$$= 635.62$$

while over 5 years the future value would be

$$FV(5) = 500e^{0.6}$$
$$= 911.06$$

For any continuously compounded interest rate there is a unique equivalent discretely compounded rate. For example, the equivalent rate to 12 per cent continuously compounded with once yearly compounding is 100y per cent where

$$(1 + y)^2 = 1.2712$$

which solves for $y = 1.1275$ or 12.75 per cent per annum. Further information on the variety of ways in which compound interest calculations can be carried out can be found in Wilkes (1983).

EXERCISES 9.1

1. In a certain country, the level of GDP in 1989 was £350b. Growth of GDP is at the rate of 3 per cent per annum.

    (i) Express this relationship as an exponential function.
    (ii) On the assumption that the relationship remains valid, use the function to state GDP in

    (a) 1991   (b) 1995

2. A car is bought new for a price of £10,000. By the end of each year of its life, the car has lost 25 per cent of its value at the start of that year.

    (i) Express this relationship as an exponential function.
    (ii) Use the function to find the value of the car after

    (a) 3 years   (b) 6 years

3. (i) A sum of £750 is deposited in an account in which interest is compounded at 8 per cent per annum. With no withdrawals, what is the value after

    (a) 5 years   (b) 10 years   (c) 20 years

    (ii) Now suppose that the interest on the account was 16 per cent. Find the value of the account after

    (a) 5 years   (b) 10 years   (c) 20 years

4. A finance company charges a nominal rate of interest of 24 per cent per annum. If a sum of £5000 is borrowed, calculate the outstanding debt after a year with the frequency of compounding

    (a) once a year   (b) twice a year
    (c) four times a year   (d) monthly

5. With interest compounded continuously at 10 per cent, find the value of £800 invested for

    (a) 2 years   (b) 10 years   (c) 20 years

9.2 THE NATURAL EXPONENTIAL FUNCTION AND ITS DERIVATIVE

The function $y = e^x$ is the *natural exponential function*. An alternative notation for e^x is $\exp(x)$, which is easier to print (particularly where the exponent is a function) and is often preferred. Although a much larger base than e could be selected, following an analogy in Finney and Thomas (1990), we can use the values of e^x to illustrate the phenomenal growth exhibited by exponential functions. In Figure 9.3 both axes are measured in centimetres and co-ordinates for $x = 1$ and $x = 2$ are shown. It is not quite possible to continue the graph to $x = 3$, when its height would be just over 20 cm.

Table 9.5 gives the height reached by the y axis (measuring e^x in centimetres) for selected values of x. As can be seen, it is unlikely that values of x in excess of 65 will have much practical significance!

Returning to earthbound discussion, the more general form

$$y = Ae^{mx} \tag{9.7}$$

describes a wide variety of processes of continuous change. The value of A locates the y intercept while the value of m modifies the slope. The values of A and m are chosen to best fit the situation under study. For example, the formula for calculating the future value of a principal sum invested under continuous compounding is an application of (9.7). Now consider the

Exponential and logarithmic functions 623

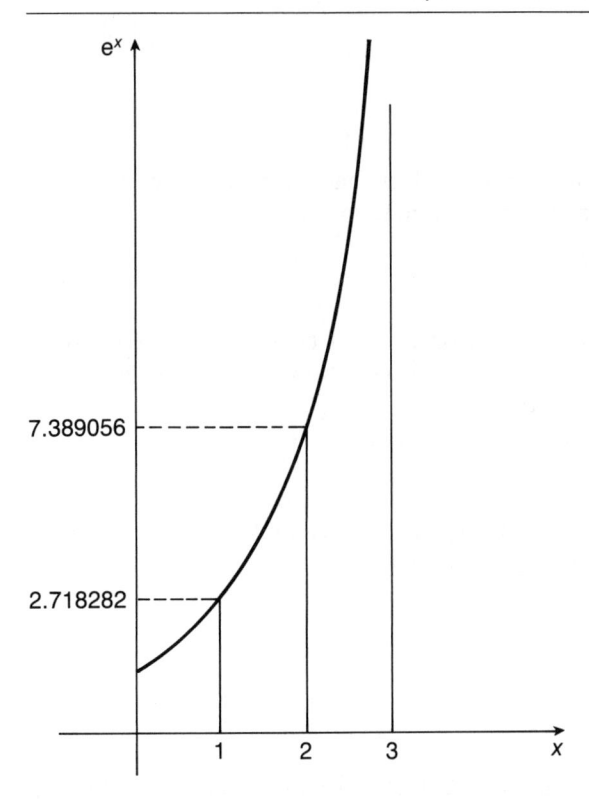

Figure 9.3

Table 9.5

x	e^x	Comment
5 cm	1.48 m	Shoulder height
10 cm	220 m	Height of respectable skyscraper
25 cm	720,000 km	Well beyond the orbit of the moon
50 cm	>5000 light years	One-sixth of the way to the Galactic rim
65 cm	>15 bn light years	To the edge of the Universe...

derivative of the function $y = e^x$. Recall that e^x was the limit of $(1 + x/n)^n$ as n increased. We can proceed by using the chain rule to find the derivative of $(1 + x/n)^n$ and then find the limit of the expression obtained. If

$$y = \left(1 + \frac{x}{n}\right)^n$$

then, by the chain rule,

$$\frac{dy}{dx} = n\left(1 + \frac{x}{n}\right)^{n-1} \frac{1}{n} = \left(1 + \frac{x}{n}\right)^{n-1}$$

but as n increases without bound, the fact that the exponent of $(1 + x/n)$ is $n - 1$ rather than n makes no difference, so that

$$\frac{d(e^x)}{dx} = \lim_{n \to \infty} \left(1 + \frac{x}{n}\right)^{n-1} = e^x \qquad (9.8)$$

The function $y = e^x$ has the unique property of being its own derivative.[3] It follows that *all* derivatives of e^x are also e^x. Now consider the derivative of $y = e^{mx}$. It can be shown that

$$\frac{d(e^{mx})}{dx} = me^{mx} \qquad (9.9)$$

The chain rule or the product rule can be used to establish (9.9).[4] Using the product rule to confirm the derivative when $m = 2$,

$$e^{2x} = e^x e^x$$

for which, using the product rule,

$$\frac{d(e^x e^x)}{dx} = e^x e^x + e^x e^x = 2e^{2x}$$

Note that the exponent will not alter as higher derivatives are taken, but the coefficient will increase by a factor of m each time. Thus for the nth derivative

$$\frac{d^n(e^{mx})}{dx^n} = m^n e^{mx} \qquad (9.9)$$

Consider some numerical examples for specific values of m. First take m as a positive integer. For

$$f(x) = e^{3x}$$

the derivative is

$$f'(x) = 3e^{3x}$$

Note that the second derivative here is

$$f''(x) = 9e^{3x}$$

Now applying the rule to the case where m is a negative integer,

$$f(x) = e^{-2x}$$
$$f'(x) = -2e^{-2x}$$

Exponential and logarithmic functions 625

In this case the second derivative is

$$f''(x) = 4e^{-2x}$$

Note that where the exponent is negative, derivatives of progressively higher order will alternate in sign. Now consider a rational exponent

$$f(x) = e^{0.1x}$$
$$f'(x) = 0.1e^{0.1x}$$

and irrational and transcendental exponents

$$f(x) = e^{\sqrt{2}x}$$
$$f'(x) = \sqrt{2}e^{\sqrt{2}x}$$
$$f(x) = e^{\pi x}$$
$$f'(x) = \pi e^{\pi x}$$

The derivative of $y = Ae^{mx}$ is mAe^{mx}. Thus for

$$y = 15e^{6x}$$

$$\frac{dy}{dx} = 90e^{6x}$$

In general, a function may contain terms involving both x^n and e^x. The simplest case of such a combination is

$$y = xe^x$$

for which the derivative is obtained by use of the product rule:

$$\frac{dy}{dx} = xe^x + e^x$$

$$= e^x(x+1)$$

And given

$$y = x^5 e^{2x}$$

the derivative may also be obtained using the product rule:

$$\frac{dy}{dx} = 5x^4 e^{2x} + 2x^5 e^{2x}$$

The derivative of the function

$$y = e^{4x}(2x + x^3)^2$$

may be obtained by combined use of the product rule and the chain rule:

$$\frac{dy}{dx} = 4e^{4x}(2x+x^3)^2 + e^{4x}2(2x+x^3)(2+3x^2)$$

This unsimplified form of the derivative is sufficient for many purposes – for example when the slope of the function is required for specific values of x. Thus, at $x = 1$, the derivative evaluates as

$$\left(\frac{dy}{dx}\right)_{x=1} = 66e^4$$

If further analysis is necessary (for example, finding higher order derivatives) the time required to simplify dy/dx may be well spent. It can be rearranged as [5]

$$\frac{dy}{dx} = 2x(x^2 + 2)(2x^3 + 3x^2 + 4x + 2)e^{4x}$$

The exponential function $f(x) = e^{mx}$ can be seen as a special case of the exponent being a function of x, as in

$$y = e^{g(x)} \qquad (9.10)$$

in which it will be assumed that the exponent $g(x)$ is a differentiable function of x. The first derivative of (9.10) is

$$\frac{dy}{dx} = g'(x)e^{g(x)} \qquad (9.11)$$

Note that, once again, the exponent does not change in the derivative. As the first numerical example, consider a linear exponent

$$y = e^{mx+c}$$

for which, since $g'(x) = m$, from (9.11) the derivative is

$$\frac{dy}{dx} = me^{mx+c}$$

Now consider the simplest form of non-linear exponent where

$$g(x) = x^2$$

so that the function to be differentiated is

$$y = e^{x^2}$$

In this case, $g'(x) = 2x$ and

$$\frac{dy}{dx} = 2xe^{x^2}$$

Now consider the function

$$y = e^{x^2 + 7x + 6}$$

for which, using (9.11), since $g'(x) = 2x + 7$ the first derivative is

$$\frac{dy}{dx} = (2x + 7)e^{x^2 + 7x + 6}$$

Obtaining the derivative of the function

$$y = (x^3 - 5x^4)e^{x^2}$$

requires the use of the product rule, and is

$$\frac{dy}{dx} = 2x(x^3 - 5x^4)e^{x^2} + e^{x^2}(3x^2 - 20x^3)$$

and, as usual, the slope of the function at any point can be found by evaluating the derivative. So, at $x = 1$ the slope will be

$$\frac{dy}{dx} = 2(1 - 5)e + e(3 - 20)$$

$$= -25e$$

As a further example of the use of (9.11), consider the function

$$f(x) = e^{x^4 - x^3 - x^2 - x}$$

for which the derivative is

$$f'(x) = (4x^3 - 3x^2 - 2x - 1)e^{x^4 - x^3 - x^2 - x}$$

Given

$$f(x) = e^{e^x}$$

since $g(x) = e^x$, $g'(x) = e^x$ and the first derivative is

$$f'(x) = e^x e^{e^x}$$

Now suppose also that the second derivative is also required in this case. Using the product rule as well as (9.11) produces the result

$$f''(x) = e^x e^x e^{e^x} + e^{e^x} e^x$$
$$= (e^x + 1)e^x e^{e^x}$$

The derivatives of exponential functions with bases other than e involve logarithms, and these cases will be considered after logarithmic functions have been introduced. Following the end of section exercises, we examine the integration of natural exponential functions.

EXERCISES 9.2

1 Find the derivative dy/dx for the following exponential functions:

   (i) $y = e^{7x}$
   (ii) $y = e^{-5x}$
   (iii) $y = e^{0.2x}$
   (iv) $y = -10e^{-0.1x}$
   (v) $y = \sqrt{3}e^{\sqrt{3}x}$
   (vi) $y = \dfrac{2}{x} e^{0.5\pi x}$

2 Find the second derivative, $f''(x)$, in the following cases:

   (i) $f(x) = 0.25e^{2x}$
   (ii) $f(x) = -200e^{-0.1x}$
   (iii) $f(x) = e^{4x} - e^{-4x}$

3 Find the required derivative in the following cases:

   (i) $f'''(x)$ where $f(x) = \sqrt{2}e^{\sqrt{x}}$
   (ii) $f''(x)$ where $f(x) = -e^{-x}$
   (iii) $f^{iv}(x)$ where $f(x) = 100/e^{0.5x}$

4 Find the first derivative where

   (i) $f(x) = 2xe^{2x}$
   (ii) $f(x) = x^3 e^{4x}$
   (iii) $f(x) = e^x(x + e^x)$

5 (i) Find the derivative dy/dx for
$$y = e^{5x}(3x + x^4)^3$$
      Evaluate your result at $x = 1$.
  (ii) Find the first derivative of
$$y = (2e^{3x} - e^{2x})(x^2 + x)^2$$
      Evaluate the result at $x = 2$.

6 Find the first derivative for

   (i) $y = e^{20-10x}$
   (ii) $y = e^{0.5x^2}$
   (iii) $y = -e^{-2x^2}$
   (iv) $y = e^{10x^2 - 30x - 40}$
   (v) $y = e^{\sqrt{x}}$
   (vi) $y = e^{2x^4 - 8x^3}$

7 (i) Find the first derivative of
$$f(x) = (2x^4 + 4x^5)e^{0.5x^2}$$
and evaluate the expression at $x = 1$.
(ii) Find an expression for the second derivative of
$$f(x) = e^{x^n}$$
and evaluate the expression where $n = 3$ and $x = 2$.

9.3 INTEGRATION OF NATURAL EXPONENTIAL FUNCTIONS

This section considers the integration of exponential functions in which the base is e. We begin with the simplest case:

$$\int e^x \, dx = e^x + k$$

This result would be expected from what we have already learnt about e^x. Given that e^x is its own derivative, it must also be its own indefinite integral (disregarding the arbitrary constant). Now consider the modest generalization, $y = e^{mx}$. The indefinite integral is

$$\int e^{mx} \, dx = \frac{e^{mx}}{m} + k$$

so that, for example,

$$\int e^{3x} \, dx = \frac{e^{3x}}{3} + k$$

As three further examples of the application of this result, consider

$$\int 6e^{5x} \, dx = \frac{6e^{5x}}{5} + k$$

and

$$\int e^{-2x} \, dx = \frac{-e^{-2x}}{2} + k$$

and

$$\int e^{\pi x} \, dx = \frac{e^{\pi x}}{\pi} + k$$

It follows from the differentiation rule (9.11) that

$$\int g'(x) e^{g(x)} \, dx = e^{g(x)} + k$$

So, for example,

$$\int 2xe^{x^2} \, dx = e^{x^2} + k$$

and

$$\int (3x^2 + 4x + 9)e^{x^3+2x^2+9x} \, dx = e^{x^3+2x^2+9x} + k$$

And, as further examples,

$$\int 0.5xe^{2x^2} \, dx = 0.125e^{2x^2} + k$$

while

$$\int \frac{e^{\sqrt{x}}}{2\sqrt{x}} \, dx = e^{\sqrt{x}} + k$$

We can now also address the subject of *definite* integrals involving exponential functions. For example,

$$\int_{x=1}^{x=2} e^{2x} \, dx = (0.5e^{2x} + k)^{x=2} - (0.5e^{2x} + k)^{x=1}$$

$$= 0.5e^4 - 0.5e^2$$

$$= 23.605$$

and

$$\int_{x=1}^{x=1.5} e^{4x-4} \, dx = [0.25e^{4x-4}]^{x=1.5} - [0.25e^{4x-4}]^{x=1}$$

$$= 0.25e^2 - 0.25e^0$$

$$\approx 1.597$$

while

$$\int_{x=0}^{x=1} (0.6x + 0.7)e^{0.3x^2+0.7x-1} \, dx$$

$$= [e^{0.3x^2+0.7x-1}]^{x=1} - [e^{0.3x^2+0.7x-1}]^{x=0}$$

$$= e^0 - e^{-1}$$

$$= 1 - \frac{1}{e}$$

$$= 0.6321$$

As an example of the application of definite integrals of exponential functions, consider the *exponential probability distribution*. This is most

Exponential and logarithmic functions 631

useful in management science and describes, among various other phenomena, the distribution of the service times for customers in a simple queuing system.[6] The distribution is

$$f(t) = \mu e^{-\mu t} \qquad (9.12)$$

where the single parameter $\mu > 0$ and the domain of the function is $t \geq 0$. In the queuing theory context, t is the time taken to serve a customer and μ represents the mean rate of service. Equation (9.12) is also referred to as the *negative exponential distribution*. The graph of the distribution is shown in Figure 9.4.

Note that, given the domain of the function, $f(t)$ begins at the point $(0, \mu)$ on the vertical axis and asymptotes to the t axis as t increases. Areas under the curve correspond to probabilities, and we first use definite integration to confirm that the total area under the curve is one. The total area is given by the improper integral

$$\int_{t=0}^{t=\infty} \mu e^{-\mu t} \, dt = [-e^{-\mu t} + k]^{t=\infty} - [-e^{-\mu t} + k]^{t=0}$$

$$= [0 + k] - [-1 + k]$$

$$= 1$$

The area under the curve between two specific values of t is the probability that a customer's service time lies between those limits. If in a particular case the average service time is 2 minutes, this corresponds to the mean rate

Figure 9.4

of service $\mu = 0.5$ per minute. We can now use definite integration to answer quality of service questions such as

1. What percentage of customers will not be served in under 3 minutes?

or

2. For what percentage of customers will the service time take longer than the average?

The answer to 1 is given by the area under the curve beyond $t = 3$. This (improper) integral is:

$$\int_{t=3}^{t=\infty} 0.5e^{-0.5t} = [-e^{-0.5t}]^{t=\infty} - [-e^{-0.5t}]^{t=3}$$

$$= [0] - [-0.2231]$$

$$= 0.2231$$

so the percentage of customers not served in under 3 minutes is 22.31 per cent. Now consider question 2. The mean service time is $1/\mu$, so the probability of a service time taking longer than $1/\mu$ is

$$\int_{t=1/\mu}^{t=\infty} \mu e^{-\mu t} \, dt = [-e^{-\mu t}]^{t=\infty} - [-e^{-\mu t}]^{t=1/\mu}$$

$$= 0 + e^{-1}$$

$$= 0.3679$$

So in any simple queuing system, 36.79 per cent of customers experience a longer than average service time. Exponential functions allow a wider range of functions to be used in the types of *differential equation* discussed in Chapter 8. For example, consider

$$12x^2 + \frac{dy}{dx} - 3e^{-3x} = 0$$

given the boundary condition that $y = 11$ when $x = 0$. Rearrangement produces

$$\frac{dy}{dx} = 3e^{-3x} - 12x^2$$

so that

$$y = \int (3e^{-3x} - 12x^2) \, dx$$

and the general solution is

$$y = -e^{-3x} - 4x^3 + k$$

At the value $x = 0$,

$$y = -1 + k$$
$$= 11$$

so $k = 12$, and the particular solution is therefore

$$y = -e^{-3x} - 4x^3 + 12$$

As a further example consider the particular solution to the second-order linear differential equation

$$2 - \frac{d^2y}{dx^2} + 0.01e^{0.1x} = 0$$

given that, when $x = 0$, $y = 6$ and $dy/dx = 5.1$. The equation rearranges as

$$\frac{d^2y}{dx^2} = 2 + 0.01e^{0.1x}$$

upon which integration produces

$$\frac{dy}{dx} = 2x + 0.1e^{0.1x} + k_1$$

which equals 5.1 when $x = 0$, so that k_1 must equal 5. Insertion of this value for k_1 into the derivative and subsequent integration results in

$$y = x^2 + e^{0.1x} + 5x + k_2$$

and since $y = 6$ when $x = 0$, the value of k_2 must be 5. The particular solution is therefore

$$y = x^2 + e^{0.1x} + 5x + 5$$

We have yet to address the question of the differentiation and integration of exponential functions to bases other than e. For this it will be necessary to have some familiarity with logarithmic functions, which we consider in the following section.

EXERCISES 9.3

1 Find the following indefinite integrals:

(i) $\int e^{5x} \, dx$

(ii) $\int -e^{-0.25x} \, dx$

(iii) $\int \pi e^{\pi x} \, dx$

(iv) $\int (6x^2 + 3e^{1.5x}) \, dx$

2 Find the following indefinite integrals:

(i) $\int x e^{x^2} \, dx$

(ii) $\int (5x - 2)e^{10x^2 - 8x - 6} \, dx$

(iii) $\int \dfrac{e^{\sqrt{x}}}{\sqrt{x}} \, dx$

3 Evaluate the following definite integrals:

(i) $\int_{x=2}^{x=4} e^{0.5x} \, dx$

(ii) $\int_{x=1}^{x=3} (0.2x + 0.8)e^{0.1x^2 + 0.8x} \, dx$

(iii) $\int_{x=0}^{x=2} (0.5\pi x + 0.25)e^{\pi x^2 + x} \, dx$

(iv) $\int_{x=0}^{x=1/\mu} \mu e^{-\mu t} \, dt$

4 With the given conditions, find particular solutions to the following differential equations:

(i) $5x + 0.5 \dfrac{dy}{dx} - 2e^{2x} = 0$

where $y = 2e^2$ when $x = 1$.

(ii) $-0.18x^2 + \dfrac{dy}{dx} + 20e^{-0.2x} = 0$

where $y = 100(e^{-2} + 1)$ when $x = 10$.

9.4 NATURAL LOGARITHMIC FUNCTIONS

An exponential relationship between the variables x and y may take the form

$$x = a^y \tag{9.13}$$

which could be read either as

x is the number that results from raising the base a to the power y

or

y is the power to which the base a must be raised to give the number x

A shorthand way of making the second statement is

y is the *logarithm* to the base a of the number x.

This is written as

$$y = \log_a x \tag{9.14}$$

and is defined for $a > 0$ and $a \neq 1$. It is important to note that equations (9.13) and (9.14) describe a *single* relationship between x and y. They view different aspects of the same connection, and are *inverse functions*. Whichever form is used is determined by the requirements of the task in hand. In this section we shall be concerned with logarithmic functions and their relevance to business and economic concepts. Consider some specific examples of logarithmic relationships between two variables, and their restatements in exponential form:

$y = \log_2 1$ means $2^y = 1$ Therefore $y = 0$
$y = \log_3 3$ means $3^y = 3$ Therefore $y = 1$
$y = \log_{10} 100$ means $10^y = 100$ Therefore $y = 2$
$y = \log_2 0.25$ means $2^y = 0.25$ Therefore $y = -2$
$y = \log_e 31.5$ means $e^y = 31.5$ Therefore $y = 3.45$
$y = \log_{1/2} 8$ means $\frac{1}{2}^y = 8$ Therefore $y = -3$

Logarithms to the base e are called *natural logarithms*, and there are two slightly different forms of notation. $y = \log_e x$ is also written as

$$y = \ln x$$

Logs to the base 10 are called *common logarithms*. It was common logarithms that were most frequently used in manual arithmetic calculations before the advent of the electronic calculator. Now consider the graphs of logarithmic functions. Figure 9.5 shows the graph of $y = \log_a x$ for $a > 1$.

$\log_a(x)$

a > 1

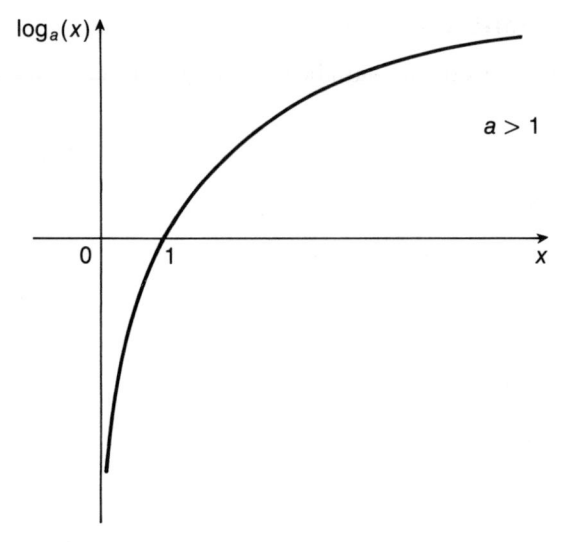

Figure 9.5

Note that the domain of the function is $x > 0$. The necessity of this can be seen from the form of function (9.13). Given that a is positive, there is no number y to which power a can be raised to give a negative outcome (x). The graph is always increasing, but at an ever decreasing rate. Although there is no limit to the value that y can take, the declining slope is such that 'astronomic' increases in x are needed, at an early stage, to produce even minute changes in y. To illustrate in the case of the natural logarithm function, with a scale of centimetres on each axis, in order that y reaches a height of 50 cm the corresponding x co-ordinate is in excess of 5000 light years distant along the horizontal axis! This phenomenal 'flatness' is the view through the inverse function of the steepness of $y = e^x$. Now consider the graph of $y = \log_a x$ when $0 < a < 1$. This is shown in Figure 9.6.

Note that the logarithm function when $0 < a < 1$, defined only for $x > 0$, is always decreasing at an always decreasing rate. The following properties of logarithms are an alternative means of stating some of the properties of exponents that we saw in Chapter 1. Where m and n are positive numbers

(i) $\log_a(mn) = \log_a m + \log_a n$ (9.15)

Equation (9.15) states that the logarithm (to any base) of the product of two numbers is the sum of the individual logarithms. With the product of 4 and 8, for example,

$$\log_2(4 \times 8) = \log_2 4 + \log_2 8$$
$$= 2 + 3 = 5$$

log$_a(x)$

$0 < a < 1$

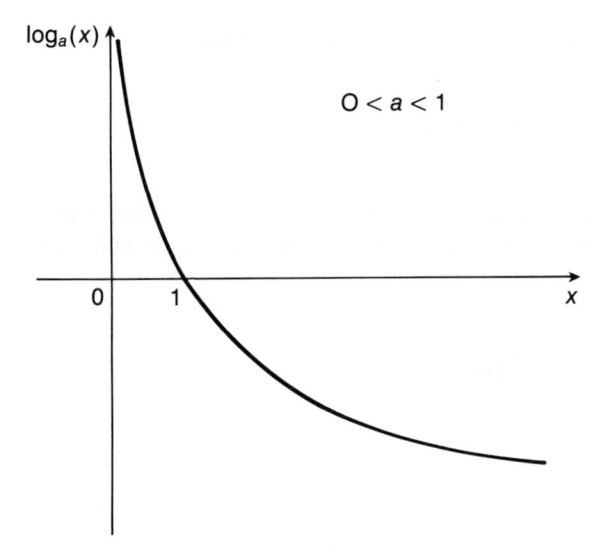

Figure 9.6

which is another way of saying that

$$2^2 \times 2^3 = 2^5$$

Equation (9.15) was the basis of calculation with logarithms – the use of the properties of exponents to transform multiplication into the simpler operation of addition.

(ii) $\quad \log_a\left(\dfrac{m}{n}\right) = \log_a m - \log_a n \qquad (9.16)$

Equation (9.16) states that the logarithm of the quotient is the difference between the logarithms of the two numbers and enables division to be transformed into subtraction by the use of logarithms. For example,

$$\log_3\left(\dfrac{243}{27}\right) = \log_3 243 - \log_3 27$$

$$= 5 - 3 = 2$$

which is another way of saying that

$$\dfrac{3^5}{3^3} = 3^2$$

(iii) $\quad \log_a m^p = p \log_a m \qquad (9.17)$

Equation (9.17) can be verified from (9.15). For example:

$$\log_a m^3 = \log_a mmm$$
$$= \log_a m + \log_a m + \log_a m$$
$$= 3 \log_a m$$

This rule, in conjunction with (9.15), is of value in re-expressing a non-linear relationship in linear form. For example, suppose that x and y are related by the equation

$$y = x^n$$

then, taking logs of both sides (to the base e) gives

$$\ln y = \ln(x^n)$$
$$= n \ln x$$

So the cubing function

$$y = x^3$$

can be replaced with the linear relationship

$$\ln y = 3 \ln x$$

The simplification achieved shows up in a side by side comparison of the two graphs, as shown in Figures 9.7(a) and 9.7(b).

The connection between the phenomena represented by x and y is the same in both cases, but the form linear in logarithms is more convenient for purposes of measurement and testing. This point is reinforced for more complicated functions such as

$$z = Ax^\alpha y^\beta \qquad (9.18)$$

This is the form taken by the *Cobb–Douglas production function* in which z represents output, x represents labour input and y is a measure of capital input at a high level of aggregation (sector or national level). While (9.18) has degree $\alpha + \beta$, it is linear in the logarithms of the variables, i.e. *log-linear*. Taking logs on both sides produces

$$\ln z = \ln A + \alpha \ln x + \beta \ln y$$

which is much more convenient for estimation purposes since linear regression can be used. The last property of logarithms to be considered here is as follows

(iv)   Change of base: $\log_a m = \dfrac{\log_b m}{\log_b a}$ \qquad (9.18)

(a)　　　　　　　　　　　(b)

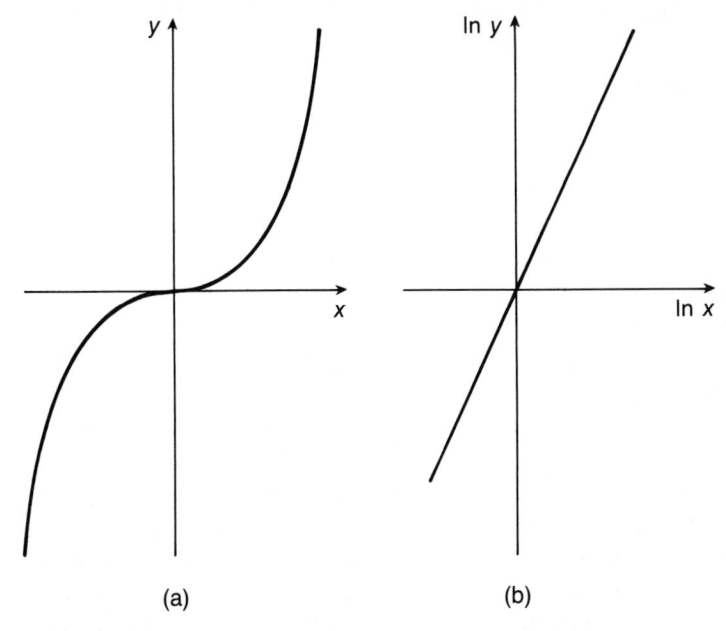

Figure 9.7

The change of base formula is useful in the differentiation of exponential functions with bases other than e. An example of the use of this formula is

$$\log_{10} 100 = \frac{\log_e 100}{\log_e 10} = \frac{4.6052}{2.3025} = 2$$

Now consider the application of logarithms to compound growth. If a population is growing at 2 per cent per annum how many years does it take for the population to double? The answer is *not* 50 years. Suppose that the original population is N. Then after x years of growth at 2 per cent the population is

$$N(1 + 0.02)^x \tag{9.19}$$

and the required value of x — the *doubling time* — is found by equating (9.19) to $2N$. So:

$$N(1 + 0.02)^x = 2N$$

Therefore, after division by N,

$$(1 + 0.02)^x = 2$$

Using property (iii) and taking the logarithms of both sides with (say) a base of 10

$$\log_{10}(1 + 0.02)^x = \log_{10} 2$$

so that

$$x \log_{10}(1 + 0.02) = \log_{10} 2$$

i.e.

$$x = \frac{\log_{10} 2}{\log_{10}(1.02)}$$

$$= \frac{0.3010}{0.0086}$$

$$\approx 35$$

So a population growing at an annual rate of 2 per cent compound will double every *35* years. Note that the doubling time is independent of the size of the population.[7] At the same rate of compound growth, a population of 5 billion will add 5 billion people in the same time that a population one-tenth the size adds 500 million. Similar questions can be addressed in other areas. If an economy has a sustained real growth rate of 3 per cent, with a stable population, the per capita gross domestic product will double in x years where

$$(1.03)^x = 2$$

Using base e logarithms:

$$x = \frac{\ln 2}{\ln 1.03}$$

$$= \frac{0.6931}{0.0296}$$

$$\approx 23.4 \text{ years}$$

Suppose that an investor places a sum on deposit at a guaranteed net 10 per cent. How long will it take for the investment to double? This will be x years where

$$x = \frac{\ln 2}{\ln 1.1}$$

$$= \frac{0.6931}{0.0953}$$

$$\approx 7.27 \text{ years}$$

Now consider a borrower who takes out a capital loan at 30 per cent compound interest per annum. In the absence of repayments (and further borrowings) the outstanding liability will double in x years where

$$x = \frac{\ln 2}{\ln 1.3}$$

$$= \frac{0.6931}{0.2624}$$

$$\approx 2.64 \text{ years}$$

So at such an interest rate, a repayment 'holiday' of under two years eight months would see a doubling of indebtedness.[8] Similar calculations can be performed under continuously compounded interest. For example, the doubling time at an interest rate of 15 per cent is given by

$$e^{0.15x} = 2$$

and on taking logs to base e

$$0.15x = \ln 2$$

so that

$$x = \frac{0.6931}{0.15}$$

$$= 4.62 \text{ years}$$

The calculations can also be applied to decay models. For example, suppose that a company currently has 20 per cent of the market for a particular product range, but that each year its market share is 97 per cent of the share in the preceding year. How long will it be before the company's market share is down to 10 per cent? The relevant equation is

$$20(0.97)^x = 10$$

so that

$$(0.97)^x = 0.5$$
$$x \ln 0.97 = \ln 0.5$$

$$x = \frac{\ln 0.5}{\ln 0.97}$$

$$= \frac{-0.6931}{-0.0305}$$

$$\approx 22.7 \text{ years}$$

So in the unlikely event of stable market conditions for that length of time, the company's share of the market would take almost 23 years to halve. Note that the time to halving at 3 per cent per annum contraction is not the same as the time to doubling at 3 per cent per annum expansion. As a final example, suppose that the quantity of a radioactive contaminant in an environment decays exponentially at the rate of one-hundredth of 1 per cent per annum. How many years must elapse before only one-half of the original quantity of the pollutant remains? The answer is the *half-life* of the substance, and is given by x where

$$e^{-0.0001x} = 0.5$$

Taking logs to base e

$$-0.0001x = \ln 0.5$$

so that

$$x = \frac{-0.6931}{-0.0001}$$

$$= 6931 \text{ years}$$

So the half-life of the radioactive material is almost 7000 years. Note that, in contrast to a situation of exponential growth, exponential decay to half of an original quantity takes significantly *longer* than would be the case under a linear process where a fixed percentage of the *original* amount is subtracted each period.[9] For example, in the above case if the volume of the contaminant was reduced by one-hundredth of 1 per cent of its original quantity each year, then the linear equivalent 'half-life' would be given by

$$n = \frac{0.5}{0.0001}$$

$$= 5000 \text{ years}$$

Exponential (true) half-life is therefore longer than the linear equivalent by the ratio

$$\frac{-\log_e 0.5}{0.5} = 1.3863$$

i.e. by some 38.63 per cent.

EXERCISES 9.4

1 Express the following logarithmic functions in terms of an exponential relationship between the variables:

(i) $y = \log_{10} x$
(ii) $x = \log_5 y$
(iii) $x = \log_e y$

2 Express the following three exponential relationships as logarithmic functions:

   (i) $y = 10^x$
   (ii) $x = a^y$
   (iii) $p = 2^q$

3 Find the value of y when

   (i) $y = \log_5 25$
   (ii) $y = \log_4 8$
   (iii) $y = \log_2 0.125$
   (iv) $y = \log_7 1$
   (v) $y = \log_{0.25} 16$
   (vi) $y = \ln e^2$

4 Evaluate:

   (i) $\log_2 8 + \log_2 16$
   (ii) $\log_3 \left(\dfrac{729}{81}\right)$
   (iii) $\log_{0.5} 8^2$

5 What is the doubling time for

   (i) a population growing at 4 per cent per annum;
   (ii) an investment returning 8 per cent per annum compound;
   (iii) a debt on which interest is charged at 20 per cent per annum?

6 What is the doubling time under continuous compounding at

   (i) 10%   (ii) 20%

7 How long would it take for a company's market share to halve when, in any one year, its market share as a percentage of the previous year is

   (i) 98%   (ii) 95%   (iii) 90%

8 What is the half-life of a radioactive substance which decays exponentially at the rate of a thousandth of 1 per cent per annum?

9.5 THE DERIVATIVE OF NATURAL LOGARITHMIC FUNCTIONS

We shall work towards the derivative of $y = \log_e x$ by making use of the facts that

1 we know how to differentiate the natural exponential function, and

2 the *inverse function* rule applies since the logarithm function is the inverse of the exponential function.

Now

$$y = \log_e x$$

can be re-expressed as

$$x = e^y$$

for which

$$\frac{dx}{dy} = e^y$$

and therefore, from the inverse function rule,

$$\frac{dy}{dx} = \frac{1}{dx/dy}$$
$$= \frac{1}{e^y}$$

so therefore

$$\frac{d(\log_e x)}{dx} = \frac{1}{x} \tag{9.20}$$

It is surprising that the derivative of $\log_e x$ is the reciprocal function x^{-1}. But the result answers the question of what is the integral of $1/x$. We now know that

$$\int x^{-1} dx = \log_e x + k.$$

Integration will be returned to later on; for the moment we take up some consequences of, and variations on, the theme of (9.20).

Higher derivatives of $y = \ln x$: since

$$\frac{dy}{dx} = \frac{1}{x}$$

$$\frac{d^2y}{dx^2} = \frac{-1}{x^2}$$

Higher order derivatives are then obtained in the normal way.

For $y = m \log_e x$

$$\frac{dy}{dx} = \frac{m}{x}$$

Thus, recalling the alternative notational schemes, given

$$f(x) = 17 \ln x$$
$$f'(x) = 17x^{-1}$$

For $y = \ln x^m$

$$\frac{dy}{dx} = \frac{m}{x}$$

This result follows from property (9.17) above. Therefore, since $\log_e x^m = m \log_e x$, their derivatives must also be identical. Consequently, if

$$y = \ln x^3$$

then

$$\frac{dy}{dx} = \frac{3}{x}$$

Now consider the derivative of

$$y = x \ln x \qquad (9.21)$$

This can be obtained by use of the product rule.

$$\frac{dy}{dx} = x \frac{1}{x} + (\ln x)1$$

$$= 1 + \ln x$$

Equation (9.21) is a special case of the product

$$y = f(x) \ln x$$

for which the derivative is

$$\frac{dy}{dx} = f(x) \frac{1}{x} + (\ln x) f'(x) \qquad (9.22)$$

so that for example if

$$y = x^3 \ln x$$

then the derivative will be

$$\frac{dy}{dx} = x^3 \frac{1}{x} + 3x^2 \ln x$$

$$= x^2(1 + 3 \ln x)$$

The quotient rule can also be employed in obtaining the derivative of expressions with a logarithmic component. For example, if

$$y = \frac{\log_e x}{2x}$$

using the quotient rule

$$\frac{dy}{dx} = \frac{2x(1/x) - 2\log_e x}{4x^2}$$

$$= \frac{1 - \log_e x}{2x^2}$$

Now consider an expression having both an exponential and a logarithmic component:

$$y = e^{m \ln x}$$

Since this function is a particular example of the form $y = e^{f(x)}$ for which the derivative is $f'(x)e^{f(x)}$, we could correctly write

$$\frac{d(e^{m \ln x})}{dx} = \frac{m}{x} e^{m \ln x}$$

But a dramatic simplification is possible in this case. Consider the original function. Taking logarithms of both sides produces the result

$$\ln y = \ln(e^{m \ln x})$$
$$= m \ln x$$
$$= \ln x^m$$

so that the relationship between y and x is, in fact,

$$y = x^m$$

for which the derivative is, of course, mx^{m-1} and to which our original derivative will simplify.[10] We now examine the derivative of composite functions involving logarithms: specifically

$$y = \log_e g(x)$$

Using the *chain rule*:

$$\frac{dy}{dx} = \frac{dy}{dg}\frac{dg}{dx}$$

$$= \frac{1}{g(x)} g'(x)$$

$$= \frac{g'(x)}{g(x)} \tag{9.23}$$

Exponential and logarithmic functions 647

In respect of the dy/dg component of the derivative, note that whatever the complexity of the expression of which the logarithm is taken, the derivative of the 'log$_e$' part is the reciprocal of the expression itself. Then, as in any instance of the chain rule, this derivative is multiplied by the derivative of the 'inner' function $g(x)$. Consider a numerical example. Suppose that

$$y = \log_e x^3$$

$$\frac{dy}{dx} = \frac{3x^2}{x^3} = \frac{3}{x}$$

This outcome confirms what we already knew about the derivative of $y = \log_e x^m$ and had obtained by other means. Now consider the derivative of

$$y = \log_e 6x$$

Application of the chain rule produces

$$\frac{dy}{dx} = \frac{6}{6x}$$

$$= \frac{1}{x}$$

It is clear from this example that the derivative of $y = \log_e mx$ is $1/x$ for any real value of m. However, the fact that the function $y = \log_e 6x$ has the same rate of change throughout the domain as $y = \log_e x$, is not altogether obvious. It may help to give an intuitive feel for the result by using property (9.15), from which we may write

$$\log_e 6x = \log_e 6 + \log_e x$$

of which the derivative of the right-hand side is

$$\frac{d}{dx} = 0 + \frac{1}{x}$$

And as a further numerical example consider

$$y = \ln(x^2 + 1)$$

for which function

$$\frac{dy}{dx} = \frac{2x}{x^2 + 1}$$

The ratio of the derivative of a function to the function itself, as in (9.23), is called the *logarithmic derivative*. Given $y = g(x)$, then

$$\frac{g'(x)}{g(x)} = \frac{dy/dx}{y} = \frac{1}{y}\frac{dy}{dx} = \frac{d(\ln y)}{dx}$$

There is an important relationship between logarithmic derivatives and the concept of the *elasticity* of a function. This is important in economics, and we shall consider elasticity in the following section.

EXERCISES 9.5

1 Find the derivative dy/dx where

  (i) $y = 20 \log_e x$
  (ii) $y = \log_e x^4$
  (iii) $y = x^2 \ln x$
  (iv) $y = (4x - 2x^2)\ln x$

2 Find $f'(x)$ where

  (i) $f(x) = \ln x^4$
  (ii) $f(x) = \ln 100x$
  (iii) $f(x) = \log_e(5x^4 - 6x^3 - 7x^2)$

3 (i) Find the derivative of

$$y = \frac{\log_e x}{x^2 + x}$$

  (ii) Find the derivative of

$$y = \frac{x}{\log_e x}$$

4 Find the required derivative in the following cases:

  (i) The third derivative of $y = \ln x$
  (ii) d^2y/dx^2 given that $y = x \log_e x$
  (iii) $f'(x)$ given that $f(x) = (\log_e x)^2$

9.6 ELASTICITY

The derivative of a function measures its rate of change in terms of x and y. The elasticity of a function measures the *proportional* rate of change. In finite terms, the elasticity of a function is the ratio of the proportionate change in the dependent variable to the proportionate change in the independent variable. That is, if x changes by an amount Δx and the resultant change in y is Δy, then the proportionate changes in the two variables are

$$\frac{\Delta x}{x} \quad \text{and} \quad \frac{\Delta y}{y}$$

Exponential and logarithmic functions 649

So the increments in the two variables are taken as a proportion of their current values. Elasticity is then calculated as the ratio of these ratios, i.e.

$$\text{elasticity} = \frac{\frac{\Delta y}{y}}{\frac{\Delta x}{x}}$$

which two-tier ratio reduces to

$$\text{elasticity} = \frac{x}{y}\frac{\Delta y}{\Delta x} \tag{9.24}$$

The *point elasticity* of the function replaces the ratio of increments in (9.24) with the derivative

$$E_{yx} = \frac{x}{y}\frac{dy}{dx} \tag{9.25}$$

What use is elasticity? In practical application, its principal advantage over slope as a measure of the responsiveness in y to changes in x is that, unlike the slope, it is *independent of the units of measure* of either x or y. This is most easily seen in the ratio (9.24). For example, any change in the units of y by some factor c also scales the increment in y by the same factor. The factor therefore cancels out in the elasticity ratio, whereas it would *not* cancel in the ratio of increments alone, the limiting value of which is the slope of the function. The elasticity ratio (9.24) can be seen to be homogeneous of degree zero in x and its increment and in y and its increment. The value of the dimensionless property of E_{yx} is evident in respect of the measurement of the sensitivity of the quantity demanded of a product (y) to changes in the product's own price (x). We speak here of the *elasticity of demand*, for which it is the usual, though not universal, practice to attach a minus sign to the formula. This is done so that a positive value results for E_d, the *own-price* demand elasticity. With p for price and q for quantity the formula therefore is

$$E_d = -\frac{p}{q}\frac{dq}{dp} \tag{9.26}$$

It is meaningless to say that the demand for a product is price sensitive because the slope of the demand function $y = f(x)$ is $f'(x) = -50$. The value of slope could be reduced to -0.5 by a change in the units of measure of price from pounds to pence. The absolute value of slope can be made to take any value by appropriate selection of the units in which the variables are measured. It *is* meaningful, however, to say that the quantity demanded of a product is sensitive to price because the elasticity of demand is greater than one – which is the watershed value for elasticity. Unity is a significant figure for E_d, since values greater than one mean that the proportionate change in quantity demanded exceeds the proportionate change in price. Consider own-price elasticity in the case of a linear demand curve. There is a minor inconvenience here, as the relationship between price and quantity is conventionally stated with price as the dependent variable. We therefore start with the equation

$$p = a - bq$$

in which $a > 0$ and $b > 0$. Note that when $q = 0$, $p = a$, and when $q = a/b$, $p = 0$. Demand elasticity at any point on the curve is given by

$$E_d = -\frac{p}{q}\frac{dq}{dp}$$

where

$$\frac{dq}{dp} = \left(\frac{dp}{dq}\right)^{-1} = -\frac{1}{b}$$

so that

$$E_d = \frac{1}{b}\frac{p}{q} = \frac{a-bq}{bq} \qquad (9.27)$$

Note from (9.27) that, while slope is invariant, elasticity is not constant along a straight line demand curve. The range of variation is from infinity when $q = 0$ to zero when $p = 0$. The graph of the demand curve is shown in Figure 9.8(a) and that of demand elasticity in Figure 9.8(b).

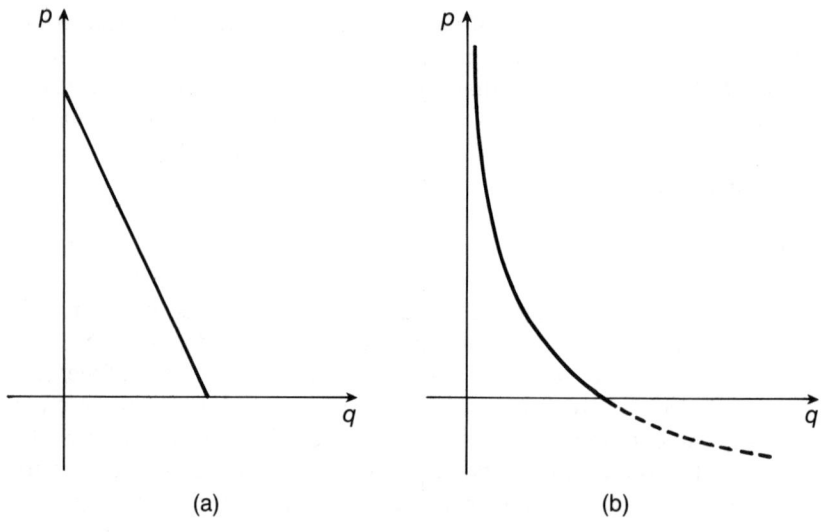

Figure 9.8

The graph of the elasticity function is one branch of the rectangular hyperbola with asymptotes at $x = 0$ and $y = -1$. Elasticity of demand is a valuable concept in assessing the effect of volume or price changes on the value of sales. For example, with quantity as the decision variable

$$R = pq$$

in which R is the value of sales (total revenue) and from which

$$\frac{dR}{dq} = p + q\frac{dp}{dq}$$

$$= p - \frac{p}{E_d}$$

$$= p\left(1 - \frac{1}{E_d}\right) \quad (9.28)$$

From (9.28) it is clear that the necessary condition for a maximum of revenue is satisfied at the output level where $E_d = 1$, i.e. where demand is *unit elastic*. Conducting the workings with price as the decision variable gives

$$\frac{dR}{dp} = q + p\frac{dq}{dp}$$

$$= -qE_d + q$$

$$= q(1 - E_d)$$

from which form it is clear that where demand is *elastic* ($E_d > 1$) a price rise, other things being equal, will result in a revenue decrease. Conversely, when demand is *inelastic* ($E_d < 1$) the strategy of generating more revenue by increasing price will succeed. The question arises as to what point on the linear demand curve corresponds to unit elasticity. From (9.27) this will be where

$$a - bq = bq$$

i.e. where

$$q = a/2b$$

which is the value of q half-way to the q intercept and which corresponds to a point half-way along the demand curve. Consider a numerical example. Let demand conditions be given by

$$p = 100 - 2q$$

for which we shall now find

(a) the elasticity of demand at $q = 20$
(b) the point where elasticity $= 0.25$

Addressing (a), first note that when $q = 20$, $p = 60$. Restating the demand equation with q in terms of p gives

$$q = 50 - 0.5p$$

from which

$$\frac{dq}{dp} = -0.5$$

so that

$$E_d = -\frac{60}{20}(-0.5) = 1.5$$

The most convenient way to find (b) is to use equation (9.27) from which

$$E_d = \frac{100 - 2q}{2q} = 0.25$$

so that

$$100 - 2q = 2q(0.25)$$
$$2.5q = 100$$

and so

$$q = 40$$

We saw that a straight line demand curve had varying elasticity along its length. It is interesting to note that constant, unit elasticity corresponds to the demand curve given by the rectangular hyperbola

$$p = \frac{A}{q}$$

From the rearrangement of this relationship we obtain

$$\frac{dq}{dp} = -\frac{A}{p^2}$$

so that

$$E_d = \frac{p}{q}\frac{A}{p^2}$$

$$= \frac{A}{pq}$$

so therefore

$$E_d = 1$$

Along such a hypothetical demand curve, price and quantity decisions have no bearing on total revenue, which is fixed at £A. What is the relationship of logarithmic derivatives and elasticity? To establish the link, first observe that given the relationship

$$p = f(q)$$

we know that

$$\frac{d \ln p}{dq} = \frac{f'(q)}{f(q)} = \frac{dp/dq}{p}$$

and that

$$\frac{d \ln q}{dq} = \frac{1}{q}$$

So, as a rather cumbersome ratio,

$$\frac{\frac{d \ln q}{dq}}{\frac{d \ln p}{dq}} = \frac{\frac{1}{q}}{\frac{dp/dq}{p}}$$

for which the left-hand side mercifully reduces to

$$\frac{d \ln q}{d \ln p}$$

while the right-hand side becomes

$$\frac{p}{q} \frac{dq}{dp}$$

so that the equation of the two means that

$$E_d = -\frac{d \ln q}{d \ln p} \qquad (9.29)$$

This result is useful since there are demand curves for which it is easier to find elasticity through (9.29) rather than using (9.26). For example, if

$$p = Aq^{-2}$$

taking logs of both sides produces

$$\log_e p = \log_e A - 2 \log_e q$$

from the derivative of which the elasticity of the original function is readily obtained since

$$\frac{d \log_e p}{d \log_e q} = -2$$

The negated reciprocal of this is demand elasticity and so

$$E_d = \tfrac{1}{2}$$

This outcome may of course be confirmed at somewhat greater length by the use of (9.26). Following the exercises we consider the integration of logarithmic functions to the base e.

EXERCISES 9.6

1 Given the demand curve relationship

$$p = 500 - 2.5q$$

where p is price, find the value of elasticity of demand where

(i) $p = 50$ (ii) $p = 100$ (iii) $p = 150$

2 Given the demand relation

$$p = 1000 - 4q$$

(i) Find the level of output for which elasticity of demand is 3.
(ii) Find the price at which the elasticity of demand is 1.5.
(iii) Find the value of price for which total revenue is maximized.

3 Find the elasticity along the demand curve

$$p = \frac{100}{q^{2.5}}$$

4 Suppose that a demand curve is given by

$$\frac{1000}{q + 10}$$

Find the elasticity of demand when

(i) $q = 2.5$ (ii) $q = 10$ (iii) $q = 40$

9.7 NATURAL LOGARITHMIC FUNCTIONS AND INTEGRATION

In this section we examine straightforward cases of integrals involving logarithmic expressions in the antiderivative or the integrand. We begin with a result familiar from earlier work:

(i) $\quad \int \dfrac{1}{x} \, dx = \log_e x + k$

Exponential and logarithmic functions 655

for which further elaboration is not required. We know also that

(ii) $\int \frac{m}{x} \, dx = m \log_e x + k$

$= \log_e x^m + k$

so that, for example,

$\int \frac{9}{x} \, dx = 9 \ln x + k$

$= \ln x^9 + k$

From the work of Section 9.5, it should also be no surprise that, for a rational function in which the numerator is the derivative of the denominator, the integral is the logarithm of the denominator:

(iii) $\int \frac{f'(x)}{f(x)} \, dx = \ln f(x) + k$

So, for example,

$\int \frac{3x^2 - 4x + 4}{x^3 - 2x^2 + 4x} \, dx = \log_e(x^3 - 2x^2 + 4x) + k$

Now consider the indefinite integral of the natural logarithm function. This is

(iv) $\int \log_e x \, dx = x \log_e x - x + k$

$= x(\log_e x - 1) + k$

We shall not prove the above result, but it can at least be confirmed by differentiation. Starting from the function

$y = x \log_e x - x + k$

and using the product rule on the first term, the derivative is obtained as

$\frac{dy}{dx} = x \frac{1}{x} + (\log_e x)1 - 1$

$= \log_e x$

which confirms the result. An extended form of (iv) can be obtained as follows:

$\int \log_e x^n \, dx = \int n \log_e x \, dx$

$= n \int \log_e x \, dx$

The rule may therefore be stated as

(v) $\quad \int \log_e x^n \, dx = nx(\log_e x - 1) + k$

so that, for example,

$$\int \ln x^7 \, dx = 7x(\ln x - 1) + k$$

Finally in this section, note that

(vi) $\quad \int \dfrac{dx}{x \ln x} = \ln(\ln x) + k$

which can be confirmed by differentiation using the chain rule. Letting

$$y = \ln(\ln x) \quad \text{and} \quad \ln x = u$$

so that

$$y = \ln(u)$$

and

$$\frac{dy}{dx} = \frac{dy}{du} \frac{du}{dx}$$

then

$$\frac{dy}{dx} = \frac{1}{u} \frac{1}{x}$$

$$= \frac{1}{x \ln x}$$

Thus, for example, using result (vi):

$$\int \frac{5}{x \ln x} \, dx = 5 \ln(\ln x) + k$$

We shall work through two examples of definite integrals. First consider the definite integral

$$\int_{x=3}^{x=10} \frac{6x - 2}{3x^2 - 2x} \, dx$$

The indefinite integral is

$$\log_e(3x^2 - 2x) + k$$

so that the definite integral evaluates as

$$[\log_e(3x^2 - 2x)]^{x=10} - [\log_e(3x^2 - 2x)]^{x=3}$$
$$= \log_e 280 - \log_e 21$$

Note, by way of revision, that this result will simplify to

$$\log_e \left(\frac{280}{21}\right) = \log_e \left(\frac{40}{3}\right)$$

$$\approx 2.59$$

Now consider

$$\int_{x=2.5}^{x=5} \log_e x^8 \, dx$$

The indefinite integral is

$$8x(\log_e x - 1) + k$$

so that the required definite integral is

$$[8x(\log_e x - 1)]^{x=5} - [8x(\log_e x - 1)]^{x=2.5}$$
$$= 40(\log_e 5 - 1) - 20(\log_e 2.5 - 1)$$
$$\approx 24.3775 - (-1.6742)$$
$$\approx 26.0517$$

Needless to say, there are far more complicated cases of integration involving logarithmic functions than the examples provided in this section. Discussion of the integration of logarithmic functions to bases other than e is contained in Section 9.9.

EXERCISES 9.7

1 Find the following indefinite integrals:

(i) $\int \dfrac{100}{x} \, dx$

(ii) $\int \dfrac{\pi}{x} \, dx$

(iii) $\int \dfrac{-dx}{x}$

2 Find:

(i) $\int \dfrac{4x^3 - 10}{x^4 - 10x} \, dx$

(ii) $\int \dfrac{ax + 0.5bx}{ax^2 + bx + c} \, dx$

3 Obtain the following indefinite integrals:

(i) $\displaystyle\int \ln x^6 \, dx$

(ii) $\displaystyle\int \ln x^\pi \, dx$

4 Find:

(i) $\displaystyle\int \frac{10}{x \log_e x} \, dx$

(ii) $\displaystyle\int \frac{dx}{2x \ln x}$

5 (i) $\displaystyle\int_{x=2}^{x=5} \frac{6x^2 - 5x}{4x^3 - 5x^2} \, dx$

(ii) $\displaystyle\int_{x=4}^{x=10} \log_e x^{10} \, dx$

9.8 INTEGRATION BY PARTS

In Sections 9.3 and 9.7 we saw how integrals can be found for certain exponential and logarithmic functions. In Chapter 8 we concentrated on the integration of power functions. In this section we consider a technique of integration that is useful for functions containing terms which are the product of power and exponential components.[11] The method is the equivalent procedure in integration to the product rule for differentiation, and is known as *integration by parts*. The product rule for differentiation stated that if the dependent variable, y, could be expressed as the product of two functions of x, as in

$$y = f(x)g(x)$$

then one way to proceed with the differentiation was to use the fact that

$$\frac{dy}{dx} = \frac{d(fg)}{dx} = g\frac{df}{dx} + f\frac{dg}{dx}$$

To see how an equivalent approach can prove to be useful in the process of integration, first integrate both sides with respect to x. The result is

$$\int \frac{dy}{dx} \, dx = \int \frac{d(fg)}{dx} \, dx = \int g \frac{df}{dx} \, dx + \int f \frac{dg}{dx} \, dx$$

Exponential and logarithmic functions 659

so that

$$y = fg = \int \left(g \frac{df}{dx}\right) dx + \int \left(f \frac{dg}{dx}\right) dx$$

This relationship is used to best advantage in the rearranged form

$$\int g \, df = fg - \int f \, dg \qquad (9.30)$$

where (9.30) is known as the *by-parts* formula. The expression to be integrated is interpreted as $\int g \, df$ and the by-parts rule is used to express this in a way that is easier to integrate. Examples will illustrate the use of the rule. Consider

(i) $\quad \int x \, e^x \, dx$

This expression will be taken as the $g \, df$ component of (9.30), the use of which will also require expressions for f and dg. Thus with

$$\frac{df}{dx} = e^x$$

it is clear that

$$f(x) = e^x$$

and with $g(x) = x$

$$\frac{dg}{dx} = 1$$

So, therefore, using (9.30)

$$\int x \, e^x \, dx = e^x(x) - \int e^x \, dx$$

$$= xe^x - e^x + k$$

$$= e^x(x - 1) + k$$

Now consider

$$\int 0.5 x^2 e^x \, dx$$

for which we shall again set

$$\frac{df}{dx} = e^x$$

It follows from this choice that
$$f(x) = e^x$$
and with
$$g(x) = 0.5x^2$$
it is therefore the case that
$$\frac{dg}{dx} = x$$
So, in the form of (9.30), we now have
$$\int 0.5x^2 e^x \, dx = 0.5x^2 e^x - \int xe^x \, dx$$
$$= 0.5x^2 e^x - [e^x(x-1)] + k$$
$$= e^x(0.5x^2 - x + 1) + k$$

As a further example consider
$$\int xe^{0.5x} \, dx$$
for which we shall set
$$\frac{df}{dx} = e^{0.5x}$$
As a result
$$f(x) = 2e^{0.5x}$$
and since
$$g(x) = x$$
it is the case that
$$\frac{dg}{dx} = 1$$
The integral can now be written as
$$\int xe^{0.5x} \, dx = 2xe^{0.5x} - \int 2e^{0.5x} \, dx$$
$$= 2xe^{0.5x} - 4e^{0.5x} + k$$
$$= e^{0.5x}(2x - 4) + k$$

Exponential and logarithmic functions 661

Now consider a further example in which the method of integration by parts is used twice. The integral required is

$$\int 2x^2 e^{0.25x} \, dx$$

Here again setting the df derivative equal to the exponential component of the integrand

$$\frac{df}{dx} = e^{0.25x}$$

Thus

$$f(x) = 4e^{0.25x}$$

and with

$$g(x) = 2x^2$$

it follows that

$$\frac{dg}{dx} = 4x$$

The integral can therefore be written as

$$\int 2x^2 e^{0.25x} \, dx = 8x^2 e^{0.25x} - \int 16x e^{0.25x} \, dx$$

in which integration by parts is applied again to the $\int f \, dg$ term on the right. In this case

$$\int 16x e^{0.25x} \, dx = 64x e^{0.25x} - \int 64 e^{0.25x} \, dx$$

$$= 64x e^{0.25x} - 256 e^{0.25x}$$

so the required integral is

$$8x^2 e^{0.25x} - 64x e^{0.25x} + 256 e^{0.25x} + k$$

which is more neatly rearranged as

$$8(x^2 - 8x + 32) e^{0.25x} + k$$

Now consider an example of definite integration:

$$\int_{x=2}^{x=12} 10x e^{0.4x - 0.8} \, dx$$

First find the indefinite integral. Setting

$$\frac{df}{dx} = e^{0.4x - 0.8}$$

then

$$f(x) = 2.5e^{0.4x - 0.8}$$

and, with

$$g(x) = 10x$$

$$\frac{dg}{dx} = 10$$

The indefinite integral is then

$$\int 10xe^{0.4x - 0.8} \, dx = 25xe^{0.4x - 0.8} - \int 25e^{0.4x - 0.8} \, dx$$

$$= 25xe^{0.4x - 0.8} - 62.5e^{0.4x - 0.8} + k$$

Ignoring the constant of integration, the definite integral is

$$[25xe^{0.4x - 0.8} - 62.5e^{0.4x - 0.8}]^{x=12} - [25xe^{0.4x - 0.8} - 62.5e^{0.4x - 0.8}]^{x=2}$$
$$= 300e^4 - 62.5e^4 - [50e^0 - 62.5e^0]$$
$$= 237.5e^4 + 12.5 \approx 12{,}979.56$$

EXERCISES 9.8

1 Use integration by parts to find the following integrals:

(i) $\int 5xe^x \, dx$

(ii) $\int 2x^2 e^x \, dx$

2 Find:

(i) $\int xe^{2x} \, dx$

(ii) $\int 5x^2 e^{0.5x} \, dx$

3 Evaluate the following definite integral:

$$\int_{x=3}^{x=5} 200xe^{0.5x - 0.5} \, dx$$

9.9 LOGARITHMIC AND EXPONENTIAL FUNCTIONS TO BASES OTHER THAN e

It is convenient to consider logarithmic functions first. Consider

$$y = \log_a x$$

There are two stages to the process of obtaining the derivative of a logarithmic function to a base other than e. These are

1 change the base to e;
2 differentiate.

We shall now put this process into effect. From the change of base formula (9.18)

$$\log_a x = \frac{\log_e x}{\log_e a}$$

It follows that, since $1/\log_e a$ is simply a constant,

$$\frac{d(\log_a x)}{dx} = \frac{1}{\log_e a} \frac{1}{x}$$

So for example given

$$y = \log_2 x$$

then

$$\frac{dy}{dx} = \frac{1}{x \ln 2}$$

$$= \frac{1}{0.6931 x}$$

The second derivative can be found in the following way: write the first derivative as

$$\frac{dy}{dx} = \left(\frac{1}{\ln 2}\right) x^{-1}$$

from which

$$\frac{d^2 y}{dx^2} = -\left(\frac{1}{\ln 2}\right) x^{-2}$$

$$= -\frac{1}{x^2 \ln 2}$$

For a further example consider

$$y = \log_{10} x^3$$

$$= \frac{\log_e x^3}{\log_e 10}$$

so that

$$\frac{dy}{dx} = \frac{1}{\log_e 10} \frac{3x^2}{x^3}$$

$$= \frac{3}{x \log_e 10}$$

$$\approx \frac{1.3}{x}$$

Now consider

$$f(x) = \log_8 (8x^2 + 7x + 6)$$

$$= \frac{\log_e (8x^2 + 7x + 6)}{\log_e 8}$$

so the derivative is

$$f'(x) = \frac{1}{\log_e 8} \frac{16x + 7}{8x^2 + 7x + 6}$$

Integration of logarithmic functions to other bases can proceed along similar lines to differentiation – first change the base to e, then integrate. For example:

$$\int \log_{10} x \, dx = \int \frac{\log_e x}{\log_e 10} \, dx$$

$$= \frac{1}{\log_e 10} \int \log_e x \, dx$$

$$= 0.4343 \int \log_e x \, dx$$

$$= 0.4343 x (\log_e x - 1) + k$$

As an illustration involving a definite integral consider

$$\int_{x=5}^{x=15} 5 \log_2 x^4 \, dx$$

Changing the base produces

$$\int_{x=5}^{x=15} \frac{5}{\log_e 2} \log_e x^4 \, dx$$

which will be

$$\left[\frac{5}{\log_e 2} 4x(\log_e x - 1)\right]^{x=15} - \left[\frac{5}{\log_e 2} 4x(\log_e x - 1)\right]^{x=5}$$

$$= \left[\frac{20x(\log_e x - 1)}{\log_e 2}\right]_{x=5}^{x=15}$$

$$= \frac{300(\log_e 15 - 1) - 100(\log_e 5 - 1)}{\log_e 2}$$

$$\approx \frac{451.4713}{0.6931} \approx 651.34$$

Now consider exponential functions to bases other than e. Consider the function

$$y = a^x$$

How may the derivative of this function be found? It will be recalled that

$$y = a^x \quad \text{means that} \quad x = \log_a y$$

and, from what we have already seen,

$$\frac{dx}{dy} = \frac{1}{y \log_e a}$$

Invoking the inverse function rule

$$\frac{dy}{dx} = y \log_e a$$

$$= a^x \log_e a \tag{9.30}$$

So that, as with the natural exponential function, the exponent of the base remains unchanged in the derivative. And each successive higher derivative will simply multiply (9.30) by a further $\log_e a$. Thus

$$\frac{d^n y}{dx^n} = a^x (\log_e a)^n$$

So, applying this result to the case of,

$$y = 2^x$$

$$\frac{dy}{dx} = 2^x \log_e 2$$

$$= (0.6931)2^x$$

and

$$\frac{d^2y}{dx^2} = (0.6931)^2 2^x$$

$$= (0.4805)2^x$$

As further examples consider

$$f(x) = 10^x$$

for which

$$f'(x) = 10^x \log_e 10$$
$$\approx (2.3026)10^x$$

and

$$f''(x) \approx (2.3026)^2 10^x$$
$$\approx (5.3019)10^x$$

Consider also the function

$$f(x) = (\tfrac{1}{2})^x$$

for which

$$f'(x) = (\tfrac{1}{2})^x \log_e (\tfrac{1}{2})$$
$$\approx -0.6931(\tfrac{1}{2})^x$$

from which it can be seen that the slope of the function is always negative but is, in absolute terms, a decreasing function of x. Note also that where the base, a, is less than one, the higher derivatives will alternate in sign. In the present case

$$f''(x) = [\log_e (\tfrac{1}{2})]^2 (\tfrac{1}{2})^x$$
$$\approx 0.4805(\tfrac{1}{2})^x$$

and

$$f'''(x) = -0.3330(\tfrac{1}{2})^x$$

Exponential and logarithmic functions 667

Now we consider an example of a derivative that requires the use of the product rule. If

$$f(x) = (x^3 + 5x^2)4^x$$

then

$$f'(x) = (x^3 + 5x^2)(\ln 4)4^x + 4^x(3x^2 + 10x)$$

Now consider the derivative of an exponential function with a base other than e, in which the exponent of the base is a function of x, i.e.

$$y = a^{f(x)}$$

The derivative can be approached in the following way. First take logarithms to the base a of both sides:

$$\log_a y - f(x)$$

Using the change of base formula for logarithms

$$\log_a y = \frac{\log_e y}{\log_e a} = f(x)$$

so that

$$\log_e y = (\log_e a) f(x)$$

Differentiating with respect to x:

$$\frac{d(\log_e y)}{dx} = \frac{1}{y}\frac{dy}{dx} = (\log_e a) f'(x)$$

so that

$$\frac{dy}{dx} = (\log_e a) f'(x) y$$

$$= (\log_e a) f'(x) a^{f(x)} \qquad (9.31)$$

So with the exception of the coefficient $\log_e a$, the outcome is as for the derivative of e^x. Therefore, given

$$y = 5^{4x^2 - 7x + 8}$$

using (9.31)

$$\frac{dy}{dx} = (\log_e 5)(8x - 7)5^{4x^2 - 7x + 8}$$

$$= (1.6094)(8x - 7)5^{4x^2 - 7x + 8}$$

$$= (12.8755x - 11.2661)5^{4x^2 - 7x + 8}$$

so the slope of this function when (say) $x = 2$ is approximately 141,454,102.

As a further example of the use of (9.31), suppose that it is required to find the slope of the function

$$f(x) = 12^{x^{1/2}}$$

when $x = 4$. The derivative is

$$f'(x) = (\ln 12)(\tfrac{1}{2}x^{-1/2})12^{x^{1/2}}$$

which evaluates at $x = 4$ as

$$f'(x) \approx 2.4849(0.25)144$$
$$\approx 89.4566$$

A derivative that requires the use of the product rule is

$$y = (x^5)4^{x^2}$$

for which

$$\frac{dy}{dx} = (x^5)[(\ln 4)(2x)4^{x^2}] + (5x^4)4^{x^2}$$

Now consider the integration of exponential functions to bases other than e. To begin with, take the simplest form for the integrand:

$$\int a^x \, dx = \frac{a^x}{\log_e a} + k \qquad (9.32)$$

The result (9.32) can be confirmed from the differentiation formula (9.30). For an example of the use of (9.32) consider the indefinite integral

$$\int 10^x = \frac{10^x}{\log_e 10} + k$$
$$= (0.4343)10^x + k$$

Putting (9.31) into reverse gives the conclusion

$$\int f'(x) a^{f(x)} \, dx = \frac{a^{f(x)}}{\log_e a} + k$$

So, for example,

$$\int [12x + 8] 2^{6x^2 + 8x - 9} \, dx = \frac{2^{6x^2 + 8x - 9}}{\log_e 2} + k$$
$$= (1.4427) 2^{6x^2 + 8x - 9} + k$$

And, as an example of a definite integral,

$$\int_{x=0}^{x=2} (4x - 1) 4^{2x^2 - x + 1} \, dx$$

The indefinite integral is

$$\frac{4^{2x^2-x+1}}{\ln 4} + k$$

and the definite integral will be:

$$\frac{1}{\ln 4}(4^7 - 4) \approx 11{,}816$$

In Sections 9.1 and 9.3 we saw examples that highlighted one sense of the term 'exponential growth'. Let us now look more closely at what exponential growth really implies. Exponential growth means that the rate of change of a quantity is proportional to the value of the quantity at any point.[12] That is,

$$\frac{dy}{dx} = ky \qquad (9.33)$$

Equation (9.33) is a differential equation the solution of which, the constant of integration apart, is an exponential function. In the case of the natural exponential function, $k = 1$, while in the case of $y = a^x$, $k = \log_e a$. So long as $k > 0$, not only does y increase with x, but the rate at which it is increasing also rises with x. It can be shown that any quantity growing exponentially will ultimately outgrow any quantity growing in proportion to fixed powers of x. It is an important property of an algorithm that the time taken to solution grows as a power of the problem size. That is, it is desirable that a computation procedure should be a *polynomial time* algorithm rather than an *exponential time* algorithm.

EXERCISES 9.9

1 Find the first derivative of the following functions:

(i) $y = \log_5 x$
(ii) $y = \log_{0.2} x$

2 Find the required derivative in the following cases:

(i) $f'(x)$ where $f(x) = \log_{60} x$
(ii) $f''(x)$ where $f(x) = \log_{10} x$
(iii) $f'(x)$ where $f(x) = \log_{e^2} x$

3 Find $f'(x)$ where

(i) $f(x) = \log_8 x^5$
(ii) $f(x) = \log_{0.5} x^{0.5}$

4 Find the derivative of the following functions:
   (i) $y = \log_{10}(5x^2 - 4x)$
   (ii) $y = \log_2(2.5x^4 - x^3 + 5x^2)$

5 Find the following indefinite integrals:
   (i) $\int \log_2 x \, dx$

   (ii) $\int 12 \log_{1.5} x^5 \, dx$

6 Evaluate
$$\int_{x=4}^{x=12} 2 \log_{10} x^3 \, dx$$

7 Find the required derivative for
   (i) $f'(x)$ where $f(x) = 10^x$
   (ii) $f''(x)$ where $f(x) = 5^x$
   (iii) $f'(x)$ where $f(x) = 2x^5 - x^3$

8 Find the first derivative of the following functions:
   (i) $y = 10^{x^2}$
   (ii) $y = 16^{5x^2 - 0.5x - 1}$
   (iii) $y = (x^3)5^{x^2}$

9 Find the following indefinite integrals:
   (i) $\int 5^x \, dx$

   (ii) $\int (\pi)6^x \, dx$

   (iii) $\int (10x - 8)10^{5x^2 - 8x} \, dx$

10 Evaluate
$$\int_{x=1}^{x=2} (4x - 1)12^{2x^2 - x} \, dx$$

9.10 AGGREGATE SALES CURVES

Many models in economics and business studies are based on exponential growth or decay, of which the simplest form is

$$y = a^x$$

where a value of a greater than one corresponds to growth, while values in the interval $0 < a < 1$ represent decay. But the function $y = a^x$ itself has only a single *parameter* (a) to adapt to the situation being modelled. This does not give sufficient flexibility to match conditions where the rate of growth itself changes over time. Take a marketing example. In the case of a new consumer product from an established company with a well resourced launch, sales may get off to a very rapid start after which growth slackens. For fashion items, beyond a certain point, sales per period time may approach zero. Examples of products likely to achieve a rapid start to sales are the latest recording by an established pop group, and an 'instant' book on a sensational but not lasting issue. Figure 9.9 shows the graph of possible aggregate sales of such products, plotted against time since launch.

Figure 9.9 is called a *modified growth curve* and has the equation

$$y = T(1 - a^{-bx}) \qquad (9.34)$$

for which the parameters must be such that

$$T > 0 \qquad a > 1 \qquad b > 0$$

In (9.34) the variable x can represent time from launch, or some measure of *marketing effort*.[13] As can be seen from the figure, the parameter T is an upper bound on total sales and, since

$$a^{-bx} \to 0 \quad \text{as} \quad x \to \infty$$

y has this value as an asymptote. T is known as the *market potential* or *saturation level* of demand. Given the ceiling set by the value of T, the values of a and b will determine how slowly or rapidly this ceiling is approached. A selection of values of (9.34) is shown in Table 9.6. While for each pair of values of a and b the curves have the same general shape as in Figure 9.9, the initial steepness and the sharpness of the 'bend' depend on the values of a and b.[14] When the base $a = e$ the function is sometimes called the *modified exponential*. The values of the parameters a and b would be selected to match as closely as possible the sales characteristics of the product under study.

But even the most careful selection of parameters in the modified growth model will not enable it to fit a common pattern of aggregate sales known as the *S curve demand function*,[15] as shown in Figure 9.10. The S curve pattern of demand sees sales at first responding slowly to increases in the independent variable but with, up to a point, an increasing rate of growth.

y↑

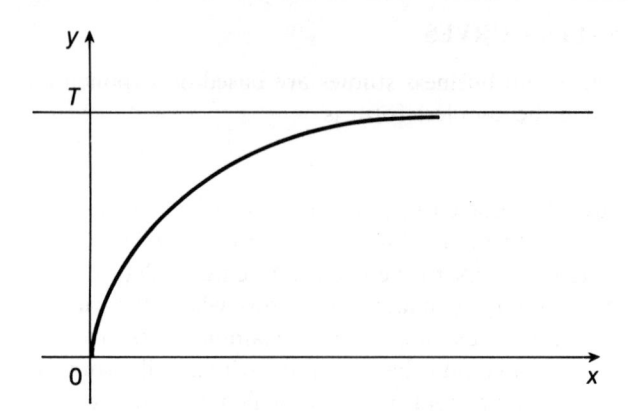

T

0 → x

Figure 9.9

Table 9.6

				x		
a	b	1	10	50	100	1000
1.02	0.2	0.40	3.88	17.97	32.70	98.09
1.2	0.02	0.36	3.58	16.67	30.56	97.39
1.2	0.2	3.58	30.56	83.85	97.39	100
1.2	0.6	10.36	66.51	99.58	100	100
1.2	1.0	16.67	83.85	99.99	100	100
1.6	0.2	8.97	60.94	99.09	99.99	100
2.0	0.2	12.94	75.00	99.90	100	100
e	0.2	18.12	86.47	100	100	100
e	1.0	63.21	100	100	100	100

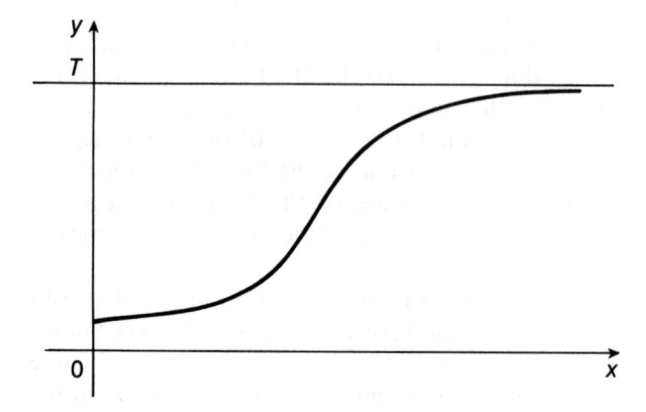

y↑

T

0 → x

Figure 9.10

Exponential and logarithmic functions 673

The maximum rate of growth of aggregate sales occurs at the point of inflection. Subsequently, growth of aggregate sales declines, with the market potential as asymptote – as with the modified growth curve. If x measures sales promotion effort (measured by duration of campaign, expenditure etc.) the y intercept represents the sales level with zero[16] effort. Then for positive x, the first part of the S curve shows *increasing marginal returns* to sales promotion input. Diminishing returns set in after the inflection point. With x representing time, an S curve pattern of aggregate sales against time might be exhibited by a good popular music recording by a relatively less well known group. Initially the record buying public becomes aware of the record gradually, but sales begin to take off after sufficient 'plays'. With x as time, the intercept may represent orders taken prior to launch. If this is not applicable, the whole function could be shifted down. The characteristic S curve shape can be produced in a number of ways.

In a marketing context two particularly relevant expressions are the *Gompertz function* and the *logistic curve*. The Gompertz function is

$$y = Ta^{b^x} \qquad (9.35)$$

in which expression the exponent of a itself has the exponent x, and where, in order to generate the S curve shape,

$$0 < a < 1 \quad \text{and} \quad 0 < b < 1$$

Results comparable with practical situations are produced when the value of b is near the upper end of its range, while a is less than 0.5. The upper asymptote is at T (representing market saturation, so $T > 0$) and the y intercept occurs at $y = Ta$. Observe also that the function has the x axis as an asymptote as the value of x becomes large and negative. In view of the fact that the decision variable, x, is the exponent of an exponent, it may be more convenient to work with the logarithm of sales. The result of taking logs on both sides of (9.35) would be

$$\log_e y = \log_e T + b^x \log_e a$$

from which, if required, the derivative dy/dx could be more easily found in the following way:

$$\frac{d(\ln y)}{dx} = b^x (\ln b)(\ln a)$$

$$= \frac{1}{y}\frac{dy}{dx}$$

so that

$$\frac{dy}{dx} = Ta^{b^x}[b^x(\ln b)(\ln a)]$$

Now consider an example. Suppose that the marketing department of a firm is considering the length of an advertising campaign for a new product. With x representing the length of campaign in weeks, it is estimated that aggregate sales y are given by

$$y = 20{,}000(0.25)^{(0.93)^x}$$

Suppose that the department has narrowed its options to five possible lengths of campaign, including the no-campaign possibility. Aggregate sales as given by the Gompertz function would be as shown in Table 9.7.

While the example has not been set in the wider context of profits maximization, from the data of Table 9.7 it would be unlikely that, with the market already at 92 per cent of saturation level, the campaign would run beyond 39 weeks. S-shaped patterns of aggregate demand are also produced by the *logistic curve*. In this model, aggregate sales y are given by the relationship

$$y = \frac{T}{1 + ae^{-bx}} \qquad (9.36)$$

In (9.36) T again represents market potential or saturation,[17] so $T > 0$ and also $a > 0$ and $b > 0$. The y intercept will be at $T/(1 + a)$. As with the Gompertz curve,[18] the lower asymptote is the x axis. The inflection point of (9.36) is of interest. This is the point where sales are growing most rapidly, and is obtained by setting the second derivative of (9.36) to zero. After some manipulation it emerges that the value of x at the inflection point is

$$x = -\frac{1}{b}\ln\left(\frac{1}{a}\right) \qquad (9.37)$$

Consider an example. For purposes of comparison with the Gompertz curve, let

$$T = 20{,}000 \quad \text{and} \quad a = 3$$

These values give the same upper asymptote (20,000) and y intercept (5000) as in the Gompertz example. A value of b of about 0.1 produces similar

Table 9.7

Length of campaign (x)	Aggregate sales (y)	Market potential (%)
0	5,000	25.00
13	11,659	58.29
26	16,210	81.05
39	18,429	92.15
52	19,373	96.87

Exponential and logarithmic functions 675

Table 9.8

| | y | |
x	Logistic curve	Gompertz curve
0	5,000	5,000
13	11,004	11,659
26	16,356	16,210
39	18,855	18,429
52	19,674	19,373

overall results. In Table 9.8, corresponding values from the logistic and Gompertz curves are presented.

The inflection point of the logistic curve is obtained from (9.37) as

$$x = -\frac{1}{0.1} \ln\left(\frac{1}{3}\right) \approx 11$$

and so the greatest rate of growth of aggregate sales would occur in the eleventh week of the campaign. Logistic curves are used to model a variety of other situations, one example being the diffusion of information through a population.[19] The 'S curve' profile occurs in a variety of business and financial situations, e.g. the graph of the total expenditure on a project against time typically follows this pattern.

9.11 BASIC PROGRAMS

The first program in this section, *E1.bas*, estimates the value of e from the definition

$$e = \lim_{n \to \infty} \left(1 + \frac{1}{n}\right)^n$$

This program will run in QuickBasic, Q-Basic, G-W Basic (with the addition of line numbers) and other versions of BASIC. The listing is as follows.

Program 9.1

```
CLS
PRINT 'n', 'e'
PRINT
FOR n% = 50000 TO 1000000 STEP 50000
    e = (1 + 1 / n%)^n%
    PRINT n%, e
NEXT n%
END
```

The beginning and endpoint of the FOR...NEXT loop can be varied as desired by editing that line of the program code. The % symbol in the variable label n% indicates an integer value. Making this distinction when possible speeds up the running of the program.

The second program, *Modlog.bas*, relates to the marketing application of Section 9.10. The program produces the values of the modified growth and logistic functions for input parameter values. The program output makes an interesting comparison of the values produced by the two functions. Try running the program with, say,

$a = 2 \quad b = 0.2 \quad T = 100 \quad 50$ time periods

The LPRINT lines mean that output is sent to the printer as the program runs. If no printer is connected and ready, these lines should be edited out of the code if you intend to run the program. The program will run in QuickBasic or Q-Basic. The program listing is as follows.

Program 9.2

```
'                        MODLOG.BAS

CLS

PRINT 'Program shows values for the modified growth and the'
PRINT 'logistic curves for the parameters specified.'
PRINT
PRINT '_ _ _ _ _ _ _ _ _ _ _ _ _ _ _ _ _ _ _ _ _ _ _ _ _ _'
PRINT
PRINT 'Please input parameters as requested'
PRINT

INPUT 'Value of a'; a
PRINT
INPUT 'Value of b'; b
PRINT
INPUT 'Value of T'; T
PRINT
INPUT 'Maximum number of time periods (x)'; n%

CLS

PRINT 'Values given by the modified growth and logistic curves'
PRINT 'for the following parameter values...'
PRINT

PRINT 'a = '; a, 'b = '; b, 'T = '; T, n%; ' time periods'

PRINT
PRINT '_ _ _ _ _ _ _ _ _ _ _ _ _ _ _ _ _ _ _ _ _ _ _ _ _ _ _';
PRINT '_ _ _ _ _'

PRINT 'x'; TAB(20); 'Modified growth'; TAB(50); 'Logistic curve'
PRINT

FOR x% = 1 TO n%

    ymod = T * (1 – a^(– b * x%))

    ylog = T / (1 + a * EXP(– b * x%))

    PRINT x%;

        PRINT USING '# #.# #'; TAB(20); ymod; TAB(50); ylog

NEXT x%

PRINT
PRINT '_ _ _ _ _ _ _ _ _ _ _ _ _ _ _ _ _ _ _ _ _ _ _ _ _ _';
PRINT '_ _ _ _ _'

END
```

ADDITIONAL PROBLEMS

1. (i) What is the future value after 10 years of a sum of £2500 invested under annual compounding at 12 per cent?

   (ii) What is the value after 5 years of a car which depreciates at 22 per cent per annum and was originally purchased for £12,500?

2. A sum of £950 is borrowed at a nominal rate of 24 per cent per annum. What is the debt outstanding:

   (i) after 1 year, with compounding

       (a) annually   (b) monthly   (c) daily

   (ii) after 5 years with

       (a) annual compounding   (b) continuous compounding

3. Find the first derivative of the following functions:

   (i) $y = 5e^{2x}$
   (ii) $y = -2.5e^{-4x}$
   (iii) $y = 4e^{3x^2 - 5x - 4}$
   (iv) $y = 2x^5 e^{4x}$

4. Find the following indefinite integrals:

   (i) $\int 5e^{0.2x} \, dx$

   (ii) $\int -6e^{-3x} \, dx$

   (iii) $\int (8x - 14)e^{2x^2 - 7x + 3} \, dx$

5. Evaluate

   (i) $\int_{x=10}^{x=20} e^{0.2x} \, dx$

   (ii) $\int_{x=2}^{x=4} (2x + 1)e^{0.1x^2 + 0.1x + 1} \, dx$

6. (i) Find the doubling time for a population growing at $3\frac{1}{2}$ per cent per annum.

   (ii) How long would it take a population growing at 5 per cent per annum to treble in size?

   (iii) A contaminant decays exponentially at the rate of one-half of 1 per cent per annum. What is the half-life of the substance?

7 Find the first derivative of the following functions:
   (i) $y = \pi^2 \log_e x$
   (ii) $y = \log_e x^7$
   (iii) $y = (x^4 - x^3)\ln x^5$
   (iv) $y = \ln(x^5 + x^4)$

8 Demand for a product is given by
$$p = 2000 - 5q$$
   Find the elasticity of demand when
   (a) $q = 50$ (b) $q = 100$
   (c) $q = 200$ (d) $q = 250$

9 Find the following indefinite integrals:
   (i) $\int \dfrac{250}{x} \, dx$

   (ii) $\int \dfrac{x - 7}{0.5x^2 - 7x} \, dx$

   (iii) $\int \log_e x^8 \, dx$

   (iv) $\int \dfrac{10}{x \ln x} \, dx$

   (v) $\int 7x e^{0.3x} \, dx$

10 Find the first derivative where
   (i) $y = \log_{20} x$
   (ii) $y = \log_4 x^5$
   (iii) $y = \log_5(x^3 - x^2)$

11 Find the following indefinite integrals:
   (i) $\int 2 \log_{12} x \, dx$

   (ii) $\int 2 \log_2 x^2 \, dx$

12 Find the derivative of
$$y = 8^{0.5x^2 - 7x}$$

13 With the boundary conditions given, find the particular solution to the following differential equation:

$$8x^3 - 0.5\frac{dy}{dx} + 4e^{-2x} = 0$$

when it is known that

when $x = -2$, $y = 40$

14 Evaluate

$$\int_{x=1}^{x=2} (2x - 0.4)3^{x^2 - 0.4x + 2}\, dx$$

REFERENCES AND FURTHER READING

1 Bishir, J. W. and Drewes, D. W. (1970) *Mathematics in the Behavioural and Social Sciences*, Harcourt, Brace and World.
2 Chiang, A. C. (1984) *Fundamental Methods of Mathematical Economics* (Third Edition), McGraw-Hill.
3 Finney, R. L. and Thomas, G. B., Jr (1990) *Calculus*, Addison-Wesley.
4 Rich, A., Rich, J. and Stoutemyer, D. (1989) *Derive: User Manual* (Third Edition), Soft Warehouse.
5 Weber, J. E. (1982) *Mathematical Analysis: Business and Economic Applications* (Fourth Edition), Harper and Row.
6 Wilkes, F. M. (1983) *Capital Budgeting Techniques* (Second Edition), Wiley.

SOLUTIONS TO EXERCISES

Exercises 9.1

1 (i) GDP$(t) = 350(1.03)^t$

   where $t = 0$ for 1989.

   (ii) (a) For 1991, $t = 2$, so that

   $$\text{GDP}(2) = 350(1.03)^2 = 371.315$$

   (b) For 1995, $t = 6$, so that

   $$\text{GDP}(6) = 350(1.03)^6 \approx 417.918$$

2 (i) Letting $V(t)$ represent the value of the car after t years, the exponential function is

   $$V(t) = 10{,}000(0.75)^t$$

   (ii) (a) For $t = 3$, $V(t) \approx 4219$
   (b) For $t = 6$, $V(t) \approx 1780$

3 (i) Future value is given by

$$FV(t) = 750(1.08)^t$$

so that

(a) with $t = 5$, $FV(5) \approx 1102$
(b) with $t = 10$, $FV(10) \approx 1619$
(c) with $t = 20$, $FV(20) \approx 3496$

(ii) Here

$$FV(t) = 750(1.16)^t$$

so that

(a) with $t = 5$, $FV(5) \approx 1575$
(b) with $t = 10$, $FV(10) \approx 3309$
(c) with $t = 20$, $FV(20) \approx 14,596$

Note the dramatic difference in the results over the 20 year horizon between the two interest rates.

4 Letting y be debt outstanding:

(a) $y = 5000(1 + 0.24) = 6200$
(b) $y = 5000(1 + 0.12)^2 = 6272$
(c) $y = 5000(1 + 0.06)^4 = 6312$
(d) $y = 5000(1 + 0.02)^{12} = 6341$

5 With a principal of £800, the future value is

$$FV(t) = 800e^{0.1t}$$

so that

(a) when $t = 2$, $FV(2) \approx £977$
(b) when $t = 10$, $FV(10) \approx £2715$
(c) when $t = 20$, $FV(20) \approx £5911$

Exercises 9.2

1 (i) $\dfrac{dy}{dx} = 7e^{7x}$

(ii) $\dfrac{dy}{dx} = -5e^{-5x}$

(iii) $\dfrac{dy}{dx} = 0.2e^{0.2x}$

(iv) $\dfrac{dy}{dx} = e^{-0.1x}$

(v) $\dfrac{dy}{dx} = 3e^{\sqrt{3}x}$

(vi) $\dfrac{dy}{dx} = e^{0.5\pi x}$

2(i) $f''(x) = e^{2x}$
(ii) $f''(x) = -2e^{-0.1x}$
(iii) $f''(x) = 16(e^{4x} - e^{-4x})$

3(i) $f'''(x) = 4e^{\sqrt{2}x}$
(ii) $f''(x) = -e^{-x}$

Note that $-e^{-x}$ is its own *second* derivative.

(iii) $f^{iv}(x) = \dfrac{6.25}{e^{0.5x}} = 6.25e^{-0.5x}$

4(i) Using the product rule

$$f'(x) = 2e^{2x} + 2x2e^{2x}$$
$$= 2e^{2x}(1 + 2x)$$

(ii) Using the product rule

$$f'(x) = 3x^2e^{4x} + x^3 4e^{4x}$$
$$= x^2 e^{4x}(3 + 4x)$$

(iii) Using the product rule

$$f'(x) = e^x(1 + e^x) + (x + e^x)e^x$$
$$= e^x(1 + x + 2e^{2x})$$

5 (i) In this case, both the product rule and chain rule are involved. The derivative is

$$\dfrac{dy}{dx} = e^{5x}[3(3x + x^4)^2(3 + 4x^3)] + 5e^{5x}(3x + x^4)^3$$

After considerable expansion and rearrangement, this can be written as

$$x^2(x^3 + 3)^2(5x^4 + 12x^3 + 15x + 9)e^{5x}$$

which, at $x = 1$, evaluates as

$$656e^5$$

(ii) Again using the product and chain rules

$$\frac{dy}{dx} = 2(2x+1)(x^2+x)(2e^{3x} - e^{2x}) + (x^2+x)^2(6e^{3x} - 2e^{2x})$$

which, at $x = 2$, evaluates as

$$336e^6 - 132e^4$$

This derivative can be reorganized as

$$2x(x+1)(3x^2 + 7x + 2)e^{3x} - 2x(x+1)(x^2 + 3x + 1)e^{2x}$$

6 (i) $\dfrac{dy}{dx} = -10e^{20-10x}$

(ii) $\dfrac{dy}{dx} = xe^{0.5x^2}$

(iii) $\dfrac{dy}{dx} = 4xe^{-2x^2}$

(iv) $\dfrac{dy}{dx} = (20x - 30)e^{10x^2 - 30x - 40}$

(v) $\dfrac{dy}{dx} = \dfrac{e^{\sqrt{x}}}{2\sqrt{x}}$

(vi) $\dfrac{dy}{dx} = (8x^3 - 24x^2)e^{2x^4 - 8x^3}$

7 (i) Using the product rule

$$f'(x) = (2x^4 + 4x^5)xe^{0.5x^2} + e^{0.5x^2}(8x^3 + 20x^4)$$

which, at $x = 1$, evaluates as

$$34e^{0.5}$$

(ii) $f'(x) = nx^{n-1}e^{x^n}$

$$f''(x) = nx^{n-1}(nx^{n-1}e^{x^n}) + e^{x^n}(n(n-1)x^{n-2})$$

and, where $n = 3$ and $x = 2$,

$$f''(x) = 3(2)^2(3(2)^2 e^8) + e^8(6(2))$$
$$= 156e^8$$

Exercises 9.3

1 (i) $0.2e^{5x} + k$
(ii) $4e^{-0.25x} + k$

(iii) $e^{\pi x} + k$

(iv) $2x^3 + 2e^{1.5x} + k$

2 (i) $0.5e^{x^2} + k$

(ii) $0.25e^{10x^2 - 8x - 6} + k$

(iii) $2e^{\sqrt{x}} + k$

3 (i) The indefinite integral is

$$2e^{0.5x} + k$$

so that the required definite integral is

$$[2e^{0.5x}]^{x=4} - [2e^{0.5x}]^{x=2} = 2(e^2 - e)$$
$$\approx 9.34$$

(ii) The definite integral is

$$[e^{0.1x^2 + 0.8x}]^{x=3} - [e^{0.1x^2 + 0.8x}]^{x=1} = e^{3.3} - e^{0.9}$$
$$\approx 24.65$$

(iii) The definite integral is

$$[0.25e^{\pi x^2 + x}]^{x=2} - [0.25e^{\pi x^2 + x}]^{x=0} = 0.25e^{4\pi + 2} - 0.25e^0$$
$$= 0.25(e^{4\pi + 2} - 1)$$
$$\approx 529{,}705$$

(iv) The definite integral is

$$[-e^{-\mu t}]^{t=1/\mu} - [-e^{-\mu t}]^{t=0} = e^{-1} + 1$$
$$\approx 0.6321$$

4 (i) The equation rearranges as

$$\frac{dy}{dx} = 4e^{2x} - 10x$$

so

$$y = 2e^{2x} - 5x^2 + k$$

and, since $y = 2e^2$ when $x = 1$,

$$2e^2 = 2e^2 - 5 + k$$

Therefore $k = 5$, and the particular solution is

$$y = 2e^{2x} - 5x^2 + 5$$

(ii) The equation reorganizes as

$$\frac{dy}{dx} = -20e^{-0.2x} + 0.18x^2$$

for which the integral is
$$y = 100e^{-0.2x} + 0.06x^3 + k$$
and, when $x = 10$,
$$y\ 100e^{-2} + 60$$
so that $k = 40$, and the required particular solution is therefore
$$y = 100e^{-0.2x} + 0.06x^3 + 40$$

Exercises 9.4

1. (i) $x = 10^y$
   (ii) $y = 5^x$
   (iii) $y = e^x$

2. (i) $x = \log_{10} y$
   (ii) $y = \log_a x$
   (iii) $q = \log_2 p$

3. (i) $y = 2$
   (ii) $y = 1.5$
   (iii) $y = -3$
   (iv) $y = 0$
   (v) $y = -2$
   (vi) $y = 2$

4. (i) $3 + 4 = 7$
   (ii) $\log_3 729 - \log_3 81 = 6 - 4 = 2$
   (iii) $2 \log_{0.5} 8 = 2(3) = 6$

5. The doubling time is x where
   (i) $x = \dfrac{\ln 2}{\ln 1.04} = \dfrac{0.6931}{0.0392} = 17.67$ years

   (ii) $x = \dfrac{\ln 2}{\ln 1.08} = \dfrac{0.6931}{0.0770} = 9.01$ years

   (iii) $x = \dfrac{\ln 2}{\ln 1.2} = \dfrac{0.6931}{0.1893} = 3.80$ years

6. The doubling time is x where
   (i) $x = \dfrac{\ln 2}{0.1} = \dfrac{0.6931}{0.1} = 6.93$ years

   (ii) $x = \dfrac{\ln 2}{0.2} = \dfrac{0.6931}{0.2} = 3.47$ years

686 Mathematics for business, finance and economics

7 The 'halving time' is x where

(i) $x = \dfrac{\ln 0.5}{\ln 0.98} = \dfrac{-0.6931}{-0.0202} = 34.31$ years

(ii) $x = \dfrac{\ln 0.5}{\ln 0.95} = \dfrac{-0.6931}{-0.0513} = 13.51$ years

(iii) $x = \dfrac{\ln 0.5}{\ln 0.9} = \dfrac{-0.6931}{-0.1054} = 6.58$ years

8 The half-life is x years where

$$x = \dfrac{\ln 0.5}{-0.00001} = 69{,}315 \text{ years}$$

Exercises 9.5

1 (i) $\dfrac{dy}{dx} = \dfrac{20}{x}$

(ii) $\dfrac{dy}{dx} = \dfrac{4}{x}$

(iii) $\dfrac{dy}{dx} = x^2 \dfrac{1}{x} + 2x \ln x$

$= x(1 + 2 \ln x)$

(iv) $\dfrac{dy}{dx} = (4x - 2x^2) \dfrac{1}{x} + (4 - 4x)\ln x$

$= 4 - 2x + (4 - 4x)\ln x$

2 (i) $f'(x) = \dfrac{4x^3}{x^4} = \dfrac{4}{x}$

(ii) $f'(x) = \dfrac{100}{100x} = \dfrac{1}{x}$

(iii) $f'(x) = \dfrac{20x^3 - 18x^2 - 14x}{5x^4 - 6x^3 - 7x^2}$

$= \dfrac{20x^2 - 18x - 14}{5x^3 - 6x^2 - 7x}$

Exponential and logarithmic functions

3 (i) Using the quotient rule
$$\frac{dx}{dy} = \frac{(x^2+x)(1/x) - (2x+1)\log_e x}{(x^2+x)^2}$$
$$= \frac{(x+1) - (2x+1)\log_e x}{(x^2+x)^2}$$

At $x = 5$, the value of the derivative is approximately -0.013.

(ii) Again using the quotient rule
$$\frac{dy}{dx} = \frac{\log_e x - x(1/x)}{(\log_e x)^2}$$
$$= \frac{1}{\log_e x} - \frac{1}{\log_e (x)^2}$$

4 (i) $\dfrac{dy}{dx} = \dfrac{1}{x}$

so that
$$\frac{d^2y}{dx^2} = -\frac{1}{x^2}$$

so that
$$\frac{d^3y}{dx^3} = -\frac{2}{x^3}$$

(ii) $\dfrac{dy}{dx} = 1 + \ln x$

so
$$\frac{d^2y}{dx^2} = \frac{1}{x}$$

(iii) Using the composite function rule
$$f'(x) = 2(\log_e x)^1 \frac{1}{x}$$

so
$$f'(x) = \frac{2\log_e x}{x}$$

Exercises 9.6

1 $\dfrac{dp}{dq} = -2.5$

so that, using the inverse function rule,

$$\dfrac{dq}{dp} = \dfrac{1}{-2.5} = -0.4$$

Thus the elasticity of demand, E_d, will in this case be

$$E_d = -\dfrac{p}{q}\dfrac{dq}{dp} = -\dfrac{p}{q}(-0.4) = \dfrac{0.4p}{q}$$

Therefore

(i) with $q = 50$, $p = 375$ and therefore

$$E_d = \dfrac{0.4(375)}{50} = 3$$

(ii) with $q = 100$, $p = 250$ and therefore

$$E_d = \dfrac{0.4(250)}{100} = 1$$

(iii) with $q = 150$, $p = 125$ and therefore

$$E_d = \dfrac{0.4(125)}{150} = 0.33$$

2 Here $dp/dq = -4$ so that $dq/dp = -0.25$ and

$$E_d = -\dfrac{p}{q}(-0.25)$$

$$= \dfrac{0.25p}{q}$$

$$= \dfrac{0.25(1000 - 4q)}{q}$$

$$= \dfrac{250 - q}{q}$$

so that

(i) when $\dfrac{250 - q}{q} = 3$

$$250 - q = 3q$$
$$250 = 4q$$

and therefore
$$q = 65$$

(ii) when $\dfrac{250-q}{q} = 1.5$

$$250 - q = 1.5q$$
$$250 = 2.5q$$

and therefore

$$q = 100 \quad \text{and} \quad p = 600$$

(iii) This occurs when $E_d = 1$, i.e. where

$$\dfrac{250-q}{q} = 1$$

so

$$q = 125 \quad \text{and} \quad p = 500$$

3 We have $p = 100q^{-2.5}$ so that

$$\ln p = \ln 100 - 2.5 \ln q$$

and

$$\dfrac{\mathrm{d}\ln p}{\mathrm{d}\ln q} = -2.5$$

the negated reciprocal of which is elasticity of demand, so that

$$E_d = \dfrac{-1}{-2.5} = 0.4$$

4 Rearrangement produces

$$q = \dfrac{1000 - 10p}{p}$$

from which, using the quotient rule,

$$\dfrac{\mathrm{d}q}{\mathrm{d}p} = -\dfrac{1000}{p^2}$$

so that elasticity of demand is given by

$$E_d = -\dfrac{p}{q}\left(-\dfrac{1000}{p^2}\right) = \dfrac{1000}{pq}$$

so therefore

(i) when $q = 2.5$, $p = 80$ and $E_d = \dfrac{1000}{2.5(80)} = 5$

(ii) when $q = 10$, $p = 50$ and $E_d = \dfrac{1000}{10(80)} = 2$

(iii) when $q = 40$, $p = 20$ and $E_d = \dfrac{1000}{40(20)} = 1.25$

5 (i) The indefinite integral is

$$0.5 \ln(4x^3 - 5x^2) + k$$

so that the definite integral is

$$[0.5 \ln(4x^3 - 5x^2)]^{x=5} - [0.5 \ln(4x^3 - 5x^2)]^{x=2}$$
$$= [0.5 \ln 375] - [0.5 \ln 12]$$
$$\approx 2.9635 - 1.2425$$
$$\approx 1.7210$$

(ii) The indefinite integral is

$$10x(\ln x - 1) + k$$

so that the definite integral is:

$$[10x(\ln x - 1)]^{x=10} - [10x(\ln x - 1)]^{x=4}$$
$$= 100[\ln 10 - 1] - 40[\ln 4 - 1]$$
$$\approx 130.2585 - 15.4518$$
$$\approx 114.8067$$

Exercises 9.7

1 (i) $100 \log_e x + k$
   (ii) $\pi \log_e x + k$
   (iii) $-\log_e x + k$

2 (i) $\ln(x^4 - 10x) + k$
   (ii) $0.5 \ln(ax^2 + bx + c) + k$

3 (i) $6x(\ln x - 1)$
   (ii) $\pi x(\ln x - 1)$

4 (i) $10 \log_e(\log_e x)$
   (ii) $0.5 \ln(\ln x)$

Exercises 9.8

1. (i) With $df/dx = e^x$, then $f(x) = e^x$. With $g(x) = 5x$, then $dg/dx = 5$ and therefore

$$\int 5xe^x \, dx = 5xe^x - \int 5e^x \, dx$$

$$= 5xe^x - 5e^x + k$$
$$= 5(x-1)e^x + k$$

(ii) With $df/dx = e^x$, then $f(x) = e^x$. With $g(x) = 2x^2$, then $dg/dx = 4x$ and therefore

$$\int 2x^2 e^x \, dx = 2x^2 e^x - \int 4xe^x \, dx$$

$$= 2x^2 e^x - 4(x-1)e^x + k$$
$$= 2(x^2 - 2x + 2)e^x + k$$

2. (i) With $df/dx = e^{2x}$, then $f(x) = 0.5e^{2x}$. With $g(x) = x$, then $dg/dx = 1$ and therefore

$$\int xe^{2x} \, dx = 0.5xe^{2x} - \int 0.5e^{2x} \, dx$$

$$= 0.5xe^{2x} - 0.25e^{2x} + k$$
$$= 0.25(2x-1)e^{2x} + k$$

(ii) With $df/dx = e^{0.5x}$, then $f(x) = 2e^{0.5x}$. With $g(x) = 5x^2$, then $dg/dx = 10x$ and therefore

$$\int 5x^2 e^{0.5x} \, dx = 10x^2 e^{0.5x} - \int 20xe^{0.5x} \, dx$$

Applying the by-parts approach to the second integral gives

$$\int 20xe^{0.5x} \, dx = 40xe^{0.5x} - \int 40e^{0.5x} \, dx$$

$$= 40xe^{0.5x} - 80e^{0.5x} + k$$

so that the original integral is

$$10x^2 e^{0.5x} - 40xe^{0.5x} + 80e^{0.5x} = 10(x^2 - 4x + 8)e^{0.5x} + k$$

3 Set

$$\frac{df}{dx} = e^{0.5x-0.5}$$

with the result that

$$f(x) = 2e^{0.5x-0.5}$$

and with

$$g(x) = 200$$

therefore

$$\frac{dg}{dx} = 200$$

so that the indefinite integral is

$$\int 200xe^{0.5x-0.5}\,dx = 400xe^{0.5x-0.5} - \int 400e^{0.5x-0.5}\,dx$$
$$= 400xe^{0.5x-0.5} - 800e^{0.5x-0.5} + k$$

and the definite integral is

$$[400xe^{0.5x-0.5} - 800e^{0.5x-0.5}]^{x=5} - [400xe^{0.5x-0.5} - 800e^{0.5x-0.5}]^{x=3}$$
$$= 2000e^2 - 800e^2 - [1200e - 800e]$$
$$= 1200e^2 - 400e$$
$$\approx 7779.55$$

Exercises 9.9

1 (i) $\dfrac{dy}{dx} = \dfrac{1}{x \ln 5} = \dfrac{1}{1.6094x}$

(ii) $\dfrac{dy}{dx} = \dfrac{1}{x \ln 0.2} = \dfrac{1}{-1.6094x}$

2 (i) $f'(x) = \dfrac{1}{x \ln 60} = \dfrac{1}{4.0943x}$

(ii) $f''(x) = -\dfrac{1}{x^2 \ln 10}$

(iii) $f'(x) = \dfrac{1}{2x}$

3 (i) $f'(x) = \dfrac{5}{x \log_e 8} \approx \dfrac{2.4}{x}$

(ii) $f'(x) = \dfrac{0.5}{x \log_e 0.5} \approx \dfrac{0.72}{x}$

4 (i) $\dfrac{dy}{dx} = \dfrac{1}{\ln 10} \dfrac{10x - 4}{5x^2 - 4x}$

(ii) $\dfrac{dy}{dx} = \dfrac{1}{\ln 2} \dfrac{10x^2 - 3x + 10}{2.5x^3 - x^2 + 5x}$

5 (i) The indefinite integral is

$$\dfrac{1}{\ln 2} \int \ln x \, dx \approx 1.4427 x (\ln x - 1) + k$$

(ii) The indefinite integral is

$$\dfrac{12}{\ln 1.5} \int \ln x^5 \, dx \approx \dfrac{60 x (\ln x - 1)}{\ln 1.5} + k$$

6 The indefinite integral is

$$\int 2 \log_{10} x^3 \, dx = \dfrac{2}{\ln 10} 3 x (\ln x - 1) + k$$

$$\approx 2.6058 x (\ln x - 1) + k$$

so that the definite integral is

$[2.6058 x (\ln x - 1)]^{x=12} - 2.6058 x (\ln x - 1)]^{x=4}$

$\approx 2.6058 [12(\ln 12 - 1) - 4(\ln 4 - 1)]$

$\approx 2.6058 (17.8189 - 1.5452) \approx 42.41$

7 (i) $f'(x) = 10^x \ln 10 \approx (2.3026) 10^x$
(ii) $f''(x) = 5^x (\ln 5)^2 \approx (2.5903) 5^x$
(iii) $f'(x) = (2x^5 - x^3)(\ln 6) 6^x + 6^x (10 x^4 - 3 x^2)$

8 (i) $\dfrac{dy}{dx} = (\ln 10)(2x) 10^{x^2}$

$\approx (4.6x) 10^{x^2}$

(ii) $\dfrac{dy}{dx} = (\ln 16)(10x - 0.5) 10^{5x^2 - 0.5x - 1}$

(iii) $\dfrac{dy}{dx} = x^3 [(\ln 5)(2x) 5^{x^2}] + (3x^2) 5^{x^2}$

Note: The product rule is used here.

694 Mathematics for business, finance and economics

9 (i) $\dfrac{5^x}{\ln 5} + k$

   (ii) $\dfrac{(\pi)6^x}{\ln 6} + k$

   (iii) $\dfrac{10^{5x^2 - 8x}}{\ln 10} + k$

10 The indefinite integral is

$$\int (4x - 1)12^{2x^2 - x}\, dx = \dfrac{12^{2x^2 - x}}{\ln 12} + k$$

so that the definite integral is

$$[(12^{2x^2 - x})/(\ln 12)]^{x=2} - [(12^{2x^2 - x})/(\ln 12)]^{x=1}$$
$$\approx 1{,}201{,}648.36 - 4.83 \approx 1{,}201{,}643.53$$

Additional problems

1 (i) $FV_{10} = 2500(1 + 0.12)^{10}$
         $= 2500(3.1058) = 7765$
   (ii) $V_5 = 12{,}500(1 - 0.22)^5$
        $= 12{,}500(0.2887) = 3609$

2 (i) (a) $950(1.24) = 1178$
       (b) $950(1.02)^{12} = 1205$
       (c) $950(1 + 0.24/365)^{365} = 1208$
   (ii) (a) $950(1.24)^5 = 2785$
        (b) $950e^{5(0.24)} = 950e^{1.2} = 3154$

3 (i) $\dfrac{dy}{dx} = 10e^{2x}$

   (ii) $\dfrac{dy}{dx} = 10e^{-4x}$

   (iii) $\dfrac{dy}{dx} = (24x - 20)e^{3x^2 - 5x - 4}$

   (iv) $\dfrac{dy}{dx} = (2x^5)4e^{4x} + (e^{4x})(10x^4)$
        $= 2(4x + 5)x^4 e^4 x$

4 (i) $25e^{0.2x} + k$
   (ii) $2e^{-3x} + k$
   (iii) $2e^{2x^2 - 7x + 3} + k$

5 (i) $[5e^{0.2x}]^{x=20} - [5e^{0.2x}]^{x=10}$
    $\approx 272.99 - 36.95 \approx 236.05$

   (ii) $[10e^{0.1x^2 + 0.1x + 1}]^{x=4} - [10e^{0.1x^2 + 0.1x + 1}]^{x=2}$
    $= 10(e^3 - e^{1.6}) \approx 151.33$

6 (i) doubling time $= \dfrac{\log_e 2}{\log_e (1.035)} \approx 20.15$ years

   (ii) trebling time $= \dfrac{\log_e 3}{\log_e (1.5)} \approx 22.52$ years

   (iii) The half-life is given by

   $\dfrac{\log_e 0.5}{-0.005} \approx 138.63$ years

7 (i) $\dfrac{dy}{dx} = \dfrac{\pi^2}{x}$

   (ii) $\dfrac{dy}{dx} = \dfrac{7}{x}$

   (iii) $\dfrac{dy}{dx} = (x^4 - x^3)\dfrac{5}{x} + \ln x^5(4x^3 - 3x^2)$
    $= 5x^2(x-1) + \ln x^5(4x^3 - 3x^2)$

   (iv) $\dfrac{dy}{dx} = \dfrac{5x^4 + 4x^3}{x^5 + x^4}$
    $= \dfrac{5x + 4}{x^2 + x}$

8 $dp/dq = -5$ so that $dq/dp = -0.2$ and the elasticity of demand is, in this case, given by

   $E_d = 0.2 \dfrac{p}{q}$

   $= \dfrac{400 - q}{q}$

   Therefore:
   (a) when $q = 50$ $E_d = 7$
   (b) when $q = 100$ $E_d = 3$
   (c) when $q = 200$ $E_d = 1$
   (d) when $q = 250$ $E_d = 0.8$

9 (i) $250 \ln x + k$
   (ii) $\log_e(0.5x^2 - 7x) + k$
   (iii) $8x(\log_e x - 1) + k$
   (iv) $10 \ln(\ln x) + k$
   (v) Using integration by parts

$$\int 16xe^{0.4x} \, dx = 40xe^{0.4x} + \int 40e^{0.4x}$$
$$= 40xe^{0.4x} + 100e^{0.4x} + k$$
$$= 20(2x - 5)e^{0.4x} + k$$

10 (i) $\dfrac{dy}{dx} = \dfrac{1}{x \ln 20}$

   (ii) $\dfrac{dy}{dx} = \dfrac{5}{x \ln 4}$

   (iii) $\dfrac{dy}{dx} = \dfrac{1}{\ln 5} \dfrac{3x - 2}{x^7 - x}$

11 (i) $\dfrac{2}{\ln 12} x(\ln x - 1) + k$

   (ii) $\dfrac{2}{\ln 2} 2x(\ln x - 1) + k$

12 $\dfrac{dy}{dx} = \ln 8(x - 7)8^{0.5x^2 - 7x}$

13 The equation rearranges as

$$\dfrac{dy}{dx} = 16x^3 + 8e^{-2x}$$

so that the general solution is

$$y = 4x^4 - 4e^{-2x} + k$$

and, since $y = 40$ when $x = -2$, we can write

$$40 = 64 - 4e^4 + k$$

so that

$$k = 4e^4 - 24$$

and the required particular solution is

$$y = 4x^4 - 4e^{-2x} + 4e^4 - 24$$

14 The indefinite integral is

$$\frac{3^{x^2 - 0.4x + 2}}{\ln 3} + k$$

so the definite integral is

$$\left[\frac{3^{x^2 - 0.4x + 2}}{\ln 3}\right]^{x=2} - \left[\frac{3^{x^2 - 0.4x + 2}}{\ln 3}\right]^{x=1} \approx 259.70$$

NOTES

1 Usually called the *base year* — not to be confused in the present context with the base of the exponential function.
2 Credit card companies normally quote interest rates per month. The equivalent annual rate is *more* than 12 times the monthly rate. For example, the annual rate equivalent to 2 per cent per month is given by $(1.02)^{12} = 1.2682$; i.e. 26.82 per cent.
3 Apart, that is, from the trivial case of $y = 0$.
4 In the use of the product rule, irrational m would involve passage to the limit.
5 Considerable manipulation is required to achieve this simplification. Packages such as *Derive* can make such rearrangements in less than a second of CPU time and free of error!
6 The term *simple queue* describes a particular type of queuing system for which the exponential distribution also describes the times between arrivals into the system.
7 Small populations will not show proportional year on year growth.
8 Far higher rates of interest are applied to personal loans from fringe financial institutions and outright 'loan sharks'.
9 As is the case under *straight line depreciation*.
10 Note that the case of $m = 1$ shows that $e^{\ln x} = x$.
11 The usefulness of the rule is not, however, confined to this situation.
12 With implications for variables such as population size.
13 The two interpretations can coincide — as when x represents the number of days for which the marketing campaign runs.
14 And on the effectiveness of the company's marketing.
15 The term 'demand functions' is used in the general sense of quantity demanded expressed as some function of any independent variable, rather than the specific economists' usage of $q = f(p)$ in a given time period.
16 Or in general some default level of promotional activity.
17 The curves are also known as *saturation curves*.
18 For the parameter ranges specified.
19 See for example Bishir and Drewes (1970).

Chapter 10

Matrices

10.1	Introduction	699
10.2	Some fundamentals	700
10.3	Addition, subtraction and scalar multiplication	709
10.4	Matrix multiplication	713
10.5	Matrix inversion	723
10.6	Simultaneous equations (i)	731
10.7	Rank	739
10.8	Higher order systems: determinants and the inverse matrix	743
10.9	Simultaneous equations (ii)	748
10.10	Cramer's rule	752
10.11	Homogeneous systems	756
10.12	Concluding remarks	760
	Additional problems	761
	References and further reading	762
	Solutions to exercises	763

Sections 10.1 and 10.2 introduce the concept of matrices, matrix notation and other basics. Sections 10.3–10.5 then consider some operations on matrices, drawing comparisons with the arithmetic of real numbers. The idea of a determinant is introduced, and the question of linear dependence is discussed.

Attention is given to the use of matrices in solving systems of linear equations involving two variables and two unknowns and of higher order. In this context Cramer's rule is introduced in Section 10.10 while homogeneous systems are described in Section 10.11.

By the end of the chapter, you will be able to add, subtract and multiply matrices, find the inverse of a matrix where this exists, and use matrices and determinants to solve simultaneous equations.

10.1 INTRODUCTION

Many problems in business and economics require the use of large numbers of unknowns and numerical values. While in principle it would be possible to handle such masses of data *ad hoc*, this would be inconvenient, inefficient and error prone. It would make the tracking of errors difficult and reduce the opportunities for systematizing procedures and taking advantage of computer algorithms.

It often desirable to arrange material in the form of *rectangular arrays*. If this is done, conventions are needed for forming and manipulating the arrays. As an example of how a rectangular arrangement of data can suggest itself, suppose that a firm can use two scarce resources to produce two products, the Alpha model and the Omega model. Table 10.1 states the resource requirements per unit of each product produced.

Stripped of the row and column labels, the resource requirements of Table 10.1 form a rectangular array with two rows and two columns. This can be shown as

$$\begin{bmatrix} 4 & 1 \\ 2 & 3 \end{bmatrix}$$

Important properties of the system can be obtained by manipulation of such arrays. Conventionally enclosed within brackets, the array is called a *matrix*. Some, but not all, of the properties of real numbers also apply to matrices and matrices have some additional properties. Before reviewing basic concepts, it is worth remembering that matrices are *notation*. There is nothing that can be done *with* matrices which in principle could not be done *without* them. In practice, however, matrices make a world of difference to what can be accomplished because, as with all good notation, matrices

1 are convenient for stating large and complex problems;
2 facilitate the deduction of general properties;
3 allow attention to be concentrated on the numerical values in particular applications;
4 are a suitable basis for computerization.

Table 10.1

	Alpha	Omega
Materials (kg)	4	1
Labour (h)	2	3

10.2 SOME FUNDAMENTALS

It is often necessary to handle several matrices, so the first concern is to ensure that they are clearly identified and distinguished.

Labelling of matrices

Matrices are usually identified by an upper case letter, optionally in bold type. For example, the matrix of resources above might be called **R**, and written as

$$R = \begin{bmatrix} 4 & 1 \\ 2 & 3 \end{bmatrix}$$

where the bold is a reminder, where necessary, that **R** is a matrix.

Dimensions of matrices

A matrix may consist of any whole positive number of rows and any whole positive number of columns. The dimensions of the matrix are stated as the number of rows times the number of columns. The dimensions give the *order* of the matrix. Thus **R** is a matrix of order 2×2. The matrix

$$A = \begin{bmatrix} 6 & 3 & 10 \\ 4 & 2 & 5 \\ 21 & 3 & 9 \end{bmatrix}$$

is a matrix of order 3×3. In some applications matrices may be involved having *thousands* of rows and columns. The matrices **A** and **R** above each have the same number of rows as columns and are *square* matrices – a category having special properties. But matrices need not be square. For example:

$$B = \begin{bmatrix} 7 & 3 & 5 \\ 2 & 4 & 1 \end{bmatrix}$$

with two rows and three columns, is a matrix of order 2×3, while the matrix **C**

$$C = \begin{bmatrix} 1 & 5 \\ 2 & 6 \\ 3 & 7 \\ 4 & 8 \end{bmatrix}$$

is a 4 × 2 matrix. In general, a matrix with m rows and n columns is of order $m \times n$. The matrix:

$$D = \begin{bmatrix} 3 \\ 4 \\ 1 \end{bmatrix}$$

is a 3 × 1 matrix. A matrix with a single column is called a *column matrix* or, more commonly, a *column vector*. A matrix of order 1 × n such as

$$E = \begin{bmatrix} 1 & 2 & 3 & 5 & 4 \end{bmatrix}$$

is called a *row matrix* or a *row vector*. By convention, when a vector is not specified as a row or column vector, it is assumed to be a column vector. A matrix can be seen as a row vector of column vectors or a column vector of row vectors. The matrix

$$F = [6]$$

is a 1 × 1 matrix. Such 1 × 1 matrices have the same properties as real numbers regarding operations such as addition and multiplication, and have no material additional properties. For many purposes an individual number can be seen as a special case of a square matrix where $m = n = 1$. The term *scalar* is used to distinguish a single number in the context of matrices.

The elements of matrices

The *elements* of (or *entries* in) a matrix are the entities of which it is composed. The elements can take a variety of forms. They could be integers as in matrices A, B, C, D, E and F above or other numbers as in the matrix

$$G = \begin{bmatrix} 5 & 6.2 & -4 \\ 0 & 4/3 & 10^6 \\ \sqrt{2} & \pi & -e \end{bmatrix}$$

which includes irrational and transcendental numbers. The elements of a matrix can also include *variables* as in

$$\begin{bmatrix} 3 & -x \\ x^2 & 7 \end{bmatrix}$$

Sub-matrices

If we begin with the matrix H where

$$H = \begin{bmatrix} 1 & 7 & -3 & 9 \\ 2.5 & 4 & 0 & 3 \\ 6 & 8 & -7 & 4 \\ 3 & \pi & 6 & \sqrt{2} \end{bmatrix}$$

then

$$J = \begin{bmatrix} 1 & 7 & -3 \\ 2.5 & 4 & 0 \end{bmatrix} \qquad L = [6 \quad 8 \quad -7 \quad 4]$$

$$R = \begin{bmatrix} 1 & -3 \\ 3 & 6 \end{bmatrix} \qquad M = [\pi]$$

are all examples of *sub-matrices* of H. A sub-matrix is obtained by deleting some of the rows and/or columns of the original matrix. The sub-matrices obtained by deletion of the last i rows and columns of a matrix are important when expressing second-order conditions for optimization using matrices. Sub-matrices are also useful when forming a *partitioned matrix*. For example the matrix (with a particular partition indicated)

$$\left[\begin{array}{ccc|cc} 20 & 32 & 16 & 15 & 12 \\ -17 & 44 & 27 & 13 & 41 \\ \hline 15 & 34 & 26 & 57 & 12 \\ 22 & 19 & 20 & 35 & 17 \end{array} \right]$$

may for some purposes be more conveniently considered as being made up of sub-matrices

$$\begin{bmatrix} A & B \\ C & D \end{bmatrix}$$

where the sub-matrices are

$$A = \begin{bmatrix} 2 & 32 & 16 \\ -17 & 44 & 27 \end{bmatrix} \qquad B = \begin{bmatrix} 15 & 12 \\ 13 & 41 \end{bmatrix}$$

$$C = \begin{bmatrix} 15 & 34 & 26 \\ 22 & 19 & 20 \end{bmatrix} \qquad D = \begin{bmatrix} 57 & 12 \\ 35 & 17 \end{bmatrix}$$

A matrix of order $m \times n$ can always be partitioned into m row vectors or n column vectors and properties of the matrix can then be analysed in terms of these vectors. Partitioning can be helpful in simplifying operations on large matrices with special structure.

The principal diagonal

The *principal diagonal* of a square matrix consists of the elements in the top left to bottom right diagonal. Alternative terminology is *main diagonal* or *leading diagonal*. Thus in the matrix N where

$$N = \begin{bmatrix} \underline{1} & 2 & 3 & 4 \\ 5 & \underline{6} & 7 & 8 \\ 9 & 10 & \underline{11} & 12 \\ 13 & 14 & 15 & \underline{16} \end{bmatrix}$$

the underlined elements 1, 6, 11, 16 constitute the principal diagonal. The remaining elements of the matrix are known as the *off-diagonal* elements. The sum of the main diagonal elements is known as the *trace*. For the matrix N the trace, T, is given by

$$T = 1 + 6 + 11 + 16 = 34$$

Triangular matrices

A *triangular matrix* is a square matrix where only the elements on, or on one side of, the main diagonal are non-zero. An example is provided by the matrix P where

$$P = \begin{bmatrix} 1 & 0 & 0 & 0 \\ 5 & 6 & 0 & 0 \\ 9 & 10 & 11 & 0 \\ 13 & 14 & 15 & 16 \end{bmatrix}$$

P is a 4×4 *lower-triangular* matrix. All the elements *above* the main diagonal must be zero in a lower-triangular matrix. In contrast, the matrix

$$Q = \begin{bmatrix} -5 & -4 & -3 & -2 \\ 0 & -1 & 0 & 1 \\ 0 & 0 & 6 & 3 \\ 0 & 0 & 0 & 5 \end{bmatrix}$$

is *upper-triangular*. Note that *some* of the elements on or above the principal diagonal can be zero in an upper-triangular matrix; a corresponding provision applies to lower-triangular matrices.

Diagonal matrices

A *diagonal matrix* is a square matrix in which all the elements except those on the leading diagonal are zero. For example, the matrix

$$S = \begin{bmatrix} 1 & 0 & 0 & 0 \\ 0 & 4 & 0 & 0 \\ 0 & 0 & 7 & 0 \\ 0 & 0 & 0 & \sqrt{2} \end{bmatrix}$$

is a 4×4 diagonal matrix. Note that diagonal matrices satisfy the conditions for both upper and lower triangularity. A special case of a diagonal matrix in which all of the elements on the main diagonal are the same is called a *scalar* matrix. For example

$$T = \begin{bmatrix} 4 & 0 & 0 \\ 0 & 4 & 0 \\ 0 & 0 & 4 \end{bmatrix}$$

is a scalar matrix. An important special case of a scalar matrix is where all the main diagonal elements are unity. Such a matrix is called an *identity matrix*. Thus

$$I = \begin{bmatrix} 1 & 0 & 0 \\ 0 & 1 & 0 \\ 0 & 0 & 1 \end{bmatrix}$$

is a 3×3 identity matrix. Identity matrices are an important concept – of which more later.

Zero matrices

In a *zero matrix* all of the elements are zero. But zero matrices, unlike identity matrices, need not be square. Thus, for example,

$$\mathbf{0} = \begin{bmatrix} 0 & 0 & 0 \\ 0 & 0 & 0 \end{bmatrix}$$

is a 2×3 zero matrix.

Matrix notation

Specific elements in a matrix can be referenced by citing their row and column number, e.g. row 1, column 4, or, in general, row i, column j. To refer to the elements in a general way, *matrix notation* is employed.

Lower-case letters are used – usually *a*, *b* or *c* – in conjunction with a *double subscript* or *index* to indicate the row and column of the element, as in

$$A = \begin{bmatrix} a_{11} & a_{12} & a_{13} & \cdots & a_{1n} \\ a_{21} & a_{22} & a_{23} & \cdots & a_{2n} \\ a_{31} & a_{32} & a_{33} & \cdots & a_{3n} \\ \vdots & \vdots & \vdots & & \vdots \\ a_{m1} & a_{m2} & a_{m3} & \cdots & a_{mn} \end{bmatrix}$$

in which a_{ij} represents the element in the *i*th row and *j*th column of *A*. Note that *A* is of order $m \times n$. In square matrices (where $m = n$) the elements on the principal diagonal have the same row and column subscript values and appear as a_{ii}. It is sometimes convenient to refer to matrices by their typical elements. The typical element for the matrix *A* would be represented as a_{ij}, for some other matrix *B* as b_{ij}. In this manner the matrix *A* above could be described as

$$A = (a_{ij}) \quad i = 1, 2, \ldots, m; \quad j = 1, 2, \ldots, n$$

In this notation the ranges of the subscripts are not explicitly stated when the order of the matrix is understood. Vectors require only one subscript to identify their elements uniquely, as for example in the column and row vectors

$$\begin{bmatrix} b_1 \\ b_2 \\ b_3 \\ \vdots \\ b_m \end{bmatrix} \quad \text{and} \quad [c_1 \ c_2 \ c_3 \ \cdots \ c_n]$$

Equal matrices

Two matrices are *equal* when:

1 the matrices have the same dimensions;
2 all corresponding elements in the matrices are equal.

Thus, given the matrices *A* and *B* where

$$A = \begin{bmatrix} 9 & 3 \\ -1 & 6 \end{bmatrix} \quad \text{and} \quad B = \begin{bmatrix} x & 3 \\ y & 6 \end{bmatrix}$$

the matrices are equal *if and only if* $x = 9$ and $y = -1$. Note that the matrices

$$A = \begin{bmatrix} 9 & 3 \\ -1 & 6 \end{bmatrix} \quad \text{and} \quad Q = \begin{bmatrix} 9 & 3 & 0 \\ -1 & 6 & 0 \end{bmatrix}$$

do *not* satisfy the conditions for equal matrices.

Transposition

The *transpose* of a matrix A is found by interchanging its rows and columns, so that $a_{ij} = a_{ji}$. The transpose of A is indicated by A' or A^T. For example, if

$$A = \begin{bmatrix} 3 & 2 \\ 7 & -1 \\ -4 & 6 \end{bmatrix} \quad \text{then } A' = \begin{bmatrix} 3 & 7 & -4 \\ 2 & -1 & 6 \end{bmatrix}$$

and if

$$B = \begin{bmatrix} 7 \\ 8 \\ 9 \\ 10 \end{bmatrix} \quad \text{then } B' = \begin{bmatrix} 7 & 8 & 9 & 10 \end{bmatrix}$$

so the transpose of a column vector is a row vector and the transpose of a row vector is a column vector. In A' above observe that the matrix $(A')' = A$, i.e. the transpose of the transpose is the original matrix. There is no analogous operation in the arithmetic of real numbers. Consider two special cases involving transposition. First, for

$$A = \begin{bmatrix} 1 & 3 & 4 \\ 3 & 7 & 8 \\ 4 & 8 & 2 \end{bmatrix}$$

the transpose is A itself, so in this case $A^T = A$. For this to be so the elements in the mth row of the matrix must be the same as the elements in the mth column, i.e. $a_{ij} = a_{ji}$ for all i, j. Such a matrix is *symmetric*. Symmetric matrices must be square and their elements are symmetric about the main diagonal. Now consider the transpose of A where

$$A = \begin{bmatrix} 0 & 4 & -3 \\ -4 & 0 & 6 \\ 3 & -6 & 0 \end{bmatrix}$$

The transpose is

$$A^T = \begin{bmatrix} 0 & -4 & 3 \\ 4 & 0 & -6 \\ -3 & 6 & 0 \end{bmatrix}$$

In this case the elements of the transposed matrix A^T have the same *absolute* value as the elements in corresponding positions in A but are of opposite sign. For this to be so, $a_{ij} = -a_{ji}$ and the matrix must square. A matrix with this property is said to be *skew-symmetric* or *antisymmetric*. The principal diagonal elements in a skew-symmetric matrix must be zero. *Transposition* can also be used to describe the interchange of two rows or of two columns of a matrix. Such an operation leaves certain essential properties unchanged and is an *elementary matrix operation*.[1]

EXERCISES 10.2

1 State the dimensions of

$$A = \begin{bmatrix} 3 & 2 \\ 6 & 1 \end{bmatrix} \quad B = [1 \ 5 \ 6 \ 7]$$

$$C = \begin{bmatrix} -3 \\ -2 \\ -1 \\ -2 \\ -3 \end{bmatrix} \quad D = [0]$$

$E = (a_{ij})$ where $i = 1, \ldots, t,\quad j = 1, \ldots, s$

2 Given the matrix

$$A = \begin{bmatrix} 6 & 3 & 7 \\ 2 & 1 & 4 \\ 5 & 9 & 3 \end{bmatrix}$$

which of the following are sub-matrices of A?

$$B = \begin{bmatrix} 6 & 7 \\ 2 & 4 \end{bmatrix} \quad C = \begin{bmatrix} 3 & 7 \\ 1 & 4 \\ 9 & 3 \end{bmatrix}$$

$$D = [6 \ 3] \quad E = \begin{bmatrix} 2 & 4 \\ 5 & 9 \end{bmatrix}$$

3 Find the trace of

$$\begin{bmatrix} 10 & 13 & 26 & 18 \\ 16 & 27 & 13 & 22 \\ 14 & 30 & 35 & 23 \\ 20 & 19 & 21 & 29 \end{bmatrix}$$

4 Are the matrices below upper-triangular, lower-triangular, diagonal, scalar or in none of these categories:

$$A = \begin{bmatrix} 2 & 3 & 4 \\ 0 & 2 & 3 \\ 0 & 0 & 2 \end{bmatrix} \quad B = \begin{bmatrix} 2 & 0 & 0 \\ 0 & 2 & 0 \\ 0 & 0 & 2 \end{bmatrix}$$

$$C = \begin{bmatrix} 4 & 0 & 0 \\ 0 & 4 & 0 \\ 0 & 2 & 4 \end{bmatrix} \quad D = \begin{bmatrix} 4 & 0 & 0 \\ 0 & 3 & 0 \\ 0 & 0 & 2 \end{bmatrix}$$

$$E = \begin{bmatrix} 4 & 0 & 4 \\ 0 & 4 & 0 \\ 4 & 0 & 4 \end{bmatrix} \quad F = \begin{bmatrix} 1 & 0 & 1 \\ 0 & 1 & 0 \\ 0 & 0 & 1 \end{bmatrix}$$

5 What conditions are necessary for the following matrices to be equal?

(i)
$$\begin{bmatrix} 9 & x & 7 \\ y & 3 & 8 \\ 2 & 4 & z \end{bmatrix} \quad \text{and} \quad \begin{bmatrix} 9 & 0 & 7 \\ 3 & 3 & 8 \\ 2 & 4 & 1 \end{bmatrix}$$

(ii)
$$A = \begin{bmatrix} 10 & 10 & 10 \\ 10 & 10 & 10 \end{bmatrix} \quad B = (b_{ij})$$

where $b_{ij} = 10$ for $i = 1, \ldots, m$ and $j = 1, \ldots, n$.

6 Find the transpose of the following matrices:

$$A = \begin{bmatrix} 4 & -1 \\ -3 & 8 \end{bmatrix} \quad B = [2 \ \ \pi \ \ 3 \ \ -1 \ \ 0]$$

$$C = \begin{bmatrix} -8 \\ 0 \\ \sqrt{2} \\ -1 \\ 4 \end{bmatrix} \quad D = [100]$$

7 Are the following matrices symmetric, skew-symmetric or not symmetric:

$$A = \begin{bmatrix} 9 & 4 & 1 & 3 \\ 4 & 1 & 2 & 7 \\ 1 & 2 & 3 & 5 \\ 3 & 7 & 3 & 6 \end{bmatrix}$$

$$B = \begin{bmatrix} 0 & 5 & -1 & -6 \\ -5 & 0 & 3 & 0 \\ 1 & -3 & 0 & 9 \\ 6 & 0 & -9 & 0 \end{bmatrix}$$

$$C = \begin{bmatrix} 0 & 1 & 2 & 1 \\ 1 & 0 & 3 & 5 \\ 2 & 3 & 0 & 4 \\ 1 & 5 & 4 & 0 \end{bmatrix}$$

$$D = \begin{bmatrix} 8 & 8 & 1 & 7 \\ 8 & \pi & 2 & 0 \\ 1 & 2 & 0 & 6 \\ 7 & 0 & 6 & e \end{bmatrix}$$

10.3 ADDITION, SUBTRACTION AND SCALAR MULTIPLICATION

We cannot march straight in and start adding and multiplying matrices. First we have to say what is meant by these and other operations.

Matrix addition

Where matrices have the same dimensions, if elements in corresponding positions are added, the result is a matrix which is the sum of the original matrices. Thus, if $A = (a_{ij})$ and $B = (b_{ij})$ are two matrices of dimension $m \times n$, then the matrices can be added (they are said to be *conformable in addition*) and the *sum* $A + B$ is the $m \times n$ matrix with elements $(a_{ij} + b_{ij})$. So, if

$$A = \begin{bmatrix} 2 & 4 \\ 3 & 0 \\ -1 & 2 \end{bmatrix} \quad \text{and} \quad B = \begin{bmatrix} 6 & 3 \\ -2 & -1 \\ 4 & 9 \end{bmatrix}$$

then

$$A + B = \begin{bmatrix} 8 & 7 \\ 1 & -1 \\ 3 & 11 \end{bmatrix}$$

More than two matrices may be added, as in

$$\begin{bmatrix} 7 & 9 \\ 3 & 2 \end{bmatrix} + \begin{bmatrix} 2 & -4 \\ 0 & 4 \end{bmatrix} + \begin{bmatrix} -3 & 2 \\ 5 & 3 \end{bmatrix} = \begin{bmatrix} 6 & 7 \\ 8 & 9 \end{bmatrix}$$

Unknown elements may be involved in matrix addition, as in

$$A = \begin{bmatrix} b & 4a & 11 \\ -b & a & 9 \\ a+b & c & 2c \end{bmatrix} \quad B = \begin{bmatrix} a & -1 & 7 \\ a & c^2 & -c \\ -2b & b & b-c \end{bmatrix}$$

the sum of which is

$$A + B = \begin{bmatrix} b+a & 4a-1 & 18 \\ a-b & a+c^2 & 9-c \\ a-b & c+b & c+b \end{bmatrix}$$

If

$$\begin{bmatrix} 15 & -8 \\ 10 & 3x \end{bmatrix} + \begin{bmatrix} 2x & 7 \\ 7 & 2y \end{bmatrix} = \begin{bmatrix} 25 & -1 \\ z & 31 \end{bmatrix}$$

then, from the elements in the (1, 1) positions, it must be the case that

$$15 + 2x = 25$$

so therefore

$$x = 5$$

and from the elements in the (2, 1) positions, it must be the case that

$$10 + 7 = z$$

so that

$$z = 17$$

and from the elements in the (2, 2) positions, it must also be the case that

$$3x + 2y = 31$$

Therefore

$$15 + 2y = 31$$

and so

$$2y = 16$$

i.e.

$$y = 8$$

The following properties hold for matrix addition.

1. The order in which the addition of matrices is carried out is immaterial, i.e.

$$A + B = B + A$$

So matrix addition is *commutative*.

2. Any grouping together for the purpose of addition is immaterial, i.e.

$$A + (B + C) = (A + B) + C = (A + C) + B$$

So matrix addition has the *associative* property.

3. The transpose of the sum of (difference between) matrices is the sum of (difference between) the transposes, i.e.:

$$(A \pm B)^T = A^T \pm B^T$$

Subtraction of matrices

If $A = (a_{ij})$ and $B = (b_{ij})$ are both matrices of dimension $m \times n$, then B may be subtracted from A, and we may write

$$A - B = (a_{ij} - b_{ij})$$

which is the resulting matrix *difference*. For example, given the matrices

$$A = \begin{bmatrix} 2 & 3 & 4 \\ 1 & 0 & 2 \end{bmatrix} \quad \text{and} \quad B = \begin{bmatrix} 6 & 2 & -1 \\ 4 & -7 & 0 \end{bmatrix}$$

then the matrix difference is

$$A - B = \begin{bmatrix} -4 & 1 & 5 \\ -3 & 7 & 2 \end{bmatrix}$$

It is obvious from the numerical values in this example that where A and B are matrices

$$A - B \neq B - A$$

Here, the difference $B - A$ is

$$B - A = \begin{bmatrix} 4 & -1 & -5 \\ 3 & -7 & -2 \end{bmatrix}$$

so if the elements of $A - B$ are (c_{ij}) then the elements of $B - A$ are $(-c_{ij})$. Matrix subtraction, unlike matrix addition, but like the subtraction of scalars, is *not* commutative.

The identity element for matrix addition

The zero matrix is the identity matrix for the operation of matrix addition (the requirements needed to fulfil an identity role depend on the operation being performed). That is, for conformable matrices A and $\mathbf{0}$,

$A + \mathbf{0} = A$ for all matrices A

Note that for $A = (a_{ij})$

$$\mathbf{0} - A = -A \tag{10.1}$$

where the matrix $-A = (-a_{ij})$ is the *negative* of A. Thus if

$$A = \begin{bmatrix} 2 & -1 \\ -4 & 3 \end{bmatrix} \quad \text{then} \quad -A = \begin{bmatrix} -2 & 1 \\ 4 & -3 \end{bmatrix}$$

The matrix difference $A - B$ is the negative of the difference $B - A$. Finally, note that as a rearrangement of equation (10.1)

$$A + (-A) = \mathbf{0}$$

That is, a matrix plus its negative equals a zero matrix of the same dimensions.

Scalar multiplication

First consider the addition of equal matrices. For example, start with a matrix A where

$$A = \begin{bmatrix} 4 & 1 \\ 2 & 3 \end{bmatrix}$$

and add A to itself

$$\begin{bmatrix} 4 & 1 \\ 2 & 3 \end{bmatrix} + \begin{bmatrix} 4 & 1 \\ 2 & 3 \end{bmatrix} = \begin{bmatrix} 8 & 2 \\ 4 & 6 \end{bmatrix}$$

which sum can be represented in general as

$$A + A = 2A$$

where the matrix $2A = (2a_{ij})$ represents a *scalar multiple* of A. Not surprisingly,

$$\begin{bmatrix} 4 & 1 \\ 2 & 3 \end{bmatrix} + \begin{bmatrix} 4 & 1 \\ 2 & 3 \end{bmatrix} + \begin{bmatrix} 4 & 1 \\ 2 & 3 \end{bmatrix} = \begin{bmatrix} 12 & 3 \\ 6 & 9 \end{bmatrix} = 3A$$

and, in general, for c equal matrices which are added, the result will be

$$cA = (ca_{ij})$$

In scalar multiplication, for matrices A and B and scalars k and m, it is the case that

(i) $(k \pm m)A = kA \pm mA$
(ii) $k(A \pm B) = kA \pm kB$
(iii) $m(kA) = (mk)A$
(iv) $0A = 0$
(v) $-1A = -A$

In (iv) **0** is the zero matrix, and in (v) $-A$ is the negative of A.

EXERCISES 10.3

1 Find the sum of the matrices in the following cases:

(i) $\begin{bmatrix} 2 & 3 & 4 \\ 1 & -7 & 3 \\ -2 & 6 & -4 \end{bmatrix} + \begin{bmatrix} -3 & 6 & 2 \\ 7 & 9 & -1 \\ 8 & -3 & 8 \end{bmatrix}$

(ii) $\begin{bmatrix} 10 & -8 \\ 4 & 11 \end{bmatrix} + \begin{bmatrix} -2 & 16 \\ 9 & 14 \end{bmatrix} + \begin{bmatrix} -3 & 12 \\ 2 & -7 \end{bmatrix}$

(iii) $\begin{bmatrix} 4 & 2a & -b \\ a & b & 6 \\ 9 & a+c & 1 \end{bmatrix} + \begin{bmatrix} -a & 6 & a \\ 9 & -c & 24 \\ 9 & c-a & c^2 \end{bmatrix}$

2 Given that

$$\begin{bmatrix} 5x & 6 \\ 4 & -y \end{bmatrix} + \begin{bmatrix} -4y & -z \\ 2x & 2y \end{bmatrix} = \begin{bmatrix} z & 4 \\ 24 & 12 \end{bmatrix}$$

find the values of x, y and z.

3 Given the matrices

$$A = \begin{bmatrix} 3 & 7 \\ -1 & 2 \end{bmatrix} \quad \text{and} \quad B = \begin{bmatrix} 2 & -4 \\ 3 & -3 \end{bmatrix}$$

find the scalar multiple $k(A + B)$ when

(a) $k = 2$ (b) $k = -1$

10.4 MATRIX MULTIPLICATION

In the example of the two-product manufacturer from the preceding section, the per unit resource consumption data formed the 2×2 matrix

$$R \begin{bmatrix} 4 & 1 \\ 2 & 3 \end{bmatrix}$$

Suppose that we now want to find the *unit costs of production* for the two goods. Suppose also that materials cost £5 per kilogramme and that the wage rate for labour is £6 per hour. The prices can be written as a 1×2 row vector

$$P = [5 \quad 6]$$

Given the price and resource consumption information, the unit costs of the Alpha and Omega models are

Alpha: 4 kg at £5 + 2 h at £6 = £32
Omega: 1 kg at £5 + 3 h at £6 = £23 (10.2)

The unit costs for the products could be written as the 1×2 row vector

$$C = [32 \quad 23]$$

Examination of (10.2) reveals that to arrive at the 32 element of the unit cost vector we perform an operation combining the row vector of prices and the first column of the resource requirement matrix R. Successive elements in the row vector

$$[5 \quad 6]$$

are multiplied by successive elements in the column

$$\begin{bmatrix} 4 \\ 2 \end{bmatrix}$$

and the products are then summed. This process is indicated by positioning the 2×1 column vector immediately after the 1×2 row vector and showing the result as a 1×1 vector thus:

$$[5 \quad 6]\begin{bmatrix} 4 \\ 2 \end{bmatrix} = [32]$$

This outcome in which a $1 \times n$ row vector multiplies an $n \times 1$ column vector producing a scalar is called the *inner product, dot product*[2] or *scalar product* of the two vectors. The workings for the unit cost of the Omega product, the second element, 23, of the unit cost vector, can be laid out in a similar fashion

$$[5 \quad 6]\begin{bmatrix} 1 \\ 3 \end{bmatrix} = [23]$$

Taken together, these workings can be presented as

$$[5 \quad 6]\begin{bmatrix} 4 & 1 \\ 2 & 3 \end{bmatrix} = [32 \quad 23] \qquad (10.3)$$

The arrangement shown in (10.3) represents *matrix multiplication*, and is written in terms of the matrices involved as

$$PR = C \qquad (10.4)$$

Now review what is involved in this process. The first element in C is the vector P *multiplied into*[3] the first column of R. The second element of C is the vector P multiplied into the second column of R. In setting out the results again we shall now specify the dimensions of the matrices:

$$[5 \quad 6]\begin{bmatrix} 4 & 1 \\ 2 & 3 \end{bmatrix} = [32 \quad 23]$$

Dimensions: $\quad 1 \times 2 \quad 2 \times 2 \qquad 1 \times 2$

A 1×2 matrix multiplying a 2×2 matrix produces a 1×2 matrix. For the process to operate, the number of elements in the row vector P *must* be the same as the number of elements in each column of R. This means that the number of *columns* of P is the same as the number of *rows* of R. If the firm finds another supplier whose prices are £7 per kilogramme for materials and £4 per hour for labour, what would be the unit costs of the products? The new prices can be included by adding a second row to *P*. The result of this is a second row in the product matrix *C*. First we set out the results and then describe them. The outcome is

$$\begin{bmatrix} 5 & 6 \\ 7 & 4 \end{bmatrix} \begin{bmatrix} 4 & 1 \\ 2 & 3 \end{bmatrix} = \begin{bmatrix} 32 & 23 \\ 36 & 19 \end{bmatrix}$$

Dimensions: 2×2 2×2 2×2

The unit cost of the Alpha model at the new prices, £36, is in the second row, first column of the product matrix, and is the *second* row of *P* multiplied 'into' the *first* column of *R*. The new unit cost of Omegas, £19, is in the second row, second column of the product matrix and is the product of the second row of *P* into the second column of *R*. The matrix workings show that a 2×2 matrix multiplied by a 2×2 matrix produces another 2×2 matrix. Now imagine a third source for materials and labour charging £3 per kilogramme and £8 per hour. The further expanded matrix operations are

$$\begin{bmatrix} 5 & 6 \\ 7 & 4 \\ 3 & 8 \end{bmatrix} \begin{bmatrix} 4 & 1 \\ 2 & 3 \end{bmatrix} = \begin{bmatrix} 32 & 23 \\ 36 & 19 \\ 28 & 27 \end{bmatrix}$$

Dimensions: 3×2 2×2 3×2

where the Alpha cost, £28, is element c_{31} of *C* and is the third row of *P* multiplied into the first column of *R*. The Omega cost, £27, is element c_{32} of *C* and results from multiplying the third row of *P* into the second column of *R*. Note the dimensions of the matrices:

$P \times R \quad = \quad C$

$3 \times 2 \times 2 \times 2 = 3 \times 2$

It is essential that the number of columns in matrix *P* is equal to the number of rows in the second matrix *R*. In terms of the dimensions, it is as if the number of columns of *P* and the number of rows of *R* cancel to give the dimensions of the product matrix *C*.

If *P* had *m* rows and two columns, the dimensions of the product matrix *C* would be $= m \times 2$.

Now suppose that there is a third *product*, with per unit resource requirements of 3.5 kilogrammes of material and 1.5 labour hours. With these details added, the matrix R becomes

$$R = \begin{bmatrix} 4 & 1 & 3.5 \\ 2 & 3 & 1.5 \end{bmatrix}$$

and there will now be three sets of unit costs for the three products, represented in matrix form as

$$\begin{bmatrix} 5 & 6 \\ 7 & 4 \\ 3 & 8 \end{bmatrix} \begin{bmatrix} 4 & 1 & 3.5 \\ 2 & 3 & 1.5 \end{bmatrix} = \begin{bmatrix} 32 & 23 & 26.5 \\ 36 & 19 & 30.5 \\ 28 & 27 & 22.5 \end{bmatrix}$$

Dimensions: 3×2 2×3 3×3

The unit costs in the third column of the expanded C matrix are successive rows of P into the last column of R. If P had dimensions $m \times 2$ and R was $2 \times n$ then C would be $m \times n$. In general, if the number of columns of P and the number of rows of R are both p, then the matrices are said to be *conformable in multiplication*, and the dimensions of the product matrix will be $m \times n$.

Notice that the product of the matrices

$$\overset{A}{\begin{bmatrix} 5 & 6 & 2 \\ 7 & 4 & 4 \\ 3 & 8 & 6 \end{bmatrix}} \overset{B}{\begin{bmatrix} 4 & 1 \\ 2 & 3 \end{bmatrix}}$$

Dimensions: 3×3 2×2

cannot be formed as the number of columns of A is not equal to the number of rows in B.

When a matrix has the same number of rows as columns, it can be multiplied by itself. The result is defined as the *square* of the original matrix. That is,

$$AA = A^2$$

For example, if

$$A = \begin{bmatrix} 4 & 1 \\ 2 & 3 \end{bmatrix}$$

then

$$A^2 = \begin{bmatrix} 4 & 1 \\ 2 & 3 \end{bmatrix} \begin{bmatrix} 4 & 1 \\ 2 & 3 \end{bmatrix} = \begin{bmatrix} 18 & 7 \\ 14 & 11 \end{bmatrix}$$

Notice that the elements of the matrix A^2 are *not* the squares of the elements of A (i.e. $(a_{ij})^2$). A square matrix can be raised to powers higher than the second. For example,

$$A^3 = AAA = \begin{bmatrix} 4 & 1 \\ 2 & 3 \end{bmatrix} \begin{bmatrix} 4 & 1 \\ 2 & 3 \end{bmatrix} \begin{bmatrix} 4 & 1 \\ 2 & 3 \end{bmatrix} = \begin{bmatrix} 86 & 39 \\ 78 & 47 \end{bmatrix}$$

Notice also that A^3 can be obtained as A^2A or as AA^2, the result will be the same in either case.[4] Consider the 2×2 square matrix

$$B = \begin{bmatrix} 4 & 6 \\ -2 & -3 \end{bmatrix}$$

Now find the matrix B^2:

$$B^2 = \begin{bmatrix} 4 & 6 \\ -2 & -3 \end{bmatrix} \begin{bmatrix} 4 & 6 \\ -2 & -3 \end{bmatrix} = \begin{bmatrix} 4 & 6 \\ -2 & -3 \end{bmatrix} = B$$

A matrix B with the property that $B^2 = B$ is *idempotent*. Note that the identity matrix is idempotent, and that it follows from $B^2 = B$ that $B^n = B$ for idempotent matrices. For square matrices

(i) $A^m A^n = A^{m+n}$

The *product rule* for the exponents of scalars is paralleled in terms of square matrices.

(ii) $(A^m)^n = A^{mn}$

The *power rule* for exponents also carries over to square matrices.

In (i) and (ii), m and n are non-negative integers and, by definition, $A^0 = I$. Unlike scalar multiplication, in matrix multiplication the order in which the matrices are taken in forming the product is important and usually affects the result. For two square matrices A and B of the same order, in general

$$AB \neq BA$$

For example, letting

$$A = \begin{bmatrix} 4 & 1 \\ 2 & 3 \end{bmatrix} \quad \text{and} \quad B = \begin{bmatrix} 5 & 6 \\ 7 & 4 \end{bmatrix}$$

then the product AB will be

$$AB = \begin{bmatrix} 4 & 1 \\ 2 & 3 \end{bmatrix} \begin{bmatrix} 5 & 6 \\ 7 & 4 \end{bmatrix} = \begin{bmatrix} 27 & 28 \\ 31 & 24 \end{bmatrix}$$

In this case the matrix A is said to *premultiply* the matrix B, or B *postmultiplies* A. Now consider the result when B premultiplies A:

$$BA = \begin{bmatrix} 5 & 6 \\ 7 & 4 \end{bmatrix} \begin{bmatrix} 4 & 1 \\ 2 & 3 \end{bmatrix} = \begin{bmatrix} 32 & 23 \\ 36 & 19 \end{bmatrix}$$

which is an outcome different from AB. Where the product matrix is to be transposed, it can be helpful to note that

$$(AB)^T = B^T A^T$$

i.e. the transpose of the product is the product of the transposes – but in reverse order. The matrices A and B above can be used to demonstrate this result:

$$\overset{B^T}{\begin{bmatrix} 5 & 7 \\ 6 & 4 \end{bmatrix}} \overset{A^T}{\begin{bmatrix} 4 & 2 \\ 1 & 3 \end{bmatrix}} = \overset{(AB)^T}{\begin{bmatrix} 27 & 31 \\ 28 & 24 \end{bmatrix}}$$

It *can* happen for particular matrices that the product AB is the same as BA, for example if the two matrices are

$$A = \begin{bmatrix} 5 & 4 \\ -4 & 5 \end{bmatrix} \quad \text{and} \quad B = \begin{bmatrix} 3 & 2 \\ -2 & 3 \end{bmatrix}$$

with the result that in this case

$$AB = \begin{bmatrix} 7 & 22 \\ -22 & 7 \end{bmatrix} = BA$$

When $AB = BA$ the matrices are *commutative*. AB is commutative with either A or B. While particular examples of commutative matrices can be formed, matrix multiplication unlike the multiplication of real numbers is *not commutative*. There are other properties of matrix multiplication which differ from the properties of scalar multiplication. For example, in scalar arithmetic, if the product ab is zero, then either a or b or both must be zero. But in terms of matrices A and B, if

$$AB = 0$$

it does not follow that A or B must be a zero matrix. For example if

$$A = \begin{bmatrix} 4 & 6 \\ 2 & 3 \end{bmatrix} \quad \text{and} \quad B = \begin{bmatrix} 3 & -9 \\ -2 & 6 \end{bmatrix} \quad \text{then } AB = \begin{bmatrix} 0 & 0 \\ 0 & 0 \end{bmatrix}$$

In the arithmetic of real numbers for given scalars a, c and d, if $ac = ad$ for $a \neq 0$, then it must be the case that $c = d$. However, in the case of matrices A, C and D, if

$$AC = AD \quad \text{where} \quad D \neq C$$

it does not follow that A must necessarily be a zero matrix. For example, given the matrices

$$A = \begin{bmatrix} 4 & 6 \\ 2 & 3 \end{bmatrix} \quad C = \begin{bmatrix} 2 & 3 \\ 4 & 1 \end{bmatrix} \quad D = \begin{bmatrix} 5 & 3 \\ 2 & 1 \end{bmatrix}$$

then

$$AC = \begin{bmatrix} 32 & 18 \\ 16 & 9 \end{bmatrix} = AD$$

A further surprising property is that the square of a matrix not itself the zero matrix can be a zero matrix. That is,

$$A^2 = 0 \quad \text{for} \quad A \neq 0$$

For example, if

$$A = \begin{bmatrix} 1 & 2 \\ -0.5 & -1 \end{bmatrix} \quad \text{then } A^2 = \begin{bmatrix} 0 & 0 \\ 0 & 0 \end{bmatrix}$$

These phenomena occur only when certain of the matrices involved have a hidden aspect of 'zeroness' about them.[5] Overall, the parallels between matrix multiplication and the multiplication of common arithmetic are limited, and there is no matrix equivalent of the operation of division amongst real numbers. Matrix multiplication may involve more than two matrices. For example, three matrices A, B and C can be multiplied together to form the product ABC if the dimensions are appropriate to multiplication in this sequence. If A is of order $m \times n$, B is $n \times p$ and C is $p \times q$, then ABC has dimensions $m \times q$. Consider some numerical examples.

(i) $\begin{array}{ccc} A & B & C \end{array}$

$$[2 \ \ 3] \begin{bmatrix} 1 & 4 & 1 \\ 3 & 0 & 5 \end{bmatrix} \begin{bmatrix} 2 \\ 3 \\ 1 \end{bmatrix} = [11 \ \ 8 \ \ 17] \begin{bmatrix} 2 \\ 3 \\ 1 \end{bmatrix} = [63]$$

(ii) $\begin{array}{ccc} A & B & C \end{array}$

$$\begin{bmatrix} 2 & 1 \\ 1 & 3 \\ 0 & 5 \end{bmatrix} \begin{bmatrix} 3 \\ 2 \end{bmatrix} [4 \ \ 2] = \begin{bmatrix} 8 \\ 9 \\ 10 \end{bmatrix} [4 \ \ 2] = \begin{bmatrix} 32 & 16 \\ 36 & 18 \\ 40 & 20 \end{bmatrix}$$

(iii) $\begin{bmatrix} 2 & 1 & 3 \\ -1 & 4 & 0 \\ 3 & 1 & 2 \end{bmatrix} \begin{bmatrix} 5 & 4 \\ 2 & 3 \\ 1 & 0 \end{bmatrix} \begin{bmatrix} 1 & 2 & 3 & 0 \\ 4 & 1 & -1 & 2 \end{bmatrix}$

$$= \begin{bmatrix} 15 & 11 \\ 3 & 8 \\ 19 & 15 \end{bmatrix} \begin{bmatrix} 1 & 2 & 3 & 0 \\ 4 & 1 & -1 & 2 \end{bmatrix} = \begin{bmatrix} 59 & 41 & 34 & 22 \\ 35 & 14 & 1 & 16 \\ 79 & 53 & 42 & 30 \end{bmatrix}$$

However, in the case

$$\begin{matrix} A & B & C \\ \begin{bmatrix} 4 & 5 \\ 1 & 7 \end{bmatrix} & \begin{bmatrix} 3 \\ 2 \\ 1 \end{bmatrix} & [6 \quad 9] \end{matrix}$$

the matrices A and B are not conformable, so the product cannot be formed. Although not commutative, matrix multiplication is *associative*. Thus, for conformable matrices A, B, C and D,

$$(AB)CD = A(BC)D = AB(CD)$$

So long as the ordering $A\ B\ C\ D$ is preserved, it does not matter how the matrices are bracketed together. For conformable matrices A, B and C

$$A(B + C) = AB + AC$$
$$(B + C)A = BA + CA$$

Matrix multiplication distributes over matrix addition. For example, if the matrices involved are

$$\begin{matrix} A & B & C \\ \begin{bmatrix} 5 & 4 \\ -1 & 9 \end{bmatrix} & \begin{bmatrix} \begin{bmatrix} 3 & 7 \\ 0 & 5 \end{bmatrix} & + & \begin{bmatrix} 2 & 1 \\ 4 & 6 \end{bmatrix} \end{bmatrix} \end{matrix}$$

the postmultiplication of A by the sum of B and C gives the result

$$\begin{matrix} A & (B+C) \\ \begin{bmatrix} 5 & 4 \\ -1 & 9 \end{bmatrix} & \begin{bmatrix} 5 & 8 \\ 4 & 11 \end{bmatrix} \end{matrix} = \begin{bmatrix} 41 & 84 \\ 31 & 91 \end{bmatrix}$$

and $AB + AC$ works out as

$$\begin{matrix} AB & + & AC \\ \begin{bmatrix} 15 & 55 \\ -3 & 38 \end{bmatrix} & + & \begin{bmatrix} 26 & 29 \\ 34 & 53 \end{bmatrix} \end{matrix} = \begin{bmatrix} 41 & 84 \\ 31 & 91 \end{bmatrix}$$

In the multiplication of real numbers, the number 1 is the *identity element* for multiplication, leaving unchanged any number by which it is multiplied. In matrix multiplication, the equivalent role is fulfilled by the *identity matrix*. An appropriately dimensioned identity matrix, I, must be such that for any matrix A

$$AI = A$$

and[6]

$$IA = A$$

Identity matrices are *diagonal* matrices in which the non-zero elements $a_{ii} = 1$. Examples are

$$\begin{bmatrix} 1 & 0 \\ 0 & 1 \end{bmatrix} = I_2 \qquad \begin{bmatrix} 1 & 0 & 0 \\ 0 & 1 & 0 \\ 0 & 0 & 1 \end{bmatrix} = I_3 \qquad \begin{bmatrix} 1 & 0 & 0 & 0 \\ 0 & 1 & 0 & 0 \\ 0 & 0 & 1 & 0 \\ 0 & 0 & 0 & 1 \end{bmatrix} = I_4$$

Note that the order of the identity matrix can be indicated by a subscript and $I_1 = [1]$. To illustrate the role of I as the identity element in multiplication, suppose that

$$A = \begin{bmatrix} 4 & 1 & 6 \\ 2 & 3 & 5 \end{bmatrix}$$

and observe that, if the 2×3 matrix A is premultiplied by an identity matrix, I must be 2×2,

$$\begin{bmatrix} 1 & 0 \\ 0 & 1 \end{bmatrix} \begin{bmatrix} 4 & 1 & 6 \\ 2 & 3 & 5 \end{bmatrix} = \begin{bmatrix} 4 & 1 & 6 \\ 2 & 3 & 5 \end{bmatrix}$$

while if A is postmultiplied, I_3 will be involved:

$$\begin{bmatrix} 4 & 1 & 6 \\ 2 & 3 & 5 \end{bmatrix} \begin{bmatrix} 1 & 0 & 0 \\ 0 & 1 & 0 \\ 0 & 0 & 1 \end{bmatrix} = \begin{bmatrix} 4 & 1 & 6 \\ 2 & 3 & 5 \end{bmatrix}$$

The identity matrix must be of the appropriate order but frequently this is presumed and the subscript on I omitted.

In this section we have examined rules for matrix multiplication. These rules can be seen as *conventions* for restoring an original order of things. For example, given the expression

$$3x_1 - 5x_2$$

and given also that this is separated out as

$$[3 \quad -5] \begin{bmatrix} x_1 \\ x_2 \end{bmatrix}$$

then in order to restore the original expression, x_1 must be multiplied by 3 and added to x_2 multiplied by -5. The rules for manipulating matrices reflect the conventions of the notation. In conjunction with appropriately framed rules, it would be possible to rewrite the original expression in the forms

$$\begin{bmatrix} 3 \\ -5 \end{bmatrix} \begin{bmatrix} x_1 \\ x_2 \end{bmatrix} \qquad \text{or} \qquad \begin{bmatrix} x_2 \\ x_1 \end{bmatrix} [3 \quad -5]$$

EXERCISES 10.4

1 Carry out the following matrix multiplications:

(i) $[3 \quad 5] \begin{bmatrix} 8 & 3 \\ -1 & 4 \end{bmatrix}$

(ii) $\begin{bmatrix} 2 & 1 \\ 3 & -2 \end{bmatrix} \begin{bmatrix} 6 & 9 \\ 1 & 4 \end{bmatrix}$

(iii) $\begin{bmatrix} 8 & 2 \\ 1 & 6 \\ -1 & 4 \end{bmatrix} \begin{bmatrix} 1 & 3 \\ 2 & -1 \end{bmatrix}$

(iv) $\begin{bmatrix} 2 & 5 \\ -3 & 4 \\ 6 & -1 \end{bmatrix} \begin{bmatrix} 5 & 4 & 3 \\ 1 & -2 & 6 \end{bmatrix}$

2 Given the matrix:

$$A = \begin{bmatrix} 2 & 1 \\ -1 & 3 \end{bmatrix}$$

find

(i) A^2
(ii) A^3
(iii) A^4

3 Which of the following matrices are idempotent?

(i) $\begin{bmatrix} 2 & 4 \\ -0.5 & -2 \end{bmatrix}$

(ii) $\begin{bmatrix} 8 & 7 \\ -8 & -7 \end{bmatrix}$

4 Show that the following matrices are commutative:

$$A = \begin{bmatrix} 3 & 7 \\ -7 & 3 \end{bmatrix} \quad \text{and} \quad B = \begin{bmatrix} -2 & -4 \\ 4 & -2 \end{bmatrix}$$

5 Carry out the following matrix multiplications:

(i) $[4 \quad 3] \begin{bmatrix} 2 & 5 & -1 \\ -2 & 6 & 4 \end{bmatrix} \begin{bmatrix} 4 \\ 3 \\ 2 \end{bmatrix}$

(ii) $\begin{bmatrix} 3 & 2 \\ 1 & 3 \\ 4 & -1 \end{bmatrix} \begin{bmatrix} 4 \\ 5 \end{bmatrix} [3 \quad 6]$

(iii) $\begin{bmatrix} 3 & 2 & 0 \\ 1 & 2 & -1 \\ 2 & 5 & 1 \end{bmatrix} \begin{bmatrix} 1 & 3 \\ 2 & 1 \\ 0 & 4 \end{bmatrix} \begin{bmatrix} 2 & 1 & 0 & 1 \\ 2 & 3 & 5 & -1 \end{bmatrix}$

10.5 MATRIX INVERSION

For two square matrices A and B of the same order, then if (and *only if*)

$$AB = BA = I$$

the matrix B is said to be the *inverse* of A, and A is the inverse of B. The inverse relationship is written as

$$B = A^{-1} \quad \text{and} \quad A = B^{-1}$$

It follows that

$$[A^{-1}]^{-1} = A$$

i.e. a matrix is the inverse of its own inverse. Note also that a matrix and its inverse are *commutative*. Consider two examples of inverses.

If $A = \begin{bmatrix} 2 & 3 \\ 1 & 4 \end{bmatrix}$ then $\begin{bmatrix} 0.8 & -0.6 \\ -0.2 & 0.4 \end{bmatrix}$ is A^{-1}

As a check, the multiplications AA^{-1} and $A^{-1}A$ may be carried out to confirm that the result in both cases is I_2. In the former case the workings are

$$\begin{bmatrix} 2 & 3 \\ 1 & 4 \end{bmatrix} \begin{bmatrix} 0.8 & -0.6 \\ -0.2 & 0.4 \end{bmatrix} = \begin{bmatrix} 1 & 0 \\ 0 & 1 \end{bmatrix}$$

For the second example consider the following:

If $A = \begin{bmatrix} 1 & -1 \\ -1 & 2 \end{bmatrix}$ then $\begin{bmatrix} 2 & 1 \\ 1 & 1 \end{bmatrix}$ is A^{-1}

If the inverse of a matrix A exists, then it is unique and A is said to be *invertible*. But in apparent contrast to the situation with real numbers, not all matrices have inverses. As may be deduced from the commutative property of inverses, only *square* matrices have inverses, but *not all* square matrices are invertible; certain conditions need to be satisfied. Consider the problem of finding the inverse of a 2×2 matrix A from scratch. The inverse, B, with elements b_{ij}, must be of order 2×2 and such that

$$\begin{bmatrix} a_{11} & a_{12} \\ a_{21} & a_{22} \end{bmatrix} \begin{bmatrix} b_{11} & b_{12} \\ b_{21} & b_{22} \end{bmatrix} = \begin{bmatrix} 1 & 0 \\ 0 & 1 \end{bmatrix}$$

Each element of the product matrix, I, gives a constraint on the elements of A and B that make it up. Thus the elements of the inverse, b_{ij}, must satisfy four 'constraints' as follows:

$$a_{11}b_{11} + a_{12}b_{21} = 1$$
$$a_{11}b_{12} + a_{12}b_{22} = 0$$
$$a_{21}b_{11} + a_{22}b_{21} = 0$$
$$a_{21}b_{12} + a_{22}b_{22} = 1$$

When the above requirements are 'solved' to state the b_{ij} in terms of the a_{ij}, the results are

$$b_{11} = \frac{a_{22}}{a_{11}a_{22} - a_{12}a_{21}} \qquad b_{12} = \frac{-a_{12}}{a_{11}a_{22} - a_{12}a_{21}}$$

$$b_{21} = \frac{-a_{21}}{a_{11}a_{22} - a_{12}a_{21}} \qquad b_{22} = \frac{a_{11}}{a_{11}a_{22} - a_{12}a_{21}}$$

The denominator in all these expressions:

$$a_{11}a_{22} - a_{12}a_{21}$$

plays a vital role in establishing the inverse and an early task is to find this value. But note that the b_{ij} will be undefined and the inverse will not exist if this number is zero. The number

$$a_{11}a_{22} - a_{12}a_{21}$$

is the *determinant* of the 2 × 2 matrix, because its value (crucially, non-zero or zero) *determines* whether or not the inverse exists, and therefore whether the system of linear equations giving rise to the A matrix has a unique solution.[7] The B matrix could be shown as the scalar product of the reciprocal of the determinant and a matrix comprising the numerators of the ratios above:

$$A^{-1} = B = \frac{1}{a_{11}a_{22} - a_{12}a_{21}} \begin{bmatrix} a_{22} & -a_{12} \\ -a_{21} & a_{11} \end{bmatrix}$$

Presented in this way, two new concepts used to find the inverse are distinguished: the determinant and the matrix of rearranged and re-signed elements of A making up the remainder of B called the *adjugate matrix* or *adjoint* of A. Consider these concepts in more detail.

The determinant

The determinant of

$$A = \begin{bmatrix} a_{11} & a_{12} \\ a_{21} & a_{22} \end{bmatrix}$$

is written either as

$$|A| \quad \text{or} \quad \det A$$

and is the product of the main diagonal elements less the product of the two remaining elements. Determinants of matrices of order higher than 2×2 require more substantial calculations and will be considered in a later section. For the moment we shall develop the notion and illustrate its uses in the context of 2×2 matrices. Consider some examples:

(i) $A = \begin{bmatrix} 2 & 3 \\ 1 & 4 \end{bmatrix}$ $\quad |A| = 2(4) - 1(3) = 5$

(ii) $A = \begin{bmatrix} 1 & -1 \\ -1 & 3 \end{bmatrix}$ $\quad |A| = 1(3) - (-1)(-1) = 2$

(iii) $A = \begin{bmatrix} 0.5 & 7 \\ 3 & 8 \end{bmatrix}$ $\quad \det A = 0.5(8) - (7)(3) = -17$

(iv) $A = \begin{bmatrix} 2 & 3 \\ 4 & 6 \end{bmatrix}$ $\quad \det A = 2(6) - (3)(4) = 0$

(v) $A = \begin{bmatrix} 9 & 4.5 \\ -4 & -2 \end{bmatrix}$ $\quad \det A = 9(-2) - (4.5)(-4) = 0$

The value of the determinant can be positive, negative or zero. In cases (iv) and (v), where $|A| = 0$, the matrices are *singular*. In contrast, cases (i)–(iii) represent non-singular matrices. This is an important distinction since only non-singular matrices have inverses. In singular matrices of order 2, one row is always a multiple of the other[8] – and the same statement must also be true of columns. So, if in terms of the rows

$$A = \begin{bmatrix} a_{11} & a_{12} \\ ka_{11} & ka_{12} \end{bmatrix}$$

then the determinant of A must be zero since

$$\det A = a_{11}ka_{12} - ka_{11}a_{12} = 0$$

The adjugate matrix

Given the matrix A where

$$A = \begin{bmatrix} a_{11} & a_{12} \\ a_{21} & a_{22} \end{bmatrix}$$

the matrix

$$\begin{bmatrix} a_{22} & -a_{12} \\ -a_{21} & a_{11} \end{bmatrix}$$

is the adjugate matrix or adjoint of A and is written as

adj A

We shall look at the adjoints of larger matrices later on. First consider some examples of the adjoints of 2×2 matrices.

(i) $A = \begin{bmatrix} 2 & -3 \\ 1 & 4 \end{bmatrix}$ adj $A = \begin{bmatrix} 4 & -3 \\ -1 & 2 \end{bmatrix}$

(ii) $A = \begin{bmatrix} 1 & -1 \\ -1 & 3 \end{bmatrix}$ adj $A = \begin{bmatrix} 3 & 1 \\ 1 & 1 \end{bmatrix}$

We are reminded of the role of the adjugate matrix in forming the inverse when we examine the products

A adj A and   adj AA

The order of multiplication is immaterial as A and adj A are commutative so these products are always the same. Take the matrices above as illustrations. In the case of example (i) A adj A works out as

$$\begin{bmatrix} 2 & 3 \\ 1 & 4 \end{bmatrix} \begin{bmatrix} 4 & -3 \\ -1 & 2 \end{bmatrix} = \begin{bmatrix} 5 & 0 \\ 0 & 5 \end{bmatrix}$$

The product adj AA gives the same result:

$$\begin{bmatrix} 4 & -3 \\ -1 & 2 \end{bmatrix} \begin{bmatrix} 2 & 3 \\ 1 & 4 \end{bmatrix} = \begin{bmatrix} 5 & 0 \\ 0 & 5 \end{bmatrix}$$

In the case of (ii), the products A adj A and adj AA work out as

$$\begin{bmatrix} 1 & -1 \\ -1 & 3 \end{bmatrix} \begin{bmatrix} 3 & 1 \\ 1 & 1 \end{bmatrix} = \begin{bmatrix} 2 & 0 \\ 0 & 2 \end{bmatrix}$$

The product of a matrix and its adjoint is a diagonal matrix with elements $a_{ii} = |A|$. In the case of matrices of order 2×2:

$$A \text{ adj } A = \begin{bmatrix} |A| & 0 \\ 0 & |A| \end{bmatrix}$$

Confirmation of this outcome is obtained by forming the product

$$\begin{bmatrix} a_{11} & a_{12} \\ a_{21} & a_{22} \end{bmatrix} \begin{bmatrix} a_{22} & -a_{12} \\ -a_{21} & a_{11} \end{bmatrix} = \begin{bmatrix} a_{11}a_{22} - a_{12}a_{21} & 0 \\ 0 & a_{11}a_{22} - a_{12}a_{21} \end{bmatrix}$$

The only modification needed to A adj A to obtain the inverse is division of its elements by $|A|$. The inverse is the *scalar product* of the reciprocal of the determinant and the adjugate matrix:

$$\frac{1}{|A|} \text{ adj } A = A^{-1} \tag{10.5}$$

In considering examples in finding the inverses of 2×2 matrices, first complete the process for the two examples above:

(i) If
$$A = \begin{bmatrix} 2 & 3 \\ 1 & 4 \end{bmatrix}$$
then since
$$|A| = 5 \quad \text{and} \quad \text{adj } A = \begin{bmatrix} 4 & -3 \\ -1 & 2 \end{bmatrix}$$
then
$$A^{-1} = \frac{1}{|A|} \text{adj } A = \begin{bmatrix} 4/5 & -3/5 \\ -1/5 & 2/5 \end{bmatrix}$$

(ii) If
$$A = \begin{bmatrix} 1 & -1 \\ -1 & 3 \end{bmatrix}$$
then since
$$|A| = 2 \quad \text{and} \quad \text{adj } A = \begin{bmatrix} 3 & 1 \\ 1 & 1 \end{bmatrix}$$
then
$$A^{-1} = \frac{1}{|A|} \text{adj } A = \begin{bmatrix} 3/2 & 1/2 \\ 1/2 & 1/2 \end{bmatrix}$$

In using this method for obtaining the inverse of a matrix it is advisable to begin by finding the determinant rather than the adjugate matrix since if the determinant is zero the inverse matrix does not exist. Consider some further examples.

(iii) If
$$A = \begin{bmatrix} 6 & 4 \\ 2 & 3 \end{bmatrix}$$
then since
$$|A| = 10 \quad \text{and} \quad \text{adj } A = \begin{bmatrix} 3 & -4 \\ -2 & 6 \end{bmatrix}$$
then
$$A^{-1} = \frac{1}{|A|} \text{adj } A = \begin{bmatrix} 0.3 & -0.4 \\ -0.2 & 0.6 \end{bmatrix}$$

(iv) If
$$A = \begin{bmatrix} 7 & 1 \\ 9 & 2 \end{bmatrix}$$
then since
$$|A| = 5 \quad \text{and} \quad \text{adj } A = \begin{bmatrix} 2 & -1 \\ -9 & 7 \end{bmatrix}$$
then
$$A^{-1} = \frac{1}{|A|} \text{adj } A = \begin{bmatrix} 0.4 & -0.2 \\ -1.8 & 1.4 \end{bmatrix}$$

(v) If
$$A = \begin{bmatrix} 2 & 1 \\ 5 & 2 \end{bmatrix}$$
then since
$$|A| = -1 \quad \text{and} \quad \text{adj } A = \begin{bmatrix} 2 & -1 \\ -5 & 2 \end{bmatrix}$$
then
$$A^{-1} = \frac{1}{|A|} \text{adj } A = \begin{bmatrix} -2 & 1 \\ 5 & -2 \end{bmatrix}$$

(vi) If
$$A = \begin{bmatrix} 4 & 6 \\ 2 & 3 \end{bmatrix}$$
then since
$$|A| = 0$$

A is singular and the inverse matrix in this case does not exist. Now consider two interesting special cases. A square matrix is *orthogonal* if

$$A'A = I$$

i.e. if the transpose of the matrix is the inverse. For example,

$$\begin{bmatrix} 0.8 & 0.6 \\ -0.6 & 0.8 \end{bmatrix}$$

is orthogonal. For orthogonal matrices, $|A| = \pm 1$ and if A and B are orthogonal matrices of the same order then the product AB is also orthogonal. An intriguing special case is

(vii) $A = \begin{bmatrix} 5 & -4 \\ 6 & -5 \end{bmatrix}$ $\quad A^{-1} = \dfrac{1}{|A|} \operatorname{adj} A = \begin{bmatrix} 5 & -4 \\ 6 & -5 \end{bmatrix}$

Here the matrix A is *its own inverse* and A is *involutional*. For involutional matrices $A^2 = I$, and the rule for real numbers that if $a^2 = 1$ then $a = \pm 1$ does not carry over to matrices.[9] The following properties of inverses can be helpful in analysis and in minimizing the extent of manual calculations.

1 The inverse of the inverse of a matrix is the original matrix. That is,

$$(A^{-1})^{-1} = A$$

2 The determinant of the inverse of a matrix A is the inverse (reciprocal) of the determinant of A:

$$|A^{-1}| = (|A|)^{-1}$$

3 The inverse of the transpose of a matrix is the transpose of the inverse. That is,

$$(A^T)^{-1} = (A^{-1})^T$$

4 The inverse of the product of two matrices is, in reverse order, the product of their two inverses:

$$(AB)^{-1} = B^{-1}A^{-1}$$

Only square matrices can have an inverse, but non-square matrices may have a *left inverse* or a *right inverse*. B is the left inverse of A if

$$BA = I$$

C is the right inverse of A if

$$AC = I$$

Both left and right inverses exist *only* if A is square, in which case

$$B = C = A^{-1}$$

We shall use the term *inverse* to refer to the multiplicative inverse defined above, but an inverse can be defined in relation to operations other than multiplication. A given matrix when combined with its inverse produces the identity matrix for that operation. The zero matrix is the identity element for matrix addition and the matrix $-X$ represents the *additive inverse* of X since

$$X + (-X) = 0$$

so if

$$X = \begin{bmatrix} 3 & -2 \\ 5 & 5 \end{bmatrix}$$

then the additive inverse is

$$-X = \begin{bmatrix} -3 & 2 \\ -5 & -5 \end{bmatrix}$$

EXERCISES 10.5

1 Establish whether or not B is the inverse of A where

(i) $A = \begin{bmatrix} 5 & 3 \\ 5 & 4 \end{bmatrix}$ $B = \begin{bmatrix} 0.8 & -0.6 \\ -1 & 1 \end{bmatrix}$

(ii) $A = \begin{bmatrix} 5 & 5 \\ 2 & 4 \end{bmatrix}$ $B = \begin{bmatrix} 0.4 & -0.5 \\ -0.2 & 0.5 \end{bmatrix}$

(iii) $A = \begin{bmatrix} 3 & 2 \\ 5 & 5 \end{bmatrix}$ $B = \begin{bmatrix} 1 & -0.4 \\ -1 & 0.2 \end{bmatrix}$

2 Find the determinant of

(i) $A = \begin{bmatrix} 9 & 7 \\ 2 & 3 \end{bmatrix}$

(ii) $A = \begin{bmatrix} -3 & -2 \\ -1 & -4 \end{bmatrix}$

(iii) $A = \begin{bmatrix} 2 & 5 \\ 3 & 4 \end{bmatrix}$

(iv) $A = \begin{bmatrix} 0.4 & 0.3 \\ 0.3 & 0.2 \end{bmatrix}$

(v) $A = \begin{bmatrix} 3 & \pi \\ 2\pi & -2 \end{bmatrix}$

(vi) $A = \begin{bmatrix} 8 & 2 \\ 12 & 3 \end{bmatrix}$

3 Find the adjugate matrices for

(i) $A = \begin{bmatrix} 3 & 2 \\ 1 & 5 \end{bmatrix}$

(ii) $A = \begin{bmatrix} 2 & -6 \\ 3 & -7 \end{bmatrix}$

(iii) $A = \begin{bmatrix} 3 & -\pi \\ 2 & e \end{bmatrix}$

(iv) $A = \begin{bmatrix} 5 & 0 \\ 0 & 5 \end{bmatrix}$

4 Find the inverse matrix, if it exists, for

(i) $A = \begin{bmatrix} 4 & 5 \\ 2 & 3 \end{bmatrix}$

(ii) $A = \begin{bmatrix} 4 & 2 \\ 3 & 4 \end{bmatrix}$

(iii) $A = \begin{bmatrix} 3 & 2 \\ 10 & 5 \end{bmatrix}$

(iv) $A = \begin{bmatrix} 1.5 & -0.5 \\ -0.4 & 0.2 \end{bmatrix}$

(v) $A = \begin{bmatrix} 8 & -4 \\ -6 & 2 \end{bmatrix}$

5 Which of the matrices below are orthogonal or involutional?

(i) $A = \begin{bmatrix} 7 & -6 \\ 8 & -7 \end{bmatrix}$

(ii) $A = \begin{bmatrix} 0.6 & -0.8 \\ 0.8 & 0.6 \end{bmatrix}$

(iii) $A = \begin{bmatrix} 0.8 & 0.2 \\ 1.8 & -0.8 \end{bmatrix}$

(iv) $A = \begin{bmatrix} -0.6 & 0.8 \\ 0.8 & 0.6 \end{bmatrix}$

10.6 SIMULTANEOUS EQUATIONS (i)

Consider the system

$7x_1 + 1x_2 = 34$
$9x_1 + 2x_2 = 48$

which in matrix form is

$$\begin{bmatrix} 7 & 1 \\ 9 & 2 \end{bmatrix} \begin{bmatrix} x_1 \\ x_2 \end{bmatrix} = \begin{bmatrix} 34 \\ 48 \end{bmatrix}$$

The equations can be seen as a 2×2 matrix of coefficients (A) premultiplying a two-element column vector of unknowns (x), the result being the 2×1 vector of right-hand side values (b). Such a system of simultaneous linear equations can be written in terms of matrices as

$$Ax = b$$

in which $b \neq 0$.[10] Suppose that the system consisted of one equation in one unknown:

$$ax = b$$

The solution for x is

$$x = \frac{b}{a}$$

i.e.

$$x = a^{-1}b \tag{10.6}$$

To show the equivalent of equation (10.6) for larger systems, given

$$Ax = b$$

premultiply both sides by the inverse of A giving

$$A^{-1}Ax = A^{-1}b \tag{10.7}$$

Since the product of a matrix and its inverse is the identity matrix (10.7) becomes

$$Ix = A^{-1}b$$

from which I can be suppressed, leaving

$$x = A^{-1}b \tag{10.8}$$

In terms of our current example, (10.8) is

$$\begin{bmatrix} x_1 \\ x_2 \end{bmatrix} = \begin{bmatrix} 7 & 1 \\ 9 & 2 \end{bmatrix}^{-1} \begin{bmatrix} 34 \\ 48 \end{bmatrix}$$

$$= \begin{bmatrix} 0.4 & -0.2 \\ -1.8 & 1.4 \end{bmatrix} \begin{bmatrix} 34 \\ 48 \end{bmatrix} = \begin{bmatrix} 4 \\ 6 \end{bmatrix}$$

That is, $x_1 = 4$ and $x_2 = 6$. The two main steps are

1 obtain the inverse (if it exists) of the matrix of coefficients;
2 premultiply the vector of right-hand side values by the inverse.

Consider some examples.

(i) The system

$$2x_1 + 3x_2 = 16$$
$$x_1 + 4x_2 = 13$$

can be expressed as

$$\begin{bmatrix} 2 & 3 \\ 1 & 4 \end{bmatrix} \begin{bmatrix} x_1 \\ x_2 \end{bmatrix} = \begin{bmatrix} 16 \\ 13 \end{bmatrix}$$

so therefore

$$\begin{bmatrix} x_1 \\ x_2 \end{bmatrix} = \begin{bmatrix} 2 & 3 \\ 1 & 4 \end{bmatrix}^{-1} \begin{bmatrix} 16 \\ 13 \end{bmatrix}$$

and so

$$\begin{bmatrix} x_1 \\ x_2 \end{bmatrix} = \begin{bmatrix} 0.8 & -0.6 \\ -0.2 & 0.4 \end{bmatrix} \begin{bmatrix} 16 \\ 13 \end{bmatrix} = \begin{bmatrix} 5 \\ 2 \end{bmatrix}$$

(ii) If

$$\begin{bmatrix} 3 & -1 \\ 4 & 2 \end{bmatrix} \begin{bmatrix} x_1 \\ x_2 \end{bmatrix} = \begin{bmatrix} 23 \\ 24 \end{bmatrix}$$

then

$$\begin{bmatrix} x_1 \\ x_2 \end{bmatrix} = \begin{bmatrix} 0.2 & 0.1 \\ -0.4 & 0.3 \end{bmatrix} \begin{bmatrix} 23 \\ 24 \end{bmatrix} = \begin{bmatrix} 7 \\ -2 \end{bmatrix}$$

(iii) If

$$\begin{bmatrix} 0.4 & -0.2 \\ -1.8 & 1.4 \end{bmatrix} \begin{bmatrix} x_1 \\ x_2 \end{bmatrix} = \begin{bmatrix} -50 \\ 250 \end{bmatrix}$$

then

$$\begin{bmatrix} x_1 \\ x_2 \end{bmatrix} = \begin{bmatrix} 7 & 1 \\ 9 & 2 \end{bmatrix} \begin{bmatrix} -50 \\ 250 \end{bmatrix} = \begin{bmatrix} -100 \\ 50 \end{bmatrix}$$

(iv) But if the system is

$$\begin{bmatrix} 3 & 1 \\ 6 & 2 \end{bmatrix} \begin{bmatrix} x_1 \\ x_2 \end{bmatrix} = \begin{bmatrix} 20 \\ 40 \end{bmatrix}$$

the determinant of the coefficient matrix is zero and the inverse does not exist. The equations are either *inconsistent* or, as here, *linearly dependent*.

The matrix of coefficients exhibits linear dependence in either case. If the right-hand side elements are in the same relationship as the rows in the A matrix, then the equations are consistent (but linearly dependent). Otherwise the equations are inconsistent – in the two-variable case producing parallel straight lines. Here the second equation is twice the first and there are infinitely many solutions.[11]

(v) In the case of

$$\begin{bmatrix} 2 & 3 \\ 6 & 9 \end{bmatrix} \begin{bmatrix} x_1 \\ x_2 \end{bmatrix} = \begin{bmatrix} 10 \\ 40 \end{bmatrix}$$

the A matrix is again singular and so is non-invertible. In this case, however, the equations are inconsistent – the relationship between the rows of A is not replicated between the elements of b. The straight lines corresponding to the equations are parallel and there are *no* solutions. The value of the inverse matrix in the solution of linear equations is apparent when the system has to be solved more than once using different right-hand side values. For example, suppose that solutions are required as follows:

$$\begin{bmatrix} 7 & 1 \\ 9 & 2 \end{bmatrix} \begin{bmatrix} x_1 \\ x_2 \end{bmatrix} \text{ for } b = \begin{bmatrix} 34 \\ 48 \end{bmatrix}, \begin{bmatrix} 3 \\ 11 \end{bmatrix}, \begin{bmatrix} 10 \\ 0 \end{bmatrix} \text{ and } \begin{bmatrix} 21 \\ 27 \end{bmatrix}$$

The inverse need be obtained only once, solutions being found by postmultiplying the inverse by the respective right-hand side vectors. The results are

$$\begin{bmatrix} 0.4 & -0.2 \\ -1.8 & 1.4 \end{bmatrix} \begin{bmatrix} 34 \\ 48 \end{bmatrix} = \begin{bmatrix} 4 \\ 6 \end{bmatrix}$$

$$\begin{bmatrix} 0.4 & -0.2 \\ -1.8 & 1.4 \end{bmatrix} \begin{bmatrix} 3 \\ 11 \end{bmatrix} = \begin{bmatrix} -1 \\ 10 \end{bmatrix}$$

$$\begin{bmatrix} 0.4 & -0.2 \\ -1.8 & 1.4 \end{bmatrix} \begin{bmatrix} 10 \\ 0 \end{bmatrix} = \begin{bmatrix} 4 \\ -18 \end{bmatrix}$$

$$\begin{bmatrix} 0.4 & -0.2 \\ -1.8 & 1.4 \end{bmatrix} \begin{bmatrix} 21 \\ 27 \end{bmatrix} = \begin{bmatrix} 3 \\ 0 \end{bmatrix}$$

The use of the inverse matrix as outlined above is not the most rapid means of obtaining a once only solution to a linear system, but advantages follow if the inverse of the coefficient matrix is available. For example it can be useful to know the range of values of right-hand side elements for which the solution values are non-negative – a situation which arises in linear programming. Suppose that x_1 and x_2 represent production levels and the two right-hand side values are resource availabilities. Suppose also that 48 units of the second resource are obtainable, but that availability of the first resource is subject to variability; for what range of values of the first resource level are both x_1 and x_2 non-negative? To answer this question, let the level of the first resource be b_1. The system is then

$$\begin{bmatrix} 7 & 1 \\ 9 & 2 \end{bmatrix} \begin{bmatrix} x_1 \\ x_2 \end{bmatrix} = \begin{bmatrix} b_1 \\ 48 \end{bmatrix}$$

and the solution values of x_1 and x_2 – along with the statement of sign requirements – will be

$$\begin{bmatrix} x_1 \\ x_2 \end{bmatrix} = \begin{bmatrix} 0.4 & -0.2 \\ -1.8 & 1.4 \end{bmatrix} \begin{bmatrix} b_1 \\ 48 \end{bmatrix} \geqslant \begin{bmatrix} 0 \\ 0 \end{bmatrix}$$

We can read directly from this that

$$x_1 = 0.4b_1 - 0.2(48) \geqslant 0$$

which means that

$$b_1 \geqslant 24$$

Similarly, as regards x_2,

$$x_2 = -1.8b_1 + 1.4(48) \geqslant 0$$

which means that

$$b_1 \leqslant 37.\overline{33}$$

So, with the other data at their original levels, for the solution values of both variables to be non-negative the amount of the first resource must lie in the range

$$24 \leqslant b_1 \leqslant 37.\overline{33}$$

This range of values is called the *tolerance interval* or *range of feasibility* for the parameter b_1. As a check on this result, note that

$$\begin{bmatrix} 0.4 & -0.2 \\ -1.8 & 1.4 \end{bmatrix} \begin{bmatrix} 24 \\ 48 \end{bmatrix} = \begin{bmatrix} 0 \\ 24 \end{bmatrix}$$

and

$$\begin{bmatrix} 0.4 & -0.2 \\ -1.8 & 1.4 \end{bmatrix} \begin{bmatrix} 37.\overline{33} \\ 48 \end{bmatrix} = \begin{bmatrix} 5.\overline{33} \\ 0 \end{bmatrix}$$

Similar calculations can be carried out for the level of the second resource. Suppose that the amount of the first resource available is fixed at 34 units. Then permissible values for the second resource must be such that

$$\begin{bmatrix} x_1 \\ x_2 \end{bmatrix} = \begin{bmatrix} 0.4 & -0.2 \\ -1.8 & 1.4 \end{bmatrix} \begin{bmatrix} 34 \\ b_2 \end{bmatrix} \geqslant \begin{bmatrix} 0 \\ 0 \end{bmatrix}$$

i.e.

$$x_1 = 0.4(34) - 0.2b_2 \geq 0$$

from which

$$b_2 \leq 68$$

The required non-negativity of x_2

$$x_2 = -1.8(34) + 1.4b_2 \geq 0$$

implies that

$$b_2 \geq \frac{1.8(34)}{1.4} \approx 43.71$$

In summary:

$$43.71 \leq b_2 \leq 68$$

which represents the tolerance interval for the b_2 parameter. We have considered the scope for variation in b_1 and b_2 alone in examples of *single-parameter analysis*. It is also possible to conduct a *joint analysis*. In this case, statements about the *relative* sizes of b_1 and b_2 will result. It is required that

$$\begin{bmatrix} x_1 \\ x_2 \end{bmatrix} = \begin{bmatrix} 0.4 & -0.2 \\ -1.8 & 1.4 \end{bmatrix} \begin{bmatrix} b_1 \\ b_2 \end{bmatrix} \geq \begin{bmatrix} 0 \\ 0 \end{bmatrix}$$

from which it emerges that, in order to ensure the non-negativity of x_1, b_1 and b_2 must be such that

$$0.4b_1 - 0.2b_2 \geq 0$$

i.e.

$$b_2 \leq 2b_1$$

To make sure that x_2 does not fall below zero, it is necessary that

$$-1.8b_1 + 1.4b_2 \geq 0$$

i.e. (approximately)

$$b_2 \geq 1.29b_1$$

In summary, neither x_1 nor x_2 will become negative so long as the relative sizes of the two parameters b_1 and b_2 satisfy the condition

$$1.29b_1 \leq b_2 \leq 2b_1$$

This result is graphed in Figure 10.1.

Matrices 737

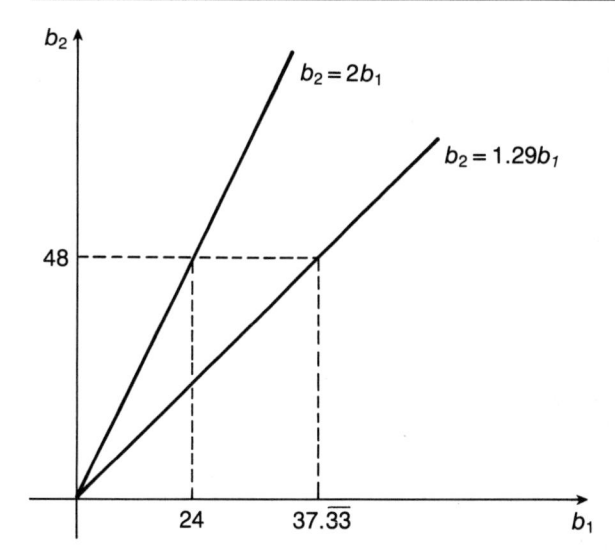

Figure 10.1

In Figure 10.1, the two limiting relationships between the values of b_2 and b_1 are shown as lines through the origin with respective equations

$$b_2 = 2b_1 \quad \text{and} \quad b_2 = 1.29b_1$$

The permitted region is on and between the lines. The individual range of variation for b_1 corresponding to a specified value of b_2 is obtained by taking a horizontal line into the diagram at the given height. The points of intersection with the rays then give the endpoints of the individual range for b_1.

EXERCISES 10.6

1 Solve, if possible, the following systems of simultaneous linear equations using the inverse of the matrix of coefficients:

(i) $\begin{bmatrix} 4 & 3 \\ 1 & 2 \end{bmatrix} \begin{bmatrix} x_1 \\ x_2 \end{bmatrix} = \begin{bmatrix} 55 \\ 20 \end{bmatrix}$

(ii) $\begin{bmatrix} 2 & -4 \\ 1 & 3 \end{bmatrix} \begin{bmatrix} x_1 \\ x_2 \end{bmatrix} = \begin{bmatrix} 120 \\ 10 \end{bmatrix}$

(iii) $\begin{bmatrix} 5 & 9 \\ 2 & 4 \end{bmatrix} \begin{bmatrix} x_1 \\ x_2 \end{bmatrix} = \begin{bmatrix} 45 \\ 20 \end{bmatrix}$

(iv) $\begin{bmatrix} 9 & 6 \\ 6 & 4 \end{bmatrix} \begin{bmatrix} x_1 \\ x_2 \end{bmatrix} = \begin{bmatrix} 30 \\ 20 \end{bmatrix}$

(v) $\begin{bmatrix} 2 & -7 \\ 2 & -2 \end{bmatrix} \begin{bmatrix} x_1 \\ x_2 \end{bmatrix} = \begin{bmatrix} -18 \\ -8 \end{bmatrix}$

(vi) $\begin{bmatrix} 9 & -6 \\ -6 & 4 \end{bmatrix} \begin{bmatrix} x_1 \\ x_2 \end{bmatrix} = \begin{bmatrix} 30 \\ 20 \end{bmatrix}$

(vii) $\begin{bmatrix} 8 & 10 \\ 4 & 6 \end{bmatrix} \begin{bmatrix} x_1 \\ x_2 \end{bmatrix} = \begin{bmatrix} 189 \\ 106 \end{bmatrix}$

(viii) $\begin{bmatrix} -0.6 & 0.4 \\ 0.8 & -0.2 \end{bmatrix} \begin{bmatrix} x_1 \\ x_2 \end{bmatrix} = \begin{bmatrix} 2 \\ 4 \end{bmatrix}$

2 For the system

$$\begin{bmatrix} 4 & 7 \\ 2 & 6 \end{bmatrix} \begin{bmatrix} x_1 \\ x_2 \end{bmatrix} = \begin{bmatrix} b_1 \\ b_2 \end{bmatrix}$$

solve for the following right-hand side values:

(i) $\begin{bmatrix} b_1 \\ b_2 \end{bmatrix} = \begin{bmatrix} 67 \\ 46 \end{bmatrix}$

(ii) $\begin{bmatrix} b_1 \\ b_2 \end{bmatrix} = \begin{bmatrix} 37 \\ 36 \end{bmatrix}$

(iii) $\begin{bmatrix} b_1 \\ b_2 \end{bmatrix} = \begin{bmatrix} 8 \\ -6 \end{bmatrix}$

(iv) $\begin{bmatrix} b_1 \\ b_2 \end{bmatrix} = \begin{bmatrix} 143 \\ 109 \end{bmatrix}$

(v) $\begin{bmatrix} b_1 \\ b_2 \end{bmatrix} = \begin{bmatrix} 133 \\ 114 \end{bmatrix}$

(vi) $\begin{bmatrix} b_1 \\ b_2 \end{bmatrix} = \begin{bmatrix} -230 \\ -340 \end{bmatrix}$

(vii) $\begin{bmatrix} b_1 \\ b_2 \end{bmatrix} = \begin{bmatrix} 220 \\ 160 \end{bmatrix}$

3 For the following system:

$$\begin{bmatrix} 4 & 5 \\ 2 & 3 \end{bmatrix} \begin{bmatrix} x_1 \\ x_2 \end{bmatrix} = \begin{bmatrix} b_1 \\ b_2 \end{bmatrix}$$

(i) Solve the system for $b_1 = 90$ and $b_2 = 50$.

(ii) With $b_2 = 50$, for what range of values of b_1 are both x_1 and x_2 non-negative?

(iii) With $b_1 = 90$, for what range of values of b_2 are both x_1 and x_2 non-negative?

(iv) For what range of relative sizes of b_1 and b_2 are the values of both x_1 and x_2 non-negative?

10.7 RANK

The system of equations

$$\begin{bmatrix} 2 & 3 \\ 4 & 6 \end{bmatrix} \begin{bmatrix} x_1 \\ x_2 \end{bmatrix} = \begin{bmatrix} b_1 \\ b_2 \end{bmatrix}$$

does not have a unique solution, since the matrix of coefficients

$$A = \begin{bmatrix} 2 & 3 \\ 4 & 6 \end{bmatrix}$$

is singular. But if the equations are consistent it may be possible to solve for one of the variables *in terms of the other*. For example, suppose that

$$\begin{bmatrix} 2 & 3 \\ 4 & 6 \end{bmatrix} \begin{bmatrix} x_1 \\ x_2 \end{bmatrix} = \begin{bmatrix} 12 \\ 24 \end{bmatrix}$$

Then it can be said that

$$x_2 = 4 - \tfrac{2}{3} x_1$$

But note that in the case of

$$\begin{bmatrix} 0 & 1 \\ 0 & 1 \end{bmatrix} \begin{bmatrix} x_1 \\ x_2 \end{bmatrix} = \begin{bmatrix} 1 \\ 1 \end{bmatrix}$$

then x_1/x_2 can have any value. The original system of equations is, in a sense, not as big as it seems. While the matrix A is of *order* 2×2, it is of *rank* 1. The rank of a matrix is the largest number of its rows or columns which together exhibit *linear independence*. If the rank of a square matrix is less than its order, there will not be a unique solution to the corresponding linear equations. Consider some examples involving the rank of matrices.

(i) The matrix

$$\begin{bmatrix} 2 & -1 & 4 \\ 6 & 5 & 3 \\ 8 & 4 & 7 \end{bmatrix}$$

is of *order* 3×3 but has *rank* 2. The linear dependence is in this case most easily seen in terms of the rows; the third row is the sum of the first two.

This does not mean that it is the third row in particular which is 'dependent'. *Any* one of the rows can be expressed as a linear combination of the other two (e.g. row 1 = row 3 − row 2) and can be dropped from the system without losing information. There must be the same degree of linear dependence between the columns as between the rows of a square matrix. In case (i), the linkage between the columns is less obvious, but inspection reveals that

$$\text{col } 3 = \frac{23}{16} \text{ col } 1 - \frac{18}{16} \text{ col } 2$$

If the rows (or columns) of a matrix are represented by the vectors

$$a_1, a_2, \ldots, a_m$$

and there can be found m numbers

$$w_1, w_2, \ldots, w_m$$

not all of which are zero and such that

$$w_1 a_1 + w_2 a_2 + \cdots + w_m a_m = \mathbf{0}$$

then the vectors of which the matrix is composed are linearly dependent. If no such numbers w_i exist, then the rows (or columns) are linearly independent. A collection of linearly independent vectors forms the *basis* of a set of vectors if every vector in the set is a linear combination of vectors in the collection.[12] This may seem obscure, but the idea of a basis is important in linear programming – in which context the term *basic solutions* may be familiar. The columns of coefficients of variables in a basic solution can be used to generate the columns of coefficients of variables at present zero. For example, given the three vectors

```
3  1  4
2  0  2
4  1  5
```

then the two vectors

```
3    and    1
2           0
4           1
```

form a basis for the three-vector system as a whole, since the third vector can be formed from a linear combination of the first two (it is simply their sum in this case). Of course, bases are also formed by the other possible pairings of vectors. Thus, if the vectors

```
3    and    4
2           2
4           5
```

are selected, the second vector is the difference between the third and the first. If the vectors selected are

$$\begin{matrix} 1 & & 4 \\ 0 & \text{and} & 2 \\ 1 & & 5 \end{matrix}$$

then from this basis, the first vector is obtainable by taking the difference between the third and the second vectors. Consider two further examples of matrices where the order is greater than the rank.

(ii) The matrix

$$\begin{bmatrix} 2 & 1 & 7 & 4 \\ 6 & 8 & 4 & 6 \\ -1 & 5 & 9 & 2 \\ 3 & 4 & 2 & 3 \end{bmatrix}$$

is of order 4×4 but has rank 3. In this case the last row is half the second row. Given consistency it would be possible to state three of the variables in terms of the fourth. This situation arises quite commonly in economics, where the variable used as the parameter is determined outside of the system.[13] The in-system (*endogenous*) variables can then be given specific values. Now consider the following

(iii) The matrix

$$\begin{bmatrix} 9 & 7 & 8 & 6 \\ 2 & 1 & 0 & 3 \\ 7 & 6 & 8 & 3 \\ 11 & 8 & 8 & 9 \end{bmatrix}$$

is of rank 2. In this case, any collection of more than *two* of the four rows (or columns) will show linear dependence. The relationship between the rows can be expressed as

row 3 = row 1 − row 2

row 4 = row 1 + row 2

Two of the variables could, given consistency, be expressed in terms of the other two. For a system to have a definite solution, rank must equal order, but there is no easy test to establish the rank. Without prior information, the fact that the rank of a matrix is less than its order is often found through the determinant being zero. Rank does not only apply to square matrices. For example, the matrix

$$\begin{bmatrix} 2 & 1 & 4 & 0.5 & 3 \\ 4 & 2 & 8 & 1 & 6 \end{bmatrix}$$

is of order 2 × 5 and rank 1. In terms of the rows, the second row is twice the first. This is the *row* rank. The *column* rank must be the same as the row rank, which means that in this case there is only one independent column of which all others are multiples – a fact discernible here by inspection. The matrix

$$\begin{bmatrix} 2 & 3 & 4 & 1.5 \\ -1 & 7 & -2 & 3.5 \\ 4 & 6 & 8 & 3 \end{bmatrix}$$

has order 3 × 4 but is of rank 2, there being only two linearly independent rows or columns.

EXERCISES 10.7

1 In the following systems, state the rank of the matrix of coefficients, A. What can be said about the relationship, if any, between the solution values of the variables in each case?

(i) $\begin{bmatrix} 2 & -1 \\ -1 & 0.5 \end{bmatrix} \begin{bmatrix} x_1 \\ x_2 \end{bmatrix} = \begin{bmatrix} 20 \\ -10 \end{bmatrix}$

(ii) $\begin{bmatrix} 9 & 16 \\ 5 & 10 \end{bmatrix} \begin{bmatrix} x_1 \\ x_2 \end{bmatrix} = \begin{bmatrix} 11 \\ 5 \end{bmatrix}$

(iii) $\begin{bmatrix} -6 & 12 \\ 4 & -8 \end{bmatrix} \begin{bmatrix} x_1 \\ x_2 \end{bmatrix} = \begin{bmatrix} 24 \\ -18 \end{bmatrix}$

2 What is the rank of the following matrices:

(i) $\begin{bmatrix} 1 & 2 & 3 \\ 2 & 4 & 6 \\ 4 & 8 & 12 \end{bmatrix}$

(ii) $\begin{bmatrix} 2a & a & -0.5a \\ 2b & b & -0.5b \\ 2c & c & -0.5c \end{bmatrix}$

for $a, b, c \neq 0$?

(iii) $\begin{bmatrix} a & b & c \\ 2a & 3b & 4c \\ 2a & 2b & 2c \end{bmatrix}$

for $a, b, c \neq 0$?

3 (i) Do the vectors

$$\begin{matrix} 6 \\ -2 \\ 3 \end{matrix} \quad \text{and} \quad \begin{matrix} 1 \\ 4 \\ -1 \end{matrix}$$

form a basis for

$$A = \begin{bmatrix} 6 & 1 & 7 & 5 \\ -2 & 4 & 2 & -6 \\ 3 & -1 & 2 & 4 \end{bmatrix}$$

(ii) What is the rank of A?

4 What is the rank of the matrix

$$\begin{bmatrix} 4 & 2 & -1 & 0.5 & -2 \\ -8 & -4 & 2 & -1 & 4 \end{bmatrix}$$

10.8 HIGHER ORDER SYSTEMS: DETERMINANTS AND THE INVERSE MATRIX

Applications may involve systems with anything from three to several thousand equations. Computer software is available to solve systems of linear equations and to obtain matrix inverses. While it is quicker and less error prone to use packages such as *Derive* for larger matrices than to work out the inverse by hand there is value in knowing something about larger systems. Consider for example

$$\begin{bmatrix} 3 & 2 & 9 \\ 3 & 1 & 5 \\ 2 & 1 & 4 \end{bmatrix} \begin{bmatrix} x_1 \\ x_2 \\ x_3 \end{bmatrix} = \begin{bmatrix} 49 \\ 33 \\ 26 \end{bmatrix}$$

We shall find the solution to this system using the inverse of the matrix of coefficients (A). In obtaining the inverse of A, the first step is to find the determinant.[14] There are various ways in which the determinant of a matrix may be found. Here we shall use the technique of *Laplace expansion*, which involves the evaluation of a determinant as a weighted sum of lower order determinants. To use this approach we shall need two new concepts. A *minor* is the determinant of a sub-matrix of A.[15] In particular, we shall need the minors associated with the individual elements, a_{ij}, of A. For the element a_{ij}, this is the determinant of the sub-matrix that is obtained by striking out the ith row and the jth column of A.[16] For example, the minor associated with a_{11} is m_{11} where

$$m_{11} = \begin{vmatrix} 1 & 5 \\ 1 & 4 \end{vmatrix} = -1$$

The minors associated with the elements a_{12} and a_{13} are respectively

$$m_{12} = \begin{vmatrix} 3 & 5 \\ 2 & 4 \end{vmatrix} = 2 \quad \text{and} \quad m_{13} = \begin{vmatrix} 3 & 1 \\ 2 & 1 \end{vmatrix} = 1$$

A *cofactor* is a *signed* minor. To obtain the cofactor, the minor associated with an element is multiplied by $+1$ or -1 depending on the position of the element in the matrix, the sign for element a_{ij} being given by

$$(-1)^{i+j}$$

Let the cofactors of elements be represented by c_{ij}. The cofactors then are

$$c_{ij} = (-1)^{i+j} m_{ij}$$

The point of this is the following result:

> The determinant of a matrix is the sum of products of the elements in any row or column and their associated cofactors.

So, in the case of matrix A, using the first row

$$\begin{aligned} |A| &= a_{11}c_{11} + a_{12}c_{12} + a_{13}c_{13} \\ &= a_{11}m_{11} - a_{12}m_{12} + a_{13}m_{13} \\ &= 3(-1) - 2(2) + 9(1) \end{aligned}$$

so the determinant is

$$|A| = 2$$

We can gain practice and confirm this result by expanding the determinant using another row or column. For example, using the second row elements and cofactors, the resulting calculations are

$$m_{21} = \begin{vmatrix} 2 & 9 \\ 1 & 4 \end{vmatrix} = -1 \text{ so } c_{21} = (-1)^{2+1} m_{21} = 1$$

$$m_{22} = \begin{vmatrix} 3 & 9 \\ 2 & 4 \end{vmatrix} = -6 \text{ so } c_{22} = (-1)^{2+2} m_{22} = -6$$

$$m_{23} = \begin{vmatrix} 3 & 2 \\ 2 & 1 \end{vmatrix} = -1 \text{ so } c_{23} = (-1)^{2+3} m_{23} = 1$$

So the determinant is

$$|A| = a_{21}c_{21} + a_{22}c_{22} + a_{23}c_{23}$$
$$= 3(1) + 1(-6) + 5(1) = 2$$

The same value for the determinant is obtained by use of the elements of the third row and their cofactors. Similarly, the value of the determinant could have been obtained by using the elements and cofactors in any one of the columns.[17] It is useful to know this, as the amount of work involved is reduced if there are zeros in a row or column. For example, the determinant of the matrix

$$B = \begin{bmatrix} 2 & 4 & 0 \\ 5 & 7 & 0 \\ 3 & 6 & 5 \end{bmatrix}$$

is most easily found by use of the third column:

$$|B| = 5(-1)^{3+3} \begin{vmatrix} 2 & 4 \\ 5 & 7 \end{vmatrix} = -30$$

The adjugate matrix is the transpose of the matrix of cofactors. For a matrix of order 3×3 the adjoint would be

$$\begin{bmatrix} c_{11} & c_{21} & c_{31} \\ c_{12} & c_{22} & c_{32} \\ c_{13} & c_{23} & c_{33} \end{bmatrix}$$

The inverse is the scalar product of the reciprocal of the determinant and the adjugate matrix. That is,

$$A^{-1} = \frac{1}{|A|} \begin{bmatrix} c_{11} & c_{21} & c_{31} \\ c_{12} & c_{22} & c_{32} \\ c_{13} & c_{23} & c_{33} \end{bmatrix}$$

In the present case, the only missing information is the cofactors of elements in the third row of A. These are obtained as follows:

$$m_{31} = \begin{vmatrix} 2 & 9 \\ 1 & 5 \end{vmatrix} = 1 \text{ so } c_{31} = (-1)^{3+1} m_{31} = 1$$

$$m_{32} = \begin{vmatrix} 3 & 9 \\ 3 & 5 \end{vmatrix} = -12 \text{ so } c_{32} = (-1)^{3+2} m_{32} = 12$$

$$m_{33} = \begin{vmatrix} 3 & 2 \\ 3 & 1 \end{vmatrix} = -3 \text{ so } c_{33} = (-1)^{3+3} m_{33} = -3$$

So the inverse should be

$$A^{-1} = \frac{1}{2}\begin{bmatrix} -1 & 1 & 1 \\ -2 & -6 & 12 \\ 1 & 1 & -3 \end{bmatrix} = \begin{bmatrix} -0.5 & 0.5 & 0.5 \\ -1 & -3 & 6 \\ 0.5 & 0.5 & -1.5 \end{bmatrix}$$

Confirmation is obtained by postmultiplication of A by A^{-1}:

$$\begin{bmatrix} 3 & 2 & 9 \\ 3 & 1 & 5 \\ 2 & 1 & 4 \end{bmatrix}\begin{bmatrix} -0.5 & 0.5 & 0.5 \\ -1 & -3 & 6 \\ 0.5 & 0.5 & -1.5 \end{bmatrix} = \begin{bmatrix} 1 & 0 & 0 \\ 0 & 1 & 0 \\ 0 & 0 & 1 \end{bmatrix}$$

We can now solve the original equations. Using the inverse,

$$\begin{bmatrix} x_1 \\ x_2 \\ x_3 \end{bmatrix} = \begin{bmatrix} -0.5 & 0.5 & 0.5 \\ -1 & -3 & 6 \\ 0.5 & 0.5 & -1.5 \end{bmatrix}\begin{bmatrix} 49 \\ 33 \\ 26 \end{bmatrix} = \begin{bmatrix} 5 \\ 8 \\ 2 \end{bmatrix}$$

The inverse matrix is valuable when solutions are needed for various right-hand side values. Incidentally, note that for the order 2 matrix

$$A = \begin{bmatrix} a_{11} & a_{12} \\ a_{21} & a_{22} \end{bmatrix}$$

the matrix of cofactors will be

$$\begin{bmatrix} a_{22} & -a_{21} \\ -a_{12} & a_{11} \end{bmatrix}$$

the transpose of which produces the adjoint. Determinants have a number of properties, which it is useful to know. Here are ten:

1   The determinant of a matrix is unchanged if the matrix is transposed. That is,

    $$|A| = |A^T|$$

2   Switching two rows or two columns of a matrix leaves the absolute value of the determinant unchanged, but changes its sign.
3   If any row (column) of a matrix is a weighted average of other rows (columns) the determinant is zero.
4   Multiplication of all elements in a row or column of a matrix by a constant multiplies the determinant by that constant.
5   If to any row of a matrix is added a multiple of any other row, the determinant is unchanged – so also for columns.
6   The determinant of an upper- or lower-triangular matrix is the product of the elements on the main diagonal.

7 If A is of order n, then the determinant of the scalar product of the matrix is the determinant of A multiplied by the scalar raised to the nth power:[18]

$$|kA| = k^n |A|$$

8 For square matrices of the same order, the determinant of the product of the matrices is the product of the individual determinants. That is,

$$|AB| = |A| \, |B|$$

9 Where A is invertible:

$$|A^{-1}| = (|A|)^{-1}$$

10 If the elements of any row (column) are multiplied by the cofactors of any other row[19] (column), the result is zero.

Properties 1–10 are useful in operations involving matrices and their associated determinants.

EXERCISES 10.8

1 Given the matrix

$$A = \begin{bmatrix} 7 & 5 & 9 \\ 3 & 8 & 4 \\ 6 & 2 & 1 \end{bmatrix}$$

(i) Find the minors of the elements

(a) a_{33} (b) a_{11} (c) a_{22}
(d) a_{32} (e) a_{21}

(ii) Find the cofactors of the minors in (i).

2 (i) Find the determinant of the matrix

$$A = \begin{bmatrix} 2 & 4 & 3 \\ 4 & 5 & 1 \\ 9 & 2 & 4 \end{bmatrix}$$

using the elements and cofactors of the first row.

(ii) Confirm the outcome of (i) by expanding the determinant by the third column.

3 Find the determinants of the following matrices:

(i)
$$A = \begin{bmatrix} 7 & 9 & 8 \\ 1 & 0 & 2 \\ 5 & 3 & 4 \end{bmatrix}$$

(ii)
$$A = \begin{bmatrix} 4 & 3 & 2 \\ 8 & 5 & 8 \\ 0 & -1 & 4 \end{bmatrix}$$

4 Find the inverse of the matrix
$$A = \begin{bmatrix} 1 & 1 & 2 \\ 5 & 4 & 9 \\ 3 & 2 & 3 \end{bmatrix}$$

5 Find the inverse of
$$A = \begin{bmatrix} 7 & -3 & 2 \\ 0 & 1 & 0 \\ 5 & 1 & 1.5 \end{bmatrix}$$

10.9 SIMULTANEOUS EQUATIONS (ii)

In this section the work of the preceding section is applied and the use of determinants and inverse matrices in the solution of linear equations is further demonstrated. For a numerical example, we shall use the inverse matrix approach to solve the system

$$4x_1 + 7x_3 = b_1$$
$$7x_1 + 3x_2 + 5x_3 = b_2$$
$$4x_1 + 2x_2 + 3x_3 = b_3$$

for the following sets of right-hand side values:

$$\begin{bmatrix} b_1 \\ b_2 \\ b_3 \end{bmatrix} = \begin{bmatrix} 47 \\ 58 \\ 35 \end{bmatrix}, \quad \begin{bmatrix} 34 \\ 19 \\ 12 \end{bmatrix} \quad \text{and} \quad \begin{bmatrix} 14 \\ -2 \\ -2 \end{bmatrix}$$

The matrix of coefficients is

$$A = \begin{bmatrix} 4 & 0 & 7 \\ 7 & 3 & 5 \\ 4 & 2 & 3 \end{bmatrix}$$

There being no obvious dependences between the rows of A in this case, we first attempt to find the determinant. The first row, which contains a zero, is the most convenient by which to expand. The determinant will be

$$|A| = 4c_{11} + 0c_{12} + 7c_{13}$$

The minors and cofactors are

$$m_{11} = \begin{vmatrix} 3 & 5 \\ 2 & 3 \end{vmatrix} = -1 \text{ so } c_{11} = (-1)^{1+1}m_{11} = -1$$

$$m_{12} = \begin{vmatrix} 7 & 5 \\ 4 & 3 \end{vmatrix} = 1 \text{ so } c_{12} = (-1)^{1+2}m_{12} = -1$$

Although not needed for the determinant, if $|A|$ is not zero, we shall need c_{12} later on.

$$m_{13} = \begin{vmatrix} 7 & 3 \\ 4 & 2 \end{vmatrix} = 2 \text{ so } c_{13} = (-1)^{1+3}m_{13} = 2$$

so that the determinant is given by

$$|A| = a_{11}c_{11} + a_{12}c_{12} + a_{13}c_{13}$$
$$= 4(-1) + 0(-1) + 7(2) = 10$$

The matrix is non-singular and each vector of right-hand side values gives a unique solution. Proceeding towards the inverse, the remaining cofactors are found as follows. For the second row

$$m_{21} = \begin{vmatrix} 0 & 7 \\ 2 & 3 \end{vmatrix} = -14 \text{ so } c_{21} = (-1)^{2+1}m_{21} = 14$$

$$m_{22} = \begin{vmatrix} 4 & 7 \\ 4 & 3 \end{vmatrix} = -16 \text{ so } c_{22} = (-1)^{2+2}m_{22} = -16$$

$$m_{23} = \begin{vmatrix} 4 & 0 \\ 4 & 2 \end{vmatrix} = 8 \text{ so } c_{23} = (-1)^{2+3}m_{23} = -8$$

and for the third row

$$m_{31} = \begin{vmatrix} 0 & 7 \\ 3 & 5 \end{vmatrix} = -21 \text{ so } c_{31} = (-1)^{3+1}m_{31} = -21$$

$$m_{32} = \begin{vmatrix} 4 & 7 \\ 7 & 5 \end{vmatrix} = -29 \text{ so } c_{32} = (-1)^{3+2}m_{32} = 29$$

$$m_{33} = \begin{vmatrix} 4 & 0 \\ 7 & 3 \end{vmatrix} = 12 \text{ so } c_{33} = (-1)^{3+3}m_{33} = 12$$

so the inverse is

$$A^{-1} = \frac{1}{10} \begin{bmatrix} -1 & 14 & -21 \\ -1 & -16 & 29 \\ 2 & -8 & 12 \end{bmatrix} = \begin{bmatrix} -0.1 & 1.4 & -2.1 \\ -0.1 & -1.6 & 2.9 \\ 0.2 & -0.8 & 1.2 \end{bmatrix}$$

The result can be checked by premultiplication or postmultiplication of A by A^{-1}. In the latter case

$$\begin{bmatrix} 4 & 0 & 7 \\ 7 & 3 & 5 \\ 4 & 2 & 3 \end{bmatrix} \begin{bmatrix} -0.1 & 1.4 & -2.1 \\ -0.1 & -1.6 & 2.9 \\ 0.2 & -0.8 & 1.2 \end{bmatrix} = \begin{bmatrix} 1 & 0 & 0 \\ 0 & 1 & 0 \\ 0 & 0 & 1 \end{bmatrix}$$

We can now use the inverse to solve the system of linear equations for the three sets of right-hand side values. For the first,

$$\begin{bmatrix} x_1 \\ x_2 \\ x_3 \end{bmatrix} = \begin{bmatrix} -0.1 & 1.4 & -2.1 \\ -0.1 & -1.6 & 2.9 \\ 0.2 & -0.8 & 1.2 \end{bmatrix} \begin{bmatrix} 47 \\ 58 \\ 35 \end{bmatrix} = \begin{bmatrix} 3 \\ 4 \\ 5 \end{bmatrix}$$

For the second set of right-hand side values, the solution will be

$$\begin{bmatrix} x_1 \\ x_2 \\ x_3 \end{bmatrix} = \begin{bmatrix} -0.1 & 1.4 & -2.1 \\ -0.1 & -1.6 & 2.9 \\ 0.2 & -0.8 & 1.2 \end{bmatrix} \begin{bmatrix} 34 \\ 19 \\ 12 \end{bmatrix} = \begin{bmatrix} -2 \\ 1 \\ 6 \end{bmatrix}$$

while for the third group of b values, the resulting x_i are

$$\begin{bmatrix} x_1 \\ x_2 \\ x_3 \end{bmatrix} = \begin{bmatrix} -0.1 & 1.4 & -2.1 \\ -0.1 & -1.6 & 2.9 \\ 0.2 & -0.8 & 1.2 \end{bmatrix} \begin{bmatrix} 14 \\ -2 \\ -2 \end{bmatrix} = \begin{bmatrix} 0 \\ -4 \\ 2 \end{bmatrix}$$

The inverse matrix is a powerful instrument for assessing the impact of changing circumstances. The right-hand side values may be resource availabilities, and the x_i may represent production levels. If the current levels of the resources effectively limit the achievement of the company's objective, then it will be someone's job to increase availability. In many cases, the model used will generate information as to the effect on profit or other objectives of unit changes in resource levels. These are the *shadow prices* of Chapter 3. But the shadow prices apply only so long as the variables in the current solution remain at feasible (usually non-negative) values. So it will be important to know how wide a variation in the b_i the current solution can tolerate.[20] Consider the first set of right-hand side values in the example above, and b_1 in particular. The range of values of b_1 for which all the x_i are non-negative satisfies

$$\begin{bmatrix} x_1 \\ x_2 \\ x_3 \end{bmatrix} = \begin{bmatrix} -0.1 & 1.4 & -2.1 \\ -0.1 & -1.6 & 2.9 \\ 0.2 & -0.8 & 1.2 \end{bmatrix} \begin{bmatrix} b_1 \\ 58 \\ 35 \end{bmatrix} \geq \begin{bmatrix} 0 \\ 0 \\ 0 \end{bmatrix}$$

The requirement on b_1 for the non-negativity of x_1 is found by multiplying the first row of the inverse by the right-hand side vector including the unknown b_1. This produces the following requirement. From x_1,

$$-0.1b_1 + 1.4(58) - 2.1(35) \geq 0$$

i.e.

$$-0.1b_1 + 81.2 - 73.5 \geq 0$$

which implies that $b_1 \leq 77$. From x_2,

$$-0.1b_1 - 1.6(58) + 2.9(35) \geq 0$$

i.e.

$$-0.1b_1 - 92.8 + 101.5 \geq 0$$

which implies that $b_1 \leq 87$. From x_3,

$$0.2b_1 - 0.8(58) + 1.2(35) \geq 0$$

i.e.

$$0.2b_1 - 46.4 + 42 \geq 0$$

which implies that $b_1 \geq 22$. Of the two upper bounds, 77 is the least and so gives the tightest restriction, and in conjunction with the single lower bound the overall requirement on b_1 for sign feasibility of all variables is that

$$22 \leq b_1 \leq 77 \tag{10.9}$$

The range given by (10.9) is the *tolerance interval* or *range of feasibility* for the parameter b_1. The tolerance interval for x_2 can be obtained by similar means from the second row of the inverse into the right-hand side vector in which b_2 is substituted for the original specific value of 58. The resulting tolerance interval is

$$55.86 \leq b_2 \leq 60.5$$

Note that this is a much narrower range than for b_1. The feasibility of the solution is more sensitive to variations in b_2 than b_1. This is likely to be an important concern for management if b_2 is *not* a control variable – being, perhaps, the subject of random variation, external regulation or the actions of competitors. For the right-hand side element b_3, the tolerance interval works out as

$$33.62 \leq b_3 \leq 36.43$$

So here again the permissible range of variation of the resource level is much narrower than for b_1.

EXERCISES 10.9

1 Solve the system

$$\begin{bmatrix} 2 & 1 & 0 \\ 3 & 2 & 4 \\ 0.5 & 1 & 5 \end{bmatrix} \begin{bmatrix} x_1 \\ x_2 \\ x_3 \end{bmatrix} = \begin{bmatrix} 35 \\ 140 \\ 120 \end{bmatrix}$$

2 Given the system

$$\begin{bmatrix} 1 & 1 & 3 \\ 4 & 7 & 2 \\ 3 & 5 & 4 \end{bmatrix} \begin{bmatrix} x_1 \\ x_2 \\ x_3 \end{bmatrix} = \begin{bmatrix} b_1 \\ b_2 \\ b_3 \end{bmatrix}$$

solve for the values of x_i when

(i) $\begin{bmatrix} b_1 \\ b_2 \\ b_3 \end{bmatrix} = \begin{bmatrix} 85 \\ 185 \\ 185 \end{bmatrix}$

(ii) $\begin{bmatrix} b_1 \\ b_2 \\ b_3 \end{bmatrix} = \begin{bmatrix} 70 \\ 50 \\ 90 \end{bmatrix}$

(iii) $\begin{bmatrix} b_1 \\ b_2 \\ b_3 \end{bmatrix} = \begin{bmatrix} -4 \\ 139 \\ 78 \end{bmatrix}$

(iv) $\begin{bmatrix} b_1 \\ b_2 \\ b_3 \end{bmatrix} = \begin{bmatrix} 9 \\ 63 \\ 45 \end{bmatrix}$

3 For the data of exercise 1(i), find the ranges of variation, for each right-hand side element individually, for which x_1, x_2 and x_3 are all non-negative.

10.10 CRAMER'S RULE

Another way of finding the solution to a system of linear equations is by use of *Cramer's rule*.[21] In Cramer's rule the solution values of the x_i are found as the ratios of determinants. The method proceeds as follows. For the system

$$\begin{bmatrix} a_{11} & a_{12} & a_{13} \\ a_{21} & a_{22} & a_{23} \\ a_{31} & a_{32} & a_{33} \end{bmatrix} \begin{bmatrix} x_1 \\ x_2 \\ x_3 \end{bmatrix} = \begin{bmatrix} b_1 \\ b_2 \\ b_3 \end{bmatrix}$$

the values of the x_i are given by

$$x_1 = \frac{\begin{vmatrix} b_1 & a_{12} & a_{13} \\ b_2 & a_{22} & a_{23} \\ b_3 & a_{32} & a_{33} \end{vmatrix}}{|A|}$$

$$x_2 = \frac{\begin{vmatrix} a_{11} & b_1 & a_{13} \\ a_{21} & b_2 & a_{23} \\ a_{31} & b_3 & a_{33} \end{vmatrix}}{|A|}$$

$$x_3 = \frac{\begin{vmatrix} a_{11} & a_{12} & b_1 \\ a_{21} & a_{22} & b_2 \\ a_{31} & a_{32} & b_3 \end{vmatrix}}{|A|}$$

In each case the numerator is the determinant obtained by *replacing the column of coefficients of the variable to be obtained, by the vector of right-hand side values*. The denominator in each ratio is the determinant of A itself. Now apply Cramer's rule to the system of Section 10.9. We have already found that $|A| = 10$. Using the first set of right-hand side values, the value of x_1 is given by Cramer's rule as

$$x_1 = \frac{\begin{vmatrix} 47 & 0 & 7 \\ 58 & 3 & 5 \\ 35 & 2 & 3 \end{vmatrix}}{10}$$

Expanding the numerator by the second column, this determinant works out as

$$3\begin{vmatrix} 47 & 7 \\ 35 & 3 \end{vmatrix} + -2\begin{vmatrix} 47 & 7 \\ 58 & 5 \end{vmatrix}$$
$$= 3(-104) - 2(-171) = 30$$

so that

$$x_1 = \tfrac{30}{10} = 3$$

The value of x_2 will be

$$x_2 = \frac{\begin{vmatrix} 4 & 47 & 7 \\ 7 & 58 & 5 \\ 4 & 35 & 3 \end{vmatrix}}{10}$$

Again expanding by the second column, the numerator determinant is

$$-47\begin{vmatrix} 7 & 5 \\ 4 & 3 \end{vmatrix} + 58\begin{vmatrix} 4 & 7 \\ 4 & 3 \end{vmatrix} - 35\begin{vmatrix} 4 & 7 \\ 7 & 5 \end{vmatrix}$$
$$= -47(1) + 58(-16) - 35(-29) = 40$$

so that

$$x_2 = 4$$

And the value of x_3 is

$$x_3 = \frac{\begin{vmatrix} 4 & 0 & 47 \\ 7 & 3 & 58 \\ 4 & 2 & 35 \end{vmatrix}}{10}$$

Using the second column by which to expand, the determinant works out as

$$3\begin{vmatrix} 4 & 47 \\ 4 & 35 \end{vmatrix} - 2\begin{vmatrix} 4 & 47 \\ 7 & 58 \end{vmatrix}$$
$$= 3(-48) - 2(-97) = 50$$

Therefore

$$x_3 = 5$$

and the solution to the system is

$$x_1 = 3 \qquad x_2 = 4 \qquad x_3 = 5$$

For an example of the use of Cramer's rule in a 2×2 case consider

$$\begin{bmatrix} 5 & 1 \\ 3 & 2 \end{bmatrix}\begin{bmatrix} x_1 \\ x_2 \end{bmatrix} = \begin{bmatrix} 20 \\ 12 \end{bmatrix}$$

for which the determinant of the coefficient matrix is

$$\begin{vmatrix} 5 & 1 \\ 3 & 2 \end{vmatrix} = 7$$

Using Cramer's rule:

$$x_1 = \frac{\begin{vmatrix} 20 & 1 \\ 12 & 2 \end{vmatrix}}{7}$$
$$= 28/7 = 4$$

and
$$x_2 = \frac{\begin{vmatrix} 5 & 20 \\ 3 & 12 \end{vmatrix}}{7}$$
$$= 0/7 = 0$$

So the solution is
$$x_1 = 7 \quad \text{and} \quad x_2 = 0$$

Cramer's rule works because for each variable individually the rule is equivalent to use of the inverse matrix but, significantly, *does not require the complete inverse to find the values of particular variables*. This can be advantageous. Expanding the ratio of determinants given by the rule gives the equivalent result for each variable as the vector of right-hand side elements premultiplied by the inverse of the coefficient matrix. This is most easily seen in the 2×2 case. Let

$$\begin{bmatrix} a_{11} & a_{12} \\ a_{21} & a_{22} \end{bmatrix} \begin{bmatrix} x_1 \\ x_2 \end{bmatrix} = \begin{bmatrix} b_1 \\ b_2 \end{bmatrix}$$

So by the inverse matrix method

$$\begin{bmatrix} x_1 \\ x_2 \end{bmatrix} = \begin{bmatrix} a_{11} & a_{12} \\ a_{21} & a_{22} \end{bmatrix}^{-1} \begin{bmatrix} b_1 \\ b_2 \end{bmatrix}$$

$$= \frac{1}{|A|} \begin{bmatrix} a_{22} & -a_{12} \\ -a_{21} & a_{11} \end{bmatrix} \begin{bmatrix} b_1 \\ b_2 \end{bmatrix}$$

$$= \frac{1}{|A|} \begin{bmatrix} a_{22}b_1 & -a_{12}b_2 \\ -a_{21}b_2 & +a_{11}b_2 \end{bmatrix}$$

By Cramer's rule,

$$x_1 = \frac{\begin{vmatrix} b_1 & a_{12} \\ b_2 & a_{22} \end{vmatrix}}{|A|}$$

$$= \frac{a_{22}b_1 - a_{12}b_2}{|A|}$$

and x_2 is given by

$$x_2 = \frac{\begin{vmatrix} a_{11} & b_1 \\ a_{21} & b_2 \end{vmatrix}}{|A|}$$

$$= \frac{a_{11}b_2 - a_{21}b_1}{|A|}$$

which is, in two parts, the result obtained by use of the matrix inverse. Cramer's rule is a useful means of solving systems of simultaneous linear equations. Because the rule finds solution values for the variables individually, it can be an attractive method when the values of all unknowns are not required.

EXERCISES 10.10

1 Use Cramer's rule to solve the following linear equations:

(i)
$$\begin{bmatrix} 9 & 1 \\ 6 & 2 \end{bmatrix} \begin{bmatrix} x_1 \\ x_2 \end{bmatrix} = \begin{bmatrix} 53 \\ 46 \end{bmatrix}$$

(ii)
$$\begin{bmatrix} 4 & 5 \\ 1 & 3 \end{bmatrix} \begin{bmatrix} x_1 \\ x_2 \end{bmatrix} = \begin{bmatrix} 26 \\ 3 \end{bmatrix}$$

2 Solve the following system by Cramer's rule:

$$\begin{bmatrix} 2 & 3 & 1 \\ 4 & 1 & 5 \\ 2 & 6 & 3 \end{bmatrix} \begin{bmatrix} x_1 \\ x_2 \\ x_3 \end{bmatrix} = \begin{bmatrix} 16 \\ 34 \\ 30 \end{bmatrix}$$

3 Use Cramer's rule to find the value of x_2 satisfying the simultaneous equations

$$x_1 - 22x_2 + 7x_3 = 107$$
$$x_1 + 29x_2 + 8x_3 = 61$$
$$-2x_1 + 12x_2 + 4x_3 = -92$$

10.11 HOMOGENEOUS SYSTEMS

We have concentrated on linear systems with the same number of equations as unknowns. Regardless of the size of such systems, the number of solutions will be 1, 0 or ∞. In the last case, there are fewer *independent* equations than unknowns, but such systems will differ in respect of the number of variables that can be assigned arbitrary values.[22] In the *n*-equation, *n*-unknown system written in matrix form as

$$Ax = b$$

in which $b \neq 0$, there will be a unique solution to the system *if and only if*

$$|A| \neq 0$$

Where the elements of **b** are all zero, the system is said to be *homogeneous*. In general, a homogeneous system of m linear equations in n unknowns takes the form

$$a_{11}x_1 + a_{12}x_2 + \cdots + a_{1n}x_n = 0$$
$$a_{21}x_1 + a_{22}x_2 + \cdots + a_{2n}x_n = 0$$
$$a_{31}x_1 + a_{32}x_2 + \cdots + a_{3n}x_n = 0$$
$$\vdots \qquad \vdots \qquad \vdots$$
$$a_{m1}x_1 + a_{m2}x_2 + \cdots + a_{mn}x_n = 0$$

In matrix form this is written as

$$Ax = 0$$

If the vector x is a solution of a homogeneous system, then so is any scalar multiple of x, kx. One solution of a homogeneous system is always

$$x = 0$$

which is the *trivial solution* to the system. The interesting question is whether the system possesses any *non-trivial* solutions. It can be shown that, if $m = n$, the system will have a non-trivial solution *only if* $|A| = 0$. Consider some examples. The following is a two-equation, two-unknown, homogeneous system where the A matrix is non-singular:

$$10x + 2y = 0$$
$$5x + 2y = 0$$

There will therefore be no non-trivial solutions to this system. The first of the equations can be restated as

$$y = -5x$$

while the second translates to

$$y = -2.5x$$

With no constant terms, the two equations give lines going through the origin as shown in Figure 10.2(a).

The lines or planes corresponding to homogeneous systems pass through the origin. Regarding non-trivial solutions, the question is whether the intersection of the lines or surfaces is a single point (trivial solution) or a continuum of points (non-trivial solutions). Consider a homogeneous system where the coefficient matrix is singular:

$$10x + 2y = 0$$
$$5x + y = 0$$

With $|A| = 0$, there *will* be non-trivial solutions to this system, since both equations have the implication that $y = -5x$. The two lines corresponding

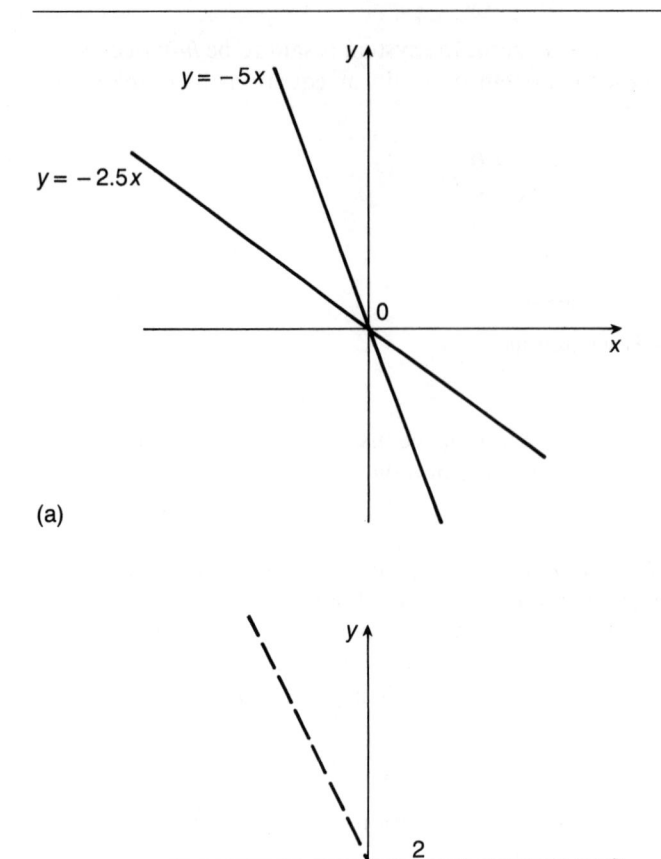

(a)

(b)

Figure 10.2

to the equations are identical and are as shown in Figure 10.2(b). One example of the continuum of solutions in this case is

$$x = 2 \quad y = -10$$

while another is

$$x = -10 \quad y = 50$$

As an example of a homogeneous system where the number of unknowns exceeds the number of equations, consider the two-equation, three-unknown case

$$10x + 4y + 2z = 0$$
$$5x + y - z = 0$$

From the second equation, the relationship of z to x and y can be stated explicitly as

$$z = 5x + y$$

Substitution into the first equation produces

$$10x + 4y + 10x + 2y = 0$$

so that

$$20x + 6y = 0$$

for which there is again a continuum of non-trivial solutions, one example of which is

$$x = 1 \qquad y = -\frac{10}{3}$$

while another is

$$x = -3 \qquad y = 10$$

EXERCISES 10.11

1 Which of the following homogeneous systems *do not* have non-trivial solutions?

(i) $\begin{bmatrix} -6 & 8 \\ -3 & 4 \end{bmatrix} \begin{bmatrix} x_1 \\ x_2 \end{bmatrix} = \begin{bmatrix} 0 \\ 0 \end{bmatrix}$

(ii) $\begin{bmatrix} 2 & 5 \\ -6 & 15 \end{bmatrix} \begin{bmatrix} x_1 \\ x_2 \end{bmatrix} = \begin{bmatrix} 0 \\ 0 \end{bmatrix}$

(iii) $\begin{bmatrix} 1 & 0 \\ 0 & 1 \end{bmatrix} \begin{bmatrix} x_1 \\ x_2 \end{bmatrix} = \begin{bmatrix} 0 \\ 0 \end{bmatrix}$

(iv) $\begin{bmatrix} 1 & 1 \\ 1 & 1 \end{bmatrix} \begin{bmatrix} x_1 \\ x_2 \end{bmatrix} = \begin{bmatrix} 0 \\ 0 \end{bmatrix}$

2 What relationship between x_1 and x_2 must be satisfied by non-trivial solutions to the following systems?

(i) $\begin{bmatrix} 6 & -4 \\ -3 & 2 \end{bmatrix} \begin{bmatrix} x_1 \\ x_2 \end{bmatrix} = \begin{bmatrix} 0 \\ 0 \end{bmatrix}$

(ii) $8x_1 - 4x_2 + 2x_3 = 0$
$4x_1 + x_2 - x_3 = 0$

(iii) $9x_1 - 7x_2 + 3x_3 = 0$
$2x_1 + 2x_2 - x_3 = 0$

3 In the systems 2(ii) and 2(iii) above:
   (i) What is x_3 in the non-trivial solutions when $x_1 = 5$?
   (ii) In case 2(iii), is a value of $x_3 = -8$ consistent with a non-trivial solution to the system?

10.12 CONCLUDING REMARKS

In this brief introduction to matrices, the primary focus has been on systems of simultaneous linear equations. In this chapter and in Chapter 2 we have discussed other methods for solving such systems. No single method is the most appropriate for all applications, but there are circumstances in which each technique is likely to work to advantage. The less systematized approaches, such as elimination or the use of substitution as described in Chapter 2, are a quick way to solve 2×2 systems. They can also be practicable in 3×3 cases where a convenient feature of the system can be exploited. But informal approaches are not recommended for larger systems. The use of determinants, as in Cramer's rule, has advantages if the values of only some of the unknowns are required. Gauss–Jordan elimination, or a variant making use of elementary row operations, is highly efficient if a one-off, complete solution is required. If a solution is required for more than one set of right-hand side values, or if tolerance intervals for right-hand side values are needed, then the inverse matrix is valuable.

The fact that the number of arithmetic operations needed to find the solution to a system of n equations in n unknowns varies approximately with the *cube* of n reveals the laborious calculations required in solving larger systems. In comparison with a system of order 3, over twice as much work is involved in solving a system of order 4, and about five times as much work is required to solve a system of order 5. So it is as well that there are software packages (such as *Derive*) for evaluating determinants, finding inverse matrices and solving systems of linear simultaneous equations. They are well worth investigating if you expect to be doing much work in this area.

ADDITIONAL PROBLEMS

1 Find x and y where

(i) $\begin{bmatrix} 2 & x & -3 \\ y & 4 & 2 \end{bmatrix} - \begin{bmatrix} -5 & 2 & -4 \\ -3 & 1 & 2 \end{bmatrix} = \begin{bmatrix} 7 & 7 & 1 \\ 9 & 3 & 0 \end{bmatrix}$

(ii) $\begin{bmatrix} x & 4 \\ 2 & y \end{bmatrix} \begin{bmatrix} 3 \\ -1 \end{bmatrix} = \begin{bmatrix} 11 \\ 10 \end{bmatrix}$

2 Find the product

$$\overset{A}{\begin{bmatrix} 5 & 2 \\ 3 & 1 \\ 4 & 2 \end{bmatrix}} \overset{B}{\begin{bmatrix} 10 \\ -2 \end{bmatrix}} \overset{C}{[6 \quad 3]}$$

3 Find the determinant of the following matrices:

(i) $\begin{bmatrix} 3 & 5 \\ 7 & 6 \end{bmatrix}$

(ii) $\begin{bmatrix} 2 & 1 & 3 \\ 4 & 3 & 5 \\ 6 & -2 & 7 \end{bmatrix}$

(iii) $\begin{bmatrix} 4 & -6 & 8 \\ 7 & 5 & 2 \\ -2 & 3 & -4 \end{bmatrix}$

4 Find the inverse of the following matrices:

(i) $\begin{bmatrix} 6 & 8 \\ 5 & 10 \end{bmatrix}$

(ii) $\begin{bmatrix} 2 & 1 & 3 \\ 3 & -1 & 3 \\ 4 & 3 & 5 \end{bmatrix}$

5 Solve the following systems using the inverse matrix:

(i) $\begin{bmatrix} 3 & 4 \\ 1 & 2 \end{bmatrix} \begin{bmatrix} x_1 \\ x_2 \end{bmatrix} = \begin{bmatrix} 5 \\ 3 \end{bmatrix}$

(ii) $\begin{bmatrix} 2 & 1 & 3 \\ 3 & -1 & 3 \\ 4 & 3 & 5 \end{bmatrix} \begin{bmatrix} x_1 \\ x_2 \\ x_3 \end{bmatrix} = \begin{bmatrix} 100 \\ 200 \\ 300 \end{bmatrix}$

6 Use Cramer's rule to find:

   (i) the values of x_1 and x_2 in the system

   $$\begin{bmatrix} 5 & 2 \\ 10 & 5 \end{bmatrix} \begin{bmatrix} x_1 \\ x_2 \end{bmatrix} = \begin{bmatrix} 160 \\ 350 \end{bmatrix}$$

   (ii) the value of x_3 in the system

   $$\begin{bmatrix} 3 & -2 & 3 \\ 4 & 6 & 5 \\ 4 & 4 & 6 \end{bmatrix} \begin{bmatrix} x_1 \\ x_2 \\ x_3 \end{bmatrix} = \begin{bmatrix} 13 \\ 29 \\ 30 \end{bmatrix}$$

7 What can be said about the existence of non-trivial solutions to the homogeneous systems below? Where a non-trivial solution does exist, give the values of x_2 and x_3 corresponding to $x_1 = 4$.

   (i) $\begin{bmatrix} 2 & 0 & -1 \\ 3 & 4 & 2 \\ 6 & 8 & 4 \end{bmatrix} \begin{bmatrix} x_1 \\ x_2 \\ x_3 \end{bmatrix} = \begin{bmatrix} 0 \\ 0 \\ 0 \end{bmatrix}$

   (ii) $\begin{bmatrix} 5 & 1 & 0 \\ 2 & 4 & 7 \\ 3 & 2 & 2 \end{bmatrix} \begin{bmatrix} x_1 \\ x_2 \\ x_3 \end{bmatrix} = \begin{bmatrix} 0 \\ 0 \\ 0 \end{bmatrix}$

8 Given

   $$A = \begin{bmatrix} 2 & 4 & 3 \\ 6 & -5 & 4 \\ 3 & 1 & 6 \end{bmatrix}$$

   find the value of the determinant using signed products.

REFERENCES AND FURTHER READING

1 Borowski, E. J. and Borwein, J. M. (1989) *Dictionary of Mathematics*, Collins.
2 Chiang, A. C. (1984) *Fundamental Methods of Mathematical Economics* (Third Edition), McGraw-Hill.
3 Finney, R. L. and Thomas, G. B., Jr (1990) *Calculus*, Addison-Wesley.
4 Mizrahi, A. and Sullivan, M. (1988) *Mathematics for Business and Social Sciences* (Fourth Edition), Wiley.
5 Rich, A., Rich, J. and Stoutemyer, D. (1989) *Derive: User Manual* (Third Edition), Soft Warehouse.
6 Weber, J. E. (1982) *Mathematical Analysis: Business and Economic Applications* (Fourth Edition), Harper and Row.
7 Wisniewski, M. (1991) *Introductory Mathematical Methods in Economics*, McGraw-Hill.

SOLUTIONS TO EXERCISES

Exercises 10.2

1 A is 2×2 B is 1×4 C is 5×1 D is 1×1 E is $t \times s$

2 B, C and D are sub-matrices of A. The matrix E, however, cannot be obtained by deleting rows and columns of A and is therefore not a sub-matrix of A.

3 The trace, T, is the sum of the elements of the main diagonal of the matrix, i.e.
$$T = 10 + 27 + 35 + 29 = 101$$

4 A is upper-triangular; B is a scalar matrix; C is lower-triangular; D is a diagonal matrix; E does not fall within any of the categories; F is upper-triangular.

5 (i) $x = 0$ $y = 3$ $z = 1$ (ii) $m = 2$ $n = 3$

6
$$A^T = \begin{bmatrix} 4 & -3 \\ -1 & 8 \end{bmatrix} \qquad C^T = [-8 \quad 0 \quad \sqrt{2} \quad -1 \quad 4]$$

$$B^T = \begin{bmatrix} 2 \\ \pi \\ 3 \\ -1 \\ 0 \end{bmatrix} \qquad D^T = [100]$$

7 A is not symmetric; B is skew-symmetric; C is symmetric; D is symmetric.

Exercises 10.3

1 (i) $\begin{bmatrix} -1 & 9 & 6 \\ 8 & 2 & 2 \\ 6 & 3 & 4 \end{bmatrix}$

(ii) $\begin{bmatrix} 5 & 20 \\ 15 & 18 \end{bmatrix}$

(iii) $\begin{bmatrix} 4-a & 6+2a & a-b \\ a+9 & b-c & 30 \\ 11 & 0 & 1+c^2 \end{bmatrix}$

2 From the elements in the (1, 1) positions
$$5x - 4y = z$$
and from those in the (1, 2) positions
$$6 - z = 4$$

so that

$$z = 2$$

From the elements in the (2, 1) positions

$$4 + 2x = 24$$

so that

$$x = 10$$

Hence

$$50 - 4y = 2$$

gives

$$y = 12$$

which value is confirmed by the elements in the (2, 2) positions in the matrix.

3
$$A + B = \begin{bmatrix} 5 & 3 \\ 2 & -1 \end{bmatrix}$$

so that

(i) when $k = 2$,

$$k(A + B) = \begin{bmatrix} 10 & 6 \\ 4 & -2 \end{bmatrix}$$

(ii) when $k = -1$,

$$k(A + B) = \begin{bmatrix} -5 & -3 \\ -2 & 1 \end{bmatrix}$$

Exercises 10.4

1 (i) [19 29]

(ii) $\begin{bmatrix} 13 & 22 \\ 16 & 19 \end{bmatrix}$

(iii) $\begin{bmatrix} 12 & 22 \\ 13 & -3 \\ 7 & -7 \end{bmatrix}$

(iv) $\begin{bmatrix} 15 & -2 & 36 \\ -11 & -20 & 15 \\ 29 & 26 & 12 \end{bmatrix}$

2 (i)
$$A^2 = \begin{bmatrix} 3 & 5 \\ -5 & 8 \end{bmatrix}$$

(ii)
$$A^3 = \begin{bmatrix} 1 & 18 \\ -18 & 19 \end{bmatrix}$$

(iii)
$$A^4 = \begin{bmatrix} -16 & 55 \\ -55 & 39 \end{bmatrix}$$

3 (i) In this case
$$A^2 = \begin{bmatrix} 2 & 0 \\ 0 & 2 \end{bmatrix}$$

so that $A^2 \neq A$ and the matrix A is therefore *not* idempotent.

(ii) In this case
$$A^2 = \begin{bmatrix} 8 & 7 \\ -8 & -7 \end{bmatrix}$$

so that $A^2 = A$ and the matrix A therefore *is* idempotent.

4 With A *pre*multiplying B, the result is
$$AB = \begin{bmatrix} 22 & -26 \\ 26 & 22 \end{bmatrix}$$

With A *post*multiplying B, the result is
$$BA = \begin{bmatrix} 22 & -26 \\ 26 & 22 \end{bmatrix}$$

so that the matrices A and B are commutative.

5 (i)
$$\begin{bmatrix} 4 & 3 \end{bmatrix} \begin{bmatrix} 2 & 5 & -1 \\ -2 & 6 & 4 \end{bmatrix} \begin{bmatrix} 4 \\ 3 \\ 2 \end{bmatrix}$$

$$= \begin{bmatrix} 2 & 38 & 8 \end{bmatrix} \begin{bmatrix} 4 \\ 3 \\ 2 \end{bmatrix}$$

$$= [138]$$

(ii) $\begin{bmatrix} 3 & 2 \\ 1 & 3 \\ 4 & -1 \end{bmatrix} \begin{bmatrix} 4 \\ 5 \end{bmatrix} [3 \quad 6]$

$= \begin{bmatrix} 22 \\ 19 \\ 11 \end{bmatrix} [3 \quad 6]$

$= \begin{bmatrix} 66 & 132 \\ 57 & 114 \\ 33 & 66 \end{bmatrix}$

(iii) $\begin{bmatrix} 3 & 2 & 0 \\ 1 & 2 & -1 \\ 2 & 5 & 1 \end{bmatrix} \begin{bmatrix} 1 & 3 \\ 2 & 1 \\ 0 & 4 \end{bmatrix} \begin{bmatrix} 2 & 1 & 0 & 1 \\ 2 & 3 & 5 & -1 \end{bmatrix}$

$= \begin{bmatrix} 7 & 11 \\ 5 & 1 \\ 12 & 15 \end{bmatrix} \begin{bmatrix} 2 & 1 & 0 & 1 \\ 2 & 3 & 5 & -1 \end{bmatrix}$

$= \begin{bmatrix} 36 & 40 & 55 & -4 \\ 12 & 8 & 5 & 4 \\ 54 & 57 & 75 & -3 \end{bmatrix}$

Exercises 10.5

1 (i) Since

$$\begin{bmatrix} 5 & 3 \\ 5 & 4 \end{bmatrix} \begin{bmatrix} 0.8 & -0.6 \\ -1 & 1 \end{bmatrix} = \begin{bmatrix} 1 & 0 \\ 0 & 1 \end{bmatrix}$$

then B is the inverse matrix to A (and A is the inverse of B).

(ii) Since

$$\begin{bmatrix} 5 & 5 \\ 2 & 4 \end{bmatrix} \begin{bmatrix} 0.4 & -0.5 \\ -0.2 & 0.5 \end{bmatrix} = \begin{bmatrix} 1 & 0 \\ 0 & 1 \end{bmatrix}$$

then B is the inverse matrix to A (and A is the inverse of B).

(iii) Since

$$\begin{bmatrix} 3 & 2 \\ 5 & 5 \end{bmatrix} \begin{bmatrix} 1 & -0.4 \\ -1 & 0.2 \end{bmatrix} = \begin{bmatrix} 1 & -0.8 \\ 0 & -1 \end{bmatrix}$$

then B and A are *not* inverses.

2 (i) $|A| = 13$
 (ii) $|A| = 10$
 (iii) $|A| = -7$

(iv) $|A| = -0.1$
(v) $|A| = -(6 + 2\pi^2)$
(vi) $|A| = 0$

3 (i)
$$\text{adj } A = \begin{bmatrix} 5 & -2 \\ -1 & 3 \end{bmatrix}$$

(ii)
$$\text{adj } A = \begin{bmatrix} -7 & 6 \\ -3 & 2 \end{bmatrix}$$

(iii)
$$\text{adj } A = \begin{bmatrix} e & \pi \\ -2 & 3 \end{bmatrix}$$

(iv)
$$\text{adj } A = \begin{bmatrix} 5 & 0 \\ 0 & 5 \end{bmatrix}$$

4 (i) $\begin{bmatrix} 1.5 & -2.5 \\ -1 & 2 \end{bmatrix}$

(ii) $\begin{bmatrix} 0.4 & -0.2 \\ -0.3 & 0.4 \end{bmatrix}$

(iii) $\begin{bmatrix} -1 & 0.4 \\ 2 & -0.6 \end{bmatrix}$

(iv) $\begin{bmatrix} 2 & 5 \\ 4 & 15 \end{bmatrix}$

(v) No inverse: determinant is zero.

5 (i) Involutional: $A^{-1} = A$
(ii) Orthogonal: $A^T = A^{-1}$
(iii) Involutional: $A^{-1} = A$
(iv) Involutional: $A^{-1} = A$

Exercises 10.6

1 (i) $\begin{bmatrix} x_1 \\ x_2 \end{bmatrix} = \begin{bmatrix} 0.4 & -0.6 \\ -0.2 & 0.8 \end{bmatrix} \begin{bmatrix} 55 \\ 20 \end{bmatrix} = \begin{bmatrix} 10 \\ 5 \end{bmatrix}$

(ii) $\begin{bmatrix} x_1 \\ x_2 \end{bmatrix} = \begin{bmatrix} 0.3 & 0.4 \\ -0.1 & 0.2 \end{bmatrix} \begin{bmatrix} 120 \\ 10 \end{bmatrix} = \begin{bmatrix} 40 \\ -10 \end{bmatrix}$

(iii) $\begin{bmatrix} x_1 \\ x_2 \end{bmatrix} = \begin{bmatrix} 2 & -4.5 \\ -1 & 2.5 \end{bmatrix} \begin{bmatrix} 45 \\ 20 \end{bmatrix} = \begin{bmatrix} 0 \\ 5 \end{bmatrix}$

(iv) The determinant is zero and the inverse does not exist. In this case, the equations are consistent; the first and second right-hand side elements are in the same relationship as the first and second rows of the coefficient matrix. There is therefore an infinite number of solutions to the system.

(v) $\begin{bmatrix} x_1 \\ x_2 \end{bmatrix} = \begin{bmatrix} -0.2 & 0.7 \\ -0.2 & 0.2 \end{bmatrix} \begin{bmatrix} -18 \\ -8 \end{bmatrix} = \begin{bmatrix} -2 \\ 2 \end{bmatrix}$

(vi) The determinant is zero and the inverse does not exist. In this case, the equations are inconsistent; the first and second right-hand side elements are not in the same relationship as the first and second rows of the coefficient matrix. There are therefore no solutions to the system.

(vii) $\begin{bmatrix} x_1 \\ x_2 \end{bmatrix} = \begin{bmatrix} 0.75 & -1.25 \\ -0.5 & 1 \end{bmatrix} \begin{bmatrix} 189 \\ 106 \end{bmatrix}$

$= \begin{bmatrix} 9.25 \\ 11.5 \end{bmatrix}$

(viii) $\begin{bmatrix} x_1 \\ x_2 \end{bmatrix} = \begin{bmatrix} 1 & 2 \\ 4 & 3 \end{bmatrix} \begin{bmatrix} 2 \\ 4 \end{bmatrix} = \begin{bmatrix} 10 \\ 20 \end{bmatrix}$

2 (i) $\begin{bmatrix} x_1 \\ x_2 \end{bmatrix} = \begin{bmatrix} 8 \\ 5 \end{bmatrix}$

(ii) $\begin{bmatrix} x_1 \\ x_2 \end{bmatrix} = \begin{bmatrix} -3 \\ 7 \end{bmatrix}$

(iii) $\begin{bmatrix} x_1 \\ x_2 \end{bmatrix} = \begin{bmatrix} 9 \\ -4 \end{bmatrix}$

(iv) $\begin{bmatrix} x_1 \\ x_2 \end{bmatrix} = \begin{bmatrix} 9.5 \\ 15 \end{bmatrix}$

(v) $\begin{bmatrix} x_1 \\ x_2 \end{bmatrix} = \begin{bmatrix} 0 \\ 19 \end{bmatrix}$

(vi) $\begin{bmatrix} x_1 \\ x_2 \end{bmatrix} = \begin{bmatrix} 100 \\ -90 \end{bmatrix}$

(vii) $\begin{bmatrix} x_1 \\ x_2 \end{bmatrix} = \begin{bmatrix} 20 \\ 20 \end{bmatrix}$

3 (i) $\begin{bmatrix} x_1 \\ x_2 \end{bmatrix} = \begin{bmatrix} 1.5 & -2.5 \\ -1 & 2 \end{bmatrix} \begin{bmatrix} 90 \\ 50 \end{bmatrix} = \begin{bmatrix} 10 \\ 10 \end{bmatrix}$

(ii) Using the inverse matrix, it is required that

$$\begin{bmatrix} x_1 \\ x_2 \end{bmatrix} = \begin{bmatrix} 1.5 & -2.5 \\ -1 & 2 \end{bmatrix} \begin{bmatrix} b_1 \\ 50 \end{bmatrix} \geq \begin{bmatrix} 0 \\ 0 \end{bmatrix}$$

so that, from the sign requirement on x_1,

$$x_1 = 1.5 b_1 - 2.5(50) \geq 0$$

which implies that

$$b_1 \geq 83.\overline{33}$$

and, from the sign requirement on x_2,

$$x_2 = -b_1 + 2(50) \geq 0$$

which implies that

$$b_1 \leq 100$$

Therefore, the tolerance interval for b_1 is

$$83.\overline{33} \leq b_1 \leq 100$$

(iii) In terms of b_2, it is required that

$$\begin{bmatrix} x_1 \\ x_2 \end{bmatrix} = \begin{bmatrix} 1.5 & -2.5 \\ -1 & 2 \end{bmatrix} \begin{bmatrix} 90 \\ b_2 \end{bmatrix} \geq \begin{bmatrix} 0 \\ 0 \end{bmatrix}$$

so that, from the sign requirement on x_1,

$$x_1 = 1.5(90) - 2.5 b_2 \geq 0$$

which implies that

$$b_2 \leq 54$$

and, from the sign requirement on x_2,

$$x_2 = -1(90) + 2 b_2 \geq 0$$

which implies that

$$b_2 \geq 45$$

Therefore, the tolerance interval for b_2 is

$$45 \leq b_2 \leq 54$$

770 Mathematics for business, finance and economics

(iv) In the joint analysis, it is required that

$$\begin{bmatrix} x_1 \\ x_2 \end{bmatrix} = \begin{bmatrix} 1.5 & -2.5 \\ -1 & 2 \end{bmatrix} \begin{bmatrix} b_1 \\ b_2 \end{bmatrix} \geq \begin{bmatrix} 0 \\ 0 \end{bmatrix}$$

so that, from the sign requirement on x_1,

$$x_1 = 1.5b_1 - 2.5b_2 \geq 0$$

which implies that

$$b_2 \leq 0.6b_1$$

and, from the sign requirement on x_2,

$$x_2 = -b_1 + 2b_2 \geq 0$$

which implies that

$$b_1 \geq 0.5b_1$$

Therefore, the range of relative sizes is

$$0.5b_1 \leq b_2 \leq 0.6b_1$$

Exercises 10.7

1 (i) As revealed by the fact that $|A| = 0$, the rank of $A = 1$, the second row being the first row multiplied by -0.5. Since the same relationship holds between the right-hand side elements, the equations are consistent, and any pair of values that satisfies

$$2x_1 - x_2 = 20$$

i.e.

$$x_2 = 2x_1 - 20$$

will be a solution to the system.

(ii) Here, the rows (columns) are linearly independent, and the rank of the A matrix is 2. The equations have the unique solution given by

$$\begin{bmatrix} x_1 \\ x_2 \end{bmatrix} = \begin{bmatrix} 1 & -1.6 \\ -0.5 & 0.9 \end{bmatrix} \begin{bmatrix} 11 \\ 5 \end{bmatrix} = \begin{bmatrix} 3 \\ -1 \end{bmatrix}$$

(iii) Here, $|A| = 0$, and the rows (columns) of the A matrix are linearly dependent (row 2 can be seen to be $-\frac{2}{3}$ of row 1). But in this case the right-hand side elements do not stand in the same relationship as that between the rows of A. So the equations are inconsistent, and there is no solution to the system.

2 (i) The rank of A is 1. The second and third rows are twice and four times the first row respectively.

(ii) The rank of A is 1. Here, the linear dependence is more easily seen in the relationship between the columns. Column 1 is twice column 2, while column 3 is $-\frac{1}{2}$ column 2.

(iii) The rank of A is 2. The third row is twice the first row so the rank cannot be greater than 2. But the rank cannot be less than 2, as the second row is not a multiple of the first.

3 (i) Yes. The third column of the matrix is the sum of the two vectors making up the basis, and the fourth column of the matrix is the difference between the two basic vectors.

(ii) The rank of A is 2.

4 The rank of the matrix is 1. All columns can be seen as constant multiples of any one given column.

Exercises 10.8

1 (i) (a) $m_{33} = \begin{vmatrix} 7 & 5 \\ 3 & 8 \end{vmatrix}$

$= 56 - 15 = 41$

(b) $m_{11} = \begin{vmatrix} 8 & 4 \\ 2 & 1 \end{vmatrix}$

$= 8 - 8 = 0$

(c) $m_{22} = \begin{vmatrix} 7 & 9 \\ 6 & 1 \end{vmatrix}$

$= 7 - 54 = -47$

(d) $m_{32} = \begin{vmatrix} 7 & 9 \\ 3 & 4 \end{vmatrix}$

$= 28 - 27 = 1$

(e) $m_{21} = \begin{vmatrix} 5 & 9 \\ 2 & 1 \end{vmatrix}$

$= 5 - 18 = -13$

(ii) $c_{33} = (-1)^{3+3} m_{33} = 41$
$c_{11} = (-1)^{1+1} m_{11} = 0$
$c_{22} = (-1)^{2+2} m_{22} = -47$
$c_{32} = (-1)^{3+2} m_{32} = -1$
$c_{21} = (-1)^{2+1} m_{21} = 13$

2 (i) Using the first row
$$|A| = a_{11}c_{11} + a_{12}c_{12} + a_{13}c_{13}$$
$$= 2c_{11} + 4c_{12} + 3c_{13}$$

Now
$$m_{11} = \begin{vmatrix} 5 & 1 \\ 2 & 4 \end{vmatrix} = 20 - 2 = 18$$

so
$$c_{11} = (-1)^{1+1}m_{11} = 18$$
$$m_{12} = \begin{vmatrix} 4 & 1 \\ 9 & 4 \end{vmatrix} = 16 - 9 = 7$$

so
$$c_{12} = (-1)^{1+2}m_{12} = -7$$
$$m_{13} = \begin{vmatrix} 4 & 5 \\ 9 & 2 \end{vmatrix} = 8 - 45 = -37$$

so
$$c_{13} = (-1)^{1+3}m_{13} = -37$$

So the determinant is
$$|A| = 2(18) + 4(-7) + 3(-37)$$
$$= 36 - 28 - 111 = -103$$

(ii) Using the third column:
$$|A| = a_{13}c_{13} + a_{23}c_{23} + a_{33}c_{33}$$
$$= a_{13}m_{13} - a_{23}m_{23} + a_{33}m_{33}$$
$$= 3m_{13} - m_{23} + 4m_{33}$$

where
$$m_{13} = \begin{vmatrix} 4 & 5 \\ 9 & 2 \end{vmatrix} = 8 - 45 = -37$$
$$m_{23} = \begin{vmatrix} 2 & 4 \\ 9 & 2 \end{vmatrix} = 4 - 35 = -32$$
$$m_{33} = \begin{vmatrix} 2 & 4 \\ 4 & 5 \end{vmatrix} = 10 - 16 = -6$$

so that
$$|A| = 3(-37) - (-32) + 4(-6)$$
$$= 111 + 32 - 24 = -103$$

which confirms the result in (i).

Matrices 773

3 (i) Here, the most convenient row or column by which to expand is the second row. The value of the determinant is then

$$(-1)^{2+1}(1)m_{21} + (-1)^{2+3}(2)m_{23} = -m_{21} - m_{23}$$

where

$$m_{21} = \begin{vmatrix} 9 & 8 \\ 3 & 4 \end{vmatrix} = 36 - 24 = 12$$

and

$$m_{23} = \begin{vmatrix} 7 & 9 \\ 5 & 3 \end{vmatrix} = 21 - 45 = -24$$

so that

$$|A| = -12 - (2)(-24) = 36$$

(ii) Expanding by the third row:

$$|A| = (-1)^{3+2}(-1)m_{32} + (-1)^{3+3}(4)m_{33}$$
$$= m_{32} + 4m_{33}$$

where

$$m_{32} = \begin{vmatrix} 4 & 2 \\ 8 & 8 \end{vmatrix} = 32 - 16 = 16$$

and

$$m_{33} = \begin{vmatrix} 4 & 3 \\ 8 & 5 \end{vmatrix} = 20 - 24 = -4$$

so that

$$|A| = 16 + (4)(-4) = 0$$

The matrix is singular. The linear dependence in this case is most evident as the third row being row 2 less twice row 1.

4 Begin by evaluating the determinant by the first row and note that

$$c_{11} = m_{11} \qquad c_{12} = -m_{12} \qquad c_{13} = m_{13}$$

so

$$|A| = m_{11} - m_{12} + 2m_{13}$$

where

$$m_{11} = \begin{vmatrix} 4 & 9 \\ 2 & 3 \end{vmatrix} = 12 - 18 = -6$$

$$m_{12} = \begin{vmatrix} 5 & 9 \\ 3 & 3 \end{vmatrix} = 15 - 27 = -12$$

$$m_{13} = \begin{vmatrix} 5 & 4 \\ 1 & 2 \end{vmatrix} = 10 - 12 = -2$$

so that

$$|A| = -6 - (-12) + 2(-2) = 2$$

The minors of elements in the second row are

$$m_{21} = \begin{vmatrix} 1 & 2 \\ 2 & 3 \end{vmatrix} = 3 - 4 = -1$$

$$m_{22} = \begin{vmatrix} 1 & 2 \\ 3 & 3 \end{vmatrix} = 3 - 6 = -3$$

$$m_{23} = \begin{vmatrix} 1 & 1 \\ 3 & 2 \end{vmatrix} = 2 - 3 = -1$$

so that the cofactors for this row are

$$c_{21} = 1 \quad c_{22} = -3 \quad c_{23} = 1$$

For the third row:

$$m_{31} = \begin{vmatrix} 1 & 2 \\ 4 & 9 \end{vmatrix} = 9 - 8 = 1$$

$$m_{32} = \begin{vmatrix} 1 & 2 \\ 5 & 9 \end{vmatrix} = 9 - 10 = -1$$

$$m_{33} = \begin{vmatrix} 1 & 1 \\ 5 & 4 \end{vmatrix} = 4 - 5 = -1$$

so that the cofactors are

$$c_{31} = 1 \quad c_{32} = 1 \quad c_{33} = -1$$

The inverse matrix is then

$$A^{-1} = \frac{1}{|A|} \begin{bmatrix} c_{11} & c_{21} & c_{31} \\ c_{12} & c_{22} & c_{32} \\ c_{13} & c_{23} & c_{33} \end{bmatrix}$$

so

$$A^{-1} = \frac{1}{2} \begin{bmatrix} -6 & 1 & 1 \\ 12 & -3 & 1 \\ -2 & 1 & -1 \end{bmatrix} = \begin{bmatrix} -3 & 0.5 & 0.5 \\ 6 & -1.5 & 0.5 \\ -1 & 0.5 & -0.5 \end{bmatrix}$$

Checking the result by multiplication:

$$\begin{bmatrix} 1 & 1 & 2 \\ 5 & 4 & 9 \\ 3 & 2 & 3 \end{bmatrix} \begin{bmatrix} -3 & 0.5 & 0.5 \\ 6 & -1.5 & 0.5 \\ -1 & 0.5 & -0.5 \end{bmatrix} = \begin{bmatrix} 1 & 0 & 0 \\ 0 & 1 & 0 \\ 0 & 0 & 1 \end{bmatrix}$$

5 In this case use the second row to find the determinant:

$$|A| = \begin{vmatrix} 7 & 2 \\ 5 & 1.5 \end{vmatrix} = 10.5 - 10 = 0.5$$

The minors of the first row are

$$m_{11} = \begin{vmatrix} 1 & 0 \\ 1 & 1.5 \end{vmatrix} = 1.5 - 0 = 1.5$$

$$m_{12} = \begin{vmatrix} 0 & 0 \\ 5 & 1.5 \end{vmatrix} = 0$$

$$m_{13} = \begin{vmatrix} 0 & 1 \\ 5 & 1 \end{vmatrix} = 0 - 5 = -5$$

so the cofactors for this row are

$$c_{11} = 1.5 \qquad c_{12} = 0 \qquad c_{13} = -5$$

In the second row

$$m_{21} = \begin{vmatrix} -3 & 2 \\ 1 & 1.5 \end{vmatrix} = 4.5 - 2 = -6.5$$

$$m_{22} = \begin{vmatrix} 7 & 2 \\ 5 & 1.5 \end{vmatrix} = 10.5 - 10 = 0.5$$

$$m_{23} = \begin{vmatrix} 7 & -3 \\ 5 & 1 \end{vmatrix} = 7 + 15 = 22$$

so the second row cofactors are

$$c_{21} = 6.5 \qquad c_{22} = 0.5 \qquad c_{23} = -22$$

In the third row

$$m_{31} = \begin{vmatrix} -3 & 2 \\ 1 & 0 \end{vmatrix} = 0 - 2 = -2$$

$$m_{32} = \begin{vmatrix} 7 & 2 \\ 0 & 0 \end{vmatrix} = 0$$

$$m_{33} = \begin{vmatrix} 7 & -3 \\ 0 & 1 \end{vmatrix} = 7 + 0 = 7$$

so that the third row cofactors are

$$c_{31} = -2 \qquad c_{32} = 0 \qquad c_{33} = 7$$

The inverse is then

$$A^{-1} = \frac{1}{0.5}\begin{bmatrix} 1.5 & 6.5 & -2 \\ 0 & 0.5 & 0 \\ -5 & -22 & 7 \end{bmatrix} = \begin{bmatrix} 3 & 13 & -4 \\ 0 & 1 & 0 \\ -10 & -44 & 14 \end{bmatrix}$$

Checking the result by multiplication:

$$\begin{bmatrix} 7 & -3 & 2 \\ 0 & 1 & 0 \\ 5 & 1 & 1.5 \end{bmatrix}\begin{bmatrix} 3 & 13 & -4 \\ 0 & 1 & 0 \\ -10 & -44 & 14 \end{bmatrix} = \begin{bmatrix} 1 & 0 & 0 \\ 0 & 1 & 0 \\ 0 & 0 & 1 \end{bmatrix}$$

Exercises 10.9

1 The determinant of the coefficient matrix is -1 and its inverse is

$$A^{-1} = \begin{bmatrix} -6 & 5 & -4 \\ 13 & -10 & 8 \\ -2 & 1.5 & -1 \end{bmatrix}$$

so that the solution values for the variables are given by

$$\begin{bmatrix} x_1 \\ x_2 \\ x_3 \end{bmatrix} = \begin{bmatrix} -6 & 5 & -4 \\ 13 & -10 & 8 \\ -2 & 1.5 & -1 \end{bmatrix}\begin{bmatrix} 35 \\ 140 \\ 120 \end{bmatrix} = \begin{bmatrix} 10 \\ 15 \\ 20 \end{bmatrix}$$

2 The inverse of the coefficient matrix is

$$A^{-1} = \begin{bmatrix} 3.6 & 2.2 & -3.8 \\ -2 & -1 & 2 \\ -0.2 & -0.4 & 0.6 \end{bmatrix}$$

so that the solution values for the variables are given by

(i) $$\begin{bmatrix} x_1 \\ x_2 \\ x_3 \end{bmatrix} = \begin{bmatrix} 3.6 & 2.2 & -3.8 \\ -2 & -1 & 2 \\ -0.2 & -0.4 & 0.6 \end{bmatrix}\begin{bmatrix} 85 \\ 185 \\ 185 \end{bmatrix} = \begin{bmatrix} 10 \\ 15 \\ 20 \end{bmatrix}$$

(ii) $$\begin{bmatrix} x_1 \\ x_2 \\ x_3 \end{bmatrix} = \begin{bmatrix} 3.6 & 2.2 & -3.8 \\ -2 & -1 & 2 \\ -0.2 & -0.4 & 0.6 \end{bmatrix}\begin{bmatrix} 70 \\ 50 \\ 90 \end{bmatrix} = \begin{bmatrix} 20 \\ -10 \\ 20 \end{bmatrix}$$

(iii) $$\begin{bmatrix} x_1 \\ x_2 \\ x_3 \end{bmatrix} = \begin{bmatrix} 3.6 & 2.2 & -3.8 \\ -2 & -1 & 2 \\ -0.2 & -0.4 & 0.6 \end{bmatrix}\begin{bmatrix} -4 \\ 139 \\ 78 \end{bmatrix} = \begin{bmatrix} -5 \\ 25 \\ -8 \end{bmatrix}$$

(iv) $$\begin{bmatrix} x_1 \\ x_2 \\ x_3 \end{bmatrix} = \begin{bmatrix} 3.6 & 2.2 & -3.8 \\ -2 & -1 & 2 \\ -0.2 & -0.4 & 0.6 \end{bmatrix}\begin{bmatrix} 9 \\ 63 \\ 45 \end{bmatrix} = \begin{bmatrix} 0 \\ 9 \\ 0 \end{bmatrix}$$

3 The tolerance interval for the parameter b_1 is obtained from the requirements that all x_i are non-negative. That is, b_1 must be such that

$$\begin{bmatrix} x_1 \\ x_2 \\ x_3 \end{bmatrix} = \begin{bmatrix} 3.6 & 2.2 & -3.8 \\ -2 & -1 & 2 \\ -0.2 & -0.4 & 0.6 \end{bmatrix} \begin{bmatrix} b_1 \\ 185 \\ 185 \end{bmatrix} \geqslant \begin{bmatrix} 0 \\ 0 \\ 0 \end{bmatrix}$$

From the x_1 line:

$$3.6b_1 + 2.2(185) - 3.8(185) \geqslant 0$$

which implies that

$$b_1 \geqslant 82.\overline{22}$$

From the x_2 line:

$$-2b_1 - 185 + 2(185) \geqslant 0$$

which implies that

$$b_1 \leqslant 92.5$$

From the x_3 line:

$$-0.2b_1 - 0.4(185) + 0.6(185) \geqslant 0$$

which implies that

$$b_1 \leqslant 185$$

So the tolerance interval for this parameter is

$$82.\overline{22} \leqslant b_1 \leqslant 92.5$$

For b_2 it is required that

$$\begin{bmatrix} x_1 \\ x_2 \\ x_3 \end{bmatrix} = \begin{bmatrix} 3.6 & 2.2 & -3.8 \\ -2 & -1 & 2 \\ -0.2 & -0.4 & 0.6 \end{bmatrix} \begin{bmatrix} 85 \\ b_2 \\ 185 \end{bmatrix} = \begin{bmatrix} 0 \\ 0 \\ 0 \end{bmatrix}$$

From the x_1 line:

$$3.6(85) + 2.2b_2 - 3.8(185) \geqslant 0$$

which implies that

$$b_2 \geqslant 180.\overline{45}$$

From the x_2 line:

$$-2(85) - b_2 + 2(185) \geqslant 0$$

which implies that

$$b_2 \leqslant 200$$

From the x_3 line:

$$-0.2(85) - 0.4b_2 + 0.6(185) \geqslant 0$$

which implies that

$$b_2 \leqslant 235$$

So the tolerance interval for this parameter is

$$180.\overline{45} \leqslant b_2 \leqslant 200$$

For b_3 it is required that

$$\begin{bmatrix} x_1 \\ x_2 \\ x_3 \end{bmatrix} = \begin{bmatrix} 3.6 & 2.2 & -3.8 \\ -2 & -1 & 2 \\ -0.2 & -0.4 & 0.6 \end{bmatrix} \begin{bmatrix} 85 \\ 185 \\ b_2 \end{bmatrix} \geqslant \begin{bmatrix} 0 \\ 0 \\ 0 \end{bmatrix}$$

From the x_1 line:

$$3.6(85) + 2.2(185) - 3.8b_3 \geqslant 0$$

which implies that

$$b_3 \leqslant 187.63$$

From the x_2 line:

$$-2(85) - 185 + 2b_3 \geqslant 0$$

which implies that

$$b_3 \geqslant 177.5$$

From the x_3 line:

$$-0.2(85) - 0.4(185) + 0.6b_3 \geqslant 0$$

which implies that

$$b_3 \geqslant 151.\overline{66}$$

So the tolerance interval for b_3 is

$$177.5 \leqslant b_3 \leqslant 187.63$$

Exercises 10.10

1 (i)
$$x_1 = \frac{\begin{vmatrix} 53 & 1 \\ 46 & 2 \end{vmatrix}}{\begin{vmatrix} 9 & 1 \\ 6 & 2 \end{vmatrix}}$$

$$= \frac{106 - 46}{18 - 6} = 5$$

$$x_2 = \frac{\begin{vmatrix} 9 & 53 \\ 6 & 46 \end{vmatrix}}{\begin{vmatrix} 9 & 1 \\ 6 & 2 \end{vmatrix}}$$

$$= \frac{414 - 318}{18 - 6} = 8$$

So the solution to the system is

$$x_1 = 5 \quad \text{and} \quad x_2 = 8$$

(ii)
$$x_1 = \frac{\begin{vmatrix} 26 & 5 \\ 3 & 3 \end{vmatrix}}{\begin{vmatrix} 4 & 5 \\ 1 & 3 \end{vmatrix}}$$

$$= \frac{78 - 15}{12 - 5} = 9$$

$$x_2 = \frac{\begin{vmatrix} 4 & 26 \\ 1 & 3 \end{vmatrix}}{\begin{vmatrix} 4 & 5 \\ 1 & 3 \end{vmatrix}}$$

$$= \frac{12 - 26}{12 - 5} = -2$$

2 The determinant of the coefficient matrix (expanded by the first row is

$$2m_{11} - 3m_{12} + m_{13}$$

where

$$m_{11} = \begin{vmatrix} 1 & 5 \\ 6 & 3 \end{vmatrix} = 3 - 30 = -27$$

$$m_{12} = \begin{vmatrix} 4 & 5 \\ 2 & 3 \end{vmatrix} = 12 - 10 = 2$$

$$m_{13} = \begin{vmatrix} 4 & 1 \\ 2 & 6 \end{vmatrix} = 24 - 2 = 22$$

So

$$|A| = 2(-27) - 3(2) + 22 = -38$$

By Cramer's rule

$$x_1 = \frac{\begin{vmatrix} 16 & 3 & 1 \\ 34 & 1 & 5 \\ 30 & 6 & 3 \end{vmatrix}}{-38}$$

in which the numerator determinant is

$$16\begin{vmatrix} 1 & 5 \\ 6 & 3 \end{vmatrix} - 3\begin{vmatrix} 34 & 5 \\ 30 & 3 \end{vmatrix} + 1\begin{vmatrix} 34 & 1 \\ 30 & 6 \end{vmatrix}$$
$$= 16(-27) - 3(-48) + (174) = -114$$

so that

$$x_1 = \frac{-114}{-38} = 3$$

For x_2:

$$x_2 = \frac{\begin{vmatrix} 2 & 16 & 1 \\ 4 & 34 & 5 \\ 2 & 30 & 3 \end{vmatrix}}{-38}$$

in which the numerator determinant is

$$2\begin{vmatrix} 34 & 5 \\ 30 & 3 \end{vmatrix} - 16\begin{vmatrix} 4 & 5 \\ 2 & 3 \end{vmatrix} + 1\begin{vmatrix} 4 & 34 \\ 2 & 30 \end{vmatrix}$$
$$= 2(-48) - 16(2) + (52) = -76$$

so that

$$x_2 = \frac{-76}{-38} = 2$$

And for x_3:

$$x_3 = \frac{\begin{vmatrix} 2 & 3 & 16 \\ 4 & 1 & 34 \\ 2 & 6 & 30 \end{vmatrix}}{-38}$$

in which the numerator determinant is

$$2\begin{vmatrix} 1 & 34 \\ 6 & 30 \end{vmatrix} - 3\begin{vmatrix} 4 & 34 \\ 2 & 30 \end{vmatrix} + 16\begin{vmatrix} 4 & 1 \\ 2 & 6 \end{vmatrix}$$
$$= 2(-174) - 3(52) + 16(22) = -152$$

so that

$$x_3 = \frac{-152}{-38} = 4$$

The complete solution is therefore

$$x_1 = 3 \quad x_2 = 2 \quad x_3 = 4$$

3 The coefficient matrix is

$$A = \begin{bmatrix} 1 & -22 & 7 \\ 1 & 29 & 2 \\ -2 & 12 & 4 \end{bmatrix}$$

and the value of x_2 is given by

$$x_2 = \frac{\begin{vmatrix} 1 & 107 & 7 \\ 1 & 61 & 8 \\ -2 & -92 & 4 \end{vmatrix}}{\begin{vmatrix} 1 & -22 & 7 \\ 1 & 29 & 8 \\ -2 & 12 & 4 \end{vmatrix}}$$

Expanding by the second column, the numerator determinant is

$$107(-1)^{1+2}m_{12} + 61(-1)^{2+2}m_{22} - 92(-1)^{3+2}m_{32}$$
$$= -107m_{12} + 61m_{22} - 92m_{32}$$

where

$$m_{12} = \begin{vmatrix} 1 & 8 \\ -2 & 4 \end{vmatrix} = 4 + 16 = 20$$

$$m_{22} = \begin{vmatrix} 1 & 7 \\ -2 & 4 \end{vmatrix} = 4 + 14 = 18$$

$$m_{32} = \begin{vmatrix} 1 & 7 \\ 1 & 8 \end{vmatrix} = 8 - 7 = 1$$

so that the determinant is

$$-107(20) + 61(18) + 92(1)$$
$$= -2140 + 1098 + 92 = -950$$

The denominator determinant also expanded by the second column is

$$-22(-1)^{1+2}m_{12} + 29(-1)^{2+2}m_{22} + 12(-1)^{3+2}m_{32}$$
$$= 22(20) + 29(18) - 12 = 950$$

so that the value of x_2 is

$$x_2 = -950/950 = -1$$

Exercises 10.11

1 As case (b) and case (c) have non-singular coefficient matrices the systems cannot be satisfied for non-zero values of x_1 and x_2.

2 (i) Either equation implies that $x_2 = 1.5x_1$.

(ii) From the second equation

$$x_3 = 4x_1 + x_2$$

and substitution into the first equation produces

$$8x_1 - 4x_2 + 8x_1 + 2x_2 = 0$$

so

$$16x_1 = 2x_2$$

i.e.

$$x_2 = 8x_1$$

(iii) From the second equation

$$x_3 = 2x_1 + 2x_2$$

and substitution in the first equation produces

$$9x_1 - 7x_2 + 6x_1 + 6x_2 = 0$$

so

$$15x_1 - x_2 = 0$$

i.e.

$$x_2 = 15x_1$$

3 (i) $x_3 = 60$ when $x_2 = 5$ (in case 2(i))
$x_3 = 160$ when $x_2 = 5$ (in case 2(ii))

(ii) Yes, a scalar multiple of any non-trivial solution is also a non-trivial solution to the system. Hence, when $x_3 = -8$, the values of x_1 and x_2 are a factor -0.05 times their respective values when $x_3 = 160$. That is, another non-trivial solution is given by

$$x_1 = -0.25 \qquad x_2 = -3.75 \qquad x_3 = -8$$

Additional problems

1 (i) $x - 2 = 7$ so $x = 9$
$y + 3 = 9$ so $y = 6$

(ii) $3x - 4 = 11$ so $x = 5$
$6 - y = 10$ so $y = -4$

2 The required product is

$$\begin{bmatrix} 46 \\ 28 \\ 36 \end{bmatrix} \begin{bmatrix} 6 & 3 \end{bmatrix}$$

$$= \begin{bmatrix} 276 & 138 \\ 168 & 84 \\ 216 & 108 \end{bmatrix}$$

3 (i) $|A| = 3(6) - 7(5) = -17$
   (ii) Expansion of the determinant by the first row produces

$$2 \begin{vmatrix} 3 & 5 \\ -2 & 7 \end{vmatrix} - 1 \begin{vmatrix} 4 & 5 \\ 6 & 7 \end{vmatrix} + 3 \begin{vmatrix} 4 & 3 \\ 6 & -2 \end{vmatrix}$$

$$= 2(21 + 10) - (28 - 30) + 3(-8 - 18)$$
$$= 62 + 2 - 78 = -14$$

   (iii) The determinant is zero (row 3 is minus half of row 1).

4 (i) $\begin{bmatrix} 0.5 & -0.4 \\ -0.25 & 0.3 \end{bmatrix}$

   (ii) The determinant is 8, and the inverse is

$$\begin{bmatrix} -1.75 & 0.5 & 0.75 \\ -0.375 & -0.25 & 0.375 \\ 1.625 & -0.25 & -0.625 \end{bmatrix}$$

5 (i) $\begin{bmatrix} x_1 \\ x_2 \end{bmatrix} = \begin{bmatrix} 1 & -2 \\ -0.5 & 1.5 \end{bmatrix} \begin{bmatrix} 5 \\ 3 \end{bmatrix} = \begin{bmatrix} -1 \\ 2 \end{bmatrix}$

   (ii) $\begin{bmatrix} x_1 \\ x_2 \\ x_3 \end{bmatrix} = \begin{bmatrix} -1.75 & 0.5 & 0.75 \\ -0.375 & -0.25 & 0.375 \\ 1.625 & -0.25 & -0.625 \end{bmatrix} \begin{bmatrix} 100 \\ 200 \\ 300 \end{bmatrix} = \begin{bmatrix} 150 \\ 25 \\ -75 \end{bmatrix}$

6 (i) The determinant is 5, so by Cramer's rule

$$x_1 = \frac{\begin{vmatrix} 160 & 2 \\ 350 & 5 \end{vmatrix}}{5}$$

$$= (800 - 700)/5 = 20$$

$$x_2 = \frac{\begin{vmatrix} 5 & 160 \\ 10 & 350 \end{vmatrix}}{5}$$

$$= (1750 - 1600)/5 = 30$$

(ii) The determinant of the coefficient matrix is 32, so x_3 is

$$x_3 = \frac{\begin{vmatrix} 3 & -2 & 13 \\ 4 & 6 & 29 \\ 4 & 4 & 30 \end{vmatrix}}{32}$$

$$= 96/32 = 3$$

7 (i) The determinant is zero. From the first equation

$$2x_1 + 0x_2 - x_3 = 0$$

so that

$$x_3 = 2x_1$$

Substitution into the second equation then produces

$$3x_1 + 4x_2 + 4x_1 = 0$$

i.e.

$$7x_1 + 4x_2 = 0$$

so

$$x_2 = -\tfrac{7}{4}x_1$$

so when $x_1 = 4$, $x_2 = -7$ and $x_3 = 8$.

(ii) The determinant here is non-zero ($= -13$) so no non-trivial solutions will exist.

NOTES

1. Also included in elementary matrix operations are the *elementary row operations* described in Chapter 2.
2. This term (sometimes considered colloquial) arises because the vectors have a dot between them to represent multiplication.
3. In the sense indicated, i.e. element by corresponding element and summing. The result can be seen as the dot product of the row vector and the first column of the matrix.
4. But in terms of matrix multiplication, this instance of the order of multiplication being immaterial is the exception rather than the rule.
5. The matrices A and B are *singular*. Singularity is defined in Section 10.5 below.
6. The dimension of I will differ between premultiplication and postmultiplication if A is not a square matrix.
7. Determinants were once referred to as *eliminants* because they arose in the course of solving linear equations through progressive substitution and elimination of variables.
8. This is an example of *linear dependence* — which can take more varied forms in larger matrices.
9. But as we see below, there is an element of 'oneness' about involutional matrices.
10. See Section 10.11 for the case where b is a zero vector.
11. Recall that this does not mean that *any* values of x_1 and x_2 will suffice. Only those values satisfying $x_2 = 20 - 3x_1$ belong to the solution set.
12. A collection of r linearly independent vectors belonging to a set of vectors A forms a basis for A if and only if the rank of A is r.
13. It is said to be an *exogenous* variable.
14. For a description of the use of Gaussian elimination in matrix inversion, see Weber (1982).
15. Minors are sometimes called *minor determinants*.
16. This is sometimes called the *complementary minor* of a_{ij}.
17. Note that the sum of products of the elements of one row and the cofactors of another is zero.
18. This property follows from property (4).
19. Sometimes called *alien* cofactors.
20. *Solution* in this context means the group of non-negative variables x_1, x_2 and x_3, rather than their particular *values*.
21. The technique is named after the eighteenth-century Swiss mathematician and physicist Gabriel Cramer.
22. This is the difference between the order of the matrix of coefficients and its rank.

Appendix

Optimization: Further Considerations

A1	Second-order conditions for unconstrained maxima and minima using determinants	787
A2	Quadratic functions: stationary values	791
A3	Constrained optimization: second-order conditions using determinants	797
A4	Optimization of quadratic functions of two variables with a linear equality constraint	800

In Chapter 6, we considered the maximization and minimization of functions of two variables. In Chapter 10, matrices and determinants were introduced. In this Appendix some of the ideas are drawn together and consideration is given to how second-order conditions for unconstrained maxima and minima can be expressed using determinants.

A further look is taken at the optimization of quadratic functions of two independent variables and we examine cases in which strict inequalities do not hold in the second-order conditions.

We then reconsider the case of the optimization of a function of two or more variables, subject to a linear equality constraint, and examine the determinantal form of second-order conditions in these cases.

A1 SECOND-ORDER CONDITIONS FOR UNCONSTRAINED MAXIMA AND MINIMA USING DETERMINANTS

For a maximum of a function of two independent variables

$$F = f(x_1, x_2)$$

at the first order it is required that both partial derivatives are zero:

$$f_1 = 0 \quad f_2 = 0$$

where f_1 and f_2 are the partial derivatives of F with respect to x_1 and x_2 respectively.[1] At the second order, it is sufficient that[2]

$$f_{11} < 0 \quad f_{22} < 0 \quad \text{(A1)}$$

$$(f_{11})(f_{22}) - (f_{12}f_{21}) > 0 \quad \text{(A2)}$$

These second-order conditions for a maximum can be expressed using determinants in the following way:

$$f_{11} < 0 \quad \begin{vmatrix} f_{11} & f_{12} \\ f_{21} & f_{22} \end{vmatrix} > 0 \quad \text{(A3)}$$

System (A3) represents precisely the same requirements as (A1) and (A2) taken together. Clearly, $f_{11} < 0$ is required in both cases. Now expand the determinant. The result is

$$f_{11}f_{22} - f_{21}f_{12} > 0$$

which restates (A2). But what of the requirement that $f_{22} < 0$? This is *implicit* in (A3). It will be recalled that second-order cross partial derivatives are equal, so $f_{12} = f_{21}$. We could therefore rewrite (A2) as

$$f_{11}f_{22} - (f_{12})^2 > 0 \quad \text{(A4)}$$

In (A4), it is clear that a *positive* number, $(f_{12})^2$, is to be subtracted from the product of f_{11} and f_{22}, and the overall result is to be positive. This can only occur if the product $f_{11}f_{22}$ is itself positive; and since $f_{11} < 0$, it must be the case that $f_{22} < 0$ too. As the first of two examples, suppose that it is required to find the maximum of

$$F = 20x_1 + 105x_2 + 5x_1x_2 - 4x_1^2 - 4.5x_2^2$$

At the first order:

$$f_1 = 20 + 5x_2 - 8x_1 = 0$$
$$f_2 = 105 + 5x_1 - 9x_2 = 0$$

which conditions solve for

$$x_1 = 15 \quad \text{and} \quad x_2 = 20$$

With the first-order conditions satisfied, at the second order it is sufficient that the conditions (A3) are fulfilled. In the present case

$$f_{11} = -8 \quad f_{12} = 5 = f_{21} \quad f_{22} = -9$$

Insertion of these values into (A3) gives

$$-8 < 0 \quad \begin{vmatrix} -8 & 5 \\ 5 & -9 \end{vmatrix} > 0$$

which is as required for a maximum. The second example is a case where the second-order conditions are functions of x_1 and x_2 and need to be evaluated at the values of x_1 and x_2 which satisfy the first-order conditions. The problem is to confirm that a local maximum of

$$F = 15x_1 + 25x_2 - 0.5x_1^2 - x_2^2 - 0.01x_1^2 x_2^2$$

occurs when

$$x_1 = 5 \quad \text{and} \quad x_2 = 10$$

At the first order, it is necessary that

$$f_1 = 15 - x_1 - 0.02 x_1 x_2^2 = 0$$
$$f_2 = 25 - 2x_2 - 0.02 x_1^2 x_2 = 0$$

The values $x_1 = 5$ and $x_2 = 10$ satisfy these conditions. At the second order:

$$f_{11} = -1 - 0.02 x_2^2$$
$$f_{12} = -0.04 x_1 x_2 = f_{21}$$
$$f_{22} = -2 - 0.02 x_1^2$$

The conditions (A3) require that

$$-1 - 0.02 x_2^2 < 0 \quad \begin{vmatrix} -1 - 0.02 x_2^2 & -0.04 x_1 x_2 \\ -0.04 x_1 x_2 & -2 - 0.02 x_1^2 \end{vmatrix} > 0$$

When $x_1 = 5$ and $x_2 = 10$, these requirements are that

$$-3 < 0 \quad \begin{vmatrix} -3 & -2 \\ -2 & -2.5 \end{vmatrix} > 0$$

which conditions are met, confirming that a local maximum occurs when $x_1 = 5$ and $x_2 = 10$, where $F = 187.5$.[3] The use of determinants is an economical way to express second-order conditions for maxima and minima when more than two independent variables are involved. For example, in the case of a function of three variables

$$f(x_1, x_2, x_3)$$

to be maximized without constraint, it is required at the first order that

$$f_1 = 0 \quad f_2 = 0 \quad f_3 = 0$$

At the second order, the conditions can be expressed as

$$f_{11} < 0 \qquad \begin{vmatrix} f_{11} & f_{12} \\ f_{21} & f_{22} \end{vmatrix} > 0 \qquad \begin{vmatrix} f_{11} & f_{12} & f_{13} \\ f_{21} & f_{22} & f_{23} \\ f_{31} & f_{32} & f_{33} \end{vmatrix} < 0 \qquad (A5)$$

The largest of these determinants

$$H = \begin{vmatrix} f_{11} & f_{12} & f_{13} \\ f_{21} & f_{22} & f_{23} \\ f_{31} & f_{32} & f_{33} \end{vmatrix}$$

is a *Hessian determinant*, the elements of which are the second-order partial derivatives of F arranged as above. The smaller determinant in (A5) is that *minor* of the Hessian obtained by deleting the last row and the last column of H. In fact, the determinant

$$H_2 = \begin{vmatrix} f_{11} & f_{12} \\ f_{21} & f_{22} \end{vmatrix}$$

is a *principal minor* of H. *Principal* minors are obtained by deleting the last $n - i$ rows and the last $n - i$ columns of the array. Thus in this case, if $i = 1$, the last two of the three rows and columns are struck out and the result is f_{11}; if $i = 2$, the last row and column are deleted and the determinant resulting is H_2, while H itself follows if $i = 3$. For a maximum of the function, sufficient conditions at the second order are that

> The principal minors of the Hessian must alternate in sign, beginning with a negative.

Thus (A5) sets out the conditions in this form for the three-variable case.

For an unconstrained *minimum* of a function of several variables, the second-order conditions are that

> The principal minors of the Hessian should all be positive.

So in the case of a function of two variables, the conditions are

$$f_{11} > 0 \qquad \begin{vmatrix} f_{11} & f_{12} \\ f_{21} & f_{22} \end{vmatrix} > 0$$

from which, expansion of the determinant produces

$$f_{11}f_{22} - f_{21}f_{12} > 0$$

i.e.

$$f_{11}f_{22} - (f_{12})^2 > 0$$

which, since $f_{11} > 0$, it must follow that $f_{22} > 0$. Thus these conditions are equivalent to those expressed without determinants in Chapter 6. For an unconstrained minimum of a function of three variables, the second-order conditions are

$$f_{11} > 0 \qquad \begin{vmatrix} f_{11} & f_{12} \\ f_{21} & f_{22} \end{vmatrix} > 0 \qquad \begin{vmatrix} f_{11} & f_{12} & f_{13} \\ f_{21} & f_{22} & f_{23} \\ f_{31} & f_{32} & f_{33} \end{vmatrix} > 0$$

which is a more manageable expression of the conditions than could have been achieved without determinants. Now return to the context of maximization and consider a numerical example of the maximization of a function of three variables using second-order conditions in determinantal form. For the function

$$F = 4x_1 + 5x_2 + 6x_3 - 3x_1^2 - 1.5x_2^2 \\ - 9x_3^2 + 2x_1x_2 + 3x_2x_3$$

is there a point which satisfies both the first- and second-order conditions for a maximum? At the first order, it is required that

(i) $f_1 = 4 - 6x_1 + 2x_2 = 0$
(ii) $f_2 = 5 - 3x_2 + 2x_1 + 3x_3 = 0$
(iii) $f_3 = 6 - 18x_3 + 3x_2 = 0$

Using substitution to solve these conditions, from condition (i)

$$x_2 = 3x_1 - 2$$

which can be substituted into condition (ii) to give

$$5 - 9x_1 + 6 + 2x_1 + 3x_3 = 0$$

which upon simplification becomes

(ii)' $7x_1 - 3x_3 = 11$

and substitution of $3x_1 - 2$ for x_2 in condition (iii) produces the result

$$6 - 18x_3 + 9x_1 - 6 = 0$$

i.e.

(iii)' $9x_1 - 18x_3 = 0$

so that

$$x_1 = 2x_3$$

which relationship, when substituted into (ii)', produces

$$14x_3 - 3x_3 = 11$$

so that

$$x_3 = 1$$

It follows that $x_1 = 2$ and $x_2 = 4$, so the function has a stationary value when

$$x_1 = 2 \qquad x_2 = 4 \qquad x_3 = 1$$

The second-order conditions are now used to establish whether a maximum is produced by the above x_j values. At the second order, the partial derivatives are

$$f_{11} = -6 \qquad f_{22} = -3 \qquad f_{33} = -18 \qquad f_{12} = 2$$
$$f_{13} = 0 \qquad f_{21} = 2 \qquad f_{23} = 3 \qquad f_{31} = 0 \qquad f_{32} = 3$$

so that the Hessian determinant is

$$\begin{vmatrix} -6 & 2 & 0 \\ 2 & -3 & 3 \\ 0 & 3 & -18 \end{vmatrix}$$

and for a maximum the principal minors of the Hessian must alternate in sign beginning with a negative. So it is required that

$$f_{11} = -6 < 0 \qquad \begin{vmatrix} f_{11} & f_{12} \\ f_{21} & f_{22} \end{vmatrix} = \begin{vmatrix} -6 & 2 \\ 2 & -3 \end{vmatrix} = 14 > 0$$

and

$$\begin{vmatrix} f_{11} & f_{12} & f_{13} \\ f_{21} & f_{22} & f_{23} \\ f_{31} & f_{32} & f_{33} \end{vmatrix} = \begin{vmatrix} -6 & 2 & 0 \\ 2 & -3 & 3 \\ 0 & 3 & -18 \end{vmatrix}$$

$$= -6 \begin{vmatrix} -3 & 3 \\ 3 & -18 \end{vmatrix} - 2 \begin{vmatrix} 2 & 3 \\ 0 & -18 \end{vmatrix}$$

$$= -270 + 72$$
$$= -198 < 0$$

so all the conditions for a maximum are met.

A2 QUADRATIC FUNCTIONS: STATIONARY VALUES

Consider again the question of finding and identifying the nature of the stationary value of a quadratic expression in two independent variables of the form

$$z = ax^2 + bxy + cy^2 + dx + ey + f$$

The first-order conditions for any stationary value require that

$$z_x = 2ax + by + d = 0 \qquad (A6a)$$

and

$$z_y = bx + 2cy + e = 0 \qquad (A6b)$$

Note that the conditions mean that the function will be stationary *in the fundamental x direction* everywhere along the straight line in the x, y plane defined by (A6a), and that the function is stationary *in the fundamental y direction* everywhere along the straight line defined by (A6b). A unique stationary value requires that these equations are independent and consistent, which means that the lines defined by the equations are not parallel. The conditions (A6a) and (A6b) can be expressed using determinants as

$$\begin{bmatrix} 2a & b \\ b & 2c \end{bmatrix} \begin{bmatrix} x \\ y \end{bmatrix} = \begin{bmatrix} -d \\ -e \end{bmatrix}$$

If the determinant of the coefficient matrix in these conditions is non-zero, the conditions have a unique solution.[4] If the determinant *is* zero there will be either a continuum of solutions if the equations are consistent (when each produces the same line) or no solutions at all if they are not (when the lines are parallel and not superimposed).

Now consider what second-order conditions require. Given the satisfaction of the first-order conditions, then from (A3) for a maximum of z it is *sufficient* that

$$2a < 0 \qquad \begin{vmatrix} 2a & b \\ b & 2c \end{vmatrix} > 0$$

For a minimum of z it is *sufficient* that

$$2a > 0 \qquad \begin{vmatrix} 2a & b \\ b & 2c \end{vmatrix} > 0$$

while it is sufficient for a saddle point that

$$\begin{vmatrix} 2a & b \\ b & 2c \end{vmatrix} < 0$$

Once again, the matrix

$$\mathbf{B} = \begin{vmatrix} 2a & b \\ b & 2c \end{vmatrix}$$

is crucial in establishing the nature of the stationary value at the second order, as well as its existence and location at the first order. Before looking at the possible outcomes in terms of the above, the matrix B repays further study. First consider the degree two terms in the function z. These are

$$z = ax^2 + bxy + cy^2$$

which can be written as

$$[x \ y] \begin{bmatrix} a & \frac{1}{2}b \\ \frac{1}{2}b & c \end{bmatrix} \begin{bmatrix} x \\ y \end{bmatrix} \tag{A7}$$

from which it is evident that the matrix B is the scalar product ($\times 2$) of the matrix of the coefficients of the degree two terms in (A7). Call the latter matrix A. A crucial fact to note is that if $|A| = 0$ then $|B| = 0$. Indeed, A could be substituted for B in the first- and second-order conditions. The expression (A7) is a *quadratic form*, and $|A|$ is the *discriminant* of the quadratic form. Clearly, much hangs on the nature of the discriminant (positive, negative or zero) as regards solutions in any particular case. In general, a *form* is a polynomial expression where each term has the same degree (i.e. two, in the case of a *quadratic* form). In general, a quadratic form can be written in the following way as

$$x' \ A \ x \tag{A8}$$

where in (A8) A is a matrix of coefficients and x is the vector of variables.[5] Now consider the various possibilities for stationary values of the quadratic[6]

$$z = ax^2 + bxy + cy^2 + dx + ey + f$$

Case 1 $4ac > b^2$

With the discriminant positive, there are two possibilities for the unique stationary value:

(a) this will be a minimum if $a > 0$ or
(b) a maximum if $a < 0$

Note that $a = 0$ cannot arise in this case.

Case 2 $4ac < b^2$

When the discriminant is negative, the stationary value is a saddle point.

Case 3 $4ac = b^2$

There are several possibilities when the discriminant is zero:

1 The first-order conditions are inconsistent. This is the case if the coefficients are such that

$$eb \neq 2cd$$

or, equivalently, since $4ac = b^2$, $bd \neq 2ae$. Here the function has no stationary value. An example of such a function is

$$z = -32x^2 + 16xy - 2y^2 + x + y$$

which is described and graphed later on.

2 The first-order conditions are consistent. This occurs if $eb = 2cd$, but there are a number of distinct possibilities here.

   (a) It may not be possible to satisfy z_x *and* z_y as when

   $$a = b = c = 0 \text{ and } d \neq 0 \text{ or/and } e \neq 0$$

   The function in this case is linear, and the first order conditions would require the equation of constants to zero.

   (b) In the ultimately degenerate case, where
   $$a = b = c = d = e = 0$$
   then the zeroth degree function

   $$z = f$$

   (producing a horizontal plane at height f given the existence of x and y axes) is all that will remain.

   (c) If

   $$a = 0 \quad \text{and} \quad b = 0 \quad \text{but} \quad c \neq 0$$

   then the function becomes a quadratic in the variable y only[7] and the stationary value will be

   (i) a minimum if $c > 0$ or,
   (ii) a maximum if $c < 0$.

   (d) If

   $$a \neq 0 \quad \text{but} \quad b = 0 \quad \text{and} \quad c = 0$$

   the function is a quadratic in x only, there being

   (i) a minimum if $a > 0$ or
   (ii) a maximum if $a < 0$.

(e) If a, b and c are *all* non-zero, an interesting pair of cases results, but note that a and c cannot be of opposite signs since $b^2 > 0$ and $4ac = b^2$ here.

   (i) If $a > 0$ and $c > 0$ then the result is the 'frontier' between a minimum and a saddle point. The function will have a 'trough' over (or below) a straight line in the x, y plane. An example of such a function is

   $$z = 16x^2 - 8xy + y^2 - 24x + 6y$$

   which is discussed and plotted below.

   (ii) If $a < 0$ and $c < 0$, then the result is a position between a *maximum* and a saddle point. The function will have a 'ridge' over (or below) a straight line in the x, y plane. An example of such a function is

   $$z = -8x^2 + 4xy - 0.5y^2 + 20x - 5y$$

   which is discussed and plotted below.

Now consider some numerical examples. First of all, take Case 3, 1 where

$$z = -32x^2 + 16xy - 2y^2 + x + y$$

is shown in Figure A1. That this function is unbounded both from above and from below can be seen by making the substitution $y = 4x$. The second degree terms then cancel giving

$$z = 5x$$

as the result, which, of course, can be made arbitrarily large and positive or large and negative. Case 3, 2(e)(i) also has a discriminant of zero, but here the first-order conditions give the same relationship between x and y:

$$z_x = 32x - 8y - 24 = 0$$

and

$$z_y = -8x + 2y + 6 = 0$$

so that $z_x = -4z_y$ and all that is required for a stationary value is that

$$y = -3 + 4x$$

That the stationary value is not a maximum is evident from the fact that the direct partial derivatives at the second order are both positive,

$$z_{xx} = 32 \quad \text{and} \quad z_{yy} = 2$$

796 Appendix

[3D surface plot]

Center x: 0   y: 0   Length x: 10   y: 10   Derive 3D-plot

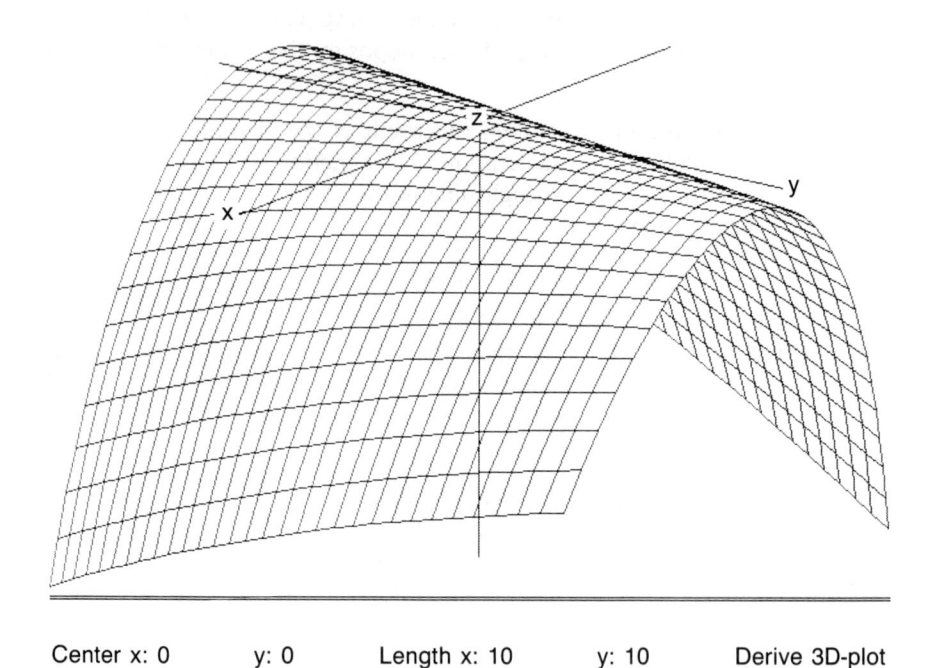

Figure A1

while the fact that the Hessian

$$\begin{vmatrix} z_{xx} & z_{xy} \\ z_{yx} & z_{yy} \end{vmatrix} = \begin{vmatrix} 32 & -8 \\ -8 & 2 \end{vmatrix} = 0$$

in this case reflects the fact that the stationary values form a 'trough' along $y = -3 + 4x$, as shown in Figure A2.

A minor change in one of the coefficients of the second degree terms would mean either a minimum (if the determinant became positive) or a saddle point (if it became negative). In case 3, 2(e)(ii), the first-order conditions both give rise to the relationship

$$y = -5 + 4x$$

along which the surface has a 'ridge' as shown in Figure A3. In this case, the fact that the knife-edge situation is between a maximum and a saddle point is evident because the direct second-order derivatives, both being negative, rule out a minimum. Finally, note that if we consider a quadratic form (no linear components) in which the discriminant is zero, such as

$$z = -32x^2 + 16xy - 2y^2$$

Center x: 0   y: 0   Length x: 10   y: 10   Derive 3D-plot

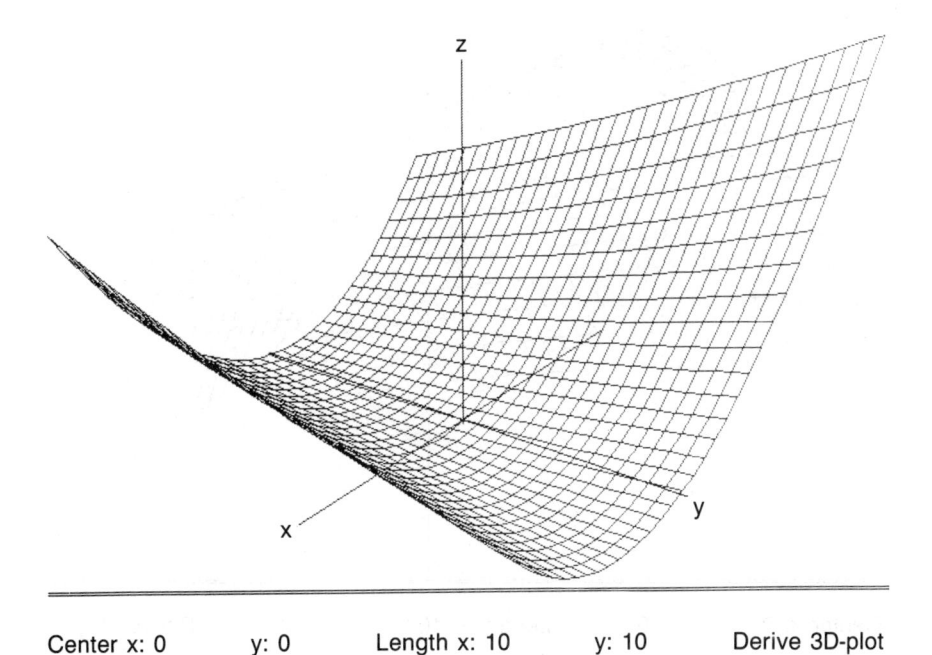

Figure A2

then the conditions at the first order

$$z_x = -64x + 16y = 0$$
$$z_y = 16x - 4y = 0$$

form a homogeneous system in which non-trivial solutions[8] satisfy

$$y = 4x$$

and these non-trivial solutions give rise to a 'ridge' rather than a 'trough' in this case.

A3 CONSTRAINED OPTIMIZATION: SECOND-ORDER CONDITIONS USING DETERMINANTS

In Chapter 7, a sufficient second-order condition was presented for the maximum of a function of two independent variables subject to a single equality constraint. That condition can be expressed conveniently using determinants. It was seen that, given the problem,

maximize $F = f(x_1, x_2)$
subject to $b - g(x_1, x_2) = 0$

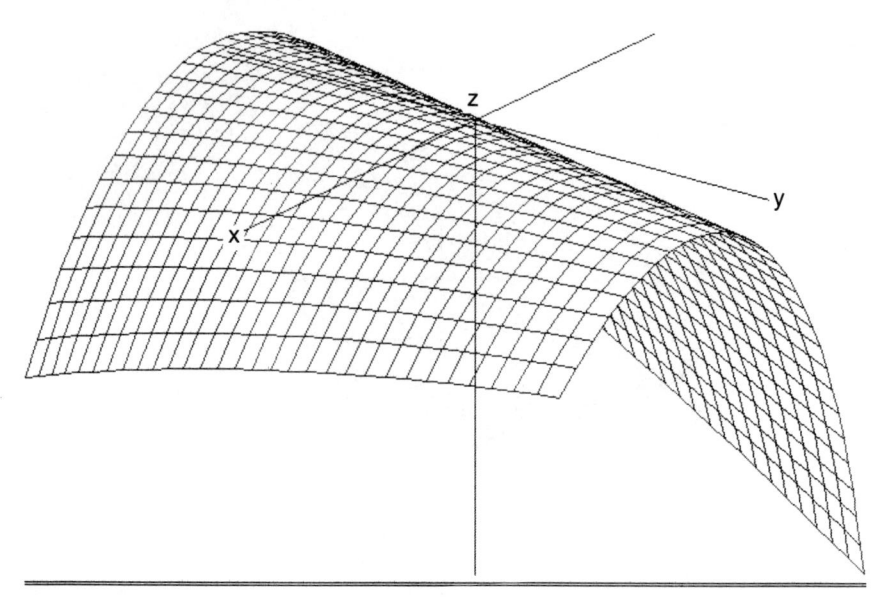

Center x: 0   y: 0   Length x: 10   y: 10   Derive 3D-plot

Figure A3

for which we form the Lagrangian

$$L = f(x_1, x_2) + j[b - g(x_1, x_2)]$$

then, given the satisfaction of conditions at the first order, for a constrained maximum it will be sufficient at the second order that

$$L_{11}g_2^2 - 2L_{12}g_1g_2 + L_{22}g_1^2 < 0 \tag{A9}$$

where, in (A9),

$$L_{11} = \frac{\partial^2 L}{\partial x_1^2} \quad L_{12} = \frac{\partial^2 L}{\partial x_1 \partial x_2} = \frac{\partial^2 L}{\partial x_2 \partial x_1} \quad L_{22} = \frac{\partial^2 L}{\partial x_2^2}$$

The condition (A9) can be expressed as

$$\begin{vmatrix} 0 & g_1 & g_2 \\ g_1 & L_{11} & L_{12} \\ g_2 & L_{21} & L_{22} \end{vmatrix} > 0 \tag{A10}$$

in which the Hessian matrix of the second-order partial derivatives of L has been bordered by a column and a row which include the first-order partial derivatives of the constraint. To confirm the equivalence of (A10) and (A9), expand the determinant by the first row, giving the requirement

$$-g_1 \begin{vmatrix} g_1 & L_{12} \\ g_2 & L_{22} \end{vmatrix} + g_2 \begin{vmatrix} g_1 & L_{11} \\ g_2 & L_{21} \end{vmatrix} > 0$$

i.e.

$$-g_1(g_1 L_{22} - g_2 L_{12}) + g_2(g_1 L_{21} - g_2 L_{11}) > 0$$
$$-g_1^2 L_{22} + g_1 g_2 L_{12} + g_1 g_2 L_{21} - g_2^2 L_{11} > 0$$

so

$$-g_1^2 L_{22} + 2 g_1 g_2 L_{12} - g_2^2 L_{11} > 0$$

to which multiplication through by -1 produces

$$L_{11} g_2^2 - 2 L_{12} g_1 g_2 + L_{22} g_1^2 < 0$$

as required. For a problem in three variables subject to a linear constraint, sufficient second-order conditions for a maximum are that

$$\begin{vmatrix} 0 & g_1 & g_2 \\ g_1 & L_{11} & L_{12} \\ g_2 & L_{21} & L_{22} \end{vmatrix} > 0 \qquad \begin{vmatrix} 0 & g_1 & g_2 & g_3 \\ g_1 & L_{11} & L_{12} & L_{13} \\ g_2 & L_{21} & L_{22} & L_{23} \\ g_3 & L_{31} & L_{32} & L_{33} \end{vmatrix} < 0$$

For a constrained minimum, both the bordered Hessian determinants should be negative. For a constrained maximum of a problem in n variables with a single equality constraint, given the satisfaction of first-order conditions, at the second order it is sufficient that the principal minors of the bordered Hessian

$$\begin{vmatrix} 0 & g_1 & g_2 & \cdots & g_n \\ g_1 & L_{11} & L_{12} & \cdots & L_{1n} \\ \vdots & \vdots & \vdots & \vdots & \vdots \\ g_n & L_{n1} & L_{n2} & \cdots & L_{nn} \end{vmatrix}$$

alternate in sign[9] beginning with a positive. For a constrained minimum in the n-variable case, the principal minors should all be negative.

For a numerical example using the bordered Hessian form of second-order condition, consider the problem

maximize $F = 240 x_1 + 80 x_2 - 4 x_1 x_2 - 8 x_1^2 - 0.5 x_2^2$
subject to $4 x_1 + 2 x_2 = 60$

Forming the Lagrangian

$$L = 240 x_1 + 80 x_2 - 4 x_1 x_2 - 8 x_1^2 - 0.5 x_2^2 + j(60 - 4 x_1 - 2 x_2)$$

then, at the first order, it is required that
$$L_1 = 240 - 4x_2 - 16x_1 - 4j = 0$$
$$L_2 = 80 - 4x_1 - x_2 - 2j = 0$$
$$L_j = 60 - 4x_1 - 2x_2 = 0$$

which conditions solve for
$$x_1 = 5 \qquad x_2 = 20 \qquad j = 20$$

Noting that the constraint is written in implicit form as
$$60 - 4x_1 - 2x_2 = 0$$
then
$$g_1 = -4 \quad \text{and} \quad g_2 = -2$$

Therefore, with the first-order conditions satisfied, at the second order it will be sufficient for a constrained maximum that

$$\begin{vmatrix} 0 & -4 & -2 \\ -4 & -16 & -4 \\ -2 & -4 & -1 \end{vmatrix} > 0$$

Expansion of this determinant by the first row gives

$$-4(-1)\begin{vmatrix} -4 & -4 \\ -2 & -1 \end{vmatrix} - 2\begin{vmatrix} -4 & -16 \\ -2 & -4 \end{vmatrix} > 0$$

i.e.
$$4(4 - 8) - 2(16 - 32) > 0$$
$$-16 + 32 = 16 > 0$$

confirming the constrained maximum.

A4 OPTIMIZATION OF QUADRATIC FUNCTIONS OF TWO VARIABLES WITH A LINEAR EQUALITY CONSTRAINT

With two variables, a quadratic objective function and a linear constraint, we can write

maximize $z = ax^2 + bxy + cy^2 + dx + ey + f$
subject to $gx + hy = i$

The Lagrangian will be

$$L = ax^2 + bxy + cy^2 + dx + ey + f + j(i - gx - hy)$$

and the first-order conditions are

$$L_x = 2ax + by + d - gj = 0$$
$$L_y = bx + 2cy + e - hj = 0 \quad \text{(A11)}$$
$$L_j = i - gx - hy = 0$$

With rearrangement the conditions can be written in matrix form as

$$\begin{bmatrix} 2a & b & -g \\ b & 2c & -h \\ -g & -h & 0 \end{bmatrix} \begin{bmatrix} x \\ y \\ j \end{bmatrix} = \begin{bmatrix} -d \\ -e \\ -i \end{bmatrix} \quad \text{(A12)}$$

A special situation arises if the determinant of the coefficient matrix is zero. The determinant is

$$\begin{vmatrix} 2a & b & -g \\ b & 2c & -h \\ -g & -h & 0 \end{vmatrix} \quad \text{(A13)}$$

Multiplication of the third row and the third column by -1 leaves the value of the determinant unchanged, as does the combined operation of repositioning column 3 as column 1 and row 3 as row 1. The result is

$$\begin{vmatrix} 0 & g & h \\ g & 2a & b \\ h & b & 2c \end{vmatrix}$$

which, in this quadratic case, is the determinant (A10) used to express the second-order condition, and for which we have so far not considered the possibility of a zero outcome. To see what happens when the determinant is zero, begin by evaluating the determinant. The result is

$$|\quad| = 2(bgh - cg^2 - ah^2)$$

Now express y in terms of x from the constraint

$$y = \frac{i - gx}{h}$$

and substitute in z. The result rearranges to

$$z = \frac{1}{h^2} [x^2(ah^2 - bgh + cg^2) + x(bhi - 2cgi + dh^2 - egh)$$
$$+ ci^2 + ehi + fh^2] \quad \text{(A14)}$$

Note that the coefficient of x^2 is a multiple of the value of the determinant:

$$-0.5 \frac{1}{h^2} |\quad|$$

So when the coefficients of the problem are such that the value of the determinant is zero: *z subject to the constraint can be reduced to a **linear** function of x*. If the substitution $(i - hy)/g$ is made for x, then z subject to the constraint can be expressed as a linear function of y. In the present case, despite the constraint, z is unbounded at either extreme provided that the coefficient of x is not also zero [10] and in the absence of sign requirements or other restrictions on the values of x and y. A numerical example of such a problem is:

$$\text{maximize } z = x^2 + 8xy + 15y^2$$
$$\text{subject to } 2x + 10y = 100 \qquad \text{(A15)}$$

Substitution for y from the constraint into the objective function produces

$$z^* = 20x + 1500$$

which can be made arbitrarily large and positive or arbitrarily large and negative. Alternatively, after the substitution for x from the constraint into the objective function, the resulting z^* function of y is

$$z^* = 2500 - 100y$$

which can also be made as high or as low as desired. If in (A14) the coefficient of x is *also* zero, then z subject to the constraint maintains a constant value. An example of such a case is

$$\text{maximize } z = x^2 + 8xy + 15y^2 - 20x$$
$$\text{subject to } 2x + 10y = 100 \qquad \text{(A16)}$$

From the constraint,

$$y = 10 - 0.2x$$

Substitution into the objective function gives

$$z^* = x^2 + 8x(10 - 0.2x) + 15(10 - 0.2x)^2 - 20x$$
$$= x^2 + 80x - 1.6x^2 + 1500 - 60x + 0.6x^2 - 20x$$

which reduces to

$$z^* = 1500$$

These are degenerate cases that arise when the bordered Hessian is zero.[11] In one case, *nowhere* is a maximum or a minimum; in the other case, *everywhere* is simultaneously a joint maximum and a joint minimum. These two outcomes correspond to the cases where (A12) represents an *inconsistent* or a *consistent* set of linear equations. Take the latter case first. In (A12), if the solution for x were sought by the use of Cramer's rule, the numerator determinant would be

$$\begin{vmatrix} -d & b & -g \\ -e & 2c & -h \\ -i & -h & 0 \end{vmatrix}$$

which evaluates as

$$bhi - 2cgi + dh^2 - egh$$

In other words, it is the coefficient of x in (A14). When this determinant is zero, the equations are at least consistent, and all permissible values of x and y produce the same value of z. This is the case in the problem (A16) where the numerator in Cramer's rule would have been

$$\begin{vmatrix} 20 & 8 & -2 \\ 0 & 30 & -10 \\ -100 & -10 & 0 \end{vmatrix} = 0$$

In this case, the linear dependence can be seen as

$$\text{row } 3 = -5(\text{row } 1) + \text{row } 2$$

This is the same relationship that obtains in the equations (A12) with these data. However, in the case of problem (A15), the equations (A12) are

$$\begin{bmatrix} 2 & 8 & -2 \\ 8 & 30 & -10 \\ -2 & -10 & 0 \end{bmatrix} \begin{bmatrix} x \\ y \\ j \end{bmatrix} = \begin{bmatrix} 0 \\ 0 \\ 100 \end{bmatrix}$$

which are inconsistent. In this case the right-hand side does not have the same relationship between its rows as does the left-hand side and there are no solutions to the equilibrium conditions.

NOTES

1 This notation for the partial derivatives is more convenient in the present context than the $\partial f/\partial x$ notation.
2 In the following discussion, for simplicity, we shall use strict inequalities in the second-order conditions.
3 This is in fact the global maximum of the function.
4 Which, in the case of $d = e = 0$, is the origin.
5 For a discussion of quadratic forms see Chiang (1984).
6 In some of the special cases considered below, z ceases to be a quadratic.
7 Note that the coefficient of x, d, must be zero here since we are still under the requirement that $eb = 2cd$.
8 Note that the trivial solution, $(0, 0)$, also satisfies $y = 4x$ and is on the ridge.
9 Starting from the 3×3 determinant (A10).
10 This case will be discussed shortly.
11 Note that no such difficult cases result in the presence of a constraint if the discriminant in the unconstrained case, $4ac - b^2$, is zero.

Index

abscissa 62
absolute maximum 316
Absolute Value Function 254
absolute values 24
addends 28
addition of matrices 709
adjacent extreme point solution practice 150, 164
adjugate matrix 725, 745
aggregate sales curves 671 ff
aggregation 30
algebraic functions 241
alien cofactors 785
allowable range 176
alternative optima *see* non-unique optima
annuities 47
antiderivative 544
antisymmetric matrix 706
antitonic function 252
arguments 28, 196, 362
arithmetic, fundamental laws of 16 ff
arithmetic progression 41
associative law 17
asymptote 208, 235
axis of symmetry 201

Base Year 697
Basic feasible solutions 164
Basic solutions 740
Basic variables 168
basis 740
binary coded decimal 26
binary notation 26
binary operation 31
biquadratic function 227, 263
blending problems 97

boundary condition 558
breakeven analysis 117 ff
buffer stock 324
by-parts formula 659

Cartesian co-ordinates 62
chain rule 295 ff, 625, 647, 656
change of base of logarithms 638
chord 359
closed form expression 44
closed interval 142
Cobb-Douglas production function 638
co-factors 744; alien 785
column vector 701
common difference 41
common logarithms 635
common multiple 267
common ratio 46
commutative law 17
commutative property 710, 723
complementary minor 785
complementary slackness conditions 493
complement of a set 11
complex conjugates 5
composite function 242
composite function rule 295, 478
compound interest 617 ff
concave function 326, 388
concurrent lines 70
conditional equation 115
consistent equations 89, 803
consols 38, 48
constant function rule: differentiation 281; integration 547
constant multiple rule 550
constant of integration 545, 596

constrained optimization 402 ff, 800 ff
constraint 152
constraint qualification conditions 498
consumption function 339
continuous compounding 619
continuous functions 245 ff
contours 155, 365, 404
control variable 196
convex function 326
convex programming problems 499
convexity property 149
co-ordinates 61 ff; Cartesian 62; rectangular 62
cost functions 225; joint 427
counting numbers 2
counting problems 14
Cramer's rule 752
critical line 408
critical points 271, 359
cross partial derivative 383
cubic equations, solution of 360; functions 221 ff
Cusp 466, 508

decay curves 615
decision variables 152, 196
defining relation 8
definite integral 571 ff, 630, 656
degenerate solution 193
degree of polynomial 229
demand curve 119, 237, 244
demand function 244, 362, 511; S-curve 671
dependent variable 80, 85, 196
depreciation 42, 46, 616
derivatives 274 ff; partial 375 ff; exponential functions 624, 665; logarithmic functions 643, 663
derived function 359
derive program 133, 227, 592, 612, 697, 760
Descartes's Rule of Signs 230 ff, 351
determinants 724, 743 ff; second order conditions using 787 ff
determinate system 104
diagonal matrix 704
differential 406, 568, 611
differential equations 555 ff, 632
differentiation 274 ff; *see also* derivatives

diophantine equation 103
discontinuous functions 247
discount, rate of 47
discrete function 249
discriminant 205, 388; of quadratic form 793, 804
disjoint sets 11
distance formula 64
Distributive Law 17
domain 196 ff
dominant term 209
dot product 714
doubling time 639
duality in LP 179 ff
dual problem 179
dual values 162, 173, 180, 413, 513, 750

e 619
elasticity 34, 648 ff
elasticity of demand 237, 339
elementary matrix operations 706
elementary row operations 99, 170, 785
eliminants 785
elimination procedure 78, 92
ellipse 219
equal sets 8
equation of a line: finding 76 ff
equilibrium price 120
equipment utilization rule 154
empty set 10
endogenous variables 110
entering variable 169
equal matrices 705
equivalent sets 8
escapable cost 288
even functions 253
Euler's formula 58
excess demand 120
excess supply 120
excise duties 134
exogenous variables 110, 785
expansion paths 408
exponential functions 267, 614 ff
exponential notation 28
exponents 19 ff
extended function 264, 267
extrapolation 82
extreme points 150, 271

factorials 39, 51

Index

factoring 204
family of lines 69
feasible set 153
finite set 8
fixed costs 239, 288
fixed point notation 27
floating point notation 27
free maximum 403
function notation 275
function of a function rule 295
functions 195 ff
functions of several variables 361 ff
fundamental directions 364, 792
future value 617

Game Theory 33, 323
Gaussian elimination 100, 112 ff, 785
Gauss-Jordan elimination 113, 760
GDP 116, 615, 617
general form of equation 85, 86
general quadratic equation 218
general solution to differential equation 558
Gompertz function 673
gradient form of equation 85
greatest lower bound 142
Gross Domestic Product 116, 615, 617

half life 642
half space 144
Hessian determinant 789
homogeneous functions 250, 373
homogeneous systems 756 ff, 797
horizontal lines 68
hyperbola 220, 235, 652
hyperbolic demand curve 237

identities 115 ff
identity matrix 704, 720
identity transformation 32
implicit differentiation 304 ff
implicit function 406
improper integral 574
income 211
inconsistent equations 94, 733, 803
indefinite integral 545
independent demand functions 511
independent equations 89
independent variable 196
indeterminate system 106
inequalities, linear 136 ff; quadratic 216

inelastic demand 651
infimum 142
infix notation 28
inflection 223, 327 ff
initial condition 559
initialization phase 166
inner product 714
input variables 362
instrument 196
integers 2
integrand 546
integration 544 ff; of exponential functions 629 ff, 668; of logarithmic functions 654 ff; by parts 658 ff; by substitution 567 ff
intercepts 68, 208
interest, compound 617 ff; simple 42
interpolation 82
intersection of sets 10
interval closed 142; open 143
interval scale 33
inventory control 82, 254, 324
inverse function rule 302, 644
inverse functions 243, 635
inversion of matrices 723 ff
involutional matrices 729
irrational numbers 2
Iso-profit lines 155, 372
isoquands 372
isotonic functions 59, 252
iteration 153
iterative phase 166

joint cost function 427
jordan elimination 113

Karush-Kuhn-Tucker conditions 498 ff
Kuhn-Tucker conditions 498 ff

Lagrange multipliers – interpretation of 461 ff
Lagrange Multiplier method 450 ff; solving first order conditions 487, 492 ff, 506 ff
Laplace expansion 743
law of scarcity 136
leading coefficient 201, 221
leading diagonal 703
lead time 82
least upper bound 141
leaving variable 169

Leibniz's notation 275
limits 619, 623
limits of integration 572
linear dependence 733, 739, 785
linear differential equation 556
linear functions of two variables 374 ff
linear inequalities 136 ff
linear regression 81, 82
Local maxima 270; minima 271
logarithmic derivative 647
logarithms 635
logical operators 31
logistic curve 674
log-linear function 638
lower envelope 324
lower triangular matrix 703

main diagonal 703
marginal cost 287, 331, 552, 576; profit 287, 377; revenue 284, 553
marginal propensity to consume 339
marginal values 162; *see also* dual values
market clearing price 120
market equilibrium 119, 214
market potential 671
market saturation 238, 671
Marshallian Demand Curve 134
matrices 699 ff; order of 700
matrix addition 709; determinant of 724, 743 ff; difference 711; inner product 714; inversion 723 ff; multiplication 713 ff; negative of 711; notation 704; rank 739 ff; scalar multiple of 712; subtraction 710
maxima, conditions for 271, 310, 330; constrained 452; global 270, 316 ff, 327, 411, 466; local 270; several variables 387 ff; sign restricted 468
Mean Absolute Deviation 254
minima, constrained 455 ff; local 271; several variables 390; sign restricted 420
minors 743, 785; principal 789
mixed derivative 383
mixed expression 236
modified growth curve 671
modulus 24
monopoly: multi product 510 ff

monotonic functions 252
multiplication of matrices 713 ff
multi product monopolist 510 ff

National Income Accounting 115
natural exponential function 622
natural logarithmic functions 635 ff
natural number 619
negative of a matrix 711
net evaluation row 168
Newton's notation for derivative 359
Newton's serpentine 236
nominal scale 33
non-negativity requirements 145
non-unique optima 159
normal lines 70
null set 10
numbers, algebraic 4; complex 5, 207; counting 2; imaginary 5, 207; irrational 2; notation 23 ff; rational 2; real 4; system 2 ff; transcendental 3, 242
numerals 2, 25
numerical integration 584 ff

objective function 152, 196
odd functions 252
off-diagonal elements 703
open interval 143
operators 28
opportunity loss 157, 174, 191
opportunity set 153
order preserving transformation 32
ordinal scale 32
ordinate 62
ordinary differential equation 556
orthants 66
orthogonal lines 70
orthogonal matrices 728
output variable 363
overdeterminate system 107

parabola 201 ff
parallel lines 70
Pareto optimality 154
parity: equality in 230
partial derivatives 375 ff; second order 383 ff
partial sum 38
periodicity 253
particular solution 558
perpendicular lines 70

Index 809

piecewise defined functions 199, 319, 578
pivotal column 169; number 169; row 169
place value notation 25
plane 110
plotting straight lines 74
point-slope form of equation 84
Points of Inflection 327 ff
Polish notation 28
polynomials 229 ff
pool resource 34
positive quadrant 65
postfix notation 28
postmultiplication 717
post-optimality analysis 175, 193
power function rule 277 ff
power rule: exponents 19, 717; integration 547
precedence rules 28
prefix notation 28
premultiplication 717
present value 47
price breaks 248
price elasticity of demand 339, 650
primal problem 179
principal diagonal 703
principal sum 617
product: infinite 39; notation 39 ff; Wallis's
production function 362; Cobb-Douglas 638
production plan 152
production possibility frontier 153
productivity curve 371
product rule: for exponents 19, 717; for differentiation 290 ff, 625, 645
progressions: arithmetic 41 ff; general term of 41; geometric 45 ff
proper subset 9

quadrants 65
quadratic equation: general 218
quadratic forms 793
quadratic formula 205
quadratic functions 201 ff, 369 ff, 791 ff, 800 ff
quadratic inequality 216
quartic functions 226 ff
quintic equation 229, 233
quotient rule: for exponents 19; for differentiation 292 ff, 646

range 196 ff
range of feasibility 142; see also tolerance interval
range of optimality 176; see also tolerance interval
rank of a matrix 739 ff, 770
rate of substitution 157, 405
Ratio Scale 31
rational functions 233
real line 61
reciprocal function 233, 644
reciprocal rule 294
rectangular co-ordinates 62
Rectangular Hyperbola 235, 652
recursion formula 48
reduced form equations 110
redundant constraint 186
reflex polygon 150
relational operators 31, 137
re-order level 83
reverse Polish notation 28
roots 58, 205 ff
row vector 701

saddle points 392 ff, 500, 540, 792
sales curves 671 ff
sales revenue 211
saturation curves 697
saturation level of demand 671
scalar 701
scalar matrix 704
scalar multiplication 712
scalar product 714, 726, 745
scientific notation 28
S curve demand function 671
secant 359
second derivative 307 ff; test 310
second order conditions using determinants 787 ff
sequences 41 ff
sensitivity analysis 175 ff, 736
series 41 ff; divergent 48; finite 43; power 43
set difference 12
sets 8 ff
sexagesimal notation 26
shadow prices 162, 750; see also dual values
sigma notation 36 ff
sign requirements 145
sign restricted variables 464 ff
simple interest 42

simple root 206
simplex 367
simplex method 153, 163 ff
Simpson's Rule 588 ff, 612
simultaneous equations 89 ff; solution using matrices 732 ff, 748 ff; solution by Cramer's rule 752 ff
simultaneous inequalities 141
singular matrices 725, 785
skew symmetric matrix 706
slack variables 156, 486
slope 64
slope-intercept form of equation 68
smooth function 267
solution set 109, 140
stationary point of infection 327
stationary points 271, 387, 398, 791 ff
step discontinuity 200
step function 248
stock control 82
stockout costs 324; defined 83
stopping rule 173
straight line depreciation 42, 697
straight lines 68 ff
strings 29
strip 147
structural variables 164
subject of equation 80, 85
sub matrices 702, 743
subsets 9
substitution method (linear equations) 90; constrained optimization 408 ff
subtraction of matrices 710
sum: partial 38; rule for summations 37; to infinity 38
sum-difference rule: differentiation 283; integration 549
summation: index of 36; limits of 36; notation 36 ff; sum rule 37
superset 9
supply curve 120
supremum 141

symmetric difference 12
symmetric matrix

technical independence 511
terminating condition 48
Theory of Equations 230
Theory of Games 33
third derivative 307, 329
tolerance interval 142, 176, 735, 750
total product curve 371
total revenue 211
transcendental functions 242
transitive relation 10
transpose of a matrix 706
trapezium rule 584 ff, 611
trapezoidal rule *see* trapezium rule
triangular matrix 703
trivial solution 757, 804
turning points 208, 270
turnover 211

undefined slope 64
units of measure 33
union of sets 10
unit costs 288, 713
universal set 10
upper triangular matrix 703
utility function 32

variable costs 288, 552, 576
Venn diagrams 12 ff
vertex 201
vertical lines 69
vinculum 30

Wallis's product 39, 51
War Loan 38, 59
Witch of Agnesi 236

zero matrix 704, 711
zero slope condition 271
zero sum game 323